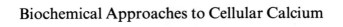

Biochemical Approaches to Cellular Calcium

METHODOLOGICAL SURVEYS IN BIOCHEMISTRY AND ANALYSIS

Series Editor: Eric Reid

Guildford Academic Associates
72 The Chase
Guildford GU2 5UL, United Kingdom

The series is divided into Subseries A: Analysis, and B: Biochemistry.
Enquiries concerning Volumes 1–18 should be sent to the above address.

Recent Titles

How to obtain future titles on publication

A standing order plan is available for this series. A standing order will bring delivery of each new volume immediately upon publication. For further information, please write to:

The Royal Society of Chemistry
Distribution Centre
Blackhorse Road
Letchworth
Herts. SG6 1HN

Telephone: Letchworth (0462) 672555

Methodological Surveys in Biochemistry and Analysis, Volume 19

Biochemical Approaches to Cellular Calcium

The Proceedings of the 11th International Subcellular Methodology Forum entitled Study of Cellular Roles of Calcium

Guildford, 6th–9th September 1988

Edited by

Eric Reid
Guildford Academic Associates, Guildford, United Kingdom

G.M.W. Cook
University of Cambridge

J.P. Luzio
Addenbrooke's Hospital, Cambridge

ROYAL
SOCIETY OF
CHEMISTRY

British Library Cataloguing in Publication Data
Study of Cellular Roles of Calcium : 11th : 1988 :
  Guildford, England
  Biochemical approaches to cellular calcium.
  1. Organisms. Cells. Metabolism. Regulation. Role of calcium
  I. Title   II. Reid, Eric, *1922*–   III. Cook, G. M. W.
  (Geoffrey Malcolm Weston) : 1938–   IIII. Luzio, J. P.
  V. Series
  574.87'61

  ISBN 0–85186–926–2

Published by the Royal Society of Chemistry, Thomas Graham House, Cambridge CB4 4WF

Printed in Great Britain by Henry Ling Ltd., at the Dorset Press, Dorchester, Dorset

Senior Editor's Preface

Amongst our array of contributors, some have heeded with alacrity the mandate that know-how and pitfalls should be the focus of the book – as for the 'Subcellular Methodology' Forum which led to it. Others have, at least in part, focussed on fascinating findings and fitted them into the field. This will benefit any reader who is not immersed in the cell-calcium field and seeks perspective. The subject-matter is wide-ranging, encompassing biochemical pharmacology, cell pathology and molecular biology, besides cell biology. Once initial perspective has been gained from Campbell's opening article, certain others warrant an 'advance look', even if designated 'Notes' (prefix '**nc**'): <u>e.g.</u> #A-3 (Rink), #A-5 (England), #ncA.4 (Spedding), #ncA.5 (Rüegg), #B-2 (Dawson), #B-3 (Crawford), #ncB.1 (Carafoli), #C-2 (Rubin), #D-2 (McCormack), #E-6 (Poole-Wilson) and #E-8 (Trump).

Assuredly the repute of the Forum-based book series will be upheld by this newest book which, having been subjected to the usual thorough editing and indexing, has escaped being a proverbial 'Proceedings' patchwork. In respect of presentation, some of the manuscripts were a joy to edit, and others entailed much editorial trouble which, however, earned compliments from some of the authors concerned.

The change to a new publisher should obviate the inordinate delay in publication and the non-optimal publicity that have hitherto plagued the book series. If the book list facing the title page raises the thought of a useful shelf run, amplification can be sought from this Editor.

Acknowledgements.- Valuable guidance in planning the Forum (held in September 1988) was given by Dr. Tony Campbell, Prof. N. Crawford, Dr. W.H. Evans and, as Forum Co-organizer, Prof. C.A. Pasternak. Since registration fees were set at an affordable level, it was a boon that many speakers 'paid their way', or at least bore the cost of travel – <u>e.g.</u> from Berlin (DDR), Brazil, Japan or the USSR. It was also gratifying that support came from certain U.K. pharmaceutical companies, <u>viz.</u> Beechams, Ciba-Geigy, Glaxo, ICI and Smith Kline & French. Throughout the text there is acknowledgement to journal publishers, <u>e.g.</u> the Biochemical Society, who have kindly allowed illustrative material to be reproduced in this book. A compliment is due to the Sigma Chemical Co. for the usefulness of their catalogue in elucidating diverse abbreviations used by authors.

Conventions and abbreviations.- $Ca^{2+}{}_i$, or $[Ca^{2+}]_i$ signifying concentration, denotes free intracellular calcium ions, usually cytosolic unless the context is an organelle. However, in the term P_i (orthophosphate) the subscript i denotes inorganic. $Ca^{2+}{}_e$ (or sometimes $Ca^{2+}{}_o$) denotes external free calcium. The term Ca^{2+} (or $^{45}Ca^{2+}$) has sometimes been substituted where, in an uptake context, the author's term was Ca (or ^{45}Ca) — and a converse change sometimes made where the context was bound or stored calcium. For any non-uniformity, blame lies with this Editor. Uniformity has prevailed in respect of inositol phosphates (InsP's, *not* IP's). InsP₃ signifies Ins(1,4,5)P₃, <u>i.e.</u> phosphate substituents at 1, 4 and 5, unless otherwise indicated (<u>cf.</u> the useful guidance early in art. #B-4). Any reader unfamiliar with $[Ca^{2+}]_i$ 'probes' (an Index entry) should consult p. 14 for types and structures (example: fura-2). For temperatures, ° always implies Celsius.

Well-known abbreviations such as HPLC and EDTA are used without definition. Generally, however, each article has a list of abbreviations. Frequent ones include the following:

Ab, antibody (MAb if monoclonal)   e.m., electron microscopy
ELISA, enzyme-linked immuno-   e.r., endoplasmic ⎱ retic-
    sorbent assay   s.r., sacroplasmic ⎰ ulum
CaM, calmodulin   p.m., plasma membrane
FI, fluorescence intensity (*not* F) ROC, receptor-operated
MP, membrane potential (ψ)   channel; *similarly* SMOC,
M_r, mol. wt. relative to 'markers'   second messenger-; VOC
    (gel; 'kDa' units imply M_r)   (POC), voltage- (potential-)

Ion transport systems: a summary excerpted from a booklet *Calcium and Cellular Function* (by Urs T. Rüegg; Sandoz, Basle):

Ion pumps (ATPases):   *Passive ion transport systems:*
    Na^+/K^+ Exchangers - Na^+/Ca^{2+}, Na^+/H^+
    Ca^{2+} - p.m., e.r. $Na^+/K^+/Cl^-$ cotransporter
                                        Ion channels - Na^+, K^+, Ca^{2+}

Guildford Academic Associates   ERIC REID
72 The Chase, Guildford,
Surrey GU2 5UL, U.K.   *24 April 1989*

Contents

The 'NOTES & COMMENTS' ('nc') items at the end of each section conclude with comments made at the Forum which led to this book, together with some reinforcing literature.

Contents

List of Authors

Primary author	Co-authors, with relevant name to be consulted in left column
A.P. Allshire – pp. 455–461 Univ. of Liverpool	M. Antolini – Luciani (ii) K.S. Authi – Crawford
C.C. Ashley – pp. 131–132 Univ. of Oxford	C.J. Barker – Kirk R.J. Barsotti – Ashley G. Bastiaens – Hilderson D. Bataille – Lotersztajn
E.M. Bevers – pp. 289–301 Limburg Univ., Maastricht, The Netherlands	A.R. Baydoun – Markham A. Beit-Or – Eilam (ii) I.K. Berezesky – Trump M.J. Berridge – Cheek D. Bingham – Dobrota
A.K. Campbell – (i) pp. 1–14; (ii) pp. 483–486; & see Morgan Welsh National Sch. of Medicine, Cardiff	A. Bock – Scheller M. Borgers – Ver Donck S. Bova – Luciani (i) P.G. Bradford – Rubin
A.M. Capponi – pp. 313–315 Univ. Cantonal Hospital, Geneva	V. Brechler – Lotersztajn G.R. Brown – Dormer R.D. Burgoyne – Cheek
E. Carafoli – (i) pp. 17–29; (ii) pp. 235–236 ETH – Zentrum, Zürich	A. Bush – Maxfield R. Busse – Lückhoff
C.A.M. Carvalho – pp. 133–136; & see Coutinho Univ. of Coimbra, Portugal	G. Cargnelli – Luciani (both) E.J. Cargoe jr. – Carvalho C. Caruso-Neves – Vieyra
T.R. Cheek – pp. 319–320 Zoology Dept., Univ. of Cambridge	A.P. Carvalho – Carvalho, Coutinho C. Cauvin – Rüegg (both)
P.H. Cobbold – pp. 211–217; & see Allshire Univ. of Liverpool	D. Chernichovsky – Eilam (i) J.G. Comerford – Dawson P. Comfurius – Bevers
O.P. Coutinho – pp. 137–138 Univ. do Minho, Braga, Portugal	P.J. Cullen – Dawson F. Cusinato – Luciani (ii) K.S.R. Cuthbertson – Cobbold
N. Crawford – pp. 179–190 Hunterian Inst., Lincoln's Inn Fields, London	
M. Crompton – pp. 357–364 Univ. Coll. London, London	I. Das – Janah E. Davies – Capponi P. Debetto – Luciani (both) K. Dengler – Scheller
A.P. Dawson – pp. 167–178; & see Hutton Univ. of E. Anglia, Norwich	R.M. Denton – McCormack, Midgley M. De Wolf – Hilderson

Primary author	*Co-authors, with relevant name to be consulted in left column*
B.F. Trump – pp. 439–452 Univ. of Maryland, Baltimore, MD	A. Wallnöfer – Rüegg (ii) D. Wermelskirchen – Wilffert
W.G. Turnell – pp. 91–101 Lab. of Molecular Biol., Cambridge	D. Wilhelm – Wilffert R.B.J. Wilson – Graham F.B.P. Wooding – Koch
L. Ver Donck – pp. 429–437 Janssen Res. Foundn., Beerse	N.M. Woods – Allshire, Cobbold
A. Vieyra – pp. 31–40 Inst. Ciências Biomédicas, Univ. Federal, Rio de Janeiro	
P.K. Wierenga – pp. 475–478 State Univ., Groningen	E. Zacharias – Scheller R. Zeh – Lückhoff
B. Wilffert – pp. 73–80 Janssen Res. Foundn., Neuss	J-F. Zuber – Rüegg (i) R.F.A. Zwaal – Bevers

A CENTURY OF INTRACELLULAR CALCIUM

Anthony K. Campbell

Department of Medical Biochemistry,
University of Wales College of Medicine,
Heath Park, Cardiff CF4 4XN, U.K.

"The calcium ion has an unusual importance in biological phenomena, and the literature concerning its effects is extremely voluminous. No other ion exerts such interesting effects on protoplasmic viscosity."

Lewis Victor
Heilbrunn
1937

A PIONEER

Calcium was the first intracellular signal to be discovered [1, 2], and one of the first scientists to recognize its wide-ranging importance in controlling cells, and in cell injury, was Lewis Victor Heilbrunn [see 3,4]. Heilbrunn was born in 1892. As a graduate student he began work at the Marine Biological Laboratory, Woods Hole, where whilst based at the University of Pennsylvania he was to have close ties throughout his life. His untimely death in a car crash in 1959 meant that he missed the opportunity of seeing the rest of the scientific community awaken to his vision of calcium.

In 1928 he observed a remarkable effect of calcium on cell protoplasm, which he called a surface precipitation reaction [2, 3, 5-8]. It was his belief that the release of calcium from internal stores, and through the cell membrane, provoked this reaction, analogous to blood clotting and that it played a key role in the events responsible for muscle contraction, amoeboid and cilate cell movement, egg fertilization, hormone and drug action, and in the protection of cells against injury to the p.m.[⊗]

The foregoing perceptive quotation comes from the first edition of his textbook [2]. Increasing documentation of the unique role of calcium both within the cells and in the fluid bathing them appeared in the edition of 1943 – and of 1952: "The sensitivity of protoplasm and its response to stimulation are believed to be due to a sensitivity to free calcium and it is believed that the freeing of calcium and the reaction of this calcium with the protoplasm within the cell is the most basic of all protoplasmic reactions".

Why, then, did his ideas fall on so many deaf ears? – for it is only since the mid–1960's that the truth of the 'Ca^{2+} hypothesis' has been widely acknowledged. Two particular reasons can be identified. Firstly, for the first half of this century there was a conceptual misunderstanding about the difference between a trigger and an energy source required for a cellular event to occur. Furthermore, the molecular basis of calcium binding was not known: purified actin + myosin could contract quite happily with ATP + Mg^{2+} and no Ca^{2+}. Secondly, only during the last 20 years have methods been available for measuring and effectively manipulating intracellular calcium. As we shall see, the techniques and reagents involved were requisite for establishing definitely a primary role for calcium in cell activation or cell injury [9].

THE PROBLEM

The biological problem we are concerned with is how primary stimuli acting at the p.m. can activate or injure a process or reaction within the cell (Fig. 1). Many such phenomena involve thresholds at the level of individual cells

---

⊗*Abbreviations*.- p.m., plasma membrane; Ins, *myo*inositol: thus InsP$_3$= 3 phosphate groups — usually Ins(1,4,5)P$_3$ isomer implied; PL, phospholipid; e.r., endoplasmic reticulum.

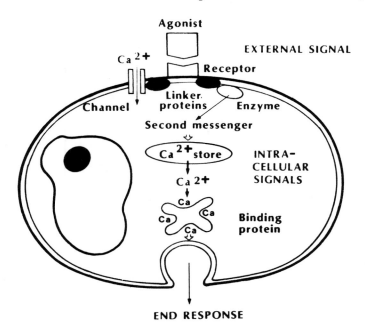

Fig. 1.
Calcium as
an intra-
cellular
signal.

or organelles [1, 9].   In other words, the time course
and dose-response of a population of cells is really a
reflection of the number of cells 'switched on' at a particular
point.   The phenomena include cell movement, cell division
and transformation, secretion, activation of intermediary metabolism, reversible cell injury and cell death.   The primary
stimuli can be physical (<u>e.g.</u> electrical or mechanical),
chemical (<u>e.g.</u> neurotransmitter or hormone) or biological
(<u>e.g.</u> bacterium or virus).   The number of cells undergoing
an end response, or the magnitude of the response in each
cell, can be modified by secondary regulators.   Three classes
of intracellular signal have been identified which can provide
the link between the initial event at the p.m. and the
ultimate cell response:
- cations: Ca^{2+}, H^+, Na^+;
- nucleotides: cyclic AMP, cyclic GMP, GTP, AMP;
- PL derivatives: InsP's and diacylglycerol.

    The five central questions regarding the role of Ca^{2+}
as an intracellular signal are:
    1. Is Ca^{2+} the primary signal responsible for initiating
the cell end-response and how does it interact with the
other intracellular signals?
    2. Where does the Ca^{2+} come from, and how?
    3. How does the Ca^{2+} work?
    4. Is the Ca^{2+} friend or foe following injury?
    5. How did the Ca^{2+} signalling system evolve?

To see how the experimental approach and methodology
to answer these questions has evolved, we must first go
back not 100 years but to the discovery of calcium itself.

GENESIS

Many Welsh miners have reason to be grateful that one
of the most important figures in 19th-century science decided
it was worthwhile finding a way to anticipate the level
of fire-damp before an explosion occurred. Early in June
1808 Humphry Davy [10] (1778-1829) at the Royal Institution
used the method he had developed for separating the positive
and negative components in salts to isolate a novel element
he named 'calcium'. But 75 years elapsed before a systematic
investigation began on the role of cations and anions in
tissue structure and function. Sydney Ringer (1835-1910)
entered University College Hospital Medical School in 1854.
A professor of medicine and dedicated physician, he also
had a passion for laboratory experiments. In 1882 he had
apparently shown that removal of calcium from the medium
bathing a beating frog heart had no effect. However, his
solutions had been made up by his technician using London
tap-water. In 1883 [11] he, somewhat embarrassed, showed
that the tap-water was rich in calcium and, through substitution
of distilled water, that the heart system did need calcium.
Not till the 1950's was an effective method found for chelating
calcium (present even in distilled water at µM levels) both
within and outside cells (see later).

Ringer, and many other distinguished physiologists at
the turn of the century, showed that removal of calcium
from the external fluid disrupted tissue structure and development, caused tissues to dissociate into isolated cells, and
initiated or blocked not only various forms of muscle contraction and cell movement but many other tissue functions.—
#Muscle contraction - heart: Ringer (1883, 1886) [11, 12];
   Mines (1910–11) [13]; Straub (1912) [14]; smooth: Stiles (1903) [15].
#Nerve impulse to muscle: Locke (1894) [16]; Overton (1904) [17].
#Amoeboid movement: Chambers & Reznikoff (1926) [18]; Pantin
   (1926) [19].
#Ciliate movement: Gray (1922, 1924) [20].
#Cell adhesion: Ringer (1890) [21]; Herbst (1900) [22].
#Egg fertilization and tissue development: Ringer (1890,1894)
   [21, 23]; Loeb (1922, citing early work) [24].
#Adrenaline effect - heart: Lawaczek (1928) [25]; Paramecium:
   Von der Wense (1934) [26].
#Digitalis effect: Loewi (1917-18) [27].
#Plant growth: Loew (1892) [28].

But what was the calcium required for? A transmission
role [16] is listed above. The discovery of, and development

of, a bioassay for acetylcholine enabled Harvey & MacIntosh [29] in 1940 to ask: "Was calcium required at the muscle endplate or for the release of acetylcholine from the nerve terminal?" They showed that calcium removal caused spontaneous excitation of nerves in a cat ganglion preparation but prevented acetylcholine secretion.

HYPOTHESIS

Ringer, and others, explained the role of calcium in vague terms involving antagonisms with ions such as potassium and synergism with others. But Heilbrunn's hypothesis [2, 8] was already clear by the early 1930's. He recognized that calcium had a role as a nutrient, as a structural component both as phosphate/oxalate/carbonate precipitates and in soft tissues, and in activation of certain extracellular events such as blood clotting. But he was most interested in the idea that calcium release inside the cell actually initiated muscle contraction and the early events of egg fertilization. By what method could he test his hypothesis?

Heilbrunn knew that citrate and oxalate bound calcium. Jacques Loeb (1859-1924) [24, 30] was one of the first to discover artificial parthenogenesis, a chemical or physical means of activating oocytes to divide without needing sperm. In 1930 Heilbrunn & Young [31] showed that oxalate-treated sea-urchin eggs failed to respond to activating doses of UV light, but did so once calcium had been restored. This approach was extended to ragworm eggs but with citrate as chelator [32]. The work of Dalcq, Heilbrunn, Pasteels and Chambers on the role of calcium in parthenogenesis was well reviewed in 1941 by Tyler [33]. It was chelation of calcium inside the cell that mattered to Heilbrunn, since he believed that calcium was released from the visibly identifiable granules of the cortex. Just [34] in 1919 and Chambers [35] in 1921 showed that egg fragments without cortex cannot be 'fertilized'. Mazia [36] in 1937 analyzed how much calcium was left in sea-urchin eggs after fertilization: there was a 15% decrease, consistent with Heilbrunn's release theory.

But it was in muscle contraction that results were most clear, through direct injection of calcium into the cell.

ZENITH AND NADIR

Chambers [37], a pioneer in microinjection of cells, with co-workers had shown in the 1920's effects of calcium on protoplasmic flow in an amoeba. Several experiments indicated that calcium itself could provoke muscle to contract

(and indeed was essential with freeze-thawed or cut muscle).-
#Applied to muscle - freeze-thawed (frog): Chambers & Hale
   (1932) [37]; cut: Heilbrunn (1940) [7].
#Micro-injected: Keil & Sichel [38]; Kamada & Kinoshita (1943) [39].
#Injected: (frog) Heilbrunn & Wiercinski (1947) [40];
   (Ca-EGTA, crab) Portzehl et al. (1964) [41].

But how specific was this effect of calcium (e.g. if micro-
injected)? In 1947 Heilbrunn and his 'student' Wiercinski
[40] published their experiments designed to test this.-
"Calcium, even in rather high dilution, causes immediate
and pronounced shortening of frog muscle...... This effect is
not shared by any one of the other cations normally present
in muscle. The results lend support to the calcium release
theory of stimulation and they are opposed to Szent-Gyorgi's
belief that potassium ion is primarily responsible for the
contraction of muscle".

     Unfortunately A.V. Hill threw a spanner into the works
by calculating that the diffusion of calcium from the muscle
surface would not be fast enough to explain the rapidity
of contraction following the action potential. Where was
this hypothetical internal calcium store? Anyway, purified
actin + myosin contract without calcium.

     Heilbrunn's theory faced two major difficulties. Firstly,
there was no direct evidence for a release of calcium within
the cell during muscle contraction, or indeed any other
type of cell activation. Secondly, it was odd that a small
release of calcium could provoke such a dramatic effect
in a cell when the calcium outside was as high as 1 mM
and the energy source for muscle had now been identified,
viz. ATP hydrolysis. The problem was that no-one knew
how low the cytosolic free Ca^{2+} was, as there was no way
of measuring it.

MEASUREMENT OF CYTOSOLIC FREE CALCIUM

     A.B. Macullum [42] wrote in a 1925 correspondence column:
"A sensitive microchemical reagent for calcium in tissues
and cells is a great desideratum". In 1928 Herbert Pollack
[43] published the first attempt to measure "calcium ions
in living protoplasm", with four criteria for "the application
of chemical tests to the living cell": (1) the reagent must
react at the pH of the cytoplasm; (2) colour tests are
preferable because precipitates in the cell are hard to
watch with ordinary illumination; and the reagent should
be (3) relatively non-toxic and (4) very sensitive.

     Pollack knew that alizarin sulphonate precipitated calcium
ions and could be used to measure blood calcium. He found

that injection into amoeba, using the 'micrurgical' technique
pioneered by Chambers, of sodium alizarin sulphonate resulted
in "purplish-red granules scattered throughout the cell" and
temporary cessation of movement.   If the amoeba tried to
send out pseudopods, purplish-red granules appeared near the
membrane and pseudopod formation stopped.- "The quiescence
which is induced after the injection of alizarin may be
due to removal of calcium of the protoplasm from the sphere
of action".   He therefore injected $CaCl_2$ into the alizarin-
injected cell.   At concentrations $>^1/_{280}$ M it caused injury,
viz. a local coagulation! But at $^1/_{280}$ M it almost immediately
caused active flowing movements, which then subsided.   In
accord with Pantin in 1926 [19], there was a "need of calcium
to effect amoeboid movement".

In conclusion Pollack wrote thus.- "The injection of
alizarin sulphonate gives a colour test which demonstrates
an appreciable amount of free calcium which serves as a
mechanism of recovering from the effects of sublethal doses
of these (phosphate, sulfate, tartrate and oxalate) anions
and of alizarin sulphonate." Yet it was to be nearly 40 years
before a really effective method, usable in any cell, for
measuring cytosolic calcium was developed.   The problem was
that no-one realized how low the cytosolic free Ca^{2+} was.

In 1957 Hodgkin & Keynes [44] injected radioactive calcium
into the axoplasm of squid giant axon.   Very little of
it diffused away from the injection site over ~2 h, unlike
other cations; so they reckoned that most of the axoplasmic
calcium was bound, and free Ca^{2+} was < 0.01 mM.   Ebashi
showed in 1963 [45] that murexide could serve as a colorimetric
indicator for Ca^{2+}, changing from orchid-purple to pink when
Ca^{2+} binds.   Jöbsis & O'Connor [46] injected it into a
toad and, in 1966, reported calcium transients correlating
with muscle contraction in isolated fibres.   But the break-
through came when Ridgway & Ashley [47] reported experiments
on injecting the bioluminescent protein aequorin into the
giant muscle fibre of the barnacle *Balanus nubilis* [see Ashley
& Lea in Vol. 6 as listed opposite title p. of this book - *Ed.*].

Macartney had reported in 1810 that extracts of luminous
jellyfish could apparently emit light in a vacuum, i.e.
without oxygen.   This was a surprise since all luminous
organisms, apart from coelenterates and radiolarians, require
oxygen to glow or flash.   The puzzle was solved by Shimomura
et al. (refs. in [9]) who isolated a jellyfish (*Aequorea*)
protein which produced a flash of blue light on adding
Ca^{2+}, but not K^+, Na^+ or Mg^{2+}.   This 'photoprotein' is
isolated as a complex containing the chemiluminescent substrate
and oxygen. Ca^{2+} binding triggers the chemiluminescent reaction.

Ridgway & Ashley observed a transient increase and then decrease in intracellular aequorin light emission, immediately following electrical excitation of the muscle. Interestingly, whilst the aequorin transient began just before contraction, in these first experiments it decayed while tension was still rising. John Blinks was also an early pioneer, using aequorin from Friday Harbor jellyfish.

Aequorin has now been used not only to demonstrate free Ca^{2+} changes in >50 cell types but also to quantify the cytosolic free Ca^{2+}. Portzehl et al. [41], using the Ca^{2+}-selective chelator EGTA synthesized by Schwartzenbach, reckoned that crab-muscle contraction needed >0.3 µM free Ca^{2+}; the level shown by aequorin was ~0.1-0.3 µM in resting cells and ~1-5 µM on activation. In 1972 I chose to work with a British relative of Aequorea, Obelia. The Ca^{2+}-activated photoprotein obelin is qualitatively very similar to aequorin; both enable free Ca^{2+} to be measured not only in activated cells but also after cell injury, when small molecules tend to leak out.

There are now 5 methods for measuring free Ca^{2+} in the cytoplasm of living cells [1, 3, 9, 48, 49] (see APPENDIX):
- (1) Ca^{2+}-activated photoproteins;
- (2) fluors;
- (3) metallochromic indicators;
- (4) NMR indicators;
- (5) Ca^{2+}-selective micro-electrodes.

During the 1970's Reynolds [50] pioneered the use of image intensification for visualizing the aequorin light signal to locate where within a cell Ca^{2+} changes occur. Up to 1980 most of the new observations about intracellular free Ca^{2+} were obtained using the photoproteins. However, at this point a brilliant innovation was made which heralded the subsequent explosion in free-Ca^{2+} measurement and localization in single cells - the development by Roger Tsien [51, 52] of fluors sensitive to Ca^{2+}, together with a way of getting them into the cytoplasm of live cells without micro-injection.

NEMESIS

Just a few years after Heilbrunn's death in 1959 the biochemical basis of internal Ca^{2+} buffering and release was discovered, together with the key molecular piece missing from Heilbrunn's puzzle - high-affinity Ca^{2+}-binding proteins. Methods became available not only for measuring and locating cytosolic free Ca^{2+} but also for manipulating cytosolic Ca^{2+} through specific buffers such as EGTA together with Ca^{2+} ionophores, antagonists and effectors of intracellular organelles.-

METHODS FOR MANIPULATING CYTOSOLIC FREE Ca^{2+}

1. *Micro-injection of calcium:* microsyringe or pressure injection; iontophoresis; electroporation.
2. *Calcium chelators (external or micro-injected):* alizarin sulphonate; oxalate; citrate; EGTA; 'BAPTA' [bis(*o*-amino-phenoxy)ethanetetra-acetate] as its acetoxy methyl ester.
3. *Ca^{2+} ionophores:* A23187; ionomycin.
4. *Membrane effectors:* p.m. channel blockers; mitochondrial, e.g. ruthenium red; endoplasmic reticulum, e.g. dantrolene.
5. *Caged compounds:* nitr 5; InsP$_3$.

A recent culmination is the ability to increase Ca^{2+} at specific sites within the cell using laser light to release Ca^{2+} from photosensitive or 'caged' substances such as nitr 5 and 'caged' InsP$_3$.

Yet during the 1950's and 1960's two groups of experimental observations caused great confusion: the uptake of Ca^{2+} by isolated mitochondria and the inhibition by Ca^{2+} of many intracellular regulatory enzymes of intermediary metabolism (refs. in [1]). Many of these turned out to be artefacts as they were obtained at very unphysiological Ca^{2+} concentrations, sometimes 10,000 times that in the resting cell's cytosol. Mitochondria may play a role in intracellular Ca^{2+} buffering, localizing Ca^{2+} transients to part of the cell, and have their own Ca^{2+}-regulated enzymes. But the releasable Ca^{2+} store is isolated from broken cells with the endoplasmic reticulum.

The Ca^{2+} store in muscle had in fact been observed by Veratti in 1902 under the light microscope and by Bennett & Porter in 1953 in the electron microscope. The crucial observation was, however, that by Marsh in 1951 (refs. in [1]) - that vesicles from muscle could 'relax' contracted myofibrils. By the early 1960's Weber [53] and Ebashi [54] had shown that this vesicle fraction could take up Ca^{2+} and that this required ATP hydrolysis. But it remained unclear how the Ca^{2+} was released from the e.r. Two theories emerged: direct electrical coupling from the action-potential opening channels, or Ca^{2+}-induced Ca^{2+} release as found in the heart. In 1955 Hokin & Hokin [55] discovered an activation of the turnover of a minor PL, phosphatidyl inositol. During the 1960's and 1970's several workers reported a similar increase in turnover when cells were activated by neurotransmitters or hormones. However, it was Michell [56] who in the mid-1970's focused attention on this puzzling biochemical phenomenon as playing a role in Ca^{2+} gating. This led to the discovery by Berridge and co-workers [57] of two new intracellular signals formed by hydrolysis of phosphatidyl

A.K. Campbell

inositol triphosphate (InsP$_3$; formerly termed triphosphatidyl inositol): diacylglycerol and Ins(1,4,5)P$_3$.

The experiments of Fatt & Katz [58] on crab muscle in 1953 led to the discovery of voltage-sensitive Ca^{2+} channels in the p.m. Yet once the paucity of cytosolic free Ca^{2+} in resting cells was realized, then it was obvious that just a small increase in p.m. permeability, or a relatively small release from intracellular stores, would provoke a large fractional rise in cytosolic free Ca^{2+}. But how was the 10,000-fold gradient of Ca^{2+} maintained across the p.m.? The Ca^{2+}-activated Mg-ATPase in erythrocytes became the first well characterized p.m. pump in the 1960's, and likewise the Ca^{2+}-uptake Mg-ATPase in muscle sarcoplasmic reticulum. In the late 1960's Baker & Blaustein identified in invertebrate excitable cells another mechanism for extruding Ca^{2+} from cells, a Na$^+$-Ca^{2+} exchange. This turned out to occur in many cells where it seemed to help restore cytosolic free Ca^{2+} after a stimulus.

Although Bailey and Needham had reported activation by Ca^{2+} of myosin ATPase in 1942 (refs. in [1]), the molecular basis of Ca^{2+} as an intracellular trigger only became clear as a result of advances in methods for purifying proteins. In 1963 Ebashi [45] discovered a complex, troponin, which restored Ca^{2+} sensitivity to isolated actomyosin. The complex was soon shown to contain three proteins named troponin T, I and C, the latter being the Ca^{2+}-binding component. This led to a search for other calcium-binding proteins. In 1967 Cheung reported a Ca^{2+} activation of cyclic-AMP phosphodiesterase. But not till the 1970's did Kakiuchi, Cheung and others isolate the Ca^{2+}-binding protein responsible for non-proteolytic activation [59, 60], viz. 'calmodulin'. Many others, some unrelated, have since been isolated. [In Vol. 13 of this series, on receptors, A.R. Means has an especially pertinent article.- *Ed.*] A crucial aspect of cell activation arising from the work of Fisher and Krebs in the late-1950's was the activation of kinases which phosphorylate key regulatory and structural proteins.

Using X-ray crystallography first on parvalbumin and then on other Ca^{2+}-binding proteins a common structural feature, the so-called 'EF hand', was highlighted by Kretskinger and others during the 1970's [cf. Turnell's art. later in this vol.- *Ed.*]. This showed how the molecular structure had evolved to provide Ca^{2+} with 7 or 8 coordinations, *vs.* 6 for Mg^{2+}, enabling the protein to bind Ca^{2+} at μM concentrations, selectively over Mg^{2+} which is ~1-5 mM in the cytosol.

Editor's note.- The author's survey of new approaches at end of the book complements the following section.

NEW HORIZONS

A century of intracellular Ca^{2+} has led us from puzzling effects of removing external calcium, through a 'doubting-Thomas' congregation, to the now accepted dogma that cytosolic free-Ca^{2+} rises are central in molecular mechanisms underlying cell activation and cell injury. The realization of Heilbrunn's vision has depended critically on the development of methods for measuring and manipulating free Ca^{2+} in living cells, and for isolating organelles and proteins responsible for regulating free Ca^{2+} in cells and for mediating its effects.

Yet there is still much to learn. The primary role of Ca^{2+} as a trigger has still not been established in many phenomena. Its interaction with other intracellular signals is still poorly understood. The molecular basis of receptor- and voltage-operated channels, $InsP_3$-stimulated release and Ca^{2+} action is not known. No complete molecular scheme from stimulus to cellular event, e.g. secretion, exists. I have contended for some 15 years [1,9] that to fully understand the molecular biology of intracellular signals we first need to re-examine the cell biology.

Many phenomena in cell activation and injury are quantal, not graded as appears from biochemical studies of millions of cells. A cell moves or stays still, a luminous cell flashes or remains invisible, a cell aggregates or remains free, a vesicle fuses and secretes or it does not, a cell divides or remains dormant, a cell dies or remains alive. It is now cardinal to establish methods for single-cell analysis which will enable us to determine the chemisymbiosis necessary between intracellular signals and energy balance to provoke a threshold end-response. We have demonstrated such thresholds in reactive oxygen metabolite production in neutrophils, and in reversible cell injury by complement [61-63]. And what about the last $3\frac{1}{2} \times 10^9$ years? How did the intracellular signals evolve and become linked to threshold events associated with cell activation and injury? Perhaps the central holistic problem in biology is how step-by-step changes in DNA lead to the evolution of cellular events and structures, which only then become susceptible to the forces of Darwinian-Mendelian selection.

During the first century of intracellular calcium literally hundreds of scientists have made major contributions to our knowledge of the special biological features of this cation. In an age where winning gold medals at the scientific olympics seems more important than natural philosophy we can perhaps all receive comfort from the wisdom, quoted overleaf, of Bertrand Russell.

*"In art nothing worth doing can be done
without genius, in science even a very
moderate capacity can contribute something
to a supreme achievement."*

Bertrand Russell, 1917

Acknowledgements

I thank the Director and staff of the Marine Biological
Association, my research group over the past 16 years for much
hard work, and the MRC, SERC, ARC, MS Society, DHSS, Welsh
Office and Royal Society for financial support.

References

1. Campbell, A.K. (1983) *Intracellular Calcium: its
    Universal Role as Regulator*, Wiley, Chichester, 556 pp.
2. Heilbrunn, L.V. (1937; & 1943, 1952) *An Outline of
    General Physiology*, Saunders, Philadelphia.
3. Campbell, A.K. (1986) *Cell Calcium 6*, 287-296.
4. Steinbach, H.N. (1960) *Science 131*, 397-399.
5. Heilbrunn, L.V. (1927) *Arch. Exp. Zellforsch. 4*, 246-
    263.
6. Heilbrunn, L.V. (1928) *A Colloid Chemistry of Protoplasma*,
    Borntraeger, Berlin.
7. **Heilbrunn, L.V. (1940)** *Physiol. Zool. 13*, 88-94.
8. Heilbrunn, L.V. (1956) *The Dynamics of Living Protoplasm*,
    Academic Press, New York.
9. Campbell, A.K. (1988) *Chemiluminescence: Principles and
    Applications to Biology and Medicine*, VCH/Ellis Horwood,
    Weinheim & Chichester, 608 pp.
10. Davy, H. (1808) *Phil. Trans. Roy. Soc. 98*, 1-44 & 333-370.
11. Ringer, S. (1883) *J. Physiol. 4*, 29-43.
12. Ringer, S. (1886) *J. Physiol. 7*, 291-308.
13. Mines, G.R. (1911) *J. Physiol. 42*, 251-166 [& (1910) *40*,
    327-346].
14. Straub, W. (1912) *Verh. Ges. Dtsch. Naturforsch. Artze
    84*, 194-214.
15. Stiles, P.H. (1903) *Am. J. Physiol. 40*, 327-346.
16. Locke, F.D. (1894) *J. Physiol. 15*, 166-167.
17. Overton, E. (1902) *Pflügers Arch. Gesamte Physiol. 42*,
    346-386.
18. Chambers, R. & Reznikoff, P. (1926) *J. Gen. Physiol. 8*,
    396-401.
19. Pantin, C.F.A. (1926) *Br. J. Exp. Path. 3*, 275.
20. Gray, J. (1924) *Proc. Roy. Soc. B 96*, 95-114 [& (1922)
    93, 122-131].
21. Ringer, S. (1890) *J. Physiol. 11*, 79-84.
22. Herbst, C. (1900) *Arch. Entwicklungsmech. 9*, 424-463.
23. Ringer, S. & Sainsbury, H. (1894) *J. Physiol. 16*, 1-9.

24. Loeb, J. (1922) *Problems and Theory of Colloid Behavior*, Columbia Univ. Press, New York.
25. Lawaczek, H. (1928) *Dtsch. Arch. Klin. Med. 160*, 302-309.
26. Von der Wense [(1934) *Nauyn-Schm.Arch. Pharmakol. 176*] cited by Hanström, B. (1939) in *Hormones in Invertebrates*, Clarendon Pr.
27. Loewi, D. (1918) *Arch. Exp. Path. Pharmakol. 82*, 131-158.
28. Loew, D. (1892) *Flora 75*, 368-394.
29. Harvey, A.M. & MacIntosh, F.C. (1940) *J. Physiol. 97*, 408-416.
30. Loeb, J. (1922) *The Dynamics of Living Matter*, Columbia Univ. Press, New York.
31. Heilbrunn, L.V. & Young, R.A. (1930) *Physiol. Zool. 3*, 330-341.
32. Heilbrunn, L.V. & Wilbur, K. (1934) *Biol. Bull. 75*, 557-564.
33. Tyler, A. (1941) *Biol. Rev. 16*, 291-336.
34. Just, E.E. (1919) *Biol. Bull. 36*, 1.
35. Chambers, R. (1921) *Biol. Bull. 41*, 318.
36. Mazia, D. (1937) *J. Cell. Comp. Physiol. 10*, 291.
37. Chambers, R. & Hale, H.P. (1932) *Proc. Roy. Soc. B 110*, 336-352.
38. Keil, E.M. & Sichel, F.J.M. (1936) *Biol. Bull. 41*, 402.
39. Kamada, T. & Kinoshita, H. (1943) *Jpn. J. Zool. 10*, 469-493.
40. Heilbrunn, L.V. & Wiercinski, F.J. (1947) *J. Cell. Comp. Physiol. 29*, 15-32.
41. Portzehl, H., Caldwell, P.C., & Rüegg, J.C. (1964) *Biochim. Biophys. Acta 79*, 581-591.
42. Macullum, A.B. (1925) *Science 62*, 511.
43. Pollack, H. (1928) *J. Gen. Physiol. 11*, 539-545.
44. Hodgkin, A.C. & Keynes, R.D. (1957) *J. Physiol. 238*, 253-281.
45. Ebashi, S. (1963) *Nature 200*, 1010.
46. Jöbsis, F.F. & O'Connor, M.J. (1966) *Biochem. Biophys. Res. Comm. 25*, 246-252.
47. Ridgway, E.B. & Ashley, C.C. (1967) *Biochem. Biophys. Res. Comm. 29*, 229-254.
48. Ashley, C.C. & Campbell, A.K. (1979) *Detection and Measurement of Free Ca^{2+} in Cells*, Elsevier/N. Holland, Amster-
49. Campbell, A.K. (1987) *Clin. Sci. 72*, 1-10.   [dam.
50. Reynolds, G.T. (1981) *Photochem. Photobiol. 27*, 405-421.
51. Tsien, R.Y. (1981) *Biochemistry 19*, 2396-2404.
52. Poenie, M., Tsien, R.Y. & Schmitt-Verhulst, A.M. (1971) *EMBO J. 6*, 2223-2232.
53. Weber, A. (1959) *J. Biol. Chem. 234*, 599-605.
54. Ebashi, S. (1961) *J. Biochem. 50*, 236-244.
55. Hokin, M.R. & Hokin, L.E. (1958) *J. Biol. Chem. 203*, 967-977.

56. Michell,R.H. (1975) *Biochim. Biophys. Acta 415*, 81-147.
57. Berridge, M.J. (1984) *Biochem. J. 220*, 345-360.
58. Fatt, P. & Katz, B. (1953) *J. Physiol. 120*, 171-194.
59. Cheung, W.Y. (1967) *Biochem. Biophys. Res. Comm. 29*, 478-482.
60. Kakiuchi, S., Yamazaki, R. & Nakajima, H. (1970) *Proc. Japan Acad. 46*, 587-592.
61. Patel, A.K., Hallett, M.B. & Campbell, A.K. (1987) *Biochem. J. 248*, 173-180.
62. Patel, A.K. & Campbell, A.K. (1987) *Immunology 60*, 135-141.
63. Morgan, B.P., Luzio, J.P. & Campbell, A.K. (1986) *Cell Calcium 7*, 399-411.

APPENDIX

with item numbers 1. to 8. corresponding to those on p. 8

Indicators for measuring intracellular free Ca^{2+}.

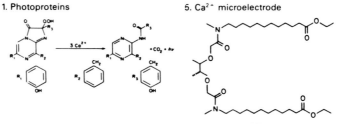

1. Photoproteins

5. Ca^{2+} microelectrode

ETH1001

2. Metallochromic dyes

Arsenazo III

Antipyrylozo III

Murexide (ammonium purpurate)

3. Fluorescent indicators

Quin 2

Chlortetracycline

Fura-2

Indo-1

4. Nuclear magnetic resonance indicators

F-bis(*o*-aminophenoxy)ethanetetra-acetate (FBAPTA)

Section #A

PLASMA MEMBRANE; Ca^{2+} MOVEMENT; MUSCLE AND NERVE

#A-1

THE CALCIUM PUMP OF THE PLASMA MEMBRANE

[1]Ernesto Carafoli, [1]Emanuel Strehler, [1]Thomas Vorherr,
[1]Peter James, [2]Anil K. Verma and [2]John T. Penniston

[1]Laboratory of Biochemistry,   [2]Department of Biochemistry
   Swiss Federal Institute of   and Molecular Biology,
   Technology (ETH),   Mayo Clinic,
   Zurich, Switzerland   Rochester, MN 55905, U.S.A.

The sequence of amino acids (1220) in the p.m. pump
of human erythrocytes has been established by protein chemistry
and DNA cloning techniques using a human teratoma library
in conjunction with oligonucleotides based on pump fragments.
Asp-475 is concerned with acyl phosphate formation, and Lys-601
binds the ATP antagonist FITC. Use of a photo-activatable
cross-linker has identified a CaM-binding stretch near the
C-terminus. Near this stretch are sequences that may have
Ca^{2+}-binding roles. Hydrophobic domains are lacking in the
mid-portion. Through phosphorylation of Ser-1178, the pump's
Ca^{2+}-affinity can be increased. Isoforms of the pump have
been demonstrated.*

The p.m. Ca^{2+} pump is an enzyme with high affinity for
Ca^{2+}, which is present in all eucaryotic p.m.'s studied thus
far. It is an ATPase of the P-type [1, 2], whose essential
properties may be summarized thus.-

Mechanism: P-type, forms aspartyl phosphate
Inhibition: vanadate
Ca^{2+} affinity: CaM (K_m ~0.5 µM)
Activators: fatty acids/acidic phospholipids
                        (K_m ~0.3 µM)
                     cAMP-dependent phosphorylation
                        (K_m ~1 µM)
Charge compensation: proton exchange, possibly neutral.

Of particular interest is the stimulation by CaM [3, 4],
due not to a phosphorylation reaction but to the direct
interaction of CaM with the ATPase. The direct interaction
has been exploited to isolate the ATPase, first from the
erythrocyte membrane [5] and then from a number of other

*Abbreviations.- CaM, calmodulin; FITC, fluorescein isothio-
cyanate; p.m., plasma membrane(s).

p.m.'s [6-8] using CaM affinity chromatography columns. The purified enzyme has M_r ~138,000 on SDA-polyacrylamide gel electrophoresis, and appears to be functionally competent [9]: *i.e.* it can be reconstituted into liposomes with optimal Ca^{2+}-transporting efficiency.

Work aimed at establishing the primary structure of the pump has been preceded in the laboratory by extensive proteolytic work on the purified erythrocyte enzyme [10, 11] which has led to the proposal of a model for the architecture of the ATPase in the membrane in which the CaM-binding domain of the molecule is located near its C-terminus, in a domain having M_r ~9,000. Subsequent work using a combination of CNBr proteolysis and specific labelling has led to the isolation of the domain and to the determination of its sequence [12], as now outlined.

THE CALMODULIN- AND Ca^{2+}-BINDING DOMAINS

CaM was labelled with a bifunctional, cleavable, radio-active and photo-activatable cross-linker [13] which conjugates to primary amines. The radioactive reagent probably labelled Lys-75/148 of CaM. Labelled CaM was then cross-linked to a purified erythrocyte ATPase (a large batch) and the azo linkage in the reagent cleaved with dithionite. The cleavage left the ^{125}I-label attached to the ATPase, very likely to its CaM-binding domain. CaM and the portion of the reagent attached to it were removed by dialysis through membranes with a cut-off at M_r 50,000, and the labelled ATPase was then fragmented by extensive CNBr cleavage. The resulting peptides were separated by HPLC, and the only water-soluble labelled peptide, which evidently contained the CaM-binding domain of the enzyme, was sequenced, yielding the following structure:

NH₂-E-L-R-R-G-Q-I-L-W-F-R-G-L-N-R-I-Q-T-Q-I-K-V-V-N-A-F-S-S-L-H-E-F

The sequence shows homology to the CaM-binding domains of a number of CaM-modulated proteins [e.g. 14, 15]. Most prominent is the abundance of positively charged amino acids, the presence of a tryptophan in the N-terminal portion of the domain, and the propensity of the domain to form an amphiphilic helix, as shown in the helix wheel model of Fig. 1. Table 1 compares some of the aligned sequences of the CaM-binding domains so far available in the literature, and stresses the points of homology mentioned above. However, the conclusion that all CaM-binding domains have a canonical primary structure would be premature: thus, with (e.g.) brain calcineurin and adenylate cyclase from *Bordetella pertussis* as recently sequenced [16, 17], besides homologies with the domains in Table 1 (notably the abundance of positively charged residues) significant differences are observed, including the presence of negatively charged amino acids. Possibly more

Table 1. A comparison of the sequences of CaM-binding domains of some CaM-modulated proteins.
MLCK-RS20: smooth muscle myosin light-chain kinase [18];
MLCK-M13: skeletal muscle MLCK [14];
PFK: skeletal muscle phosphofructokinase [15];
SPEC: human erythrocyte spectrin β-subunit [19]:
FODRIN: human brain fodrin [20];
ADC: adenylate cyclase from *Bordetella pertussis* [17].
CN: brain calcineurin subunit A [16]

Ca ATPase :	LRRGQIL WFRGLNR IQTQ IKVVNA FSSS
MLCK : Fragment RS 20	ARRK WQKTGHA VRAIGR LSSM
MLCK : Fragment M 13	MKRR WKKNFIA VSAA NRF KK ISSS
PFK : Fragment M 10 + 11	RSFMNN WEVYKLL AH IR
SPEC :	RKDNILRLWSYLQEL LQ SRRQR LETT
FODRIN :	SKTASP WKSAR LM VHTVAT FNSIK
ADC :	RERIDLL WK IARAGARSAVGTEAK
CN :	ARKEVIRNIRAIGKMARVFSVLR

than one kind of CaM-binding domain will emerge from the analysis of additional sequences, and it is interesting to imagine that different domain types will be dictated by the type of (enzyme) protein modulated by CaM.

Proteolysis work ([10]; see above) had suggested a C-terminal location for the CaM-binding domains of the Ca²⁺ pump. Conclusive proof has now come from settling the primary structure of the Ca²⁺ pump of rat-brain [21] and, as now outlined, of human teratoma [22], using protein chemistry and DNA cloning.

SEQUENCING OF THE HUMAN ENZYME

Tryptic fragments from the purified erythrocyte Ca²⁺-pump were sequenced and used to synthesize suitable oligonucleotide probes to screen a human teratoma λgt10 cDNA library: the two amino acid sequences originally used were FAMGIA and MTHPEF. The 17mer oligonucleotides obtained were ³²P-end-labelled, mixed, and used to screen ~1 × 10⁶ plaques of human teratoma library constructed in λgt10 (kindly donated by Drs. Maxine Singer and Jacek Skowronski of the National Cancer Institute, Bethesda, MD). The initial screening yielded three positive clones, but only one hybridized to both probes. This clone (t19c) was purified and analyzed by restriction mapping and nucleotide sequencing: it was ~1.2 kb long and contained two stretches that matched with the oligonucleotide probes in the only open reading frame that encompassed its entire sequence (Fig. 2).

Re-screening of the library with subfragments of the clone insert yielded several additional clones, the longest ~3.4 kb long [17]. The reading frame was found to remain open up to the 5'-end of cDNA t6, indicating that the sequence coding for the N-terminus of the ATPase had not been reached. The library was therefore re-screened with an oligonucleotide corresponding to the 5'-end sequence of cDNA clone t6 and with a mixture of oligonucleotides corresponding to a peptide of the erythrocyte pump (EGDFGC) not contained in the translated sequence of clone t6 and thus assumed to belong to the N-terminal portion of the pump. The re-screening produced a clone (t8.1) which overlapped with clone t6 at its 3'-end and extended it by >1 kb towards its 5'-end. It contained a region which matched the oligonucleotide probe made on the basis of the sequence EGDFGC in 14 of the first 15 positions (in the clone this region read DGDFGI). The combined nucleotide sequence of clones t19, t6 and t8.1 and the deduced amino acid sequence are presented in Fig. 3. The reading frame as pre-determined by the sequence of clone t6 continued into clone t8.1 and most likely began at the first in-frame Met residue (nucleotide positions 1-3 in Fig. 3).

The deduced amino acid sequence of teratoma clone t8.1 and that of several tryptic peptides of the erythrocyte pump show differences, indicating that at least two isoforms of the pump exist in humans. Analysis of the sequence of ~600 amino acids from the erythrocyte pump shows only ~86% identity with the corresponding deduced sequences of the teratoma clone(s). As expected, however, the sequences surrounding the functionally essential domains of the pump [e.g. the ATP (FITC) binding domain] are almost identical. Fig. 4 shows the translated amino acid sequence of the clones, and identifies the residues and domains. Ten hydrophobic transmembrane domains are identified tentatively by hydrophobicity plots with a window size of 18 and the Kyte-Doolittle algorithm [23]. They may correspond to residues 105-127, 156-173, 382-404, 418-438, 855-878, 885-904, 929-953, 970-990, 1004-1026 and 1040-1060.

Conclusive information on the number and location of the transmembrane domains may well come from immunological work with synthetic epitopes corresponding to the putative sequences connecting these domains. Interestingly, on

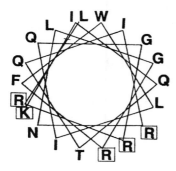

Fig. 1. A helix-wheel model of the CaM-binding domain of the p.m. Ca²⁺-ATPase. The basic residues are boxed. *From [12], courtesy of Journal of Biological Chemistry.- Figs. 2 & 3 (from [22]) likewise.*

Fig. 2 *(right)*. Partial restriction map and sequencing strategy of cDNA clones t19, 78.1 and t6 which code for the human teratoma p.m. Ca²⁺-ATPase.

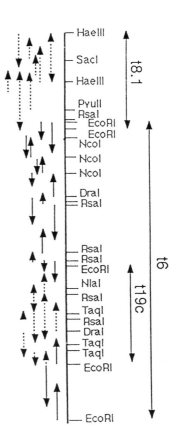

comparing the distribution of the putative hydrophobic membrane spanning domains in a number of other ion pumps of the P-class, it appears that the domains tend to be 'clustered' in the vicinity of the C-terminus and in the N-terminal region. A relatively large middle portion of the enzyme molecules is invariably free of hydrophobic domains [24-27]. However, the overall sequence identity of the p.m. Ca²⁺ pump with the other ion transporters of the P-type is rather low, and is striking only in the regions connected with the catalytic function and in the 'hinge' region which is assumed to permit the motion necessary to bring the ATP (FITC) binding site and the critical aspartic acid residue close to each other in space.

The ATP binding site as examined with the ATP antagonist FITC [28] is identified in Fig. 4 as Lys-601, and the residue forming the phosphoenzyme as Asp-475. Both residues are

[continued on p. 25

[#A-1

[continued on facing p.

[continued on right

Commencing on facing p.:

Fig. 3. Cloned nucleotide sequence of the p.m. Ca^{2+}-ATPase, with translation and aligned peptides. The *first* row of each segment contains the pointers; the *second* the nucleotide sequence, the *third* the translated sequence, and the *fourth* some erythrocyte-pump aligned tryptic peptides. The *pointers row* contains a *period* every tenth nucleotide, and an *arrowhead* (v) at each tenth amino acid. It also identifies some important residues and domains: **p**, aspartyl phosphate site; **f**, FITC-binding site; **c**, beginning of the CaM-binding domain (*panel on right, halfway down; for* **p** *see top of 2nd panel*).

Source: ref. [22], courtesy of J. Biol. Chem.

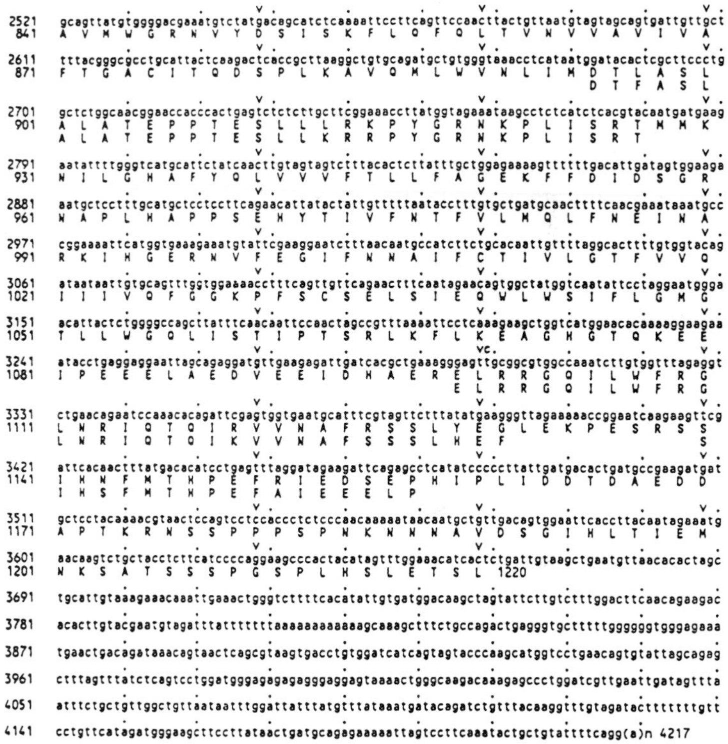

```
   1 M G D M A N N S V A Y S G V K N S L K E A N H D G D F G I T L A E L R A L M E L
  41 R S T D A L R K I Q E S Y G D V Y G I C T K L K T S P N E G L S G N P A D L E R
  81 R E A V F K K N F I P P K K P K T F L Q L V W E A L Q D V T L I L E I A A I V L
 121 S L G L S F Y Q P P E G D N A L C G E V S V G E E E G E G T G W I E G A A I L
 161 L S V V C V V L V T A F N D W S K E K Q F R G L Q S R I E Q E Q K F T V I R G G
 201 Q V I Q I P V A D I T V G D I A Q V K Y G D L L P A D G I L I Q G N D L K I D E
 244 S S L T G E S D H V K K S L D K D P L L L S G T H V R E G S G R M V V T A V G V
 281 N S Q T G I I F T L L G A A M E M Q P L K S E E G G D E K D K K K A N L P K K E K S
 321 N K A K A Q D G A A M E M Q P L K S E E G G D E K D K K K A I T V I L L V L Y F V I D T F W V Q
 361 V L Q G K L T K L A V Q I G K A G L L M S F F I I G V T V L V V A V P E G L P L A V T
 401 K R P W L A E C T P I Y I Q Y F V K F H L D A C E T M G N A T A I C S D K T G T L
 441 I S L A Y S V K K M M K D N N L V R H Y K K K V P E P E A I P P N I L S Y L V T G I S V
 481 T M N R M T V V Q A Y I N E K H Y T L V F N T F E C A L L G L L D L K R D Y
 521 N V A Y T S K I L P P E K E G G L P R H V G N K T V L K N S D G S Y R I F S
 561 Q D V R N E I P E E A L Y K V Y T F N S V R K S M S T V L K N S D G S Y R I E P M
 601 K G A S E I I L K K C F K I L S A N G E A K V F R P R D R D D I V K T V I E P M
 641 A S E G L R T I C L A F R D F P A G E P E P E W D N E N D I V T G L T C I A V V
 681 G I E D P V R P E V P D A I K K C Q R A G I T V R M V T G D N I N T A R A I A T
 721 K C G I L H P G E D F L C L E G K D F N R R I R N E K G E I E Q E R I D K I W P
 761 K L R V L A R S S P T D K H T L V K G I I D S T V S D Q R Q V V A V T G D G T N
 801 D G P A L K K A D V G F A M G I A G T D V A K E A S D I I L T D D N F T S I V K
 841 A V M W G R N V Y D S I S K F L Q F Q L T V N V V A V I V A F T G A C I T Q D S
 881 P L K A V Q M L W V N L I M D T L A S L A L A T E P P T E S L L R K P Y G R N
 921 K P L I S R T M M K N I L G H A F Y Q L V V V F T L L F A G E K F D I D S G R
 961 N A P L H A P P S E H Y T I V F N T F V L M Q L F N E I N A R K I H G E R N V F
1001 E G I F N N A I F C T I V L G T F V V Q I I V Q F G G K P F S C S E L S I E Q
1041 W L W S I F L G M G T L L W G Q L I S T I P T S R L K F L K E A G H G T Q K E
1081 I P E E L A E D V E E I D H A E R E L R R G Q I L W F R G L N R I Q T Q I R V
1121 V N A F R S S L Y E G L E K P E S R S S I H N F M T H P E F R I E D S E P H I P
1161 L I D D T D A E D D A P T K R N S S P P P S P N K N N A V D S G I H L T I E M
1201 N K S A T S S S P G S P L H S L E T S L 1220
```

Fig. 4. Amino acid sequence of the p.m. Ca²⁺-ATPase. The putative trans-membrane helices are signified by *boxed* domains, including the aspartic acid where the phosphoenzyme is formed and the lysine which binds the ATP antagonist FITC. *Also boxed:* the CaM-binding domain, and a negatively charged domain N-terminal to it (a putative Ca²⁺-binding site).

Fig. 5. A model of the
interaction of CaM with
the p.m. Ca^{2+} pump, and
of its effect on Ca^{2+}
binding. The hydrophilic
protruding C-terminal
domain contains the positively charged CaM binding
domain (marked **++**) and, N-
terminal to it, a domain
rich in acidic residues
(**--**). The 2 domains interact in the non-activated
enzyme (*top*; CaM absent).
With Ca^{2+} present CaM swings
its binding domain away from
the acidic sequence, which
is now free to bind Ca^{2+}.

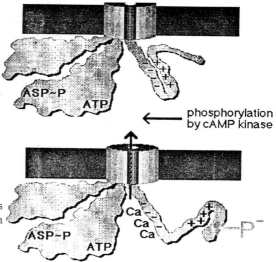

located in the enzyme's hydrophilic middle portion, which
protrudes into the intracellular space. Fig. 4 also shows
that the proposal of a C-terminal location of the CaM-binding
domain [10] was indeed correct. The domain is indeed located
next to the C-terminus of the ATPase. In Fig. 3 is identified,
immediately N-terminal to the CaM-binding domain, a 20-residue
stretch which is particularly rich in glutamic and aspartic
acids (1079-1099), and could thus function as a potential
Ca^{2+}-binding site. In the absence of CaM the acidic domains
would be Ca^{2+}-free and bound to the basic CaM-binding domain
next to it. CaM would interact with its binding domain,
removing from it the acidic sequence which would thus be
free to bind Ca^{2+}.

The scheme shown in Fig. 5 summarizes the proposal.
Although the scheme is very hypothetical, recent work in
our laboratory using synthetic peptides has provided indications in favour of the model. Initially this work showed
that the synthetic basic CaM-binding domain interacts with
CaM in a Ca^{2+}-dependent manner. Then it has further showed
that the putative Ca^{2+}-binding domain in the absence of
Ca^{2+} indeed interacts very strongly with the basic CaM-binding
domain C-terminal to it, and far less strongly in the absence
of Ca^{2+}. Very recent work has given indications that the
formation of the complex between CaM and its binding domain
in the presence of Ca^{2+} induces a conformational change in
the putative acidic Ca^{2+}-binding domain prepared synthetically.

The problem of the Ca^{2+}-binding domain of the ATPase
has another interesting aspect. A search of the translated

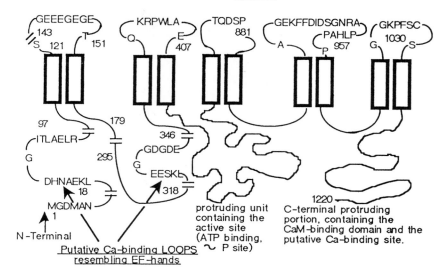

Fig. 6. A scheme of the architecture of the Ca^{2+} pump in the p.m. The boxes represent the 10 putative transmembrane domains. The small stretches of protein protruding on the outside are shown in full, whereas the large hydrophilic domains protruding into the cytosolic space are not. The cytosolic domain between transmembrane helices 3 and 4 contains >400 residues, including the catalytic site(s) and the 'hinge' region (see text). The two loops representing Ca^{2+}-binding EF-hands are formed by residues 22 through 33 and 310 through 321. The numbers shown indicate the approximate position of the domains in the structure.

sequence of the entire teratoma clone for similarities with an average Ca^{2+}-binding loop of the 'EF-hand' type found in Ca^{2+}-binding proteins [29] has identified two regions (22–33 and 310–321) which scored even higher than some classical EF-hand proteins [22, 30–35]. The two putative EF-hand loops are indicated in the scheme of Fig. 6, which summarizes the architecture of the p.m. pump. Both putative EF-hand loops show differences from the classical Ca^{2+}-coordinating loops of EF-hand structures [22] and, in addition, the flanking sequences of the regions do not show the strong propensity for α-helix formation expected of true EF-hand structures. Thus, caution is necessary in postulating a function in Ca^{2+}-binding for the two EF-hand-type regions identified here. They could, however, play a regulatory one in the process of Ca^{2+} binding.

DOMAIN FOR cAMP-DEPENDENT PUMP PHOSPHORYLATION

Very recent work in this laboratory has identified in the pump structure another domain of possible functional importance, viz. the substrate sequence for the cAMP-dependent protein kinase. The Ca²⁺ pump is stimulated by cAMP-dependent phosphorylation [36], and the stimulation is due to the phosphorylation of the pump molecule proper [37], at variance with the case of the sarcoplasmic reticulum pump, where the stimulation is linked to the phosphorylation of the accessory protein phospholamban. The stimulation corresponds to the lowering of the Km (Ca²⁺) of the pump from the normal low affinity value of ~15–20 μM to a value of ~1 μM. Added CaM, however, stimulates further, to the optimum activated level of ~0.5 μM.

Samples of purified erythrocyte ATPase were labelled with [γ-³²P]ATP and then split with CNBr [38]. The digestion products were then separated by HPLC, and the only radioactive peak found was subjected to sequence analysis. The following sequence was obtained:

-T-H-P-E-R-R-I-E-D-S-E-P-H-I-P-L-I-D-D-T-T-A-E-D-D-A-P-T-K-R-N-S-X-P-P-(P)-S-P-D-K-N.
                                                          ↑

In the pump sequence shown in Figs. 3 and 4 this domain can be easily recognized next to the C-terminus downstream from the CaM-binding domain. However, the match is not complete, showing once again that the erythrocyte pump and the pump cloned from teratoma cells are isoforms. The arrowed (↑) gap X was a sequencing cycle in which no new amino acid peak became prominent although in the preceding cycle the serine decreased. In the sequence shown in Figs. 3 and 4 the gap would correspond to a serine, and it thus appeared logical to postulate that the seemingly blank cycle corresponded to a phosphoserine.- The stretch Lys-Arg-Asn-Ser-Ser would clearly be a canonical substrate site for the cAMP-dependent protein kinase. Derivatization of the labelled peptide with ethanethiol to convert phosphoserine into ethylcysteine, which can now be identified in the normal amino acid sequencing procedures, showed that the cycle gap indeed corresponded to a phosphoserine.

Extension of the model.- The data on the cAMP phosphorylation can be used to extend the model for ATPase regulation by CaM put forward in Fig. 5. In the model shown in Fig. 7, the C-terminal domain containing the substrate sequence for the cAMP-directed phosphorylation has a strong acid character, and could thus interact in the pump in situ with the basic CaM-binding domain. The latter domain, on the other hand, probably interacts also with the acidic sequence (the putative Ca²⁺-binding site; see above) N-terminal to it. The conformational change induced by the cAMP-directed phosphorylation could swing the CaM-binding domain away from the putative Ca²⁺-binding sequence, allowing Ca²⁺ free access to it.

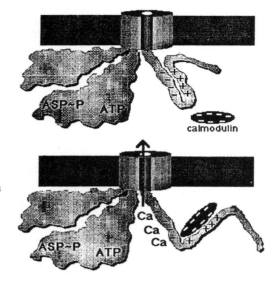

Fig. 7. A model of the effects of the phosphorylation by the cAMP-dependent kinase on the activity of the p.m. Ca^{2+} pump and on its interaction with Ca^{2+} and CaM.

References

1. Pedersen, P.L. & Carafoli, E. (1987) *Trends Biochem. Sci.* *12*, 146-160.
2. Pedersen, P.L. & Carafoli, E. (1987) *Trends Biochem. Sci.* *12*, 186-189.
3. Gopinanth, R.M. & Vincenzi, F.F. (1977) *Biochem. Biophys. Res. Comm. 77*, 1203-1209.
4. Jarrett, H.W. & Penniston, J.T. (1977) *Biochem. Biophys. Res. Comm. 77*, 1210-1216.
5. Niggli, V., Penniston, J.T. & Carafoli, E. (1979) *J. Biol. Chem. 254*,   9955-9998.   [3270.
6. Caroni, P. & Carafoli, E. (1981) *J. Biol. Chem. 256*, 3263-
7. Hakim, G., Itano, T., Verma, A.K. & Penniston, J.T. (1982) *Biochem. J. 207*, 225-231.
8. Michalak, M., Famulski, K. & Penniston, J.T. (1984) *J. Biol. Chem. 259*, 15540-15547.
9. Niggli, V., Adunyah, E.S., Penniston, J.T. & Carafoli, E. (1981) *J. Biol. Chem. 256*, 395-401.
10. Zurini, M., Krebs, J., Penniston, J.T. & Carafoli, E. (1984) *J. Biol. Chem. 259*, 618-627.
11. Benaim, G., Zurini, M. & Carafoli, E. (1984) *J. Biol. Chem. Chem. 259*, 8471-8477.
12. James, P., Maeda, M., Fischer, R., Verma, A.K., Krebs, J., Penniston, J.T. & Carafoli, E. (1980) *J. Biol. Chem. 263*, 2905-2910.
13. Jaffe, C.L., Lis, H. & Sahnon, N. (1980) *Biochemistry 19*, 4423-4429.

14. Blumenthal, D.K., Takis, K., Edelman, A.M.,
    Charbonneau, H., Titani, K., Welsch, K.A. & Krebs, E.G.
    (1985) *Proc. Nat. Acad. Sci. 82*, 3187-3191.
15. Buschmeier, B., Meyer, H.E. & Mayr, G.W. (1987) *J. Biol.
    Chem. 262*, 9454-9462.
16. Kincaid, R.L., Nightingale, M.S. & Martin, B.M. (1988)
    Proc. Nat. Acad. Sci. 85, 8983-8987.
17. Glaser, P.,Sakamoto, H., Ballabou, J., Ullmann, A. &
    Denchin, A. (1988) *EMBO J. 7*, 3997-4004.
18. Lukas, T.J., Burgess, W.H., Prendergast, F.G., Lan, W. &
    Watterson, D.M. (1986) *Biochemistry 25*, 1458-1464.
19. Anderson, J.P. & Morrow, J.S. (1988) *Work quoted in 20.*
20. Harris. A.S., Croall, D.E. & Morrow, J.S. (1988) *J. Biol.
    Chem. 263*, 15754-15761.
21. Shull, G.E. & Greeb, J. (1988) *J. Biol. Chem. 263*, 8646-8657.
22. Verma, A.K., Filoteo, A.G., Stanford, D.R., Wieben, E.D.,
    Penniston, J.T., Strehler, E.E., Fischer, R., Heim, R.,
    Vogel, G., Mathews, S., Strehler-Page, M.A., James, P.,
    Vorherr, T., Krebs, J. & Carafoli, E. (1988) *J. Biol.
    Chem. 263*, 14152-14159.
23. Kyte, J. & Doolittle, R.F. (1982) *J. Mol. Biol. 157*, 105-123.
24. MacLennan, D.H., Brandl, C.J., Korczak, B. & Green, N.M.
    (1985) *Nature 316*, 696-700.
25. Kawakami, K., Ohta, T., Nojima, H. & Nagano, K. (1986) *J.
    Biochem. (Tokyo) 100*, 389-397.
26. Serrano, R., Kielland-Brandt, M.C. & Fink, G.R. (1986)
    Nature 319, 689-692.
27. Shull, G.E. & Lingrel, J.B. (1986) *J. Biol. Chem. 261*,
    16788-16791.
28. Filoteo, A.G., Gorski, J.P. & Penniston, J.T. (1987) *J.
    Biol. Chem. 262*, 6526-6530.
29. Kretsinger, R.H. (1980) *CRC Crit. Rev. Biochem. 8*, 119-174.
30. Sasagawa, T., Ericsson, L.H., Walsh, K.A.. Schreiber, W.E.,
    Fischer, E.H. & Titani, K. (1982) *Biochemistry 21*, 2565-2569.
31. Swan, D.G., Hall, R.S., Dhillon, N. & Leadlay, P.F. (1987)
    Nature 329, 84-85.
32. Ohno, S., Emori,Y., Imajjoh, S., Kawasaki, H., Kisargi, M.
    & Suzuki, K. (1984) *Nature 312*, 566-579.
33. McManus, J.P., Watson, D.C. & Yaguchi, M. (1985) *Biochem.
    J. 235*, 566-579.
34. Hochstrasser, K., Illchmann, K. & Werle, E. (1970) *Hoppe-
    Seyler's Z. Physiol. Chem. 351*, 721-738.
35. Marashi, F., Helms, S., Shiels, A., Silverstein, S.,
    Greenspan, D., Stein, G. & Stein, J. (1985) *Biochem. Cell
    Biol. 64*, 277-298.
36. Caroni, P. & Carafoli, E. (1981) *J. Biol. Chem. 256*,
    9371-9373.
37. Neyses, L., Reinlib, L. & Carafoli, E. (1985) *J. Biol.
    Chem. 260*, 10283-10287.
38. James, P., Pruschi, M., Vorherr, T., Penniston, J.T. &
    Carafoli, E. (1989) *Biochemistry*, in press.

#A-2

REGULATION OF THE REVERSAL CYCLE OF THE CALCIUM PUMP FROM KIDNEY PROXIMAL TUBULES

[1]Adalberto Vieyra, [1]Celso Caruso–Neves and
[1,2]José Roberto Meyer–Fernandes

[1]Departamento de Bioquímica Médica,
 Instituto de Ciências Biomédicas,
 Centro de Ciências da Saúde,
 Universidade Federal do Rio de Janeiro,
 21910 Rio de Janeiro, Brazil

and [2]Departamento de Biologia Celular e Tecidual,
 Universidade do Estado do Rio de Janeiro,
 20550 Rio de Janeiro, Brazil

Ca^{2+}-ATPase from kidney proximal tubule p.m.[*] can catalyze simultaneous cycles of ATP hydrolysis and synthesis (ATP$\rightleftharpoons$$^{32}P_i$ exchange reaction). At Ca^{2+} concentrations that saturate the high-affinity, cytosolic binding site, the enzyme becomes saturated in both directions, but ATP synthesis requires a transmembrane Ca^{2+} gradient. Ca^{2+} concentrations >50 μM activate forward cycles but progressively inhibit the exchange with native vesicles. With solubilized enzyme, activation of the enzyme cycle needs high pH and mM Ca^{2+} concentrations, such that ATP hydrolysis is inhibited. The exchange reaction exhibits Michaelian behaviour with respect to the ligands P_i and Mg^{2+}, and its rate is higher the lower the ATP/ADP ratio. The osmotically active solutes urea and TMA$\rightarrow$O (found in renal tissue), whilst not affecting active Ca^{2+} transport or ATP hydrolysis, individually inhibit ATP synthesis, but TMA$\rightarrow$O offsets the inhibitory effect of urea when they are added together. In contrast, the polyol sucrose stimulates the reversal cycle. All 3 osmolytes only slightly modify the K_m for P_i. It is proposed that osmolytes interact with Ca^{2+}- ATPase during conversion of the phosphorylated intermediate (i.e. E_2-P $\longrightarrow$ E_1~P) in the direction of ATP synthesis.

[*]*Abbreviations:* p.m., plasma membrane(s); TMA$\rightarrow$O, trimethylamine oxide; E, Ca^{2+}-ATPase; E_1~P and E_2-P, phosphoenzyme forms; PEP, phospho(enol)pyruvate; P_i, inorganic orthophosphate; P_iK, a KH_2PO_4/K_2HPO_4 mixture.

In recent years remarkable progress has been attained in the knowledge of the mechanisms responsible for the transport of solutes, notably Ca^{2+}, across the membranes of kidney tubules. Renal calcium transport plays a crucial role in body calcium homeostasis since <5% of ultrafiltered calcium is lost in the urine [1]. In addition, kidney cells maintain a very low intracellular calcium concentration [2] despite the occurrence of a continuous flux of calcium through the tubular epithelium by the transcellular route [1, 3]. The maintenance of the cytosolic Ca^{2+} activity in a very narrow sub-μM range appears to be critically involved in the regulation of other epithelial functions such as NaCl and water reabsorption [4, 5].

Much information concerning renal Ca^{2+} transport mechanisms has been obtained through micropuncture of tubular segments [6], microperfusion of tubules - in situ [3] or previously isolated [5] - and techniques more recently developed for isolating p.m., as vesicles [7, 10]. The use of membrane vesicles has made it possible to characterize both active Ca^{2+} in vitro [8, 10, 11] and Ca^{2+}-ATPase activities [9, 11] without interference from other cellular structures. Thereby the basolateral membrane of the epithelial cells has been established [8, 10] as the exclusive location for the ATP-driven Ca^{2+}-pumping mechanism and the secondary active Na^{+}/Ca^{2+} exchange found in proximal tubules (in which reabsorption of two-thirds of ultrafiltered Ca^{2+} takes place [1]). The first step of transepithelial Ca^{2+} transport, i.e. the Ca^{2+} flux across the luminal membrane of the cells, is passive from a phenomenological point of view [4]. Furthermore, the solubilization and partial purification of the renal Ca^{2+}-ATPase as starting material [12] provided a useful tool for the study of the enzyme in the absence of a Ca^{2+} gradient.

In the present work we used native vesicles derived from basolateral p.m., and a Triton X-100-solubilized preparation, to study the reversal cycle of the Ca^{2+} ATPase from p.m. of kidney proximal tubules by measuring the exchange reaction between the γ-phosphoryl group of ATP and $^{32}P_i$ in the medium (ATP $\rightleftharpoons$ $^{32}P_i$ exchange). The influences of the ligands Ca^{2+}, ADP, ATP, P_i and Mg^{2+} were ascertained. Also investigated were the effects of organic molecules that are thought [13, 14] to be capable of offsetting the inhibitory effects on kidney enzymes of inorganic solutes and of urea (largely handled by kidney tubules [15]). The data indicate that the exchange is regulated by two classes of Ca^{2+}-binding sites, high–affinity located on the cytosolic surface of the membrane vesicles and low–affinity on the peritubular surface. It is further shown that sucrose (representative of the polyol molecular class [16]) stimulates the rate of exchange,

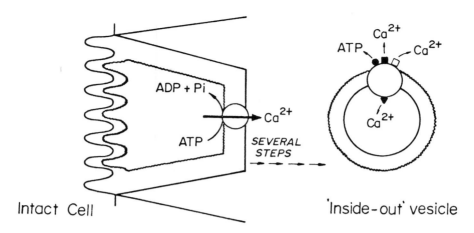

Fig. 1. Formation of 'inside-out' p.m. vesicles from kidney proximal tubules (method as in [11]). Symbols denote sidedness of ATP sites (●) and of Ca^{2+} binding sites (cytosolic, □ & ■; peritubular, ▼). Cytosolic sites for Ca^{2+} ($K_{Ca} \simeq 1 \mu M$ and $K_{Ca} \simeq 250 \mu M$) are stimulatory and the peritubular one ($K_{Ca} \simeq 1 mM$) is inhibitory towards the forward cycle [11].

and that TMA→O partially reverses the inhibitory effect of urea on the exchange reaction.

METHODS

Vesicle isolation and marker enzyme assays were as previously described [16]. This type of p.m. preparation includes membrane sheets as well as right-side-out vesicles besides inside-out vesicles [7, 8][8], but only the latter can accumulate Ca^{2+} and have the two ATP sites (catalytic and regulatory) facing the outer (cytosolic) surface of the membrane. Thus, only the inside-out vesicles sustain a transmembrane ATP-dependent Ca^{2+} gradient and can catalyze the ATP $\rightleftharpoons$ $^{32}P_i$ exchange reaction at μM Ca^{2+} concentrations (see below). In this type of vesicle the high-affinity Ca^{2+} binding sites also face the extravesicular side, whereas the lower-affinity site faces the intravesicular space [11]. The formation of the inside-out vesicles and the sidedness of ATP and Ca^{2+} sites are depicted in Fig. 1.

For enzyme solubilization and determinations of Ca^{2+} uptake and Ca^{2+}-ATPase activities, see [11]. $^{32}P_i$ was purchased from the Brazilian Institute of Atomic Energy, purified by extraction [17] and stored in dilute HCl until use. [γ-^{32}P]ATP was synthesized (method: [18]) for use in ATPase assays

---

[8] 'Sidedness' Index entries in earlier vols. are pertinent, and in Vol. 17 R.G. Price _et al._ describe brush-border p.m. isolation.— _Ed._

Fig. 2. Assays (at 37°) for Ca^{2+} uptake (o) and ATP $\rightleftharpoons$ $^{32}P_i$ exchange (●) catalyzed by native vesicles (0.2 mg wet wt./ml). Constituents (mM).- *Uptake:* Tris-HCl pH 7.4, 30; MgCl$_2$, 5; ATP, 5; P$_i$K, 20; PEP, 5; CaCl$_2$, 0.05; (mg/ml) pyruvate kinase, 0.05. *Exchange:* Tris-HCl pH 7.5, 50; MgCl$_2$, 10; ATP, 1; ADP, 0.2; P$_i$K, 4 (with $^{32}P_i$); CaCl$_2$, 0.02. *Both:* ouabain, 0.2; NaN$_3$, 10.

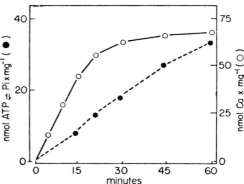

and as a standard in autoradiograms. The exchange was assayed by measuring the formation of $[\gamma\text{-}^{32}P]ATP$ from $^{32}P_i$ and ADP. Briefly, the samples were quenched with perchloric acid and the excess $^{32}P_i$ was removed as the phosphomolybdate complex by repeated extraction with acetone/butyl acetate, leaving the $[\gamma\text{-}^{32}P]ATP$ in the aqueous phase [19]. Aliquots of the molybdate extracts were counted in a liquid scintillation counter. The identification of $[\gamma\text{-}^{32}P]ATP$ as the exchange reaction product was confirmed by TLC with autoradiography [20]. All the reagents were of analytical grade.

RESULTS AND DISCUSSION

It is well established that Ca^{2+}-ATPase from sarcoplasmic reticulum vesicles [21], brain [22] and red cells [23] can function in the forward and reverse directions. To our knowledge, we are the first to describe the reversal cycle of a Ca^{2+}-ATPase from p.m. of an asymmetric cell of epithelial origin.

Ca^{2+} transport and ATP $\rightleftharpoons$ $^{32}P_i$ exchange with native vesicles.- Basolateral p.m. vesicles catalyze simultaneously ATP-dependent Ca^{2+} accumulation and this exchange in a medium containing 20 µM Ca^{2+} (Fig. 2). The amount of $[\gamma\text{-}^{32}P]ATP$ formed after 1 h was decreased by 80% in a medium without Ca^{2+} (1 mM EGTA) or when the Ca^{2+} ionophore A23187 was present (not shown). These results indicate that synthesis of ATP by the Ca^{2+} pump requires the existence of a transmembrane Ca^{2+} gradient [24] and/or the occupancy of an intravesicular and low-affinity Ca^{2+} binding site [21, 22].

Identification of $[\gamma^{32}P]ATP$.- Its Ca^{2+}-dependent formation was confirmed using two approaches. The non-extractable radioactive product completely disappeared when PEP and pyruvate

Fig. 3. TLC and autoradiography (methods as in [20])
of samples assayed for
$[\gamma\text{-}^{32}P]ATP$ formation. With
native vesicles (lane **A**) the
medium composition in the
assays for ATP $\rightleftharpoons$ $^{32}P_i$
exchange was as in the
legend to Fig. 2. With
soluble enzyme (lane **B**) the
medium was at pH 7.0 and
contained 4 mM $CaCl_2$.

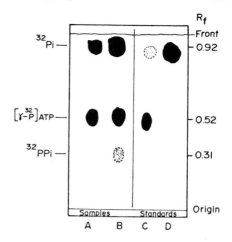

Fig. 4. Dependence of ATP $\rightleftharpoons$ $^{32}P_i$
exchange reaction, using native
vesicles, on Ca^{2+} concentration.
This was varied from zero (1 mM
EGTA) to 10 mM. Otherwise the
composition of the medium was
as in the legend to Fig. 2, except
that the pH was 7.0. The reaction
time was 15 min.

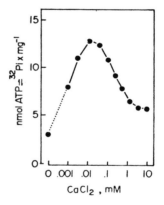

kinase were added to the reaction medium to remove all
the ADP formed during ATP hydrolysis (not shown). In addition,
the non-extractable product formed in the presence of either
native vesicles (Fig. 2) or soluble enzyme (Fig. 6, below)
co-migrated in TLC with synthetic $[\gamma\text{-}^{32}P]ATP$ (Fig. 3).

[Ca^{2+}] dependence of the exchange reaction.- Fig. 4 shows
that the formation of $[\gamma\text{-}^{32}P]$ by native vesicles was stimulated
by Ca^{2+} up to 30 µM and inhibited at higher concentrations.
Evidently the occupancy of the highest affinity site [10,
11] promotes the activation of both forward and reverse
cycles of the Ca^{2+} pump. Probably, on the other hand,
binding of Ca^{2+} to the site of intermediate affinity that

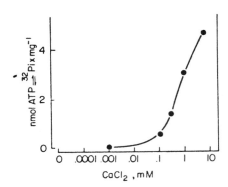

Fig. 5. Dependence of Ca^{2+}-ATPase activity (37°, 15 min) on Ca^{2+} concentration. Constituents of medium (mM): Tris–HCl pH 7.0, 30; $MgCl_2$, 10; $[\gamma^{-32}P]$-ATP, 5; ouabain, 0.2; NaN_3, 10; P_iK, 4; KCl, 20; EGTA, 1; sucrose, 160; $CaCl_2$ as shown; (mg/ml) solubilized enzyme, 0.3.

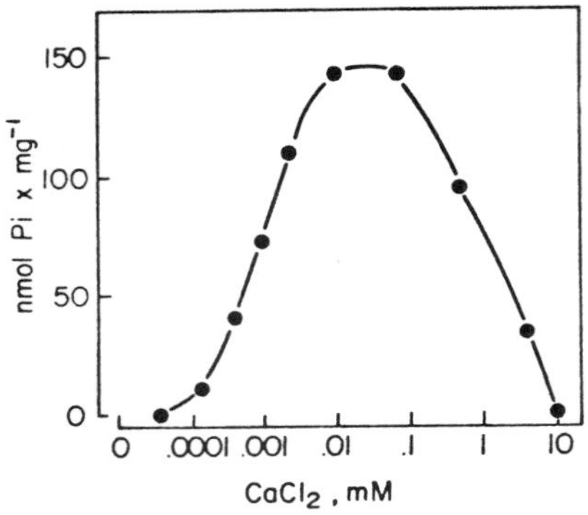

Fig. 6. Effects of varying Ca^{2+} concentration on ATP $\rightleftharpoons$ $^{32}P_i$ exchange (37°, 15 min) using solubilized Ca^{2+}-ATPase instead of native vesicles (otherwise as for Fig. 2 legend except that the pH was 7.0 and $CaCl_2$ was varied as shown).

further stimulates the forward reaction ($K_m = 250$ μM [11]) is responsible for the inhibition of the reversal cycle. This implies that as ATP hydrolysis and Ca^{2+} transport become fully activated, fewer and fewer cycles of ATP synthesis take place.

$[Ca^{2+}]$ dependence with soluble enzyme.- The Triton X-100-treated enzyme is unable to establish Ca^{2+} gradients, but ATP hydrolysis becomes highly activated in the presence of micromolar Ca^{2+} concentrations (Fig. 5). It was also observed that Ca^{2+} concentrations >100 μM progressively inhibited the rate of hydrolysis. Since this inhibition by Ca^{2+} was also observed in the presence of a constant $Ca.ATP^{2-}$ concentration (not shown), it is proposed that the binding of Ca^{2+} to a low-affinity, inhibitory site ($i_{0.5} = 1$ mM) impairs the forward cycle. The experiments shown in Fig. 6 indicate that a simultaneous activation of ATP $\rightleftharpoons$ ^{32}P exchange takes place. These results indicate that saturation of the lower affinity Ca^{2+} binding site is required for full activation of the reversal cycle of the enzyme. In addition, the formation of $[\gamma^{-32}P]$ by soluble enzyme with high Ca^{2+} was also stimulated by high pH (not shown), as for the Ca^{2+}-ATPase of sarcoplasmic reticulum [21] and brain [22].

P_i **and magnesium affinities.**- The Mg^{2+}-dependent phosphorylation by P_i is the first step in the reversal cycle
of ion-motive ATPases [21, 23]. This warranted exploration
of the affinity for these ligands in the reversal cycle
of renal p.m. Ca^{2+}-ATPase. With native vesicles the apparent
K_m for P_i in the exchange reaction was 5 mM at pH 7.0
in the presence of 10 μM Ca^{2+}. The study of $MgCl_2$ dependence
also revealed a Michaelian behaviour, with half-saturation
at 3 mM. The affinities for P_i and Mg^{2+} decreased upon
solubilization of the enzyme (the apparent K_m's were 20 mM
and 10 mM respectively). These results may indicate that
the solubilization process increases the access of water
from the medium into the hydrophobic phosphorylation domain
[19, 23, 25], thus impairing the binding of P_i and, to
a lesser extent, of Mg^{2+}.

Influence of the ATP/ADP ratio.- The apparent K_m for
ADP appears to be extremely low since the ADP derived from
ATP hydrolysis was enough to activate the ATP $\rightleftharpoons$ $^{32}P_i$ exchange
reaction. As shown by the following values for the
exchange in 15 min (units and general conditions as for
Fig. 2), further activation was observed when ADP was increased
to 1 mM, especially when the ATP concentration was lowered:

- 1 mM ATP + 0.2 mM ADP: rate 13.0 (conditions of Fig. 2);
- 1 mM ATP + 1 mM ADP: rate 18.3;
- 0.2 mM ATP + 1 mM ADP: rate 24.1.

Evidently the reversal cycle is favoured by a low ATP/ADP
ratio although a competition between P_i and ATP at the
phosphorylation site cannot be ruled out [21].

Effects of sucrose, urea and TMA→O

More than two-thirds of ultrafiltered NaCl and ~50%
of ultrafiltered urea are reabsorbed by passing through the
proximal epithelium [1]. Although high concentrations of
these solutes are found in the renal medulla rather than
in the cortex, the above fluxes challenge the cytoplasm
of proximal tubule cells by continuous exposure to solutes
that inhibit kidney enzymes [13, 16]. It has been shown
that kidney contains other osmotically active solutes such
as polyols and trimethylamines that are 'non-perturbing' [14]
and that are capable, in some conditions, of offsetting
the effects of urea on the activity of several enzymes
[13, 16].

Table 1 shows the effects of urea, TMA→O and sucrose
on the Ca^{2+} pump from proximal tubules. Evidently neither
Ca^{2+} uptake nor Ca^{2+}-ATPase activity at pH 8.5 was modified
by the addition of these osmotically active solutes. However,

Table 1. Effects of osmotically active solutes on tubule Ca^{2+} pump parameters (expressed as nmol/mg in the tabulated 37° incubation time), viz. **A**: Ca^{2+} uptake; **B**: Ca^{2+}-ATPase; **C**: ATP $\rightleftharpoons$ $^{32}P_i$ exchange. Generally the media were as stated in legend to Fig. 2 (*plus* solutes as tabulated) for *uptake* (A) or *exchange* (B, C) except for Tris-HCl pH, which was 8.5; but there were particular differences.- A, *supplement:* 20 mM KCl; *replacement:* $^{45}CaCl_2$, 0.02 mM. B, (mg/ml) soluble enzyme 0.1, *in place of native vesicles, and* asolectin, 1, *as supplement;* $MgCl_2$ *lowered to* 5.5 mM; *replacement:* [γ-^{32}P]ATP, 5 mM. **C**, *replacement:* $^{32}P_i K$, 4 mM; NaN$_3$ *omitted.*

Added solute (& final concentration, mM)	Uptake (**A**; 2 h)	ATPase (**B**; 15 min)	Exchange (**C**; 15 min)
None (control)	120.9	127.5	12.5
Sucrose (400)	110.2	129.0	21.2
Urea (800)	108.0	121.5	2.7
TMA$\rightarrow$O (400)	111.1	118.5	9.3
Urea (800) + TMA$\rightarrow$O (400)	104.9	109.5	10.9

a strong inhibition of ATP $\rightleftharpoons$ $^{32}P_i$ exchange was observed with 800 mM urea present at the same pH value, an effect that was almost completely reversed by the simultaneous presence of 400 mM TMA$\rightarrow$O, a methylamine that inhibited the exchange by 30%. The ability of methylamines to counteract the effects of urea on other enzymes when they are present conjointly in a concentration ratio of 1:2 has been reported [13, 16]. It is interesting that this ratio appears to be preserved in tissues through the evolution of living systems [16]. In contrast with the effect observed with urea or TMA$\rightarrow$O, a representative polyol, viz. sucrose, stimulated the exchange by 70% (Table 1).

Addition of the organic solutes did not significantly modify the apparent K_m for P_i (not shown). This may indicate that they induce modifications in those steps of the cycle that involve large conformational transitions, rather than the mere binding of P_i or Mg^{2+}.

The effects of these organic solutes were pH-dependent. Urea did not affect the rate of exchange at pH 6.0 and, conversely, TMA$\rightarrow$O and sucrose stimulated it by 105% and 300% respectively at this pH value (not shown).

Since neither urea nor TMA$\rightarrow$O affected the forward reaction of renal Ca^{2+}-ATPase (Table 1) or the binding of P_i and Mg^{2+} during the reversal cycle, and since they appear to

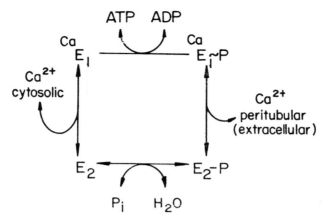

Scheme 1. Catalytic cycle of the Ca^{2+}-ATPase from kidney tubules. Ca^{2+}_{cyt}, cytosolic Ca^{2+}; Ca^{2+}_{per}, peritubular Ca^{2+}. E_1, the enzyme form phosphorylated by ATP during the forward cycle; E_2, the enzyme form phosphorylated by P_i during the reversal cycle. *Adapted from refs. [21] & [26].*

modify the solvent properties of water [16], conceivably they influence the reversal cycle by affecting the conversion of the phosphoenzyme E_2-P to the form E_1~P (Scheme 1). The conversion of the form primarily phosphorylated by P_i (low-energy phosphoenzyme, E_2-P) into another one (E_1~P) that is capable of transferring its phosphoryl group to ADP to synthesize ATP involves a hydrophobic-hydrophilic transition at the level of the phosphorylation domain of other Ca^{2+}-ATPases [19, 23, 25, 26]. Thus, it is plausible that solutes which interact with surrounding water and modify its solvent properties [16] may affect the rate of phosphoenzyme conversion during the reversal of the catalytic cycle.

Acknowledgements

The authors are deeply grateful to Dr Martha M. Sorenson for helpful suggestions and valuable discussion of the manuscript. This work was supported in part by grants from FINEP, CNPq and CPEG/UFRJ (Brazil). CAPES and CNPq made travel awards for Forum participation (by A.V.).

References

1. Suki, W.N. (1979) *Am. J. Physiol.* *237*, F1-F6.
2. Lee, C.O., Taylor, A. & Windhager, E.E. (1980) *Nature* *287*, 857-861.

3. Ullrich, K.J., Rumrich, G. & Kloss, S. (1976) *Pfluegers Arch. Eur. J. Physiol. 364*, 223-228.
4. Taylor, A. & Windhager, E.E. (1979) *Am. J. Physiol. 236*, F505-F512.
5. Friedman, P.A., Figueiredo, J.F., Maack, T. & Windhager, E.E. (1981) *Am. J. Physiol. 240*, F558-F568.
6. Lassiter, W.E., Gottschalk, C.W. & Mylle, M. (1963) *Am. J. Physiol. 204*, 771-775.
7. Kinne-Saffran, E. & Kinne, R. (1974) *J. Memb. Biol. 17*, 263-274.
8. Gmaj, P., Murer, H. & Kinne, R. (1979) *Biochem. J. 178*, 549-557.
9. Gmaj, P., Murer, H. & Carafoli, E. (1982) *FEBS Lett. 144*, 226-230.
10. Gmaj, P., Zurini, M., Murer, H. & Carafoli, E. (1983) *Eur. J. Biochem. 136*, 71-76.
11. Vieyra, A., Nachbin, L., de Dios-Abad, E., Goldfeld, M., Meyer-Fernandes, J.R. & Moraes, L. (1986) *J. Biol. Chem. 261*, 4247-4255.
12. De Smedt, H., Parys, J.B., Borghgraef, R. & Wuytack, F. (1981) *FEBS Lett. 131*, 60-62.
13. Somero, G.N. (1986) *News Physiol. Sci. 1*, 9-12.
14. Balaban, R.S. & Burg, M.B. (1987) *Kidney Int. 31*, 562-564.
15. Pitts, R.F. (1974) *Physiology of the Kidney and Body Fluids*, Year Book Medical Publishers, Chicago, pp. 71-134.
16. Yancey, P.H., Clarck, M.E., Hand, S.C., Bowlus, R.D. & Somero, G.N. (1982) *Science 217*, 1214-1222.
17. de Meis, L. (1984) *J. Biol. Chem. 259*, 6090-6097.
18. Glynn, I.M. & Chappell, J.B. (1984) *Biochem. J. 90*, 147-149.
19. de Meis, L., Martins, O.B. & Alves, E.W. (1980) *Biochemistry 19*, 4252-4261.
20. Vieyra, A., Meyer-Fernandes, J.R. & Gama, O.B.H. (1985) *Arch. Biochem. Biophys. 238*, 574-583.
21. Carvalho, M.G.C., Souza, D.G. & de Meis, L. (1976) *J. Biol. Chem. 251*, 3629-3636.
22. Trotta, E.E. & de Meis, L. (1978) *J. Biol. Chem. 253*, 7821-7825.
23. Chiesi, M., Zurini, M. & Carafoli, E. (1984) *Biochemistry 23*, 2595-2600.
24. Makinose, M. (1971) *FEBS Lett. 12*, 269-270.
25. de Meis, L., Gómez Puyou, M.T. & Gómez Puyou, A. (1988) *Eur. J. Biochem. 171*, 343-349.
26. Pedersen, P.L. & Carafoli, E. (1987) *Trends Biol. Sci. 12*, 186-189.

#A-3
RECEPTOR-MEDIATED CALCIUM ENTRY

Timothy J. Rink

Smith, Kline & French Research Ltd.,
The Frythe, Welwyn, Herts. AL6 9AR, U.K.

Ca^{2+} is still cardinal in cell signalling despite the discovery of several other second messengers and the increasing understanding of membrane transduction mechanisms such as GTP-binding proteins. For instance: the only documented signalling function for inositol phosphates is in mediating Ca^{2+} mobilization; Ca^{2+} is important, together with diacylglycerol, in regulating protein kinase-C; the major proposed functions for cGMP are in regulation of Ca^{2+} fluxes; and in many instances cAMP acts partially or mainly by enhancing Ca^{2+} movement. As for cell signalling mechanisms in general, Ca^{2+} signalling is diverse. Voltage-gated Ca^{2+} entry is subserved by many different types of channels with different kinetics, voltage dependence, ion-selectivity, physiological modulation and sensitivity to pharmacological agents [e.g. 1]. Stimulus-dependent discharge of Ca^{2+} from specialized organelles is a major source of activation in some excitable and most 'non-excitable' cells and can readily provide a generalized cellular signal. The discovery that $Ins(1,4,5)P_3$ mediates this process solved a major mystery in intracellular communication. Yet Ca^{2+} discharge in some cells, e.g. striated muscle, may well be controlled mainly by other systems, and conjecturally other inositol phosphates, Ca^{2+} itself, and possibly GTP modulate the action of $Ins(1,4,5)P_3$. However, this article considers a third general process of Ca^{2+} signalling: stimulus-dependent Ca^{2+} entry, not via voltage-gated channels. Consideration is given to receptor-mediated Ca^{2+} entry (a set of diverse phenomena), to its route, to possible transport mechanisms and their control, and to recent findings of Ca^{2+} oscillations in non-excitable cells.

Three general processes of Ca^{2+} signalling are outlined above: (1) voltage-gated Ca^{2+} entry, mediated by $Ins(1,4,5)P_3$ in many cells [2]; (2) Ca^{2+} discharge maybe involving modulation of $Ins(1,4,5)P_3$ action by other inositol phosphates, Ca^{2+} itself and perhaps GTP [3, 4]; (3) stimulus-dependent Ca^{2+} entry not <u>via</u> voltage-gated channels. It is (3) that I consider here, with the focus on receptor-mediated Ca^{2+}

entry although physical stimuli other than voltage, e.g.
mechanical deformation, probably also can gate Ca^{2+} entry,
e.g. the stretch-activated Ca^{2+} channel we have proposed
from studies of endothelial cells [5]. First I outline
some of the evidence for receptor-mediated Ca^{2+} entry and
arrive at an operational description of what is clearly
a set of diverse phenomena. Then I consider the route
of this Ca^{2+} entry, possible transport mechanisms and their
control, and the possible role of Ca^{2+} entry in regulating
Ca^{2+} oscillations in non-excitable cells.

EVIDENCE FOR RECEPTOR-MEDIATED Ca^{2+} ENTRY

The first suggestions of receptor-mediated Ca^{2+} entry
came from experiments in which a Ca-dependent event, such
as contraction, was inhibited by removal of external Ca^{2+}
in depolarized tissue, or when depolarization did not occur
[6], or in the presence of 'Ca^{2+} antagonists' such as nifedipine.
None of these results gave compelling evidence for receptor-
operated channels since neither activation by another internal
message nor the possibility that external Ca^{2+} was required
for internal Ca^{2+} release was excluded; moreover we know
that the 'Ca^{2+} antagonists' are effective only on certain
classes of voltage-gated Ca^{2+} channels. Further evidence
came from experiments with apparently non-excitable cells,
in which stimulation evoked an increase in $^{45}Ca^{2+}$ uptake
[e.g. 7]. With the introduction of the fluorescent Ca^{2+}
indicators quin-2, and later fura-2 and indo-1, it became
clear that in many cells only a transient rise in $[Ca^{2+}]_i$
could be evoked in the absence of external Ca^{2+}, whereas
with normal external Ca^{2+} the peak response was often distinctly
larger, and usually showed an extended plateau phase [e.g. 8].

The simplest explanation for these findings is that
stimulation normally evokes both internal discharge and Ca^{2+}
entry. As expected from this analysis, the difference between
the peak responses in the presence and absence of external
Ca^{2+} was greater when large amounts of quin-2 were loaded
into cells to buffer the $[Ca^{2+}]_i$ rise resulting from discharge
of a finite internal store [9]. Further support for the
presence of receptor-mediated Ca^{2+} entry came from the use
of Mn^{2+} as a surrogate for Ca^{2+}. Mn^{2+} strongly quenches
the dye fluorescence and hence stimulated entry reduces the
fluorescent signal at wavelengths where Ca^{2+} increases it,
or at the isobestic wavelength where Ca^{2+} has no effect.
In platelets [10], neutrophils [11] and endothelial cells
[12], stimulation in the presence of Mn^{2+} leads to quenching
of cytosolic indicator, demonstrating Mn^{2+} influx; the assumption then is that Mn^{2+} has entered via routes normally
used by Ca^{2+}.

Further evidence for receptor-mediated Ca^{2+} entry comes from electrical measurements. For instance, in PC12 phaeochromocytoma cells, bradykinin can evoke a Ca^{2+}-dependent depolarization (monitored with a fluorescence dye) as expected for receptor-mediated Ca^{2+} entry through some kind of pore or channel (#B-7, this vol.: Fasolato et al.). And there are now several instances of Ca^{2+} currents, evoked by agonists, measured under voltage clamp where voltage-gated Ca^{2+} entry could not occur. One example is NMDA[⊗] acting on central neurones [13]. Another is ATP acting on rabbit-ear artery smooth muscle cells [14]; very recent experiments in these cells have shown an inward current associated with elevated $[Ca^{2+}]_i$, and with Mn^{2+} entry (monitored by indo-1), evoked by ATP while the membrane potential was clamped near the resting potential [15].

OPERATIONAL DEFINITION

Significant amounts of Ca^{2+} could enter a cell either through relatively Ca^{2+}-specific transport processes, or via non-specific cation transporters. In order not to constrain our thinking, we prefer the operational concept of a receptor-mediated process that admits a biologically relevant amount of Ca^{2+} into the cell, whether in a Ca^{2+}-specific manner or along with other cations. The amount of Ca^{2+} needed to subserve a trigger function is small compared to the amount of Na^+ or Mg^{2+} needed to perturb significantly their cellular content, so that a somewhat promiscuous uptake of extracellular cations can deliver, or help to maintain, an effective Ca^{2+} signal.

ROUTES OF Ca^{2+} ENTRY

The route for receptor-mediated Ca^{2+} entry might seem quite obvious, viz. across the p.m. and into the cytosol. Yet there is growing consideration of an alternative, whereby Ca^{2+} passes from the extracellular fluid into the cytosol, indirectly via a Ca-discharging organelle [e.g. 16, 17]. This seems at first an implausible and unnecessary complication: but there turns out to be a logic and elegance to this concept, and certain experimental results are hard to explain by other mechanisms. Basically, it was found that stimulation in a Ca^{2+}-free medium produced a response and depleted the internal store, the Ca^{2+} presumably being pumped out across the p.m. After removal of the stimulus or blockade of the receptor, an exposure to external Ca^{2+} allowed refilling of the store and a restoration of responsiveness even in Ca^{2+}-free medium; however, there was no response during the

[⊗] N-methyl-D-aspartic acid
Other abbreviations: p.m., plasma membrane; SMOC, second messenger-operated channel

'refilling' period.   This lack of response was taken to indicate that refilling with no significant increase in $[Ca^{2+}]_i$ occurs <u>via</u> a route into an organelle.   However, when muscle contraction [18, 19] or $^{86}Rb^+$ efflux [16] was the 'bioassay' for $[Ca^{2+}]_i$ one could imagine a sub-threshold elevation in $[Ca^{2+}]_i$ promoting refilling into the store. But experiments with fura-2-loaded parotid cell suspensions [17], single gastric parietal cells [20] or single endothelial cells [21] confirm that such refilling of internal stores can occur with no measured rise in $[Ca^{2+}]_i$.

This experimental paradigm is shown diagrammatically in Fig. 1A.   These experiments of course create highly artificial conditions to isolate the mechanisms.   In physiological conditions this type of store-refilling would normally occur during, or shortly after, stimulation and could readily support a maintained response, or serve to refill the internal pool in the period between the $[Ca^{2+}]_i$ spikes in an oscillating cell (see below).

We would not expect this mechanism to be present in all cells, and indeed it may well co-exist with other receptor-mediated Ca^{2+} entry mechanisms in the p.m.   Since the actual existence of pathways into the internal store remains hypothetical, discussion of this mechanism might seem premature; but we proposed a simple model [17] based on the gap-junction, spanning the two relevant membranes (Fig. 1B).   This pathway is regarded as somewhat Ca-selective and regulated by the Ca^{2+} concentration   within the organelles; this idea is based on the known regulation of gap junctions by Ca^{2+} in the 100 µM range.   It is economical, simple and apparently sensible to have the state of the internal store control its own filling.

Putney [16] proposed a more complex model in which highly localized depletion of Ca^{2+} by avid sequestration led to local uptake across the p.m.   In either case one would expect close apposition of organelle membrane to p.m., as is in fact seen at points in most cell types. Interestingly, it has recently been found [24] that $Ins(1,4,5)P_3$ binding sites and the $Ins(1,4,5)P_3$-releasable Ca^{2+} pool in fractionated liver cells are indeed mostly associated with the p.m. fraction. Further suggestions for regulation of this type of pathway include a role for $Ins(1,3,4,5)P_4$ and for GTP.

Fig. 1, *opposite*. Routes of calcium entry, direct into the cytosol and indirect <u>via</u> a calcium-sequestering and -releasing organelle. A: experimental paradigm which led to the proposal for indirect entry <u>via</u> calcium-sequestering organelles.

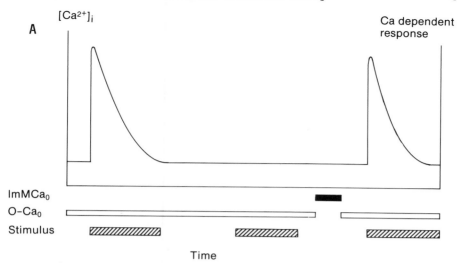

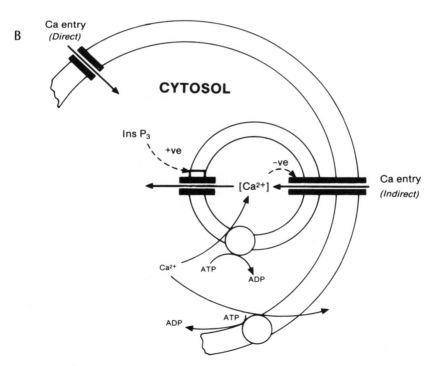

Fig. 1, *ctd.* **B** *(adapted from ref. [17]):* two possible routes for calcium entry, direct across the p.m. and indirect <u>via</u> a [Ca²⁺]-inhibited pathway linking external medium to a calcium-sequestering and -discharging organelle.

DISTINGUISHING DIRECT ENTRY INTO THE CYTOSOL FROM
ENTRY INTO A CALCIUM-DISCHARGING GRANULE

It turns out to be surprisingly difficult to determine experimentally the route of stimulated Ca^{2+} entry. Both routes will give increases in tracer Ca^{2+} uptake; both can give sustained elevation of $[Ca^{2+}]_i$ dependent on the presence of both external Ca^{2+} and agonist; both could be inhibited by similar blocking agents, e.g. Ni, Cd or La; both could give rise to Ca^{2+} currents measured by impalement voltage-clamp, or by patch-clamp. If the entry to the store carries a Ca^{2+} current, as is likely, and the $InsP_3$-gated Ca^{2+} release carries a current, as the experimental evidence suggests, then the 'indirect' route can give rise to Ca^{2+} currents and to calcium-dependent depolarization. Even in a 'detached' membrane patch experiment, one cannot distinguish between these pathways, because some cytoplasm is usually drawn up into the patch pipette and might include membrane-associated Ca^{2+}-releasing organelles. Thus Ca^{2+} channels reported to be opened by application of $InsP_3$ to inside-out patches of T-lymphocyte membranes [22] could have been due to $InsP_3$ acting on channels of calcium-discharging organelles; Ca^{2+} already available to the organelle interior through the 'indirect' pathway would then pass into the bathing medium and generate a current.

One critical piece of evidence, available in only a few cases, is the kinetics of $[Ca^{2+}]_i$ changes. If internal Ca^{2+} discharge occurs clearly before Ca^{2+} entry, as apparently occurs with thrombin-stimulated endothelial cells [13], then neither mechanism (direct entry into the cytosol or indirect entry via an organelle) is excluded, though the extra delay in the entry needs to be explained. The simplest form of the 'indirect' pathway model predicts the same delay in onset of $[Ca^{2+}]_i$ rise in the presence and absence of external Ca^{2+} - with a faster, larger, longer elevation in the presence of external Ca^{2+}, as we saw in stopped-flow measurements of fura-2-loaded parotid cells [8]. So far only one example is known where receptor-mediated influx clearly precedes internal release, viz. the ADP-evoked $[Ca^{2+}]_i$ rise in human platelets [23]. This cannot be accommodated in the 'indirect' model and indicates receptor-mediated Ca^{2+} entry directly into the cytosol.

MECHANISMS OF Ca^{2+} ENTRY

The mechanism of direct Ca^{2+} entry is usually thought of as some form of receptor-operated channel, or pore. Influx of Ca^{2+} would then depolarize the membrane, the size of this effect varying from the almost undetectable to 10's of mV, depending on the amount of influx and hence the size of the current and the other conductances of the membranes. Also the influx is expected to decrease with depolarization, diminishing to zero as the Ca^{2+} reversal potential is approached.

Ion channels are sometimes visualized as aqueous pores whose intrinsic size, shape and charge determine selectivity and permeability. However, strong and complex interactions between the permeating ions and specific binding sites within the channels appear to have a critical influence on permeability and selectivity. In the L-type voltage-operated Ca^{2+} channel where permeation has been analyzed in some detail [24], calcium-selectivity may be provided by two specific Ca^{2+}-binding sites in the 'pore'. Ca^{2+} flux occurs in a form of 'single-file' (originally proposed for K^+ channels in squid axons). If external Ca^{2+} is removed these channels become much more conductive to Na^+. One can regard the high-affinity binding of one Ca^{2+} in the channel as creating the calcium-selectivity filter, to exclude Na^+ when Ca^{2+} is present at mM concentrations in the external medium. An analysis along these lines is under way for some NMDA channels, where the Na^+ and Ca^{2+} conductance and selectivity are also markedly dependent on the external concentrations of these ions, implying a complex ion transit mechanism with specific ion binding sites within the channel [14]. A degree of interaction between Na^+ and divalent cations was seen in experiments on ATP-operated Ca^{2+} channels [15]. The available data for ligand-stimulated currents in patch-clamped T-lymphocytes appear to show selectivity for Ca^{2+} and Ba^{2+} over Na^+ [22], but a detailed analysis of permeation mechanism and selectivity has not yet been reported.

When the very existence of receptor-operated Ca^{2+} channels is still being validated, detailed consideration of the mechanism of permeation is clearly premature. However, the permeation mechanisms and selectivity processes are likely to be complex, strongly dependent on the experimental conditions, and to vary from cell to cell and from receptor to receptor, reflecting a diversity as great as that being revealed for voltage-gated Ca^{2+} channels.

Demonstration of an inward Ca^{2+} current shows that there is a transport mechanism through which Ca^{2+} can carry charge, but it does not necessarily show that transport is through a pore. An exchanger mechanism that is not tightly, electroneutrally coupled will also carry a current. This is well established for the K^+-selective antibiotic ionophore valinomycin, and for the p.m. Na^+:Ca^{2+} exchanger, which carries three Na^+ ions for each Ca^{2+} and thus carries one net charge

in the direction of the Na^+ flux at each turnover [25].
We know of no intrinsic p.m. Ca^{2+} exchangers which are
electro-neutral; but the Ca^{2+} ionophores A23187 and ionomycin
appear to mediate electro-neutral $Ca^{2+}:Mg^{2+}$ or $Ca^{2+}:2H^+$
exchange and thus the transport carries no current and is
not directly influenced by membrane potential.

COUPLING RECEPTOR–OCCUPATION TO Ca^{2+} ENTRY

How may receptors be coupled to Ca^{2+} entry? With
rather little solid experimental data, we must fall back
on analogy with modulation and regulation of other ion transporters. The simplest mechanism would be for the receptor
binding site to be on a component of the transporter and
directly gate its activation – in the way that acetylcholine
does at the nicotine receptor. We can only make guesses,
since we lack structural evidence and have no information
on the molecular basis of any receptor-mediated Ca^{2+} entry,
and moreover compelling kinetic evidence demanding direct
linking is lacking. Glutamate binding at the NMDA receptor
and ATP binding at P_{2y} receptors might work in this way.

Another form of coupling could be _via_ a G-protein which
interacts, in the plane of the membrane, with the receptor
and its transporter; this would be analogous to the muscarinic
activation of K^+ channels in the heart. In macrophages,
pertussis toxin can reduce Ca^{2+} uptake [26]; but, while
this finding implicates a G-protein, it does not exclude
an indirect coupling _via_ second-messengers whose formation
requires a G-protein coupling. In a recent report on hepatocytes, pertussis toxin preferentially blocked Ca^{2+} influx,
leaving internal release unaltered [27] – a result which
does point to a special role for a G-protein in Ca^{2+} entry.
From the circumstantial evidence of rapid kinetics of $[Ca^{2+}]_i$
signals, we suggested that the ADP-evoked Ca^{2+} entry in
platelets might be activated _via_ G-protein coupling [23].

There are numerous examples of modulation and control
of ion channels by intracellular second messengers. Ca^{2+}
itself activates several classes of channels. $Ins(1,4,5)P_3$
gates Ca^{2+} channels on specialized organelles; it has also
been claimed to do so for p.m. Ca^{2+} channels in T-lymphocytes
[22]. The experiments showing this are technically difficult
and this result is still somewhat controversial; the present
data also do not exclude entry _via_ the 'indirect' pathway
as discussed above. $Ins(1,3,4,5)P_4$ has been proposed as
a key mediator of Ca^{2+} entry, based initially on indirect
evidence, from experiments in which different inositol phosphates were microinjected into sea-urchin eggs and elevation
of the fertilization membrane was the bioassay for $[Ca^{2+}]_i$

[3]. However, more recent experiments question a major role for InsP$_4$ in activation of sea-urchin eggs [28]. A requirement for Ins(1,3,4,5)P$_4$ in addition to Ins(1,4,5)P$_3$ was reported, for activation of a K$^+$-conductance (as a bioassay for [Ca^{2+}]) in whole-cell patched lacrimal gland cells [29]. These are intriguing data, but other interpretations are possible, including a synergistic action of the two inositol phosphates on Ca^{2+} release from an organelle. In parotid gland cells, the plateau phase of Ca^{2+} entry, judged from fura-2 signals, could be roughly correlated either with Ins(1,4,5)P$_3$ levels or with Ins(1,3,4,5)P$_4$ levels [8]. Recently, however, Putney and co-workers [30] found that following an abrupt removal of the agonist (carbachol), the fall in [Ca^{2+}]$_i$ correlated with decline in Ins(1,4,5)P$_3$, while Ins(1,3,4,5)P$_4$ fell more slowly, arguing against any predominant role for Ins(1,3,4,5)P$_4$ in Ca^{2+} entry in these conditions. Thus, inositol phosphates remain plausible candidates to activate SMOC's [31], but compelling evidence is still awaited.

Cytosolic Ca^{2+} itself has been proposed to mediate Ca^{2+} entry [32]. For example, Ca^{2+}-activated, non-specific cation channels, which could pass Ca^{2+}, were identified in neutrophil membranes. Their role in intact cells is debatable; Ca^{2+} entry can be promoted by receptor occupancy even when [Ca^{2+}]$_i$ is at basal levels [17]. Again this remains a possibility to be considered; some limiting mechanism would be needed to prevent the positive feedback rapidly filling the cell with lethal amounts of Ca^{2+}.

In a sense, cAMP acts to mediate Ca^{2+} entry on behalf of β-adrenoceptors in the heart, by the special mechanism of promoting phosphorylation of voltage-gated Ca^{2+} channels to make them pass more current [33]. Also cAMP can promote mobilization of Ca^{2+} in several other cell types, but whether this is _via_ some form of SMOC is not known at present. As mentioned above, cAMP has powerful inhibitory effects on Ca^{2+} mobilization in some cells, _e.g._ platelets, though whether this is a direct effect on Ca^{2+} transport or on the receptor or on coupling systems is not clear [34]. cAMP directly activates non-specific calcium channels in vertebrate photoreceptors [34]. Whether it could, by analogy, mediate Ca^{2+} entry through SMOC's elsewhere is not known. Generally cAMP reduces cell responsiveness and is thought to reduce Ca^{2+} mobilization.

Diacylglycerol, acting _via_ protein kinase-C, can have marked effects, positive and negative, on voltage-gated Ca^{2+} channels. The main effect of C-kinase on receptor-mediated Ca^{2+} entry is inhibition [35], but the molecular basis of the action is unclear. Whether C-kinase can directly phosphorylate non-voltage-gated Ca^{2+} entry transporters is not known.

It has been proposed that $Na^+:H^+$ exchange, or increased pH_i in platelets, can mediate Ca^{2+} mobilization [36]. However, we [37] (and others) have questioned the basis for this conclusion, and can readily separate $[Ca^{2+}]_i$ changes from $Na^+:H^+$ exchange. It seems unlikely that this exchange is a requirement for receptor-mediated Ca^{2+} entry - but clearly changes in pH_i may well modulate Ca^{2+} mobilization since most biological processes show some pH dependence in the physiological range.

OSCILLATION OF CYTOSOLIC Ca^{2+}

One of the most intriguing recent results in cell signalling has been the finding that prolonged application of agonist can result in Ca^{2+} oscillations, or repetitive spikes, in non-excitable cells. Very striking records were seen by Cobbold and colleagues ([38; cf. pp. 211 this vol.) in aequorin-injected hepatocytes, where agonist concentration seemed to modulate frequency of spiking, ~0.05-0.01 Hz, rather than spike height or duration. We have found similar responses in histamine-stimulated, endothelial cells loaded with fura-2 [21]. The pattern of spiking in endothelial cells is not altered by isotonic K^+, apparently ruling out a role for membrane potential or Na^+-dependent transporters such as the $Na^+:Ca^{2+}$ or $Na^+:H^+$ exchangers. Persistent spiking requires extracellular Ca^{2+}, suggesting a role for Ca^{2+} entry, perhaps in both triggering spikes (and thus controlling spike frequency) and maintaining the filling of the internal stores. We are currently exploring the implications of models in which the primary event is a regenerative discharge of the internal store dependent in part on a slow, receptor-mediated Ca^{2+} entry direct into the cytosol, followed by refilling via the 'indirect' route during the inter-spike interval.

PHARMACOLOGICAL APPROACHES

A key question has to be posed: are there any pharmacological tools for dissection of these processes? What has been discussed so far is a set of phenomena we would like to think of as receptor-mediated Ca^{2+} entry systems. Unfortunately, there are few good tools to work with, and certainly nothing like the dihydropyridines, tetrodoxin or curare. Certain divalent cations such as Ni^{2+} and Cd^{2+} can block receptor-mediated Ca^{2+} entry, but more work is needed to see if a panel of divalent cation blockers can help us classify subtypes of Ca^{2+} entry; the different voltage-gated Ca^{2+} channels do show differential specificities to these inorganic blockers [1]. We can also explore further the ability of yet other divalent ions to substitute for Ca^{2+}. Mn^{2+} has been particularly useful, and has given one recent clear-cut result: there is no stimulus-dependent Mn^{2+} entry into parotid cells [39], whereas this is readily seen in platelets, neutrophils and endothelial cells, pointing to a marked difference in the Ca^{2+} entry mechanism. The organic Ca^{2+} channel blockers are relatively ineffective in blocking receptor-mediated Ca^{2+} entry and, at the required concentration, often show a non-specific suppression of signal transduction.

As in many other contexts, it is hard to figure out mechanism and function without specific, potent antagonists or inhibitors. We can exploit certain blocking cations and the use of Ca^{2+}-free solution in experiments on isolated cells and simple tissues, but clearly these manoeuvres are non-specific and will influence many other processes, not least voltage-gated Ca^{2+} entry. We need agents of greater specificity to pick apart the roles of the different receptor-mediated Ca^{2+} entry systems. We also need pharmacological tools for use in more complex systems, especially whole-animal experiments, to look at the physiological and pathological role of receptor-mediated Ca^{2+} entry. Possibly a molecular biology approach, _e.g._ finding mutant cell lines with defective Ca^{2+} entry, might help unravel the role of such systems and provide ways to get at the molecular basis of these transport systems. But good pharmacological probes would facilitate both molecular and physiological analysis.

There is also the prospect that selective blockers of receptor-mediated Ca^{2+} entry will show tissue and functional specificity of action with therapeutic potential, _e.g._ as anti-inflammatory agents. The effectiveness of 'Ca^{2+} antagonists' acting at voltage-operated channels shows that selective interference with Ca^{2+} signalling can have enormous clinical benefit.

Acknowledgements

I thank Beatrice Leigh and Maureen Bowden for help in preparing the manuscript, and Drs. Benham, Hallam, Jacob, Merritt and Pozzan for helpful discussion and permission to cite unpublished data.

References

1. Fox, A.P., Nowycky, M.D. & Tsien, R.Y. (1987) _J. Physiol._ _394_, 149-172.
2. Berridge, M.J. & Irvine, R.F. (1984) _Nature 312_, 315-319.
3. Irvine, R.F. & Moor, R.M. (1986) _Biochem. J._ _240_, 914-920.
4. Dawson, A.P., Comerford, J.G. & Futton, D.V. (1986) _Biochem. J. 234_, 311-315.
5. Lansman, J.B., Hallam, T.J. & Rink, T.J. (1987) _Nature 325_, 811-813.

6. Bolton, T.B. (1979) *Physiol. Rev. 59*, 607–718.
7. Foreman, J.C., Hallet, M.B. & Mongar, J.L. (1977) *J. Physiol. 271*, 193–214.   [14916.
8. Merritt, J.E. & Rink, T.J. (1987) *J.Biol. Chem. 262*, 14912–
9. Rink, T.J., Smith, S.W. & Tsien, R.Y. (1982) *FEBS Lett. 148*, 21–26.
10. Hallam, T.J. & Rink, T.J. (1985) *FEBS Lett. 186*, 175–179.
11. Merritt, J.E., Jacob, R. & Hallam, T.J. (1989) *J. Biol. Chem.*, in press.
12. Hallam, T.J., Jacob, R. & Merritt, J.E. (1988) *Biochem. J. 255*, 179–184.
13. Mayer, M.L. & Westbrook, G.C. (1987) *J. Physiol. 394*, 501–527.
14. Benham, C.D. & Tsien, R.Y. (1987) *Nature 328*, 814–817.
15. Benham, C.D. (1988) *J. Physiol. 407*, 92P.
16. Putney, J.W. (1986) *Cell Calcium 7*, 1–12.   [17369.
17. Merritt, J.E. & Rink, T.J. (1987) *J. Biol. Chem.262*, 17362–
18. Brading, A.F. & Sneddon, P. (1980) *Br. J. Pharmacol. 70*, 229–240.
19. Casteels & Droogmans, S. (1981) *J. Physiol. 317*, 263–279.
20. Negulescu, P.A. & Machen, T.E. (1988) *Am. J. Physiol. 254*, C130–C140.
21. Jacob, R., Merritt, J.E., Hallam, T.J. & Rink, T.J. (1988) *Nature 335*, 40–45.
22. Kuno, M. & Gardener, P. (1987) *Nature 326*, 301–304. [16369.
23. Sage, S.O. & Rink, T.J. (1987) *J. Biol. Chem. 262*, 16363–
24. Hess, P., Lansman, J.B. & Tsien, R.Y. (1986) *J. Gen. Physiol. 88*, 253–319.
25. Barcenas-Ruiz, L., Beulelmann, D.J. & Weir, W.G. (1987) *Science 238*, 1720–1722.
26. Murayama, T. & Ui, M. (1985) *J. Biochem. 260*, 7226–7233.
27. Hughes, B.P., Crofts, J.N., Auld, A.M., Read, L.C. & Barritt, G.J. (1987) *Biochem. J. 248*, 911–918.
28. Crossley, I., Swann, K., Chambers, E. & Whitaker, M. (1988) *Biochem. J. 252*, 252–262.
29. Morris, A.P., Gallacher, D.V., Irvine, R.F. & Peterson, L.C. (1987) *Nature 330*, 653–655.
30. Putney, J.W. (1988) *J. Exp. Biol. 139*, 135–150.
31. Meldolesi, J. & Pozzan, T. (1987) *Exp. Cell Res. 171*, 271–283.
32. von Tscharner, V., Prod'hom, B., Baggiolini, M. & Reuter, H. (1986) *Nature 324*, 369–372.
33. Tsien, R.W. (1983) *Ann. Rev. Physiol. 45*, 341–358.
34. Waldman, S.A. & Murad, F. (1987) *Pharmacol. Rev. 39*, 163–196.
35. MacIntyre, D.E., Buchfield, M. & Shaw, A.M. (1985) *FEBS Lett. 188*, 383–388.
36. Siffert, W. & Ackkerman, J.W.N. (1987) *Nature 325*, 456–458.
37. Simpson, A.W. & Rink, T.J. (1987) *FEBS Lett. 222*, 144–148.
38. Woods, W.M., Cuthbertson, K.S.R. & Cobbold, P.H. (1986) *Nature 319*, 600–602.
39. Hallam, T.J. & Merritt, J.E. (1988) *J. Biol. Chem. 263*, 6161–6164.

#A-4

CALCIUM- AND CALMODULIN-BINDING PROTEINS OF LIVER ENDOSOMAL AND PLASMA-MEMBRANE FRACTIONS

Carlos Enrich[†] and W. Howard Evans

National Institute for Medical Research,
Mill Hill, London NW7 1AA, U.K.

Advantage was taken of subcellular fractionation techniques developed to isolate specific regions of the hepatocyte's polarized p.m.*, and of parts of the endocytic compartment, to study the Ca^{2+}- and CaM-binding polypeptides present in these membranes. The studies showed that distinctive populations of Ca^{2+}- and CaM-binding polypeptides were present in the various subcellular fractions and these probably relate to the different functions accorded to the membrane networks.

As Ca^{2+} plays crucial roles in secretion, proliferation, muscle contraction, neurotransmission, etc., cells regulate intracellular Ca^{2+} levels. Various Ca^{2+}-transporting ATPases, Ca^{2+}-carriers and channels located at the p.m. and at intracellular sites ensure the fine control of Ca^{2+} levels in various parts of the cell [1]. Many Ca^{2+}-binding proteins have been identified, some of which are associated with membranes whereas others are integral membrane proteins. CaM is involved in mediating many of the actions of Ca^{2+} [2, 3], and other Ca^{2+}-binding proteins - e.g. lipocortins, calpactins, endonexin and calelectrins - have been identified recently [4].

In the present work, we have investigated the distribution of Ca^{2+}- and CaM-binding proteins in p.m. originating from the three functionally distinct regions of the hepatocyte's cell surface. Since these surface areas interface with different environments - the blood and bile spaces, and junctions attaching hepatocytes to each other - the possibility arises that different Ca^{2+}- or CaM-binding proteins may be present. We have also investigated the distribution of Ca^{2+}- and CaM-binding proteins of endocytic membranes, since 'microsomal' high-affinity pumping stores of unknown cytological nature (calciosomes) have been identified [5].

[†]*Permanent address:* Dept. de Biologia Celular y Anatomia Patologica, Univ. of Barcelona, Ave. Diagonal s/n, Barcelona 08028.
**Abbreviations.*- CaM, calmodulin; e.r., endoplasmic reticulum; p.m., plasma membrane(s). The kDa mol. wts. are, strictly speaking, M_r's.

METHODS

Subcellular fractions from liver homogenates.- The methods used for isolating domain-specific p.m., and the markers used, are well documented [6, 7] (& arts. by Evans et al., in Vols. 13 & 17, this series - Ed.). The preparation of endosomal vesicles was achieved by following the subcellular distribution in tissue homogenates of various intact radio-iodinated ligands (e.g. asialotransferrin, insulin, prolactin) at various intervals after their endocytic uptake by perfused rat livers [8]. These membranes, classified as 'early' (corresponding to vesicles encapsulating ligands 2 min after uptake from the portal vein), 'late' (vesicles encapsulating ligands 10-30 min after uptake) or 'receptor-enriched' (vesicles not enclosing ligands but with a high capacity to bind insulin or asialoglycoproteins) [9]. These endosomal membranes were shown to be impoverished in markers for the p.m., e.r., lysosomes and Golgi apparatus, but contained a monensin-activated ATPase which probably corresponds to the proton pump that acidifies the interior of the endocytic compartment [9, 10]. The fractionation procedure employed commences with a post-mitochondrial supernatant that is subjected to sequential density gradient centrifugation in sucrose and Nycodenz gradients. Lysosomes and Golgi fractions were prepared as described in [11]. Clathrin coats and clathrin were prepared from bovine cerebral cortices by standard procedures [12].

Analysis of membranes for Ca^{2+}- and CaM-binding.- Membrane fractions were electrophoresed in 10% polyacrylamide gels in one [13] or two [14] dimensions, transferred electrophoretically to nitrocellulose papers at 12 V for 12 h. They were exposed to ^{45}Ca (Amersham International) [15] or to CaM [16, 17] (Sigma) iodinated by the Bolton & Hunter procedure. Autoradiographs were then developed. In the CaM-binding experiments, control experiments carried out in the presence of 2 mM EGTA showed that no binding had occurred.

RESULTS AND DISCUSSION

Ca^{2+}-binding proteins.- Fig. 1 shows that 'early' or 'late' endosomes contained four major Ca^{2+}-binding polypeptides - 180, 92, 90 and 55 kDa. The receptor-enriched endosomes contained the same polypeptides except that only low amounts of the 180 kDa Ca^{2+}-binding polypeptides were present. In the p.m. fractions, 210 and 180 kDa Ca^{2+}-binding polypeptides were the major components present, but an additional 100 kDa polypeptide was identified in canalicular p.m. The lateral p.m. contained few Ca^{2+}-binding polypeptides, although the 180 kDa polypeptide was present in very low amount.

Fig. 1. Ca^{2+}-binding polypeptides of sub-fractions (abbreviations evident from their description in **METHODS**). *Arrows* point to the major Ca^{2+}-binding proteins of endosomes. Lchs = light polypeptide chains of bovine brain clathrin.

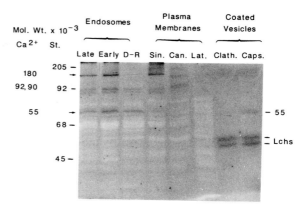

In view of the role of coated pits and vesicles in the transfer of specific populations of surface membrane proteins, especially receptors, into the endocytic compartment, clathrin-containing fractions from bovine brain were also analyzed. Purified clathrin released from coated vesicles, and clathrin cages (or caps), contained two major Ca^{2+}-binding polypeptides corresponding in electrophoretic mobility to the light chains of clathrin; a 55 kDa polypeptide - probably a clathrin-assembly factor - also bound Ca^{2+}. It was noted that the 55 kDa polypeptide in the clathrin caps was similar in electrophoretic mobility to a component present in endosomes but absent from p.m. Clathrin heavy chain (180 kDa) did not bind Ca^{2+}, indicating that the 180 kDa polypeptide in the other fractions, especially endosomes, was not clathrin.

The capacity of the clathrin light chains (36 and 33 kDa) to bind Ca^{2+} agrees with the work of Mooibroek et al. [15] and suggests that these polypeptides play a regulatory role in the formation of coated structures and thus endocytosis.

CaM-binding proteins.- Fig. 2 manifests the CaM-binding polypeptides in p.m. and endosomal subfractions. Two major ones (240 and 140 kDa) were present in the p.m., and bile-canalicular p.m. additionally showed others in the 40-60 kDa range that bound CaM weakly [11]. Endosomes contained two major CaM-binding polypeptides (140 and 115 kDa). The former probably corresponded to the polypeptide present in p.m., and the latter was near-confined to 'early' and 'late' endosomes (Fig. 3) - especially when the fractions were analyzed by 2-D electrophoresis; the absence of the 140 kDa polypeptide is due to its focussing at a pI <6.6. The 115 kDa CaM-binding polypeptide is thought to be an endosome-specific protein [18].

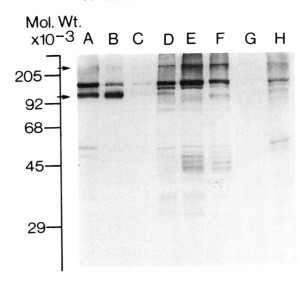

Fig. 2. CaM-binding polypeptides, resolved by polyacrylamide gel electrophoresis of various rat-liver subcellular fractions: **A**, 'late' endosomes; **B**, 'early' endosomes; **C**, 'receptor-enriched' endosomes; **D**, sinusoidal p.m.; **E**, canalicular p.m.; **F**, lateral p.m.; **G**, lysosomes; **H**, Golgi membranes. The Figure is an autoradiograph of polypeptides that bind ^{125}I-labelled CaM. *Arrows* point to the CaM-binding polypeptides specific to either endosomes or the various p.m. domains.

CONCLUSION

Clear differences in the Ca^{2+}- and CaM-binding polypeptide complements of rat-liver p.m. domains and endosomes have been demonstrated. The differences may reflect the various functions carried out at the p.m. and in the membranes forming the endocytic compartment. The Ca^{2+}- and CaM-binding polypeptides common to the cell surface and intracellular membranes may reflect the functional continuity that exists between these membrane networks as receptor-ligand complexes are internalized by the hepatocytes [cf. Evans et al., #B-7 in Vol. 13 - *Ed.*].

Acknowledgement

Carlos Enrich thanks the Wellcome Trust for a Fellowship.

Fig. 3. Comparison of the major ¹²⁵I-CaM-binding polypeptides of 'late' and 'early' endosomes with those of sinusoidal p.m. *Arrows* point to a major endosome-specific CaM-binding polypeptide (115 kDa, pI 4.3) resolved in 2 dimensions and subjected to auto-radiography.

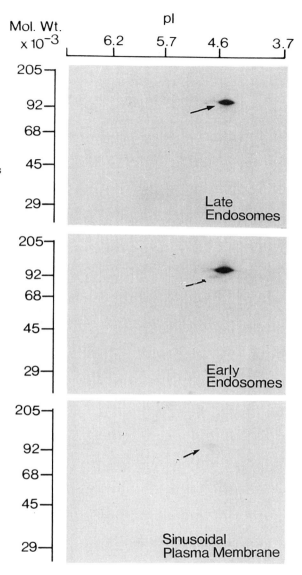

References

1. Meldolesi, J. & Pozzan, T. (1987) *Exp. Cell Res. 171*, 271-283.
2. Means, A.R. & Dedman, J.R. (1980) *Nature 285*, 73-77.
3. Cohen, P. & Klee, C.B., eds. (1988) *Calmodulin. Molecular Aspects of Cellular Regulation*, Vol. 5, Elsevier, Amsterdam.
4. Geisow, M., Fritsche, U., Hexham, J., Dash, B. & Johnson, T. (1986) *Nature 320*, 636-638.

5. Hashimoto, S., Bruno, B., Lew, D.P., Pozzan, T., Volpe, P. & Meldolesi, J. (1988) *J. Cell Biol. 107*, 2523-2531.
6. Wisher, M.H. & Evans, W.H. (1976) *Biochem. J. 146*, 375-388.
7. Evans, W.H., Flint, N. & Vischer, P. (1980) *Biochem. J. 192*, 903-910.
8. Evans, W.H. & Enrich, C. (1988) in *Cell-free Analysis of Membrane Traffic* (Morré, D.J., Howell, K.E., Cook, G.M.W. & Evans, W.H., eds.),A.R. Liss, New York, pp. 155-167.
9. Evans, W.H. & Flint, N. (1985) *Biochem. J. 232*, 25-32.
10. Saermark, T., Flint, N. & Evans, W.H. (1985) *Biochem. J. 225*, 51-88.
11. Enrich, C., Bachs, O. & Evans, W.H. (1988) *Biochem. J. 255*, 999-1005.
12. Kirchhausen, T., Harrison, S.C., Parham, P. & Brodsky, F.M. (1983) *Proc. Nat. Acad. Sci. 80*, 2481-2485.
13. Laemmli, U.K. (1970) *Nature 227*, 680-685.
14. O'Farrell, P.H. (1975) *J. Biol. Chem. 250*, 4007-4021.
15. Mooibroek, M.J., Michiel, D.G. & Wang, J.H. (1987) *J. Biol. Chem. 262*, 25-28.
16. Bachs, O. & Carafoli, E. (1987) *J. Biol. Chem. 262*, 10786-10790.
17. Hincke, M.T. (1988) *Electrophoresis 9*, 303-306.
18. Enrich, C. & Evans, W.H. (1989) *Eur. J. Cell Biol. 48*, in press.

#A-5

SITES OF ACTION OF DRUGS WHICH ALTER THE CALCIUM SENSITIVITY OF CARDIAC MYOFIBRILS

Susan J. Smith and Paul J. England

Department of Cellular Pharmacology,
Smith, Kline & French Research Ltd.,
The Frythe, Welwyn, Herts. AL6 9AR, U.K.

Drugs which increase the force of contraction of the failing heart may have therapeutic benefit in congestive heart failure, possibly due in part to enhanced Ca^{2+} sensitivity of the contractile proteins. To facilitate the development of such drugs, it is important to determine their sites of interaction within the contractile apparatus. Sites of action on the regulatory proteins Tn* (and its subunits) and Tm may be defined using systems of increasing complexity.-
1. The rates of Ca^{2+} release from sites on isolated cardiac TnC can be measured by stop-flow using the fluorescent indicator Quin 2. This identifies agents which increase the Ca^{2+} affinity of TnC directly.
2. A high-throughput +ve/-ve screen for compounds modifying Ca^{2+} binding to cTnC has been developed, needing little protein. A range of dicarbocyanine dyes bind stoichiometrically and specifically to different hydrophobic areas on the Ca^{2+}-induced conformation of TnC, producing a large change in their fluorescence which is reversed by drugs that bind to the same patch on cTnC. Thereby drugs acting at different hydrophobic sites on the protein are distinguishable.
3. A covalently bound fluorescent probe on TnI will report the rate of Ca^{2+} release from the low-affinity Ca-specific and high-affinity Ca/Mg sites of the Tn/Tm complex; hence Ca^{2+}-sensitizing compounds which act on some component of this can be assayed.
4. Effects on myofibrillar ATPase activity can be tested as a function of Ca^{2+} to determine how the agents modify the interaction of the contractile proteins in a simple, structured system.
5. Ca^{2+} affinity of TnC is modified by protein-protein interactions, including the mechanical behaviour of muscle. To

*Abbreviations.- Tn, troponin (subunits: TnC, TnI, TnT; cTnC, cardiac TnC); Tm, tropomyosin. Quin 2 & IANBD: *see over.*

test whether this alters the effects of Ca^{2+}-sensitizing agents in muscle, the Ca^{2+} dependence of force production by skinned cardiac fibres can be measured.

6.   Another approach is being developed to combine the ease of manipulation of a biochemical assay with the need for a functional working system. Observation of the motility of fluorescently-labelled regulated actin filaments over immobilized myosin may provide a model system of contraction in which the protein components can be interchanged and the effects of drugs tested.

The importance of Tn in the control of cardiac contraction, and the restriction of the cardiac isoforms of the subunits to the heart, make this protein complex an appropriate target for drug intervention [1]. The Tn complex is composed of three subunits: TnC, the Ca^{2+}-binding subunit which has considerable homology with calmodulin; TnI, the inhibitory subunit which inhibits actin-myosin interaction, and which in heart can be phosphorylated by cAMP-dependent protein kinase; TnT, which enhances the interaction of the other subunits with tropomyosin. In a beating, hormonally-unstimulated heart there is only sufficient cytoplasmic Ca^{2+} to activate 50% of the actin/myosin cross-bridges during systole. Therefore, a drug which increases the Ca^{2+} affinity of Tn may increase the interaction of actin and myosin, and hence contractility, at a given level of cytoplasmic Ca^{2+}. Methods will be described that may be used to measure the Ca^{2+} sensitivity of the cardiac contractile system at different levels of complexity. *(Section nos. as in Abstract.)*

1. MEASUREMENT OF THE RATE OF DISSOCIATION OF Ca^{2+} FROM CARDIAC TnC

Skeletal TnC has two high-affinity metal-binding sites (**III** and **IV**) in the C-terminal region which can bind Ca^{2+} or Mg^{2+}, and two low-affinity Ca^{2+} binding sites (**I** and **II**) near the N-terminus. Cardiac TnC similarly has two high-affinity C-terminal sites, but only one low-affinity N-terminal site [2].

The on-rates for Ca^{2+} binding at all sites are fast, approaching the diffusion-limited rate. Off-rates at high-affinity sites are so slow that they are irrelevant during the time-course of muscle contraction. Therefore, it is changes in the more rapid off-rates at the low-affinity sites **I** and **II** (site **II** only in the case of cardiac TnC) that are significant in the regulation of muscle contraction.

Abbreviations, CTD.- Quin 2, 2-{[2-bis(carboxymethyl)-amino-5-methylphenoxy]methyl}-6-methoxy-8-bis(carboxymethyl)-amino-quinoline; IANBD, 4-(*N*-iodoacetoxyethyl-*N*-methyl)-7-nitro-benz-2-oxa-1,3-diazole.

A drug which reduced the off-rate of Ca^{2+} from this site would enhance Ca^{2+} affinity of the complex and hence increase the force produced by the muscle at a given Ca^{2+} concentration.

The rates of Ca^{2+} dissociation can be monitored using a stopped-flow apparatus. Calcium is dissociated from TnC on reaction with a Ca^{2+} chelator [3]. The time-dependence of Ca^{2+} release can be measured directly by inclusion of the fluorescent Ca^{2+} chelator Quin 2 in the assay. This approach has been used because a large signal can be obtained; the fluorescence of Quin 2-Ca^{2+} complex with excitation at 339 nm is 6-fold greater than that of the free Quin 2.

The dissociation of Ca^{2+} from TnC sites is biphasic, corresponding to the release at the two classes of sites (rates ~160/sec and ~1/sec at 15°). The association rate of Ca^{2+} with Quin 2 is too fast to be measured and occurs within the dead time of the instrument (<3 msec) in experiments in the absence of TnC.

Bepridil has been shown to increase both the Ca^{2+} sensitivity of dog heart myofibrillar ATPase, and Ca^{2+} binding to cardiac TnC [4]. It should therefore decrease the rate of Ca^{2+} dissociation from the protein. As shown in Fig. 1, bepridil indeed causes a dose-dependent decrease in the Ca^{2+} off-rate. The decrease for the low-affinity site correlates well with an overall 5-fold increase in the total Ca^{2+} affinity of cardiac TnC in the presence of 100 μM bepridil observed by Solaro et al. [4], and demonstrates the applicability of this method.

The alternative approach of modifying the protein using a Ca^{2+}-sensitive fluorescent probe [5] might alter Ca^{2+} binding, modify protein-protein interactions, and/or interact with drug-binding sites, and hence has been avoided.

2. FORMATION OF Ca^{2+}-DEPENDENT TnC-DICARBOCYANINE DYE COMPLEXES

Measurement of the rate of Ca^{2+} release from TnC is a direct way of identifying interesting compounds, but requires a lot of protein and is time-consuming. Accordingly, a high-throughput, low-protein positive/negative screen for compounds modifying Ca^{2+} binding to cTnC has been developed.

Dicarbocyanine dyes [6] (structures: Fig. 2) have been used to assay a range of calmodulin ligands for their effects on Ca^{2+}-binding proteins [7, 8]. The relative affinities of these compounds for calmodulin and TnC were well correlated with their effects on calmodulin-stimulated phosphodiesterase activity or Ca^{2+} sensitivity of myofibrillar ATPase activity

Fig. 1. Dose-response curve for the effect of bepridil on the Ca^{2+}-off rates of cardiac TnC at 15°. Rates of Ca^{2+} dissociation from cTnC on mixing with Quin 2 were measured in the presence or absence of bepridil. The symbols distinguish data from 3 experiments; each data point is the average of 4 fluorescent transients.
A: rate of Ca^{2+} release from high-affinity sites.
B: rate of Ca^{2+} release from low-affinity site.
Conditions: excitation at 339 nm, emission at >475 nm; KCl, 100 mM; MOPS, 10 mM; $MgCl_2$, 3 mM; dithiothreitol, 0.1 mM; pH 7.0; 15°. Final concentrations: cTnC, 10 μM; Ca^{2+}, >50 μM; Quin 2, 240 μM; bepridil, 0-25 μM.
TnC was purified from bovine cardiac muscle according to Szynkiewicz *et al.* [21].

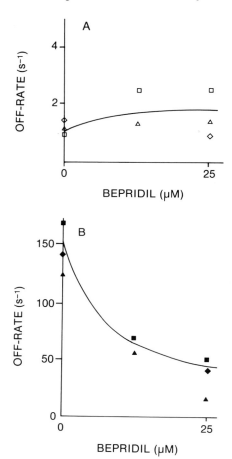

[8]. The experiments described below extend this approach. Two dyes, $diOC_6(3)$ and $diSC_3(5)$, were tested for their ability to form Ca^{2+}-dependent complexes with cTnC. The dyes are presumed to bind to hydrophobic patches exposed on the protein in the Ca^{2+}-induced conformation. This type of assay can be used to identify compounds acting selectively at the sterically-restricted drug-binding site on TnC as recognized by bepridil.

Fig. 3A shows the formation of the $cTnC.diOC_6(3)$ complex. The fluorescence of the dye is increased ~13-fold on the formation of a stoichiometric complex with cTnC. Ca^{2+} removal or titration in the absence of Ca^{2+} gives no fluorescent enhancement. Addition of bepridil (0-60 μM) had no effect on the fluorescence of the $cTnC.diOC_6(3)$ complex, irrespective of the presence or absence of Ca^{2+}. However, addition of

diY - C$_n$-(2m+1)

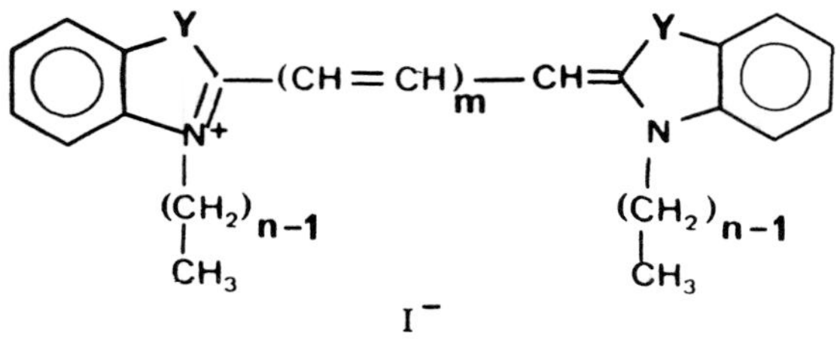

Fig. 2. Generalized structure for dicarbocyanine dyes.
Nomenclature.- Y = O, S or $(CH_3)C(CH_3)$; **n** varies from 2 to 18;
m is 1, 2 or 3; and the counter-ion is iodide. *Shorthand
notation.-* diY-C$_n$-(2m+1), where diY indicates the particular
heterocyclic nucleus of the symmetric dye, C$_n$ gives the
number of carbon atoms in the alkyl chains attached to the
nitrogen atoms of each nucleus, and the () value is the
number of methene (–CH=) groups bridging the two nuclei.

trifluoperazine, a calmodulin antagonist reported to alter
Ca^{2+} binding to skeletal TnC, causes a decrease in fluorescence
of the complex.   In the absence of cTnC, trifluoperazine
alone gives an increase in dye fluorescence.   This assay
therefore provides a means of distinguishing compounds which
bind to cTnC in a bepridil-like manner from other ligands.

Fig. 3B shows the formation of the cTnC.diSC$_3$(5) complex.
Fluorescence is decreased by 30% on formation of the stoichiometric complex.   The effect is Ca^{2+}-dependent and reversed
by addition of either bepridil or trifluoperazine. Bepridil
reverses the fluorescence change produced by formation of
cardiac TnC.diSC$_3$(5) complex (Fig. 4A), but is without effect
on the corresponding *skeletal* complex (Fig. 4B). This accords
with the specific effects of bepridil on the structure and
function of cardiac, but not skeletal, TnC (S.J. Smith,
L. MacLachlan & D. Reid, unpublished data).

Used in combination, the dicarbocyanine dye complexes
can form the basis of a rapid screen to identify compounds
showing 'bepridil-like' binding to cTnC.   The assay uses
100-fold less protein than the Quin 2 Ca^{2+}-chelation assay.

3. DEVELOPMENT OF AN ASSAY OF Ca^{2+}-OFF RATES
IN MORE COMPLEX PROTEIN ASSEMBLIES

The interactions between TnC and other components of
the Tn complex, tropomyosin and actomyosin, modify the Ca^{2+}
off-rate of the low-affinity Ca^{2+} binding sites of TnC,

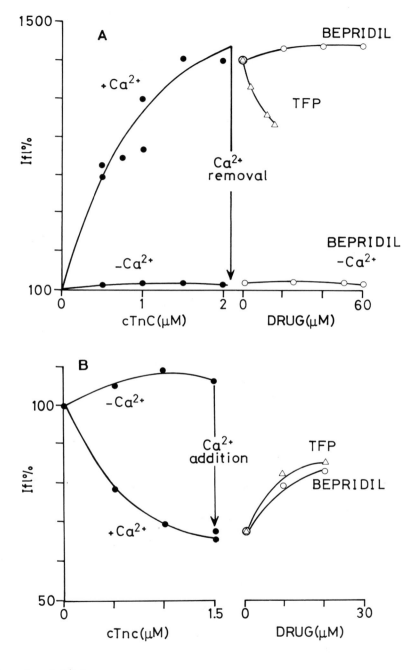

Fig. 3. Ca^{2+}-dependent cTnC complex formation with dicarbocyanine dyes: the different effects of bepridil and trifluoperazine (TFP). To the dye (fluorescence '100%') in a cuvette,

[continued opposite

Table 1. Rates of Ca^{2+} release and fluorescence transitions of
IANBD-labelled Tn/Tm (at 15°). **#1** & **#2**: Data from a stop-flow
Quin 2–Ca^{2+} chelation expt. in which protein (Tn/Tm) is mixed
with Quin 2 (240 μM final concn.); 5 μM Tn/Tm (unlabelled in **#1**;
labelled, with IANBD, in **#2**). Ca^{2+} 25 μM; Quin 2 fluorescence
excitation at 339 nm, emission at >490 nM.
#3: Data from a stop-flow expt. measuring changes in fluorescence of IANBD-labelled TnI within the Tn/Tm complex as Ca^{2+} is
released and binds to EGTA (10 mM final concn.); 1.3 μM IANBD-
labelled Tn/Tm. Ca^{2+} 50 μM; IANBD fluorescence 490_{ex} & 530_{em} nm.
Medium in **#1-#3** was as in Fig.1 legend except where indicated.

Experiment	Rates/sec	
#1: Ca^{2+} release from unlabelled Tn/Tm	23	1.2
#2: Ca^{2+} release from IANBD-labelled Tn/Tm	22	0.9
#3: Fluorescence signals from IANBD-labelled Tn/Tm	22	1.8

both *in vitro* and *in vivo* [9-11]. It is therefore important
to study the effects of drugs on the Ca^{2+} affinity of more
complex protein assemblies to determine their likely action
in vivo.   In addition, measurements of Ca^{2+} off-rates from
Tn/Tm will identify those drugs that do not interact directly
with TnC, but do modify the Ca^{2+} binding properties of the
complex.   The Ca^{2+} off-rate for a partially purified Tn/Tm
complex can be determined using Quin 2 as described above,
sect. 1.   The quality of the data is less good than that
for TnC and is more difficult to interpret, particularly
when drugs are present. It is unclear whether this is due
to   optical or biochemical properties of the system (e.g.
the presence of uncomplexed TnC).

An alternative assay using a fluorescent probe (IANBD)
[9] on TnI has been developed to overcome this problem.
Table 1 exemplifies two features of this assay.- (i) the
Ca^{2+}-binding properties of Tn/Tm are not altered by the labelling
of TnI. The rates of Ca^{2+} release from labelled and unlabelled
protein, as measured by the Quin 2 chelation assay, are

Fig. 3, *continued from opposite*
cTnC was added stepwise; then drug was added to the complex.
A, Ca^{2+}-dependent complex formation between cTnC and $diOC_6(3)$:
Ca^{2+} present (50 μM) or absent (2.5 mM EGTA); excitation 470 nm,
emission 505 nm. **B,** Ca^{2+}-dependent complex formation between
cTnC and $diSC_3(5)$: 500 μM Ca^{2+} or no added Ca^{2+}; $622_{ex}/663_{em}$ nm.
Assay conditions.- Siliconized microtubes (0.8 ml), room temp.;
constituents and pH of medium as in Fig. 1 legend; ethanolic
solution of dye added to give final concn. 1 μM/0.01% ethanol.

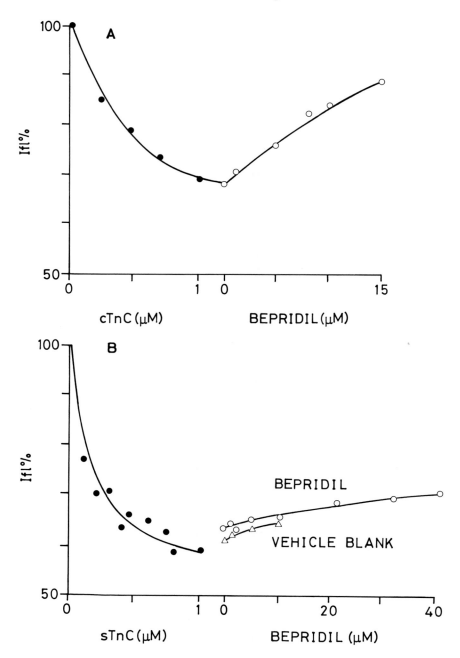

Fig. 4
TnC.diSC₃(5) binding assay discriminates between skeletal and
cardiac TnC. Conditions as for Fig. 3B. Titration of diSC₃(5)
with (**A**) cardiac TnC and bepridil, (**B**) skeletal TnC and bepridil.

the same (expts. 1 & 2). (ii) The fluorescent probe reports the rates of Ca^{2+} release from the low-affinity Ca site and the high-affinity Ca/Mg sites of TnC within the complex (expt. 3).

The amount of protein required for this assay is 5- to 10-fold less than for the Quin 2 Ca^{2+}-chelation assay, since 1 μM label can be detected. The assay can be used to determine the effects of compounds on Ca^{2+}-off rates of Tn/Tm, and can be extended to more physiologically relevant assemblies including actin and myosin cross-bridges. Fluorescence changes of IANBD-labelled skeletal TnI occur only on Ca^{2+} release from the low-affinity Ca^{2+} sites of the TnC component of the TnC/TnI complex [9]. This unexpected difference between the interactions of skeletal and cardiac contractile proteins is being further explored.

4. MYOFIBRILLAR ATPase

Calcium-sensitizing agents have been identified by their effects on the ATPase activity of washed myofibrils. These retain the basic thick and thin arrangement of the contractile proteins in the intact muscle. The hydrolysis of ATP is low in the absence of Ca^{2+}, but increases in Ca^{2+} concentration result in increased ATPase activity. This is a consequence of Ca^{2+} binding to Tn, and an increased actin-myosin interaction, analogous to contraction.

In general, compounds may modify V_{max} or basal ATPase activity in addition to any effects on Ca^{2+} sensitivity [12]. For example, Fig. 5A shows the effects of perhexiline on myofibrillar Ca^{2+} sensitivity. This compound causes an increase in the ATPase rate at submaximal Ca^{2+} concentrations. The dose-dependency of the effects is shown in Fig. 5B, where the increase in both Ca^{2+} sensitivity and basal ATPase can be seen.

Measurement of myofibrillar ATPase has been favoured as a screen for Ca^{2+}-sensitizing agents, because of its apparent simplicity and the ease of protein preparation and assay. However, it is becoming apparent on theoretical grounds that data from these assays may not predict the likely effect of the drugs in muscle. This is because the Ca^{2+} affinity of TnC is itself determined by the nature of cross-bridge interactions with the thin filament (actin and the regulatory proteins). Actively cycling cross-bridges increase the Ca^{2+} affinity of the Ca^{2+}-specific site(s) of TnC. This effect is probably large; a recent, although indirect, estimate indicated a 10-fold increase in the apparent Ca^{2+} affinity of the TnC complex as skeletal muscle fibres

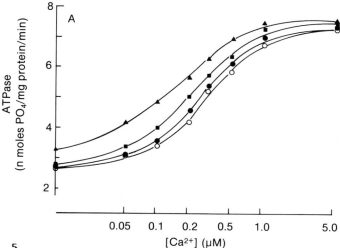

Fig. 5.
Ca^{2+} sensitivity of myofibril
ATPase: effects of varying
concentrations of perhexiline,
<u>viz.</u> (in **A**) nil (o; control),
50 μM (●), 100 μM (■) and
200 μM (▲).   Data replotted
in **B** as a function of concentration of perhexilene to
show its effects on the basal
ATPase activity, and the
increase in ATPase activity
at submaximal [Ca^{2+}]: free
Ca^{2+} nil (o), 0.1 μM (●),
0.3 μM (■) or 5.0 μM (▲).
Standard assay medium
with 0.5 ml total vol.: dog

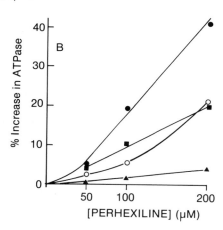

heart myofibrils, 0.5 mg protein; KCl, 60mM; imidazole, 30 mM, pH 7.0;
EGTA, 5 mM; ATP, 2.5 mM; Mg^{2+}, 3.0 mM. The amounts of CaCl$_2$ needed
to give the required free Ca^{2+} concentration in the EGTA/CaCl$_2$
buffers were calculated using a computer program (BBC Basic)
in which the interactions between EGTA, ATP, Ca^{2+}, MgCl$_2$ and
pH are all taken into account. For P$_i$ produced by ATP hydrolysis during 20 min a modified Fiske–SubbaRow assay was applied.

went from a relaxed to a fully contracted state [10]. In
comparison, the most effective reported Ca^{2+}-sensitizing agent,
calmidazolium, at ½-maximal effective dose [12] produces a
<2-fold increase in Ca^{2+} affinity of the Ca^{2+}-specific site
of cardiac myofibrils [13].

    The behaviour of the myofibrils during the ATPase assay
is complex and ill-defined.-

(i) The initial incubation of the myofibrils with the drug occurs in the absence of added ATP, leading to the formation of rigor cross-bridges. These may be analogous to the state which occurs in muscle at the end of the power stroke, but this is short-lived *in vivo* and is therefore not highly populated, nor likely to be of great significance for this mechanism of drug action.

(ii) As these cross-bridges are exposed to low concentrations of ATP diffusing from the incubation medium, Ca^{2+} regulation of contraction may be over-ridden by the effects of the remaining rigor cross-bridges.

(iii) This is followed by a short period of isotonic shortening resulting in over-contraction, in which the lattice structure is damaged. The properties of this system are those which largely determine the measured ATPase activity and its apparent Ca^{2+} sensitivity. Hence it is not clear that the effects of drugs observed in this assay will necessarily predict their behaviour in a working muscle.

5. THE Ca^{2+} DEPENDENCE OF FORCE PRODUCTION IN SKINNED CARDIAC MUSCLE FIBRES

Skinned single muscle fibres provide a system in which the effects of agents on the function of the contractile proteins can be investigated. Treatment of muscle fibres with Triton X-100 disrupts the sarcolemma and solubilizes cellular membranes including sarcoplasmic reticulum and mito-chondrial membranes [11]. The environment of the contractile proteins ($[Ca^{2+}]$, $[ATP]$, pH) can be controlled since the constituents of the bathing solutions have direct access to the myofilament space. Mechanical work is measured directly, and other contractile parameters (e.g. stiffness) can be measured with relative ease.

The approach is being increasingly used to identify Ca^{2+}-sensitizing compounds, because of its approximation to the conditions in the intact muscle. A disadvantage of the method is that it does not identify which protein in the regulatory complex is the actual drug-binding component. It is also time-consuming as an initial screen.

6. *IN VITRO* MOVEMENT ASSAY

Recently, muscle models have been developed which are intermediate in complexity between isolated contractile proteins and skinned muscle fibres. Observations are made of movement of myosin-coated beads on the array of parallel actin filaments which occur naturally in Nitella [14], or the movement of fluorescent actin filaments on microscope slides coated with myosin [15]. Investigation of the properties of these models has already provided valuable information: for example, single beads of myosin are as efficient at producing motion as myosin, implying that the two-headed nature of myosin is not fundamental to contraction [16]. The constituent proteins and ligands of such models can be easily manipulated, and could provide a controlled system to investigate the molecular action of pharmacological agents.

The simplest motility assays are not Ca^{2+}-sensitive since the regulatory components are not included. However, tropomyosin and Tn from skeletal muscle bind to Nitella actin, providing an assay of bead motion which is regulated [17]. Ca^{2+}-sensitivity of the movement of fluorescently-labelled regulated actin on myosin has also been reported [18], and the system is being investigated further. A mechanically loaded motility assay [19] would be most likely to predict the effect of a drug on cardiac contractile proteins *in vivo*, as the mechanical state of the muscle may modulate Ca^{2+} regulation [10, 11].

FINAL COMMENTS

Many different approaches are needed to search for agents which modify the Ca^{2+} sensitivity of the contractile proteins either by direct action on TnC or via other components of the regulatory complex. Recent data emphasize the dynamic effects of protein-protein interactions on the Ca^{2+} affinity of the contractile proteins. This indicates the feasibility of achieving a drug which will specifically modify cardiac contractility without altering, for example, calmodulin-dependent processes. It remains to be seen whether such agents will be of clinical significance [20].

Acknowledgements

We thank David Mills for expert technical assistance, and Nigel Carter for painstaking protein purification. S.J.S. is grateful to Dr John Sleep, MRC Unit of Cellular Biophysics, King's College, London, for collaboration on the *in vitro* movement assay.

References

1. Rüegg, J.C. (1986) *Circulation 73*, 78-84.
2. Van Eerd, J-P. & Takahashi, K. (1975) *Biochem. Biophys. Res. Comm. 64*, 122-127.
3. Rosenfeld, S.S. & Taylor, E.W. (1985) *J. Biol. Chem. 260*, 242-251.
4. Solaro, R.J., Bousquet, P. & Johnson, J.D. (1986) *J. Pharmacol. Exp. Ther. 238*, 502-507.

5. Johnson, J.D., Collins, J.H., Robertson, S.P. & Potter, J.D. (1980) *J. Biol. Chem. 255*, 9635-9640.
6. Sims, P.J., Waggoner, A.S., Hang, C.H. & Hoffmann, J.F. (1974) *Biochemistry 13*, 3315-3329.
7. Orlov, S.N., Pokudin, N.I., Ryazhskii, G.C. & Kratsov, G.M. (1984) *Biokhimya (USSR) 49*, 43-50.
8. Tkachuk, V.A., Baldenkov, G.N., Feoktisov, I.A., Mens'shikov, M.Yu., Quast, U. & Hertzig, J.W. (1987) *Arz.-Forsch./Drug Res. 37*, 1013-1017.
9. Rosenfeld, S.S. & Taylor, E.M. (1985) *J. Biol. Chem. 260*, 252-261.
10. Guth, K. & Potter, J.D. (1987) *J. Biol. Chem. 262*, 13627-13635.
11. Pan, B.S. & Solaro, R.J. (1987) *J. Biol. Chem. 262*, 7839-7849.
12. Silver, P.J., Pinto, P.B. & Dachiw, J. (1985) *Biochem. Pharmacol. 35*, 2545-2551.
13. Scheetz, M.P. & Spudich, J.A. (1983) *Nature 303*, 31-35.
14. El-Saleh, S.C. & Solaro, R.J. (1987) *J. Biol. Chem. 262*, 17240-17246.
15. Scheetz, M.P., Block, S.M. & Spudich, J.A. (1986) *Meths. Enzymol. 134*, 531-544.
16. Toyoshima, Y.Y., Kron, S.J., McNally, E.M., Niebling,K.R., Toyoshima, C. & Spudich, J.A. (1987) *Nature 328*, 536-539.
17. Vale, R., Szent-Gyorgyi, A. & Scheetz, M.P. (1984) *Biophys. J. 45*, 7a.
18. Honda, H. & Asakura, S. (1988) *9th Int. Biophysics Congr. IUPAB 29*, 52 [venue: Tel Aviv].
19. Kishino, A. & Yanagida, T. (1988) *Nature 334*, 74-76.
20. Wetzel, B. & Hauel, N. (1988) *Trends Pharmacol. Sci. 9*, 166-170.
21. Szynkiewicz, J., Stepowski, D., Brzeska, H. & Drabikowski, W. (1985) *FEBS Lett. 181*, 281-285.

#A-6

A COMPARISON BETWEEN CONTRACTION AND INCREASE IN SLOWLY EXCHANGING ^{45}Ca IN THE ISOLATED RAT AORTA

B. Wilffert, D. Wermelskirchen, P. Koch, D. Wilhelm and T. Peters

Janssen Research Foundation,
Raiffeisenstr. 8, D-4040 Neuss 21, F.R.G.

A method not needing lanthanum or EGTA, which can cause artefacts, has been developed to examine aortic strips for the fraction of slowly exchanging ^{45}Ca which is considered to be representative of contractile processes depending on Ca^{2+} influx. We label the strips for 5 min with ^{45}Ca^{2+} and measure the ^{45}Ca remaining after washing for 45 min at 4° with a normal Ca^{2+}-containing Tyrode solution.

The inhibition by Ca^{2+}-entry blockers of the K^+-induced increase in ^{45}Ca labelling in this fraction correlates well with their Ca^{2+}-entry* blocking properties determined in contraction measurements. However, parallel measurement of ^{45}Ca labelling and contraction does not always correlate even for K^+-induced contractions, and seems essential where interpretable data are needed for unknown agents. This is demonstrated with high concentrations of R 56865, a compound which, in respect of inhibition of K^+ effects, shows non-parallelism between contraction and elevation in slowly exchanging ^{45}Ca. Furthermore we compared noradrenaline and K^+ effects, and found a good correlation in respect of both contractions and the ^{45}Ca response in the lower concentration range; but for neither agent was a correlation found at higher concentrations. This might imply that either at the higher concentrations contractile processes not directly dependent on Ca^{2+}-influx predominate or the ^{45}Ca uptake in the Ca fraction studied is a limiting factor.

THE MEASUREMENT OF SLOWLY EXCHANGING ^{45}Ca

The loss of ^{45}Ca from labelled vascular smooth muscle by washing with a ^{45}Ca-free solution is characterized by a fast and a slow phase. The slowly exchanging fraction of ^{45}Ca corresponds to an intracellularly located Ca pool whose labelling reflects the ^{45}Ca concentration in

*Note by Ed.: here and elsewhere, Ca (authors' preference) altered to Ca^{2+} (but not systematically).

Fig. 1. Residual ^{45}Ca in rat aortic strips after washing with a Tyrode solution at 4° for different times. Tissues pre-labelled in a ^{45}Ca^{2+}-containing Tyrode solution (for 5 min) under control (o) or depolarizing (K$^+$; ●) conditions. Data are presented as means ±S.E.M. (n = 4); P <0.05 denoted *.

After the washing, the tissues were blotted (60 sec), weighed, and dissolved in Soluene$^{\odot}$ (Packard Insts.) for 2 h at 60°; then the residual ^{45}Ca was measured.
Adapted from ref. [8] (publ. S. Karger AG, Basel).

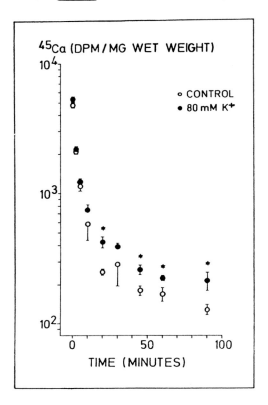

the cytosol. If vascular smooth muscle is labelled with ^{45}Ca for a short time, measurement of slowly exchanging ^{45}Ca serves to estimate the amount of ^{45}Ca which has entered the cytosol. To assess this ^{45}Ca fraction, it has to be separated from the fast-exchanging ^{45}Ca fraction. For this purpose lanthanum [1, 2] and EGTA [3] have been used. Drawbacks with lanthanum are the necessity to omit physiological phosphate and carbonate buffers [4], penetration of the membrane by lanthanum [4, 5] and a change in intracellular calcium distribution [6]. EGTA removes membrane-bound calcium and thereby induces structural changes in the plasmalemma [7]. We therefore developed the following alternative method for studying slowly exchanging ^{45}Ca^{2+} without using lanthanum or EGTA. We labelled aortic strips from Wistar rats for 5 min with ^{45}Ca under normal or depolarizing (80 mM K$^+$) conditions and washed in normal Tyrode solution at 4° or conventionally, for comparison, with lanthanum-containing (50 mM) HEPES solution at 37° or an EGTA-containing (2 mM) Ca^{2+}-free HEPES- or Tyrode-solution at 4° [8]. With each washing solution, different times were compared. As Fig. 1 illustrates (see legend for final steps before residual ^{45}Ca measurement), the K$^+$-induced increase

Fig. 2. Effect of
nifedipine on K⁺-
induced contrac-
tions and *(below)*
increases in
slowly exchanging
^{45}Ca in rat
aortas. Data are
presented as
means ±S.E.M.
(n = 4–10).
E signifies expon-
ential (thus 1E⁻5
= 10^{-5}).
Upper portion is
from [12], by
courtesy of
Elsevier
Scientific
Publications;
similarly for
Fig. 3.

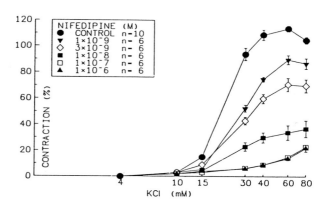

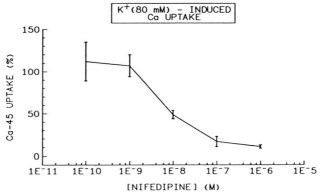

in slowly exchanging $^{45}Ca^{2+}$ could be shown by washing at 37°
with lanthanum-containing HEPES solution or else at 4° with
the normal Tyrode solution. The EGTA method failed to
demonstrate a K⁺-induced increase in slowly exchanging ^{45}Ca
in aortic strips from the rat in contrast to the rabbit [3].

Accordingly, for minimization of interference by the
fast-exchanging fraction our preferred washing procedure was
with normal Tyrode solution for 45 min, whereby unlabelled
calcium competes with ^{45}Ca; the tissue is affected only by
the temperature decrease to 4° to slow down transmembrane
exchanging processes.

INHIBITION OF THE K⁺-INDUCED CONTRACTION AND INCREASE IN SLOWLY EXCHANGING ^{45}Ca BY Ca^{2+}-ENTRY BLOCKERS AND R 56865

The K⁺-induced contraction in rat aortic rings (details
in [8]) was inhibited in a concentration-dependent way by
nifedipine   (10^{-9}–10^{-6} M;   Fig. 2),   verapamil   (10^{-8}–10^{-5} M),

Fig. 3. Effect of R 56865 on K$^+$-induced contractions and increases in slowly exchanging ^{45}Ca in rat aortas. Data are presented as means ±S.E.M. (n = 5-12). *Essentially as in ref. [12] (publ. Elsevier).*

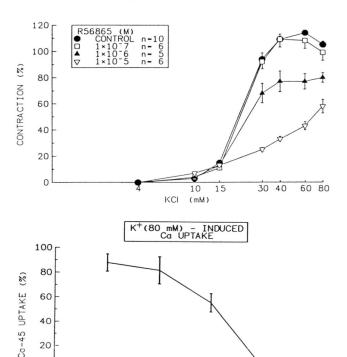

diltiazem (10^{-7}-10^{-5} M) and flunarizine (10^{-7}-10^{-5} M; data not shown). Similarly in respect of concentration range and dependency, each Ca^{2+}-entry blocker inhibited the K$^+$-induced increase in slowly exchanging ^{45}Ca (for nifedipine see Fig. 2). These data support the assumption that the slowly exchanging ^{45}Ca-fraction measured is representative for contractile processes depending on Ca^{2+}-influx. R 56865, viz. *N*-{1-[4-(4-fluorophenoxy)butyl]-4-piperidinyl} -*N*-methyl-2-benzothiazol-amine, attenuated the K$^+$-induced contraction only at high concentrations (10^{-6}-10^{-5} M) and attained a maximal inhibition of 45 ±3% at 80 mM (Fig. 3). The K$^+$-induced increase in slowly exchanging ^{45}Ca, however, was completely abolished by R 56865 (10^{-5} M; Fig. 3).

In contrast to calcium entry blockers, with R 56865 there is clearly a dissociation between inhibition of K$^+$-induced contraction and inhibition of K$^+$-induced increases in slowly

Fig. 4. The
K$^+$- and *(below)*
noradrenaline-
induced
increases in
^{45}Ca in rat
aortic strips.
Data are
presented as
means ±S.E.M.
(n = 4-8).
3E$^-$9 (e.g.)
signifies
3×10^{-9}.

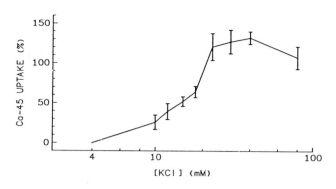

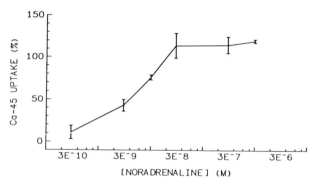

exchanging ^{45}Ca. This may be explained by an inhibition
by R 56865 of the uptake of ^{45}Ca from the cytosol into
the slowly exchanging compartment. Thus the ^{45}Ca-experiments
may overestimate Ca^{2+} entry-blocking properties of R 56865. For
a better interpretation of data with unknown agents, parallel
measurements of ^{45}Ca-labelling and contraction are evidently
necessary.

COMPARISON OF K$^+$- AND NORADRENALINE-INDUCED CONTRACTION AND INCREASE IN SLOWLY EXCHANGING ^{45}Ca

The K$^+$-induced increase in the slowly exchanging ^{45}Ca-
fraction was concentration-dependent (Fig. 4). Fig. 5 depicts
the relationship between contraction and the increase in
slowly exchanging ^{45}Ca elicited by K$^+$ at different concentrations (values with 80 mM = '100%'). A good linear correlation
was found up to a contraction of ~50%. At 30 to 80 mM
[K$^+$] contractions still increased while no further rise of
slowly exchanging ^{45}Ca was observed.

Noradrenaline also caused a concentration-dependent
increase in the slowly exchanging ^{45}Ca in the rat aorta

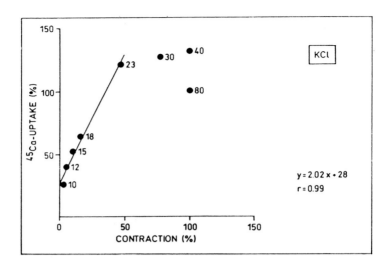

Fig. 5. Relation between contraction and increase in slowly exchanging ^{45}Ca elicited by different K^+ concentrations (mM) in rat aortas (effects '100%' for 80 mM K^+). Data are presented as means (n = 6-7).

(Fig. 4), the curve having a maximum comparable with that for K^+ ('100%' for 80 mM K^+). In the aorta of Sprague-Dawley rats Chiu <u>et al</u>. [9] found for K^+ a maximal increase in slowly exchanging ^{45}Ca twice that for noradrenaline, which may be explained by the use of a lanthanum method and a different rat strain.

The relationship between contraction and ('100%' for 80 mM K^+) ^{45}Ca uptake was characterized by a linear correlation up to a contraction of ~60%, whereafter contraction increased and the amount of slowly exchanging ^{45}Ca remained constant (Fig. 6; for contractions, '100%' = maximum of the first concentration-response curve, and the second curve is shown). The lack of correlation between slowly exchanging ^{45}Ca and contraction may indicate that at higher concentrations, for both K^+ and noradrenaline, calcium released from a pool not labelled by extracellular ^{45}Ca^{2+} in 5 min contributes to the contractile process. As far as K^+ is concerned this might be calcium- or depolarization-induced release of calcium from the sarcoplasmic reticulum (for review see [10]). For noradrenaline an increased turnover of membrane phospho-inositides by activation of the α_1-adrenoreceptors causes

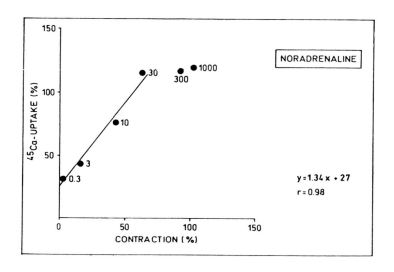

Fig. 6. As for Fig. 5, but with varied noradrenaline. For the ^{45}Ca measurements the effect of 80 mM K$^+$ represents '100%'. The contraction measurements are related to the maximum of the first concentration-response curve for noradrenaline ('100%') whereas the values of the second are shown in the Figure. Data are presented as means (n = 4-11).

an increase in Ins(1,4,5)P$_3$ which releases calcium from the sarcoplasmic reticulum (reviewed in [10]).

There is, however, another possible explanation – that ^{45}Ca uptake into the slowly exchanging compartments is saturable. In other words, at higher concentrations of K$^+$ or noradrenaline the labelling of the cytosol with extracellular ^{45}Ca may still increase but the uptake of cytosolic calcium into the slowly exchanging compartments cannot increase further.

GENERAL CONSIDERATIONS

We developed a method for measuring slowly exchanging ^{45}Ca without using lanthanum or EGTA, the washing being with a normal Tyrode solution for 45 min at 4°. With methods which measure slowly exchanging ^{45}Ca after a short labelling period with ^{45}Ca^{2+} the transport of ^{45}Ca from the extracellular space to these slowly exchanging compartments is estimated.

This method does not help decide whether the measured ^{45}Ca-fraction is derived from the extracellular space [11]. However, the short ^{45}Ca-labelling time implies a flux of ^{45}Ca out of a pool in a fast equilibrium with the extracellular space.

A good correlation between ^{45}Ca and contraction measurements is often found, but not invariably even for K^+-induced contractions. This implies that measurements of slowly exchanging ^{45}Ca can give us information about calcium movements not obtainable by contraction experiments.

Acknowledgements

The skilful technical assistance of U. Nebel and A. Leidig is gratefully acknowledged.

References

1. Van Breemen, C. & McNaughton, E. (1970) *Biochem. Biophys. Res. Comm. 39*, 567-574.
2. Godfraind, T. (1976) *J. Physiol. 260*, 21-35.
3. Meisheri, K.D., Palmer, R.F. & Van Breemen, C. (1980) *Eur. J. Pharmacol. 61*, 159-165.
4. Van Breemen, C., Hwang, O. & Siegel, B. (1977) in *Excitation-Contraction Coupling in Smooth Muscle* (Casteels, R., et al., eds.), Elsevier/North Holland Biomedical Press, Amsterdam, pp. 243-252.
5. Wendt-Gallitelli, M.F. & Isenberg, G. (1985) *Pflügers Arch. 405*, 310-322.
6. Kolbeck, R.C. & Speir, W.A. (1986) *J. Mol. Cell Cardiol. 18*, 733-738.
7. Loutzenhiser, R., Leyten, P. Saida, K. & Van Breemen, C. (1985) *Calcium and Contractility. Smooth Muscle.* (Grover, A.K. & Daniel, E.E., eds.), Humana Press, Clifton, NJ, pp. 61-92.
8. Wermelskirchen, D., Wilhelm, D., Wilffert, B., Pegram, B.L., Hunter, J., Nebel, U. & Peters, T. (1988) *Pharmacol. 36*, 84-90.
9. Chiu, A.T., McCall, D.E. & Timmermans, P.B.M.W.M. (1986) *Eur. J. Pharmacol. 127*, 1-8.
10. Karaki, H. & Weiss, G.B. (1988) *Life Sci. 42*, 111-122.
11. Lüllmann, H. (1987) in *Cardiovascular Pharmacology '87* (Papp, J. Gy., ed.), Akadémiai Kiado, Budapest, pp. 629-639.
12. Koch, P., Wilhelm, D., Wermelskirchen, D., Nebel, U., Wilffert, B. & Peters, T. (1988) *Eur. J. Pharmacol. 158*, 183-190.

#A-7

THE ROLE OF PHOSPHOLAMBAN PHOSPHORYLATION BY CYCLIC AMP- AND CYCLIC GMP-DEPENDENT PROTEIN KINASES IN REGULATING MUSCLE CALCIUM FLUXES

Katherine A. Gang, Tessa J. Mattinsley
and *John P. Huggins

Department of Cellular Pharmacology,
Smith Kline & French Research Ltd.,
The Frythe, Welwyn, Herts. AL6 9AR, U.K.

Phospholamban is an integral protein of s.r. which regulates the activity of the Ca^{2+}-ATPase. It can be phosphorylated by several protein kinases, and a physiological role is best established for the cAMP-dependent kinase (A-kinase). This article discusses methods used to examine the importance of G-kinase in phosphorylating phospholamban. The purification of kinases, of phospholamban and of heart and smooth muscle membranes is described. Evidence is given which suggests that phospholamban occurs in an s.r. subcompartment. In vitro G-kinase phosphorylates phospholamban when the latter is presented to the enzyme as a solubilized protein or when present in cardiac or smooth muscle membranes. However, we have been unable to show a role for this process in intact tissue.

Phospholamban is a low mol. wt. protein (6080 Da) found in cardiac muscle-cell s.r.[†] It can be phosphorylated *in vitro* by the action of four protein kinases stimulated respectively by cAMP (A-kinase; [1]), cGMP (G-kinase; [2]), Ca^{2+} and calmodulin [3], and diacylglycerol (protein kinase C; [4]). The physiological relevance of cAMP-induced phosphorylation is well established [5]. Phospholamban phosphorylation by A-kinase increases the Ca^{2+}-sensitivity of the Ca^{2+}-ATPase pump of cardiac s.r., such that at physiological Ca^{2+} concentrations the pump is stimulated [6]. This has led to the hypothesis that phospholamban phosphorylation may explain the increased rate at which Ca^{2+} returns to basal levels after

*addressee for any correspondence; now at Pharmacology Dept., Merrell Dow Res. Inst., 16, rue d'Ankara, 67084 Strasbourg, France.
[†]*Abbreviations.-* s.r., sarcoplasmic reticulum; cAMP, cyclic AMP, & see above for A-kinase (similarly cGMP, G-kinase). G-peptide: see overleaf.

contraction, and the increased rate of force relaxation, when cardiac muscle is exposed to agents which raise intracellular cAMP (see [5]). Indeed exposure of perfused hearts to isoprenaline causes a dramatic increase in phospholamban phosphorylation [7].

The relevance of the other three kinases in modulating phospholamban phosphorylation *in vivo* is far less clear. Small amounts of phospholamban are present in some smooth muscles [8], which suggests a role of cGMP-dependent phospholamban phosphorylation because increases in cGMP in smooth muscle are associated with relaxation [9, 10] and a reduction in intracellular Ca^{2+} [11-13].

This article describes and discusses methods which we have developed to investigate cGMP-dependent phosphorylation of phospholamban both *in vitro* and *in vivo*. The emphasis is very firmly on methods. Apart from the specific interest in the role of G-kinase in phosphorylating phospholamban, many of the methods are applicable to other systems. This applies, for example, where the effects of cGMP, subcellular fractionation of smooth muscle, or whole-tissue protein phosphorylation are involved.

PURIFICATION OF KINASES

A-kinase is activated in the presence of cAMP by the dissociation of a catalytically active subunit (C subunit) [14]. There are many advantages of using this subunit for phosphorylation experiments rather than the holoenzyme and in any case it is easier to purify. The method of Reimann & Beham [15] has been reproducibly successful in our hands and produced a protein that is virtually homogeneous. The enzyme is stored in aliquots at -70° in 100 mM phosphate, pH 7, 50% (w/v) glycerol, 1 mM EDTA, 15 mM 2-mercaptoethanol, and is stable for many months.

G-kinase

Nucleotide binding and protein kinase assays.- G-kinase activity was assayed using 10 µl of enzyme fraction in a total volume of 70 µl of 500 mM potassium phosphate, 700 mM $MgCl_2$, 50 mg/ml Sigma histone II-A, 50 µM cGMP and 50 µM $[\gamma-^{32}P]ATP$ (90 GBq/mol). Alternatively, Arg-Lys-Ser-Arg-Ala-Glu ('G-peptide'; from Peninsula Labs., St. Helens, U.K.) [16] was used as a substrate (100 µM) instead of histone. After a 1 or 2 min incubation at 30°, samples were spotted onto Whatman P81 papers, which were then washed in 37.5 mM $H_3PO_4/0.05\%$ (w/v) polyphosphoric acid. Radioactivity was determined by Cerenkov counting in water.

Cyclic nucleotide binding was determined in a buffer (pH 6.8) containing 0.7 μM [^{3}H]-cAMP or -cGMP (90 GBq/mol), 50 mM potassium phosphate, 1 mM EDTA, 2 M NaCl, 0.5 mg/ml histone II-A, and 1 μM of the competing nucleotide (cAMP for cGMP binding, and *vice versa*). After 60 min at 4°, 1 ml 10 mM potassium phosphate/1 mM EDTA, pH 6.8, was added, the entire sample was filtered onto a 0.45 μm HA Millipore filter which was washed twice, and filter radioactivity was determined by scintillation counting.

Purification.- G-kinase does not have a dissociating catalytic subunit, although a catalytically active proteolytic fragment can be prepared [17]. We have found the affinity purification method of Lincoln [18] to be a very rapid and effective technique. Ion-exchange chromatography [a critical step as most A-kinase regulatory (R) subunit is removed here] was followed by affinity chromatography using 8-2-amino-ethylamino-cAMP-Sepharose (Pharmacia). The kinase eluted from the affinity column contained a considerable amount of the A-kinase R subunit, as assessed by [^{3}H]cAMP binding.

Further purification was performed on an FPLC MonoQ column as follows: cAMP was added to the kinase eluted from the affinity column to give a final concentration of 10 mM (to displace cGMP). The enzyme was dialyzed for 12 h at 4° against 20 mM sodium phosphate/2 mM EDTA/25 mM 2-mercaptoethanol, pH 7. This material was then applied to the MonoQ column which was first washed with 20 mM Tris/1 mM EDTA/5% (w/v) glycerol/7.5 mM 2-mercaptoethanol, pH 7.5, and then eluted with a gradient of 0 to 1 M NaCl in the same buffer. The eluted kinase was concentrated by dialysis against sucrose followed by 50% (w/v) glycerol/20 mM sodium phosphate/2 mM EDTA/25 mM 2-mercaptoethanol, pH 7, and stored at -20°.

Results.- Fig. 1 shows the elution of protein (measured as in [19]) and G-kinase activity (using histone) from the MonoQ column. Of the two main protein peaks, the first corresponds to kinase activity. The two peaks (**1** and **2**) were pooled as shown and binding capacities determined: 1.27 and 0.03 nmol/ml of cGMP bound to peaks **1** and **2** respectively, compared to 0.7 and 0.86 nmol/ml for cAMP binding respectively. Scanning of an SDS-polyacrylamide gel showed the concentrated kinase of peak **1** to be ~90% pure with respect to total protein. The preparation had a maximum specific activity using G-peptide of 2.4 μmol/min per mg. The maximum stimulation of activity by cGMP was 13-fold. Hence the ion-exchange column helped to remove cGMP from the enzyme, greatly improved its purity and helped to remove the R subunit of A-kinase.

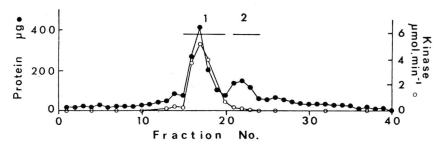

Fig. 1. Elution of G-kinase from a MonoQ column (details in text): yields of protein and of G-kinase, and pooling (———).

PHOSPHORYLATION OF PHOSPHOLAMBAN

Purification of phospholamban.- Many groups have now published purification methods, but some require denaturation by organic solvents [20] or SDS [21] or both [22, 23], or a specific monoclonal antibody [24]. An early report that phospholamban is selectively extracted from s.r. with deoxy-cholate [25] is now controversial [26, 27]. The remaining three methods [26-28] all use a combination of fairly mild detergents and gel exclusion, ion exchange and sulphydryl affinity chromatography, and all claim homogeneity (or near) for the protein.

We have used the method of Jones <u>et al</u>. [28] to purify phospholamban from porcine or bovine hearts. The final protein was only 30% pure as assessed by silver staining of SDS gels, being contaminated with some proteins of higher M_r, but this sufficed for phosphorylation experiments.

Phospholamban phosphorylation.- Incubation of purified phospholamban (60 µg/ml) in 50 mM HEPES/Tris (pH 6.8), 100 mM $MgCl_2$, 2.5 mM EGTA, 0.2% (w/v) Triton X-100, 0.5 mM dithiothrei-tol, 0.8 µM cGMP, 5 µg/ml G-kinase, 0.1 mM [γ-^{32}P]ATP resulted in its phosphorylation, manifest by a band on an autoradiograph of a 15% acrylamide SDS gel (Fig. 2a; *gel conditions: see [29]).

Studies using crude membrane vesicles.- Microsomal vesicles from smooth muscle and crude s.r. vesicles from heart are easily prepared by published methods using differen-tial centrifugation [30, 31]. For vascular smooth muscle the greatest problem is thorough homogenization in the presence of so much connective tissue. The simplest solution is to use large blood vessels which are relatively free from connec-tive tissue, and we have found sheep pulmonary artery useful in this respect. The endothelium detaches from the vessel wall after incubating for a few minutes at 4° in 0.9% NaCl.

Fig. 2b is considered later in text

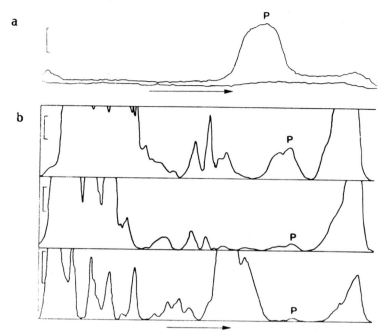

Fig. 2. Scans of autoradiographs of SDS gels showing:
a: purified phospholamban phosphorylated *in vitro* by G-kinase
(*lower trace:* phospholamban omitted); **b**: phosphorylation of
phospholamban in smooth-muscle microsomes *(top trace)* and
fractions Ca-B *(middle)* and CaP *(bottom).* Vertical bars
represent 1 a.u.; *horizontal arrow* (length represents 20 mm)
indicates direction of protein migration. **P**: position of the
monomeric form of phospholamban (5,500 Da).

Membrane phosphorylation was effected in some cases with
detergent present (as above), but in others detergent was
omitted: the reaction was stopped by adding excess EDTA (to
chelate Mg^{2+}) in 0.15 M sucrose, 10 mM HEPES/Tris, pH 7.   The
mixture was then centrifuged at 350,000 **g** for 20 min at
4° (100,000 rpm in the TL100.2 rotor of a Beckman TL100).
The pellets were dissolved for electrophoresis in 2% (w/v)
SDS, 62.5 mM Tris-HCl, 20% (w/v) sucrose, 40 mM dithiothreitol.

With phospholamban there is a fortuitous characteristic in that
it undergoes a reversible subunit dissociation in the presence
of SDS when samples are heated to 100° [32].   The resulting
change in M_r from the pentameric M_r of ~28,000 to the monomeric.
M_r of 5,500 provides a very simple means of identifying
phosphorylated phospholamban on polyacrylamide gels.   The
advantages of the detergent-free method of phosphorylation
are that the kinase reaction can be very rapidly stopped,
allowing kinetics to be analyzed, and also that Mg^{2+}, which

affects the M_r transformation of phospholamban [33], can be removed from the incubate.

When cardiac s.r. or pulmonary artery microsomes were incubated with G-kinase and $[\gamma-^{32}P]ATP$, phospholamban was phosphorylated. In addition, when cardiac s.r. was incubated with A- or G-kinase for 2 min, the rate of phosphorylation was linear with time. By altering the concentration of s.r. protein in the incubation, the kinetics of the kinases could be compared. It was found that the data fitted cooperative kinetics, that the $k_{0.5}$ and n values were comparable between kinases but that the V_{max} for G-kinase was nearly 4-fold higher than for A-kinase.

SUBFRACTIONATION OF SMOOTH MUSCLE MEMBRANES

To examine whether the phospholamban in smooth muscle is present in the same membranes as the protein that it regulates (the s.r. Ca^{2+}-ATPase), methods were developed to separate s.r. from sarcolemma [cf. pertinent arts. in Vol. 6, this series --Ed.].

Separation methods.- Techniques tried unsuccessfully for this separation included differential centrifugation, sucrose and Percoll gradient centrifugation, and phase separation using polyethylene glycol/dextran. It was therefore necessary to modify the density of one of the membranes. One reported method [30] is to incubate microsomes with digitonin, to selectively permeabilize sarcolemma, and then to separate membranes on a sucrose gradient. This method suffers from the disadvantages that theoretically proteins might be solubil- ized from the membranes by the digitonin and also that calcium transport is completely abolished, even in s.r. memb- ranes (data not shown).

We therefore tried a technique which has been used extensively for cardiac membranes [e.g. 34], viz. to load s.r. vesicles with crystals of calcium oxalate by incubating mixed membranes with ATP, Ca^{2+} and potassium oxalate, and then to separate membranes on a sucrose gradient. Some success has been obtained in separating smooth-muscle membranes by similar methods, although the membranes produced were not analyzed in terms of marker enzymes [35, 36]. We have used this technique for smooth-muscle membranes and compared the resulting fractions with those obtained using digitonin [37].

Methodological details.- Sheep pulmonary artery microsomes were prepared essentially as in ref. [30] except that the protease inhibitors phenylmethanesulphonyl fluoride (50 μM) and (each 1 μg/ml) antipain, leupeptin and pepstatin A were

Fig. 3. Fractions from gradient centrifugation after $^{45}Ca^{2+}$ loading: radioactivity before *(hatched bars)* and after *(open bars)* filtration.

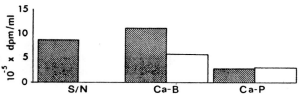

included in the homogenization buffer. For the **digitonin technique** the detergent was added to microsomes (0.25 mg/mg protein) [30] and then, in the SW41Ti rotor of a Beckman L80 centrifuge, the membranes were centrifuged at 100,000 **g** for 19 h in a linear gradient of 15-50% (w/v) sucrose in 1 mM ATP, 5 mM $MgCl_2$, 5 mM potassium oxalate, 5 mM NaN_3, 10 mM HEPES-NaOH, 0.5 M KCl, and the three inhibitors as above (pH 7.1). Of the three resulting fractions, the top fraction (D-B) and the bottom one (D-P) were relatively enriched in sarcolemma and s.r. respectively. For **Ca^{2+} loading**, 1 ml microsomes (10-15 mg protein/ml) was loaded with calcium oxalate by incubating for 30 min at 30° with 4.4 ml of 0.1 M KCl, 25 mM HEPES-NaOH, 5 mM NaN_3, 5 mM potassium oxalate, 50 mM Mg-ATP and 50 mM $CaCl_2$ (pH 6.9). The reaction mixture was cooled and then loaded onto a sucrose gradient as above. Centrifugation at 150,000 **g** for 19 h gave a diffuse band (Ca-B) and a loose pellet (Ca-P).

Separation results.- In some experiments, the $CaCl_2$ in the s.r.-loading incubation was labelled with $^{45}Ca^{2+}$ and then the distribution of radioactivity was determined after gradient centrifugation. Aliquots of each fraction were filtered through 0.45 μm Millipore HA filters (subsequently washed with 2 × 1 ml 2 mM ATP, 5 mM NaN_3, 100 mM KCl, 25 mM HEPES-NaOH, 10 mM $MgCl_2$; pH 6.9) to determine the distribution of Ca^{2+} within membrane vesicles. Fig. 3 shows that in the Ca-P, all the $^{45}Ca^{2+}$ was associated with membranes, and as this material was centrifuged to a loose pellet through 50% sucrose, the membranes presumably contained calcium oxalate and represent s.r. The less dense Ca-B fraction also contained $^{45}Ca^{2+}$, ~50% being in a membrane-bound form which, as the membranes had not entered the Ca-P fraction, clearly was not precipitated with oxalate. It is therefore assumed that the Ca^{2+} in the Ca-B membranes is that which is transported into sarcolemmal membranes during the incubation with $CaCl_2$ and ATP. The remaining $^{45}Ca^{2+}$ was observed above the Ca-B in the sucrose gradient ('S/N' in Fig. 3).

Marker enzymes were used to assess the purity of membranes prepared by either technique (Table 1). Cytosolic contamination was undetectable in all fractions after sucrose gradient

Table 1. Relative protein contents and specific activities
of marker enzymes (assayed as in [37]) in various membrane
fractions, relative to a post-nuclear supernatant (PNSN).
Besides protein, the items tabulated (*vs.* PNSN as unity) are
lactate dehydrogenase (cytosol), succinate dehydrogenase
(mitochondria), 5'-nucleotidase (5'-AMPase; sarcolemma),
NADPH-cytochrome **c** reductase (NC reductase; s.r.) and the
ratio of 5'-AMPase to NC reductase. All values are means of
3-8 preparations. N.D., not detectable.

Fraction	Protein	LDH	SDH	5'-AMPase	NC reductase	Ratio
PNSN	*1.0*	*1.0*	*1.0*	*1.0*	*1.0*	*1.0*
Microsomes	1.32	0.284	0.411	3.16	3.91	0.81
Ca-B	0.615	N.D.	0.522	5.03	16.1	0.31
Ca-P	0.156	N.D.	0.372	1.15	10.3	0.11
D-B	0.111	N.D.	0.246	4.63	12.9	0.36
D-P	0.0754	N.D.	N.D.	0.401	17.9	0.02

centrifugation, but significant mitochondrial contamination
was evident. Neither technique resulted in an absolute
separation between s.r. and sarcolemma, but the ratio of
5'-nucleotidase to NADPH-cytochrome **c** reductase shows that
for both methods s.r. is relatively enriched in the 'P'
fractions *vs.* sarcolemma. Therefore, by examining the distribution of a protein in both the 'B' and the 'P' fractions,
it should be possible to know its subcellular distribution.

  Phosphorylation studies.- Phospholamban phosphorylation
by G-kinase in smooth-muscle microsomes, Ca-B and Ca-P was
examined as described for crude membranes. A typical result
is shown in Fig. 2b. To our surprise, 35-fold more phospholamban
(after normalizing per mg protein) occurred in the Ca-B
than in the Ca-P. This result is in direct contrast to
that observed using the digitonin technique, from which phospholamban was concluded to be present in s.r. [2]. The results
reported here suggest that there are two structurally distinct
s.r. compartments in muscle, only one of which contains
phospholamban which may be phosphorylated. Similar results
to these have been observed using Ca^{2+} loading in cardiac
muscle [38]. The functional implications of these compartments
deserve further research.

PHOSPHORYLATION IN INTACT TISSUE

  Protein phosphorylation can be studied in intact tissues
by perfusing or incubating the tissue with $^{32}P_i$, to label
intracellular ATP, and then stimulating with agents which
affect kinases. For rat or guinea pig hearts a perspex-shielded

Langendorf perfusion cabinet was constructed (J.P. Huggins & P.J. England) in which up to 8 hearts may be perfused. The flow of perfusate is controlled with valves and the perfusate containing $^{32}P_i$ is recycled through the heart at constant pressure. Rabbit aortic rings were incubated in small-volume, water-jacketed organ baths containing $^{32}P_i$. Tissues were stimulated with isoprenaline (to activate A-kinase) or carbamyl-choline, nitroprusside or 8-bromo-cGMP (to stimulate G-kinase), and membranes were prepared [39]. Isoprenaline caused an 18.5-fold stimulation of phospholamban phosphorylation in the hearts, but no agent used to elevate cGMP significantly stimulated the phosphorylation of phospholamban.

CONCLUSIONS

The results obtained from these studies were as follows.-
a. Purified phospholamban is a substrate for G-kinase.
b. Phospholamban present in membranes, where it is likely to retain its native configuration, is phosphorylated by G-kinase.
c. The kinetics of cGMP-induced phospholamban phosphorylation, when compared to those for A-kinase-induced phosphorylation (where a physiological role is quite well established), show that cGMP-dependent phosphorylation occurs at a significant rate.
d. Phospholamban is not phosphorylated in the expected 's.r.' fraction of smooth muscle fractionated by Ca^{2+} loading. This suggests that phospholamban exists in a subcompartment of s.r., the specialized function of which is unclear.
e. Agents which increase intracellular cGMP in cardiac and smooth muscle do not cause phospholamban phosphorylation in whole-tissue preparations. Hence phospholamban is a good substrate for G-kinase *in vitro*, but this is apparently not the case *in vivo*.

References (BBA, Biochim. Biophys. Acta; JBC, J. Biol. Chem.)

1. Tada, M., Kirchberger, M.A., Repke, D.I. & Katz, A.M. (1974) *JBC 249*, 6174-6180.
2. Raeymaekers, L., Hofmann, F.A. & Casteels, R. (1988) *Biochem. J. 252*, 269-273.
3. Le Peuch, C.J., Haiech, J. & Demaille, J.G. (1979) *Biochemistry 18*, 5150-5157.
4. Movsesian, M.A., Nishikawa, M. & Adelstein, R.S. (1984) *JBC 259*, 8029-8032.
5. Huggins, J.P. & England, P.J. (1985) in *Molecular Advances in Transmembrane Signalling* (Cohen, P. & Houslay, M.D., eds.), Elsevier, Amsterdam, pp. 57-87.
6. Hicks, M.J., Shigekawa, M. & Katz, A.M. (1979) *Circ. Res. 44*, 384-391.
7. Lindemann, J.P., Jones, L.R., Hathaway, D.R., Henry, B.G. & Watanabe, A.M. (1983) *JBC 258*, 464-471.
8. Raeymaekers, L. & Jones, L.R. (1986) *BBA 882*, 258-265.

9. Rapoport, R.M., Draznin, M.B. & Murad, F. (1983) *Nature* *306*, 174-176.
10. Itoh, T., Kanmwa, Y., Kuriyama, H. & Sasagwi, T. (1985) *Br. J. Pharmacol. 84*, 393-406.
11. Morgan, J.P. & Morgan, K.G. (1984) *J. Physiol. 357*, 539-551.
12. Rashatwar, S.S., Cornwelt, T.L. & Lincoln, T.M. (1987) *Proc. Nat. Acad. Sci. 84*, 5685-5689.
13. Kai, H., Kanaide, H., Matsumoto, T. & Nakamura, M. (1987) *FEBS Lett. 221*, 284-288.
14. Corbin, J.D. (1983) *Meth. Enzymol. 99*, 227-232.
15. Reimann, E.M. & Beham, R.A. (1983) *Meth. Enzymol. 99*, 51-55.
16. Glass, D.B. & Krebs, E.G. (1982) *JBC 257*, 1196-1200.
17. Heil, W.G., Landgraf, W. & Hofmann, F. (1987) *Eur. J. Biochem. 168*, 117-121.
18. Lincoln, T.M. (1983) *Meth. Enzymol. 99*, 62-71.
19. Bradford, M.M. (1976) *Anal. Biochem. 72*, 605-611.
20. Capony, J-P., Rinaldi, M.L., Guilleux, F. & Demaille, J.G. (1983) *BBA 728*, 83-91.
21. Le Peuch, C.J., Le Peuch, D.A.M. & Demaille, J.G. (1980) *Biochemistry 19*, 3368-3373.
22. Holtzhauer, M., Sydow, H., Grunow, G. & Will, H. (1986) *Biomed. Biochim. Acta 45*, 719-725.
23. Collins, J.H., Kranias, E.G., Reeves, A.S., Bilezikjian, L.M. & Schwartz, A. (1981) *Biochem. Biophys. Res. Comm. 99*, 796-803.
24. Suzuki, T., Lui, P. & Wang, J.H. (1987) *Biochem. Cell Biol. 65*,
25. Bidlack, J.M., Ambudkar, I. S. & Shamoo, A.E. (1982)   [302-309. *JBC 257*, 4501-4506.
26. Jakab, G. & Kranias, E.G. (1988) *Biochemistry 27*, 3799-3806.
27. Inui, M., Kadoma, M. & Tada, M. (1985) *JBC 260*, 3708-3715.
28. Jones, L.R., Simmerman, H.K.B., Wilson, W.W., Gurd, F.R.D. & Wegener, A.D. (1985) *JBC 260*, 7721-7730.
29. Laemmli, U.K. (1970) *Nature 227*, 680-685.
30. Morel, N., Wibo, M. & Godfraind, T. (1981) *BBA 644*, 82-88.
31. Huggins, J.P. & England, P.J. (1987) *FEBS Lett. 217*, 32-36.
32. Wegener, A.D. & Jones, L.R. (1984) *JBC 259*, 1834-1841.
33. Louis, C.F., Maffitt, M. & Jarvis, B. (1982) *JBC 257*, 15182-15186.
34. Levitski, D.O., Aliev, M.K., Kuzmin, A.V., Levchenko, T.S., Smirnov, V.N. & Chazov, E.I. (1976) *BBA 443*, 468-484.
35. Raeymaekers, L., Agostini, B. & Hasselbach, W. (1980) *Histochemistry 65*, 121-129.
36. Raeymaekers, L., Wuytack, F., Eggermont, J., De Schutter, G. & Casteels, R. (1983) *Biochem. J. 210*, 315-322.
37. Gang, K.A. & Huggins, J.P. (1988) *Biochem. Soc. Trans. 16*, 356-358.
38. England, P.J., Jeacocke, S.A., Huggins, J.P., Mills, D. & Pask, H.T. (1983) *Biochem. Soc. Trans. 11*, 153.
39. Huggins, J.P., Cook, E.A., Piggott, J.R., Mattinsley, T.J. & England, P.J. (1989) *Biochem. J.*, in press.

#A-8

SEARCHES FOR Ca²⁺-BINDING MOTIFS IN PROTEIN SEQUENCES

William G. Turnell

MRC Laboratory of Molecular Biology,
Hills Road, Cambridge CB2 2QZ, U.K.

and MRC Acute Phase Protein Research Group,
Immunological Medicine Unit, Department of Medicine,
Royal Postgraduate Medical School, London W12 OHS, U.K.

*Sequences in proteins thought to contain the 'E-F hand'
Ca²⁺-binding structure(s) are compared with a linear 'template' of
residue choices. The template matches exclusively all the
actual and strongly predicted E-F hand motifs currently entered
in protein sequence data banks. It is adaptable to include
matches to some phospholipid-binding proteins.*

Protein sequence databases can be searched for matches
to patterns of amino acid residues that are found to be
characteristic of particular substructures or motifs [e.g. 1].
The pattern is represented by a weight matrix, which is
a table of scores for matching each residue type at each
positional element within the motif. Some elements of the
pattern may be invariant and so only one residue type would
score, while at other elements several different residues
may be allowed, and hence would score. Finally, remaining
elements may be completely unrestricted and so all residue
types score zero. A sparse pattern, which contains a significant
proportion of unrestricted elements is a *template*.

During a search, a match-score for each overlapping
segment of sequence is obtained by sliding the template
along the sequence. For each segment, residues are scored
according to their assigned weight in the equivalent element
of the template, and the match-score is the sum of these
residue-scores.

Many calcium-modulated proteins possess one or more characteristic, evolutionarily related, cation-binding folds known
as 'helix-loop-helix' or 'EF-hand'[†] motifs [2]. In each
motif the loop connecting the two helices is 12, or rarely
13, amino acid residues long, and provides ligands to the
calcium ion. Such a contiguous string of residues may
be represented by a template, with each element labelled

[†] EF is depictable by an appropriately oriented hand [2].
Abbreviations.- CRP, C-reactive protein; PKCα, brain protein
kinase C; SAP, serum amyloid P component; TNF, tumour necrosis
factor.

as follows: X-Y-Zg*$\overline{Y}$h$\overline{X}$--$\overline{Z}$, according to the nomenclature of
Kretsinger [3]. Here X, Y, Z, $\overline{X}$, $\overline{Y}$ and $\overline{Z}$ are residues
amongst which those that possess oxygen-bearing side-chains
are candidates for binding to the calcium ion. As a main-chain
carboxyl oxygen or a water molecule can contribute a ligand
to the calcium ion, not all these residues need be of this
type.

The remaining elements of this template are as follows:
'-' can be any residue, 'g' must be glycine, and 'h' is
a hydrophobic residue important for the integrity of the
EF-hand fold; '*' marks an allowed insertion (making the
loop 13 residues long instead of 12) that has been found
in crystallographically determined structures [4].

Structure determinations of α-lactalbumin [5] and galac-
tose-binding protein [6] have revealed novel calcium-binding
motifs that mimic to different degrees the loop in the
'helix-loop-helix' EF-hand motif, but are devoid of the flanking
helices.

As might be anticipated, the template described above
is not by itself sufficiently specific to match exclusively
to sequences of calcium-binding loops logged in the protein
sequence libraries. Employing the 'cut-off' criterion that
at least 3 of the appropriate elements must match to oxygen-
bearing side-chains, the 12-residue template matches to 46%
(~3100 sequences) of the current SWISSPROT library [7] and
with the insertion (*) to 44% (data not shown).

The following sections describe how the template may
be extended for greater specificity.

'HELIX-LOOP-HELIX' MOTIFS

In these motifs the two helices interact to different
extents. The interaction is always mediated by contacts
between hydrophobic moieties. The interacting surfaces along
each helix, together with the calcium ion liganded to oxygen
atoms, constitute the interior of the motif. This interior
of the EF-hand fold has been evolutionarily conserved, whereas
the remaining exterior is highly variable. The template
shown in Fig. 1 was constructed so as to encompass the
structural features of this interior. The residue choices
for helix/helix contact residues, at elements -10, -9, +7,
+10 and +11, were taken from those found in the hydrophobic
interiors of proteins, whereas the choices for termini of
the helices (elements -13, -6 and +14) were selected from
those commonly found at the edge of hydrophobic patches
[see 8].

Matrix (reconstructed from the rotated figure; weights of zero are left blank):

	−13	−12	−11	−10	−9	−8	−7	−6	X −5	−4	Y −3	−2	z −1	g 0	$\bar{Y}$ 1	h 2	$\bar{X}$ 3	4	5	$\bar{Z}$ 6	7	8	9	10	11	12	13	14
C	24	24	14	14				100								100					100							24
S		24		14	14				15		15		15		15		15			15			24	24	24			24
T		24		14	14		100	100	15		15		15		15	100	15			15	100		14	14	14			24
P							100	100								100					100							24
A		24		14	14																		24	24	24			
G		24		14	14									100									14	14	14			
N		24		14	14				20		20		20		20		20			20			14	14	14			
D									25		25		25		25		25			25								
E									25		25		25		25		25			25								
Q	24		14	14	14				20		20		20		20		20			20			14	14	14			24
H	24		14	14	14																		14	14	14			24
R	24		14	14	14																		14	14	14			24
K	24		14	14	14																		14	14	14			24
M	24	24	24	24	24		100	100								100					100		24	24	24			24
I	24	24	24	24	24		100	100								100					100		24	24	24			24
L	24	24	24	24	24		100	100								100					100		24	24	24			24
V	24	24	24	24	24		100	100								100					100		24	24	24			24
F	24	24	24	24	24		100	100								100					100		24	24	24			24
Y	24	24	24	24	24		100	100								100					100		24	24	24			24
W	24	24	24	24	24		100	100								100					100		24		24			24

helix (columns −13 to −9 and 9 to 14)

Fig. 1. The template used to search for sequences of EF-hand motifs, displayed in the form of a matrix. The elements of the template are the columns of weights assigned to amino acid residues, which are listed, in the one-letter code, at the start and finish of each row. Weights of zero are left blank. The elements are numbered along the top, from −13 to +14, so that the origin of the template lies at the invariant glycine (element 0). Element labels X, Y, Z, $\bar{Y}$, $\bar{X}$, $\bar{Z}$, g, h and '−' are described on the opposite page.

The results of searching the 6821 sequences in the current protein sequence library SWISSPROT with the template depicted in Fig. 1 are shown in Table 1. The template, which carries a maximum match-score of 694 (obtained by adding together the highest weight assigned to each element) and a cut-off score of 630 (whereby all template/sequence match-scores less than this are rejected) returns from the library all protein sequences with known or proposed EF-hand motifs that bind calcium with high affinity.

Most of these proteins possess two or more helix-loop-helix templates, each with different binding affinities for calcium. For example, from proteins with just two calcium-binding sites: motifs I (residues 10-38), II (82-109) and I (12-40) are, respectively, not returned from the sequences of vitamin D-dependent calcium-binding protein [9], α-parvalbumin [2] and S-100(α) protein [10].

An important result is as follows. If a non-sparse pattern is constructed from all the aligned sequences shown in Table 1, then many more non-calcium-binding sequences are returned. Probably the exterior residues of 'high-affinity motifs', whilst being variable, nevertheless form a non-random sub-set of the equivalent residues from all EF-hand motifs. A low-affinity motif may be located within a particular sequence for a segment with non-sparse homology to a high-affinity motif from the *same* sequence.

MEMBRANE-ASSOCIATED MOTIFS

The sequences listed in Table 1 are typically from intracellular proteins. In EF-hand structures that have been determined crystallographically a water molecule is often liganded to the calcium ion [e.g. 4]. Model-building (not shown) suggests that a phosphate group could replace the water molecule, but would have to be stabilized in this position by a positively charged side-chain. Arginine or lysine at the N-terminus of the second helix (element +7 in Fig. 1) could perform this function, whilst still contributing to the interior hydrophobicity of the motif.

Replacing the residue choices at element +7 in Fig. 1 with R or K (each with weight 100), and employing a cut-off score of 630, returns only the sequences shown in Table 2. All these sequences are from the extracellular (or periplasmic) domain of phospholipid membrane-associated proteins. In particular, a novel form of TNF has recently been found to be associated with the monocyte cell surface [11], and residues 292-303 of bovine PKCα have already been proposed as a calcium-binding site [12].

Needleman-Wunch [13] alignment of the TNF and PKCα sequences did not produce a significant match either of identical or of homologous equivalent residues. However, adopting the strategy described above for finding low-affinity Ca^{2+}-binding sites from high-affinity sites (i.e. aligning only residues 284-311 of PKCα to the TNF sequence) produced the same alignment between these two sequences as is shown in Table 2*. The alignment program used was ALIGN, employing the homology 'Mutation Data' matrix, a matrix-bias of 6 and·a gap-penalty of 100 [14].

POSSIBLE CALCIUM-DEPENDENT PHOSPHATE-BINDING MOTIF

Two pentraxins - a family of Ca^{2+}-binding plasma proteins [15] - occur in the human: CRP (the classical acute phase reactant) and SAP. CRP binds phosphocholine-containing molecules, e.g. phospholipids, whereas SAP is the only major Ca^{2+}-dependent DNA-binding protein in whole serum [16]. Other members of the mammalian pentraxin family include hamster female protein involved in parturition. The mammalian pentraxins contain a decapeptide, residues 133-142 (human CRP numbering), that has been proposed as an α-helical binding site to 4 phosphate groups separated by 7-11 Å [17]. Such arrays of phosphate groups occur at the surfaces of both phospholipid bilayers [18] and double-helical nucleic acids.

Commencing 13 residues beyond the phosphate-binding α-helix modelled by Turnell et al. [17], Nguyen et al. [19] have proposed a low-affinity Ca^{2+}-binding loop in pentraxins, based upon homology with the loops of EF-hand motifs, but without the adjacent helices. However, similar sequences for loops are extremely common amongst proteins that do not necessarily bind calcium (see earlier in this article).

Turnell et al. [17] used a template to search for sequences that could form phosphate-binding α-helices. Their template has been generalized (in terms of hydrophobicity and charge) to form the first template shown in Table 3. The second template in Table 3 was formed from the elements for the loop in the helix-loop-helix template shown in Fig. 1. These first and second templates were linked by an arbitrary gap of 10-15 residues. The result of searching the SWISSPROT library with the combined templates is shown in Table 3. All the mammalian pentraxin sequences logged in the library are returned, together with lipocortins I and II.

As members of the annexin family, these lipocortins are Ca^{2+}-dependent phospholipid-binding proteins. Other members of the family - p32, p68 and pII - are not currently logged in the SWISSPROT library. They contain multiple,

_____
* p. 98   [Text continues on p. 99

Table 1. The segments of protein sequences returned by searching the SWISSPROT library with the template depicted in Fig. 1. The aligned segments form columns of residues. Each column is labelled at the top with the heading of the equivalent element of the template (see Fig. 1). The residue numbers against the N-termini of the segments position them in their respective protein sequences.

```
SWISSPROT                                           helix              helix
ENTRY              PROTEIN                            ..                 ..
                                                    .   .            .      .
                                                 .         .       .          .
                                                   X-Y-ZgYhX--Z
                                                ------------- +++++++++++++
                                                1111           11111
                                                3210987654321012345678901234

AEQ1$AEQVI    AEQUORIN 1 PRECURSOR          152   CEETFRVCDIDESGQLDVDEMTRQHLGF
AEQ2$AEQVI    AEQUORIN 2 PRECURSOR          152   CEETFRVCDIDESGQLDVDEMTRQHLGF
CABI$BOVIN    VITAMIN D-DEPENDENT (CABP)     46   LDELFEELDKNGDGEVSFEEFQVLVKKI
CABI$PIG      VITAMIN D-DEPENDENT (CABP)     49   LDDLFQELDKNGNGEVSFEEFQVLVKKI
CABI$RAT      VITAMIN D-DEPENDENT (CABP)     49   LDNLFKELDKDGDGEVSYEEFEVFFKKL
CABV$BOVIN    CALBINDIN                      15   FFEIWLHFDADGSGYLEGKELQNLIQEL
                                            102   FMKTWRKYDTDHSGFIETEELKNFLKDL
                                            146   TDLMLKLFDSNNDGKLELTEMARLLPVQ
                                            190   FNKAFELYDQDGDGYIDENELDALLKDL
CABV$CHICK    CALBINDIN                      17   FFEIWHHYDSDGNGYMDGKELQNFIQEL
                                            104   FMQTWRKYDSDHSGFIDSEELKSFLKDL
                                            148   TEIMLRMFDANNDGKLELTELARLLPVQ
                                            192   FNKAFEMYDQDGNGYIDENELDALLKDL
CABV$RAT      CALBINDIN                      15   FFEIWLHFDADGSGYLEGKELQNLIQEL
                                            102   FMKTWRKYDTDHSGFIETEELKNFLKDL
                                            146   TDLMLKLFDSNNDGKLELTEMARLLPVQ
                                            190   FNKAFELYDQDGNGYIDENELDALLKDL
CALB$BOVIN    CALCINEURIN B SUBUNIT          54   VQRVIDIFDTDGNGEVDFKEFIEGVSQF
                                             91   LRFAFRIYDMDKDGYISNGELFQVLKMM
                                            132   VDKTIINADKDGDGRISFEEFSAVVGGL
CALL$CAEEL    CALMODULIN-LIKE PROTEIN        26   FREAFMMFDKDGNGTISTKELGIAMRSL
                                             62   ILEMINEVDIDGNGQIEFPEFCVMMKRM
                                             98   IREAFRVFDKDGNGVITAQEFRYFMVHM
                                            134   VDEMIKEVDVDGDGEIDYEEFVKMMSNQ
CALM$CHLRE    CALMODULIN                     15   FKEAFALFDKDGDGTITTKELGTVMRSL
                                             51   LQDMISEVDADGNGTIDFPEFLMLMARK
                                             88   LREAFKVFDKDGNGFISAAELRHVMTNL
                                            124   VDEMIREADVDGDGQVNYEEFVRMMTSG
CALM$DICDI    CALMODULIN                     14   FKEAFSLFDKDGDGSITTKELGTVMRSL
                                             50   LQDMINEVDADGNGNIDFPEFLTMMARK
                                             87   IREAFKVFDKDGNGYISAAELRHVMTSL
                                            123   VDEMIREADLDGDGQVNYDEFVKMMIVR
CALM$DROME    CALMODULIN                     12   FKEAFSLFDKDGDGTITTKELGTVMRSL
                                             48   LQDMINEVDADGNGTIDFPEFLTMMARK
                                             85   IREAFRVFDKDGNGFISAAELRHVMTNL
                                            121   VDEMIREADIDGDGQVNYEEFVTMMTSK
CALM$ELEEL    CALMODULIN                     12   FKEAFSLFDKDGDGTITTKELGTVMRSL
                                             48   LQDMINEVDADGNGTIDFPEFLTMMAKK
                                             85   IREAFRVFDKDGNGYISAAELRHVMTNL
                                            121   VDEMIREADIDGDGQVNYEEFVQMMTAK
CALM$HUMAN    CALMODULIN                     12   FKEAFSLFDKDGDGTITTKELGTVMRSL
                                             48   LQDMINEVDADGNGTIDFPEFLTMMARK
                                             85   IREAFRVFDKDGNGYISAAELRHVMTNL
                                            121   VDEMIREADIDGDGQVNYEEFVQMMTAK
CALM$PATSP    CALMODULIN                     12   FKEAFSLFDKDGDGTITTKELGTVMRSL
                                             48   LQDMINEVDADGDGTIDFPEFLTMMARK
                                             85   IREAFRVFDKDGDGFISAAELRHVMTNL
                                            121   VDEMIREADIDGDGQVNYEEFVTMMTSK
CALM$SPIOL    CALMODULIN                     12   FKEAFSLFDKDGDGCITTKELGTVMRSL
                                             48   LQDMINEVDADGNGTIDFPEFLNLMARK
                                             85   LKEAFRVFDKDQNGFISAAELRHVMTNL
                                            121   VDEMIREADVDGDGQINYEEFVKVMMAK
CALM$TETPY    CALMODULIN                     12   FKEAFSLFDKDGDGTITTKELGTVMRSL
                                             48   LQDMINEVDADGDGTIDFPEFLSLMARK
                                             85   LIEAFKVFDRDGDGLITAAELRHVMTNL
```

[continued

Table 1, *continued*

CALM$TRYBR	CALMODULIN	12	FKEAFSLFDKDGDGTITTKELGTVMRSL
		48	LQDMINEVDQDGSGTIDFPEFLTLMARK
		85	IKEAFRVFDKDGNGFISAAELRHIMTNL
		121	VDEMIREADVDGDGQINYEEFVKMMMSK
CALM$WHEAT	CALMODULIN	13	FKEAFSLFDKDGDGCITTKELGTVMRSL
		49	LQDMINEVDADGNGTIDFPEFLNLMARK
		86	LKEAFRVFDKDQDGFISAAELRHVMTNL
		122	VDEMIREADVDGDGQINYEEFVKVMMAK
CALM$YEAST	CALMODULIN	13	FKEAFALFDKDNNGSISSSELATVMRSL
		86	LLEAFKVFDKNGDGLISAAELKHVLTSI
CALS$CHICK	CALMODULIN, STRIATED MUSCLE	12	FKEAFSLFDRDGDGCITTMELGTVMRSL
		48	LQDMVGEVDADGSGTIDFPEFLSLMARK
		85	IREAFRVFDKDGNGYISAAELRHVMTNL
		121	VDEMIKEADCNNDGQVNYEEFVRMMTEK
CAP1$CHICK	CALPAIN I	610	WLTIFRQYDLDKSGTMSSYEMRMALESA
CAP1$HUMAN	CALPAIN I	620	YLSIFRKFDLDKSGSMSAYEMRMAIESA
CAP1$RABIT	CALPAIN I	208	YLAIFRKFDLDKSGSMSAYEMRMAIESA
CAP2$RABIT	CALPAIN 2	329	YQKIYREIDVDRSGTMNSYEMRKALEEA
CAPS$HUMAN	CALCIUM-DEPENDENT PROTEASE	174	WQAIYKQFDTDRSGTICSSELPGAFEAA
		209	LYNMIIRRYSDESGNMDFDNFISCLVRL
CAPS$PIG	CALCIUM-DEPENDENT PROTEASE	172	WQAIYKQFDVDRSGTIGSSELPGAFEAA
CAPS$RABIT	CALCIUM-DEPENDENT PROTEASE	172	WQAIYKQFDVDRSGTICSRELPGAFEAA
CART$CHICK	CALRETININ	14	FLDVWRHFDADGNGYIEGKELENFFQEL
		61	MKEFMHKYDKNADGKIEMAELAQILPTE
		105	FMEAWRRYDTDRSGYIEANELKGFLSDL
CAVP$BRALA	CALCIUM VECTOR PROTEIN	90	ILRAFKVFDANGDGVIDFDEFKFIMQKV
		127	VEEAMKEADEDGNGVIDIPEFMDLIKSK
CBP$STRER	CALCIUM-BINDING PROTEIN	105	VKGTWGMCDKNADGQINADEFAAWLTAL
		139	AAEAFNQVDTNGNGELSLDELLTAVRDF
CD31$YEAST	CELL DIVISION CONTROL P31	25	IYEAFSLFDMNNDGFLDYHELKVAMKAL
		134	LRAMIEEFDLDGDGEINENEFIAICTDS
CFAG$HUMAN	CYSTIC FIBROSIS ANTIGEN	51	ADVWFKELDINTDGAVNFQEFLILVIKM
ML1C$CHICK	MYOSIN CARDIAC MUSCLE	130	FVEGLRVFDKEGDGTVMGAELRHVLATL
ML1M$AQUIR	MYOSIN, LIGHT CHAIN, MUSCLE	86	YMEAFKTFDREGQGFISGAELRHVLTAL
ML1M$PACYE	MYOSIN, LIGHT CHAIN, MUSCLE	86	YMEAFKTFDREGQGFISGAELRHVLTAL
ML1S$CHICK	MYOSIN, SKELETAL MUSCLE	128	FVEGLRVFDKEGNGTVMGAELRHVLATL
ML1S$RAT	MYOSIN, SKELETAL MUSCLE	128	FVEGLRVFDKEGNGTVMGAELRHVLATL
ML2A$CHICK	MYOSIN, CARDIAC MUSCLE	27	FKEAFTIMDQNRDGFIDKADLRDTFAAL
ML2A$PATYE	MYOSIN, SMOOTH MUSCLE	25	MKEAFTMIDQNRDGFIDINDLKEMFSSL
ML2B$CHICK	MYOSIN, CARDIAC MUSCLE	26	FKEAFTIMDQNRDGFIDKADLRDTFAAL
ML2B$PATYE	MYOSIN, SMOOTH MUSCLE	20	MKEAFTMIDQNRDGFIDINDLKEMFSSL
ML2E$PATSP	MYOSIN, (EDTA LIGHT CHAIN)	20	MKEAFSMIDVDRDGFVSKDDIKAISEQL
ML2M$CHICK	MYOSIN, SMOOTH MUSCLE	33	FKEAFNMIDQNRDGFIDKEDLHDMLASM
ML2S$CHICK	MYOSIN, SKELETAL MUSCLE	27	FKEAFTVIDQNRDGIIDKDDLRETFAAM
ML2S$HALRO	MYOSIN, SMOOTH MUSCLE	17	FKEAFTMIDANRDGFIDQEDLKDTYASL
ML2S$RABIT	MYOSIN, SKELETAL MUSCLE	29	FKEAFTVIDQNRDGIIDKEDLRDTFAAM
ML2S$RAT	MYOSIN, SKELETAL MUSCLE	29	FKEAFTVIDQNRDGIIDKEDLRDTFAAM
ML3F$HUMAN	MYOSIN, FETAL MUSCLE	87	FVEGLRVFDKEGNGTVMGAELRHVLATL
ML3L$DROME	MYOSIN, LARVA AND ADULT FORM	85	FIECLKLYDKEENGTMLLAELQHALLAL
ML3M$CHICK	MYOSIN, SMOOTH MUSCLE	88	YVEGLRVFDKEGNGTVMGAEIRHVLVTL
ML3P$DROME	MYOSIN, PUPA FORM	85	FIECLKLYDKEENGTMLLAELQHALLAL
ML3S$CHICK	MYOSIN, SKELETAL MUSCLE	87	FVEGLRVFDKEGNGTVMGAELRHVLATL
ML3S$RAT	MYOSIN, SKELETAL MUSCLE	87	FVEGLRVFDKEGNGTVMGAELRHVLATL
ONCO$RAT	ONCOMODULIN	43	VKDIFRFIDNDQSGYLDGDELKYFLQKF
PRVA$AMPME	PARVALBUMIN ALPHA	43	VTKAFHILDKDRSGYIEEEELQLILKGF
PRVA$ESOLU	PARVALBUMIN ALPHA	42	VKKVFKAIDADASGFIEEEELKFVLKSF
		81	TKAFLKAADKDGDGKIGIDEFETLVHEA
PRVA$LATCH	PARVALBUMIN ALPHA	45	LKEVFGILDQDKSGYIEEEELKFVLKGF
PRVA$RABIT	PARVALBUMIN ALPHA	43	VKKVFHILDKDKSGFIEEEELGFILKGF
PRVA$RAJCL	PARVALBUMIN ALPHA	43	LAEIFNVLDGDQSGYIEVEELKNFLKCF
PRVA$RANES	PARVALBUMIN ALPHA	43	MQKVFHVLDQDQSGFIEKEELCLILKGF
PRVA$RAT	PARVALBUMIN ALPHA	43	VKKVFHILDKDKSGFIEEDELGSILKGF
PRVB$AMPME	PARVALBUMIN BETA	43	VKKVFDILDQDKSGYIEEDELQLFLKNF
PRVB$BOACO	PARVALBUMIN BETA	43	LTKVFGVIDRDKSGYIEEDELKKFLQNF
PRVB$CYPCA	PARVALBUMIN BETA	43	VKKAFAIIDQDKSGFIEEDELKLFLQNF
PRVB$ESOLU	PARVALBUMIN BETA	42	VKKAFYVIDQDKSGFIEEDELKLFLQNF
PRVB$GADCA	PARVALBUMIN BETA (ALLERGEN M)	43	LKKLFKIADEDKEGFIEEDELKLFLIAF
PRVB$GADME	PARVALBUMIN BETA	43	IKKAFVFIDQDKSGFIEEDELKLFLQVF
PRVB$GRAGE	PARVALBUMIN BETA	43	VKKIFGILDQDKSGFIEEDELQLFLQNF
PRVB$LATCH	PARVALBUMIN BETA	43	LEAIFKILDQDKSGFIEEDELELFLQNF
PRVB$RANES	PARVALBUMIN BETA	43	AKKVFEILDRDKSGFIEQDELGLFLQNF
S10A$BOVIN	S-100 PROTEIN, ALPHA CHAIN	54	VDKVMKELDENGDGEVDFQEYVVLVAAL

[continued over

Table 1, *continued*

SCP$NEIDI	SARCOPLASMIC (CBP)	8	MKTYFNRIDFDKDGAITRMDFESMAERF
SCP$PERVT	SARCOPLASMIC (CBP)	8	MKTYFNRIDFDKDGAITRKDFESMATRF
SCP1$BRALA	SARCOPLASMIC (CBP I)	11	KFTFDFFLDYNKDGSIQQEDFEEMIKRY
		107	IPFLFKGMDVSGDGIVDLEEFQNYCKNF
SCP2$BRALA	SARCOPLASMIC (CPB II)	11	KFTFDFFLDMNHDGSIQDNDFEDMMTRY
		107	IPFLFKGMDVSGDGIVDLEEFQNYCKNF
SCPA$PENSP	SARCOPLASMIC (CBP), ALPHA-B	61	WNEIAELADFNKDGEVTVDEFKMAVQKH
SCPB$PENSP	SARCOPLASMIC (CBP), BETA	9	KYIVRYMYDIDNDGFLDKNDFECLAVRV
		61	WNEIAELADFNKDGEVTVDEFKQAVQKN
SORC$CRILO	22 KD PROTEIN (SORCIN/V19)	75	CRLMVSMLDRDMSGTMGFNEFKELWAVL
		105	WRQHFISFDSDRSGTVDPQELQKALTTM
SP1C$STRPU	SPEC 1C PROTEIN	58	IDKMISDVDTDESGTIDFSEMLMGIAEQ
		96	YTKAFDDMDKDGNGSLSPQELREALSAS
		132	IKAIIQKADANKDGKIDREEFMKLIKSC
SP2A$STRPU	SPEC 2A PROTEIN	86	LTKAFDDLDKDHDGSLSPQELRTAMSAC
SP2C$STRPU	SPEC 2C PROTEIN	15	FKSSFKSIDADGDGKITPEELKAAFKSI
		86	YFKAFDALDTDKSGSLSPEELRTALSAC
TPC$HALRO	TROPONIN C, BODY WALL MUSCLE	49	LQEMIEEVDIDGSGTIDFEEFCLMMYRQ
TPCC$HUMAN	TROPONIN C, CARDIAC MUSCLE	57	LQEMIDEVDEDGSGTVDFDEFLVMMVRC
		97	LSDLFRMFDKNADGYIDLEELKIMLQAT
TPCC$RABIT	TROPONIN C, CARDIAC MUSCLE	57	LQEMIDEVDEDGSGTVDFDEFLVMMVRC
		97	LSDLFRMFDKNADGYIDLDELKIMLQAT
TPCS$CHICK	TROPONIN C, SKELETAL MUSCLE	22	FKAAFDMFDADGGGDISTKELGTVMRML
		58	LDAIIEEVDEDGSGTIDFEEFLVMMVRQ
		98	LANCFRIFDKNADGFIDIEELGEILRAT
TPCS$HUMAN	TROPONIN C, SKELETAL MUSCLE	19	FKAAFDMFDADGGGDISVKELGTVMRML
		55	LDAIIEEVDEDGSGTIDFEEFLVMMVRQ
		95	LAECFRIFDRNADGYIDAEELAEIFRAS
TPCS$PIG	TROPONIN C, SKELETAL MUSCLE	19	FKAAFDMFDADGGGDISVKELGTVMRML
		55	LDAIIEEVDEDGSGTIDFEEFLVMMVRQ
		95	LAECFRIFDRNMDGYIDAEELAEIFRAS
TPCS$RABIT	TROPONIN C, SKELETAL MUSCLE	19	FKAAFDMFDADGGGDISVKELGTVMRML
		55	LDAIIEEVDEDGSGTIDFEEFLVMMVRQ
		95	LAECFRIFDRNADGYIDAEELAEIFRAS
TPCS$RANES	TROPONIN C, SKELETAL MUSCLE	22	FKAAFDMFDTDGGGDISTKELGTVMRML
		58	LDAIIEEVDEDGSGTIDFEEFLVMMVRQ
		98	LAECFRIFDKNADGYIDSEELGEILRSS

Table 2. Sequence segments returned from the SWISSPROT library by a search with a template as shown in Fig. 1 but with the modifications made to element +7 as described in the text. The format is the same as in Table 1.

SWISSPROT ENTRY	PROTEIN		helix helix
			.. ..
			. . . .
			. . . .
			X-Y-ZgYhX--Z
			------------ +++++++++++++++
			1111 11111
			3210987654321012345678901234
BTUE$ECOLI	VITAMIN B12 TRANSPORT PERIPLASMIC PROTEIN	2	QDSILTTVVKDIDGEVTTLEKFAGNVLL
IL3$MOUSE	INTERLEUKIN-3 PRECURSOR (COLONY-STIMULATING FACTOR)	73	FRRVNLSKFVESQGEVDPEDRYVIKSNL
KPCA$BOVIN	PROTEIN KINASE C, ALPHA TYPE	284	EYYNVPIPEGDEEGNVELRQKFEKAKLG
MYS$CHICK	MYOSIN HEAVY CHAIN, FAST SKELETAL MUSCLE	315	QIQKLEARVRELEGEVDAEQKRSAEAVK
MYSP$RAT	MYOSIN HEAVY CHAIN, PERINATAL SKELETAL MUSCLE	134	QIQKLEARVRELEGEVENEQKRNAEAVK
TUNF$HUMAN	TUMOR NECROSIS FACTOR PRECURSOR	87	KPVAHVVANPQAEGQLQWLNRRANALLA

Table 3. Sequences returned by the combined template search described in the text. The residue choices for the first template were as shown, and were assigned weights of 100, making a maximum score for this template of 800. Weights for the residue choices of the second template were the same as elements −5 to +6 of the template shown in Fig. 1. Thus the second template carried a maximum score of 350. It is labelled as described early in this article, complete with the allowed insertion, marked ' * '.

The cut-off scores were 700 for the 1st template and 275 for the 2nd; hence 975 was the combined cut-off score.

SWISSPROT ENTRY	PROTEIN		1st template		2nd template
			P-SNNSMNN		
			TQQTIQQ		
			AHHALHH		
			GRRGVRR		
			MKKMFKK		
			I IW		
			L L		
			V V		X-Y-Zg*YhX--Z
CL1H$HUMAN	CALPACTIN I (LIPOCORTIN II)	222	PHLQKVFDR	YKSYSPYDMLESIRKEVKG	DLENAF
CL2$HUMAN	CALPACTIN II (LIPOCORTIN I)	230	PQLRRVFQK	YTKYSKHDMNKVLDLELKG	DIEKCL
CL2$RAT	CALPACTIN II (LIPOCORTIN I)	230	PHLRKVFQN	YRKYSQHDMNKALDLELKG	DIEKCL
CRP$HUMAN	C-REACTIVE PROTEIN PRECURSOR	133	PRVRKSLKKGYTVGAEASIILGQEQDSFGGNFEGSQ		
CRP$RABIT	C-REACTIVE PROTEIN PRECURSOR	134	PMVRKSLKKGYILGPEASIILGQDQDSFGGSFEKQQ		
FP$MESAU	FEMALE PROTEIN	112	PWVKKGLQKGYTVKNKPSIILGQEQDNYGGGFDNYQ		
SAMP$HUMAN	SERUM AMYLOID P-COMPONENT PRECURSOR	132	PLVKKGLRQGYFVEAQPKIVLGQEQDSYGGKFDRSQ		

(Text, ctd. from p. 95)
(Text, ctd. from p. 95)

duplicated domains [see 20] which bind calcium with different affinities. These duplicated, homologous sequences (not shown) fit the combined templates to different extents. Residues are conserved that would be essential for the proposed structures: an α-helix for sequences that correspond to the first template, and a Ca^{2+}-binding loop for the second. For example, residues corresponding to the sixth element of the first template (Table 3) and to element 'h' of the second are conserved as hydrophobic. Those corresponding to 'g' of the second are glycine, except in domain 2 of lipocortin II and of pII. Both these proteins bind fewer Ca^{2+} ions than their number of repeated domains.

The segments of lipocortins I and II shown in Table 3 follow immediately after a segment of 16 residues that are highly conserved throughout the annexins [21]. Taylor & Geisow [22] have proposed a model for the 3-D structure of the repeated annexin domains based upon predicted secondary structures and conserved hydrophobic residues. In their model, the highly conserved 16-residue segment in each domain contains the Ca^{2+}-binding loop. Unfortunately it is impossible to ascertain *a priori* whether protein residues are conserved for functional or for purely structural reasons, or for a mixture of both.

CONCLUSIONS

No attempt has been made to describe in detail the findings presented here, as distinct from the procedures adopted to obtain them. Evidently selective (and provocative) sequence alignments may be obtained by searching the libraries with templates that have been designed upon structural principles.

Acknowledgements

Templates were compared with sequences using the program ANALYSEPL written and maintained by Dr R. Staden [23]. I thank him for advice and encouragement, together with Dr M. Hanley and Dr S. Cooper for reading the manuscript.

References

1. Brenner, S. (1987) *Nature 329*, 21.
2. Kretsinger, R.H. & Nockolds, C.E. (1973) *J. Biol. Chem. 248*, 3313-3326.
3. Kretsinger, R.H. (1976) *Annu. Rev. Biochem. 45*, 239-266.
4. Hertzberg, O. & James, M.N.G. (1985) *Biochemistry 24*, 5298-5302.
5. Stuart, D.I., Acharya, K.R., Walker, N.P.C., Smith, S.G., Lewis, M. & Phillips, D.C. (1986) *Nature 324*, 84-87.
6. Vyas, N.K., Vyas, M.N. & Quiocho, F.A. (1987) *Nature 327*, 635-638.
7. Bairoch, A. (1988) Biochemie Médicale, Centre Medical University, 1211 Genève 4, Switzerland.[*]
8. Miller, S., Janin, J., Lesk, A.M. & Chothia, C. (1987) *J. Mol. Biol. 196*, 641-656.
9. Hofmann, T., Kawakami, M., Hitchman, A.J.W., Harrison, J.E. & Dorrington, K.J. (1979) *Can. J. Biochem. 57*, 737-748.
10. Baudier, J. & Gerard, D. (1983) *Biochemistry 22*, 3360-3369.
[*]*Access to SWISSPROT library needs a direct approach.*

11. Kriegler, M., Perez, C., DeFay, K., Albert, I. & Lu, S.D. (1988) *Cell 53*, 45-53.
12. Parker, J.P., Coussens, L., Totty, N., Rhee, L., Young, S., Chen, E., Stabel, S., Waterfield, M.D. & Ullrich, A. (1986) *Science 233*, 853-859.
13. Needleman, S.B. & Wunch, C.D. (1970) *J. Mol. Biol. 48*, 444-453.
14. Barker, W.C., Hunt, L.T., George, D.G., Yeh, L.S., Chen, H.R., Blomquist, M.C., Seibel-Ross, E.I., Elzanowski, A., Blair, J.K., Lewis, M.T., Davalos, D.P. & Ledley, R.S. (1987)   National Biomedical Research Foundation, Georgetown University Medical Center, 3900 Reservoir Road NW, Washington D.C. 2007. *(Access to 'ALIGN' program needs a direct approach.)*
15. Pepys, M.B. & Baltz, M.L. (1983) *Adv. Immunol. 34*, 141-212.
16. Pepys, M.B. & Butler, P.J.G. (1987) *Biochem. Biophys. Res. Comm. 148*, 308-313.
17. Turnell, W.G., Satchwell, S.C. & Travers, A.A. (1988) *FEBS Lett. 232*, 263-268.
18. Engelman, D.M. (1971) *J. Mol. Biol. 58*, 153-165.
19. Nguyen, N.Y., Suzuki, A., Boykins, R.A. & Liu, T-Y. (1986) *J. Biol. Chem. 261*, 10456-10465.
20. Crompton, M.R., Owens, R.J., Totty, N.F., Moss, S.E., Waterfield, M.D. & Crumpton, M.J. (1988) *EMBO J. 7*, 21-27.
21. Geisow, M.J. (1986) *FEBS Lett. 203*, 99-103.
22. Taylor, W.R. & Geisow, M.J. (1987) *Protein Engineering 1*, 183-187.
23. Staden, R. (1988) *CABIOS 4*, 53-60.

#ncA

NOTES and COMMENTS relating to

PLASMA MEMBRANE; Ca^{2+} MOVEMENT; MUSCLE AND NERVE

'COMMENTS' are on pp. 149–156, starting with Forum discussions
on the preceding main articles, then on the 'Notes'

#ncA.1

A Note on

THE POSSIBLE MODULATION BY Ca^{2+} IONS OF THE EXCIMER FORMATION OF PYRENE-LABELLED GM$_1$ AND PHOSPHATIDYLCHOLINE MOLECULES IN ARTIFICIAL MEMBRANES

[1]G. Bastiaens, [1]M. De Wolf, [2]G. Van Dessel, [1]A. Lagrou,
[1][⊗]H.J. Hilderson and [1,2]W. Dierick

[1]RUCA-Laboratory for   [2]UIA-Laboratory for
   Human Biochemistry,   Pathological Biochemistry,
          University of Antwerp,
          Groenenborgerlaan 171, B-2020 Antwerp, Belgium

The background to the present study is that ganglioside molecules might serve as modulators of plasma membrane characteristics by both fluidity changes of the bilayer and specific interactions of the carbohydrate portion with some membrane constituents, and that multivalent binding between CT[*] and GM$_1$ (its receptor) seems to require lateral movement of GM$_1$ in the plane of the membrane. In order to investigate the possible role of Ca^{2+} in the binding of CT to its receptor, a model system was tested using both pyrene-labelled GM$_1$ and PC[*] molecules. [Receptors feature in Vol. 13.- *Ed.*]

We developed a simpler, gentler and faster method than that of Ollmann et al. [1] for labelling GM$_1$ at the single double bond of the sphingosine moiety. (The fatty acid side-chains of GM$_1$ are almost completely saturated.) A solution of GM$_1$ (2 mg/ml) in CCl$_4$ was slowly (45 min) saturated with ozone at 0°. The double bond was quantitatively converted to an ozonide. A previous study [2] had demonstrated that ozonolysis has no deleterious effect on the oligosaccharide moiety of the gangliosides. The ozonide bridge was subsequently cleaved reductively by adding 1 vol. of 50% (v/v) acetic acid containing Zn powder. After stirring for 1 h at 40° a 5-fold molar excess of solid Py[*]-butyryl hydrazine (Molecular Probes Inc.) was added, giving rise to Schiff bases with the generated aldehyde groups on the GM$_1$ molecules. These bases underwent reduction to the corresponding stable hydrazine compounds. After centrifugation (20 min, 12500 rpm; Sorvall SS-34 rotor), the Py-labelled GM$_1$ was recovered from the CCl$_4$/acetic acid interface as a narrow band.

[⊗]addressee for any correspondence

[*]*Abbreviations (& see overleaf).-* CT, cholera toxin (subunit: CT-B); PC, phosphatidylcholine; Py, pyrene. *See end of art. for the term* excimer. GM = monosialoganglioside.

This band, dispersed in a minimal volume of CHCl₃/methanol/- 0.25 M CaCl₂, 55:45:10 by vol., was loaded onto a radially compressed mini-column (Sep-Pak Silica cartridge, Waters). To remove the unreacted Py and other impurities, we first eluted with CHCl₃/methanol, firstly 9:1 and then 2:1. Finally the products of interest were recovered by elution with CHCl₃/methanol/0.25 M CaCl₂, 55:45:10. Every step in both the synthesis and the purification of GM_1-PBH* was followed by HPTLC (Merck Si 60 plates).

The GM_1-probe can be incorporated into artificial membranes (LUV and SUV), wherein the excimer formation shows a linear concentration dependence up to 10 mol%†. The probe also retained its receptor capacity towards CT, indicating that the oligosaccharide moiety remained intact during the procedure. This was demonstrated by the occurrence of a fluorescence energy transfer between the Trp-88 residue of each β-polypeptide chain of CT-B and the Py label of the ganglioside.

Since published values for critical micelle concentration [3-5] vary widely, evaluation was attempted. Below 1 μM GM_1-PBH a constant but significant amount of excimer was always obtained, suggesting that these micelles could still exist at those low concentrations. However, the curve for E/M ratio *vs.* probe concentration had a discontinuity which indicated that at ~10 μM a change in the characteristics of the micelles occurs (Fig. 1).

Investigation of Ca²⁺ effects

In the literature the possible role of Ca^{2+} in the behaviour of gangliosides in artificial membranes is also still a matter of debate [1, 6]. Using our probe we were able to demonstrate that the effect of Ca^{2+} in increasing the Tm of DPPC-LUV's containing 10 mol% of GM_1-PBH was due chiefly to an effect upon the PC molecules. A similar effect was obtained by incorporating Py-labelled PC molecules (Py-PC); even in the absence of gangliosides an increase in Tm could be observed. Moreover, a possible aggregation of ganglioside molecules produced by Ca^{2+} was not demonstrable: addition of increasing amounts of Ca^{2+} did not lead to a parallel increase in the excimer formation.

At 47°, well above the Tm, Ca^{2+} did not influence the diffusion constant of GM_1-PBH (7.7×10^{-9} cm²/sec), although

---

*PBH, pyrene butyrylhydrazine; DPPC, dipalmitoyl-PC (mol% connotes mol introduced/mol DPPC); LUV & SUV, large/small unilamellar vesicle; E/M, excimer/monomer intensity ratio; Tm, transition temperature.
†mol% connotes mol introduced *vs.* mol DPPC.

Fig. 1. E/M as a function
of concentration (-log C)
pf GM_1-PBH. After each
dilution step: **o**, a short
sonication and 1-h equilibration at room temp.; **or**
•, merely left standing for
equilibration at room temp.

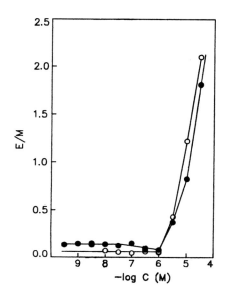

the lifetime of the label was decreased from 78 to 72 nsec
such that the E/M ratio changed (from 0.33 to 0.31).

Using GM_1-PBH, lower values of E/M are obtained than
with Py-PC at the same probe-to-lipid ratio. This could
be due either to a different level of the fluorophore in
the lipid bilayer (the deeper the label the more room there
is to accommodate it), or to the fact that the sugar moiety
can lead to a steric hindrance preventing excimer formation.

Upon addition of Ca^{2+} to 200 mM, Tm shifts to higher
values, whereas Mg^{2+} has only a minor effect: in Ca^{2+}-exposed
DPPC-LUV's containing 10 mol% Py-PC from 42° to 45°, in DPPC-
LUV's containing 10 mol% GM_1-PBH from 47° to 48°, and in
DPPC-LUV's containing 10 mol% Py-PC and 10 mol% unlabelled
GM_1 from 43° to 45°. Furthermore, the temperature region
of phase separation of DPPC-LUV's containing 10 mol% Py-PC
ranges from 32° to 42°; upon adding 10 mol% unlabelled GM_1
to the mixture the phase-separation starts at 29° and is
completed at 47.5°. GM_1 obviously widens the region of
phase-separation of DPPC in both directions. From these
data it is also clear that in a mixture of DPPC and GM_1
the two kinds of probe behave differently in response to
Ca^{2+}, unless perhaps labelled molecules do not exactly represent
the parent unlabelled molecules which they are supposed to
mimic.

Finally, in DPPC-LUV's containing 10 mol% Py-PC, Ca^{2+}

ions enhance the excimer formation below Tm. This effect is abolished completely or even slightly reversed upon addition of either GM_1 or GM_1-PBH. This effect is evidenced by Py-PC as well as GM_1-PBH.

In conclusion, use of labelled GM_1 and PC enables us to look separately at the behaviour of each of these molecules in the same physicochemical environment. This will be of great value in future studies on the effect of Ca^{2+} and other parameters on both the binding and the effect of CT upon viable cells.

References

1. Ollmann, M. Schwarzmann, G., Sandhoff, K. & Golla, H-J. (1987) *Biochemistry 26*, 5943-5952.
2. Fishman, P.H., Moss, J. & Osborne, J.C. jr. (1978) *Biochemistry 17*, 711-716.
3. Formisano, S., Johnson, M.L., Lee, G., Aloj, S.M. & Edelhoch, H. (1979) *Biochemistry 18*, 1119-1124.
4. Yohe, H.C. & Rosenbergh, A. (1972) *Chem. Phys. Lipids 9*, 279-294.
5. Corti, M., Degiorgio, V., Ghidoni, R., Sonnino, S. & Tettamonti, G. (1980) *Chem. Phys. Lipids 26*, 225-238.
6. Sharom, F.J. & Grant, C.W.M. (1978) *Biochim. Biophys. Acta 507*, 280-293.

EXPLANATION OF 'EXCIMER'
by Senior Editor, with help from G.V.C. Davies (University of Surrey Library) and H.J. Hilderson

"Excimers are molecular associates that exist only in excited electronic states. They are therefore detectable only in emission spectra, and particularly in fluorescence spectra…." [Th. Förster (1969) *Agnew.Chem. Int. Edn. 8*, 333; he cites B. Stevens & E. Hutton (1960) *Nature 186*, 1045-46]. The term is now commonly used in the narrower sense of excited molecular associates, particularly dimers.

From *Chemistry and Physics of Lipids 27* (1980) 199-219:- "Pyrene probes are used for their ability to form excimers. Excimers are molecular complexes formed by associations between an excited pyrene molecule Py^* and a ground state molecule Py. This process is diffusion controlled. The rate of excimer formation yields information on the dynamic molecular properties of artificial as well as of natural membranes."

#ncA.2

A Note on

BIPHASIC REGULATION OF THE LIVER PLASMA MEMBRANE Ca^{2+} PUMP BY GLUCAGON (19-29) IS MEDIATED BY TWO DISTINCT G PROTEINS

S. Lotersztajn, C. Pavoine, V. Brechler, [†]M. Dufour, [†]D. Le Nguyen, [†]D. Bataille and F. Pecker

INSERM-U99, hôpital Henri Mondor, 94010 Créteil, France

and [†]Centre CNRS-INSERM, Pharmaco. Endocrino., 34094 Montpellier Cedex, France

Plasma membrane Ca^{2+} pumps are responsible for active extrusion of Ca^{2+} from the cell. We have recently shown that nM concentrations of glucagon (19-29) which can arise from native glucagon by proteolytic cleavage of the dibasic doublet Arg^{17}-Arg^{18}, inhibit the Ca^{2+} pump in rat-liver plasma membranes independently of the adenylate cyclase activation [1]. We show here that GTP^{Y}s, a non-hydrolyzable analogue of GTP, reveals at 10 μM a biphasic response of the liver Ca^{2+} pump to increasing concentrations of glucagon (19-29). Low concentrations of the peptide (1 pM to 0.1 nM) potentiated inhibition of the Ca^{2+} pump, while raising the concentration of glucagon (19-29) resulted in the reversal of the Ca^{2+} pump inhibition caused by lower ones. The action of guanine nucleotides is specific since GTP^{Y}s > Gpp(NH)p > GTP ⋙ [GDPβs, App(NH)p, ATP^{Y}s, UTP]. Fig. 1 shows results for GTP^{Y}s.

After treatment of rats with cholera toxin, a 70% increase of the basal Ca^{2+} pump activity was observed, with a loss of sensitivity towards glucagon (19-29) (Fig. 1). Treatment with pertussis toxin caused a decrease in the basal activity of the Ca^{2+} pump but had no effect on its sensitivity towards glucagon (19-29).

We conclude that the Ca^{2+} pump in liver p.m. is dependent on a G protein sensitive to cholera toxin. Glucagon (19-29) can exert a biphasic effect on this system, and thus would modulate Ca^{2+} homeostasis in liver cells.

Recently it has been shown that glucagon (18-29), a potential precursor of glucagon (19-29), can be liberated from glucagon by insulin proteinase [2]. These data suggest

Fig. 1. Ca²⁺ pump activity
in relation to glucagon
(19-29) concentration, and
influence of a GTP analogue.

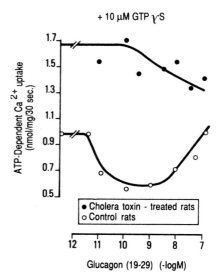

a new role for glucagon as a prohormone, distinct from its own
two previously described hormonal actions on adenylate
and phospholipase C systems.

References

1. Mallat, A., Pavoine, C., Dufour, M., Lotersztajn, S.,
   Bataille, D. & Pecker, F. (1987) *Nature 325*, 620-622.
2. Rose, K., Savoy, J.L., Muir, A.V., Davies, J.G.,
   Offord, R.E. & Turcatti, G. (1988) *Biochem. J.*, *256*, 847-
   851.

#ncA.3

A Note on

THE EFFECT OF RARE-EARTH METALS ON CALCIUM LEVELS IN ISOLATED HEPATOCYTES

D. Bingham, A.E. Stagg and [⊛]M. Dobrota

The Robens Institute of Health and Safety,
University of Surrey,
Guildford, Surrey GU2 5XH, U.K.

REPORTED EFFECTS OF LANTHANIDES

The rare-earth elements (lanthanides) interact with bio-logical systems mainly as calcium 'analogues' and thus affect numerous physiological and biochemical processes which are Ca^{2+}-dependent [1-5]. Lanthanides readily replace calcium because they have virtually the same ionic radius and a higher affinity for the binding site: being trivalent, they have a higher charge:volume ratio. The interactions of lanthanides with Ca^{2+}-dependent systems occur at molecular and cellular levels. They readily displace calcium from calmodulin, troponin C, parvalbumin, concanavalin A, and struc-tural proteins (G-actin, haemocyanin, collagen) whose functional properties are affected by the binding through promotion of polymerization; they also bind to other proteins such as transferrin, IgG and serum albumin [3].

The effect of lanthanides on Ca^{2+}-dependent enzymes may be inhibitory (phospholipase C, Ca^{2+}-ATPase, prothrombin/factor X activation) or, by stabilizing conformation, stimula-tory as with acetylcholinesterase (at a low terbium concentra-tion) and amylase. It has been suggested that whether the lanthanide inhibits or stimulates may indicate whether calcium in the protein plays a functional role or merely a structural role [6].

At the cellular level it is noteworthy that lanthanides do not normally enter cells. The major effects of the binding to the p.m.[*] are (1) blockage of Ca^{2+} channels and thus inhibition of Ca^{2+} entry, and (2) blockage of the Ca^{2+}-ATP-ase pump so that Ca^{2+} efflux from cells is inhibited. Effect (2) bears particularly on intracellular signalling by Ca^{2+} and on the inhibition of stimulus-coupled secretion in a

[⊛] addressee for any correspondence
[*]*Abbreviations.-* AAS, atomic absorption spectrometry; LDH, lactate dehydrogenase; PBS, phosphate-buffered saline; p.m., plasma membrane.

wide range of cells and tissues. Amongst numerous effects the most notable are inhibition of synaptic release of acetylcholine from nerve axons and of muscle contraction. The wide-ranging effects of 'exogenous' lanthanides on intracellular processes are well illustrated by the observations that, besides blocking Ca^{2+} efflux (<u>via</u> the Ca^{2+}-ATPase pump), lanthanum also brings about an increase in phosphorylated proteins by blocking the Mg^{2+}-dependent dephosphorylation [7].

Despite much work with rare earths as Ca^{2+} blockers, it is fair to say that our understanding of their various biological effects remains rather poor. Although free lanthanide ions apparently do not cross p.m.'s it is possible that they are taken up into cells; most of the proteins which readily bind lanthanides are also taken up by receptor-mediated endocytosis by various tissues. Lanthanides may also be taken up by the reticulo-endothelial system (<u>e.g.</u> the phagocytic Kupffer cells of the liver) following their i.v. adminstration, due to the formation of insoluble aggregates/precipitates. At present it is not known whether lanthanides internalized by such processes are contained as inactive precipitates (<u>e.g.</u> insoluble phosphates) within the endosome/lysosome compartment or redistribute to other cellular compartments or indeed other tissues. A difficulty in experimental work with lanthanides is precipitation in the presence of phosphate due to the very low solubility product.

Whilst it has been proposed that lanthanides offer potentially useful therapeutic activities and are considered as virtually non-toxic, it must be emphasized that at present we lack a clear understanding of their cellular handling and of all their biological activities.

The present preliminary study was concerned with the effects of holmium (Ho), which has a radius very close to that of calcium, on total cell calcium in isolated hepatocytes, measured after pre-incubation of the cells with different Ho^{3+} concentrations.

EXPERIMENTAL PROCEDURES (Scheme 1)

Preparation of hepatocytes, by an adaptation of the method of Reese & Byard [8].- Male Wistar albino rats (200 g) were killed by gassing with CO_2. The liver was quickly removed, washed in PBS (Flow Labs.) and the lobes separated. A slice was made in the edge of each lobe to expose large blood vessels into which a needle could be placed. The lobes were perfused initially at 25-30 ml/min at 37° with 400 ml 24 mM $NaHCO_3$/142 mM NaCl/4.4 mM KCl/1.24 mM KH_2PO_4/0.62 mM $MgSO_4$/0.62 mM $MgCl_2$/0.5 mM EDTA (pH 7.4), gassed with 95% O_2/5% CO_2. The perfusate was then changed to 100 ml pre-gassed

LIVER, *perfused with Ca²⁺-free*
bicarbonate, then collagenase; cells spun down
|
Washed HEPATOCYTES - *determine counts & viability*
| *15-min incubations:* +Ho³⁺ *(0.001-5 mM)*
| *then* +Ca²⁺ *(2.5 mM); then centrifuge*
┌──────────────────────────┴──────────────────────────┐
CELLS *(washed once)*   SUPERNATANT
Measure calcium, protein, LDH   *Measure calcium, LDH*

Scheme 1.   Outline of the procedures.

buffer with 50 mg collagenase and 5 mM Ca^{2+} in place of EDTA; the perfusate was recirculated at 10-15 ml/min.

When the liver was soft to touch, perfusion was stopped and the lobes gently dispersed in 50 ml PBS. Filtering through bolting cloth removed any large aggregates of cells. The cell suspension was pelleted by centrifugation at 50 **g** for 1 min, then resuspended in PBS; this was repeated twice. A final wash and resuspension was performed with the incubation medium to be used. Hepatocyte yield and viability in the incubation medium was determined by dye exclusion (see below).

Ho³⁺ and Ca²⁺ treatment of hepatocytes.- The suspension was diluted at 37° with incubation medium (pre-gassed as above) to give 10^6 cells/ml. The medium comprised 0.1 M mannitol, 0.1 M NaCl, 0.1 mM $MgSO_4$ and 20 mM PIPES, pH 7.4 at 37°; it was sterilized by filtration and kept at 4° till the day of use. Into 10 ml glass conical flasks, with triplication, were pipetted 2 ml of the cell suspension, and 20 µl PIPES buffer containing, except for the 3 controls, $HoCl_3$ at differing levels obtained by dilution from a stock solution of $HoCl_3 . 6 H_2O$ (0.5 M; Aldrich). After 15 min at 37° in a shaking water-bath, 20 µl of 0.255 M $CaCl_2 . 6 H_2O$ was added, and incubation continued for a further 15 min.

Separation of cells from the medium.- With a Heraeus–Christ Digifuge, the suspensions were centrifuged for 1 min at 25 **g** in 3 ml plastic test tubes. The supernatant was carefully aspirated, placed on ice, and all but 100 µl (used within 24 h for LDH assay) was frozen for later analyses. The cells were washed with fresh incubation medium (supernatant discarded) and, in 2 ml of fresh medium, placed on ice, and similarly frozen except for the LDH aliquot (usually 100 µl) to which was added an equal volume of solubilizing agent (0.9% NaCl/0.1% bovine serum albumin/0.1% v/v Triton X-100).

Analyses.- Hepatocyte initial viability and yield were estimated by adding, to 100 µl of the suspension with gentle mixing, 100 µl of 0.5% Trypan Blue (Flow Labs.): on a sample

Table 1. Hepatocyte calcium and LDH leakage after holmium treatment (**H**, hepatocytes; **S**, supernatant; ± is **S.D.**, n = 6).

Ho^{3+}, mM	H +S: nmol total Ca	H Ca: % of total	pmol/mg prot.	% leakage
0	4.73 ±0.16	1.25±0.07	19.0 ±2.7	40.7 ±9.2
0.001	4.86 ±0.20	1.24 ±0.08	18.9 ±1.1	41.0 ±8.8
0.01	4.80 ±0.17	1.24 ±0.05	18.5 ±0.5	43.1 ±6.6
0.1	5.02 ±0.23	1.31 ±0.07	19.4 ±2.5	39.1 ±6.6
1.0	4.97 ±0.26	1.49 ±0.06*	25.9 ±2.8*	41.8 ±1.8[†]
5.0	4.88 ±0.15	1.47 ±0.06*	26.8 ±3.7*	32.2 ±10.6

vs. control: $P < 0.005$. (No significant LDH effects). [†]n = 3 only

of the mixture, the numbers of total cells and dye-excluding (viable) cells were counted in a haemocytometer by phase-contrast microscopy. Cell integrity after the 30 min of incubation was assessed by LDH leakage into the medium [9]. The LDH was assayed by following the change in A_{340} over 3 min (linear part of curve) of a mixture containing 0.61 mM pyruvate and 0.32 mM NADH.

Calcium was assayed by flame AAS (Instrumentation Lab., model 353), the Ca^{2+} standards (1-5 mM) and supernatant samples being diluted 1:100 with 0.1% $LaCl_3$ (Spectrosol; BDH Chemicals) to remove phosphate interference. Cell samples were diluted 2-fold with 10% $LaCl_3$. Holmium had to be assayed by a nitrous oxide/acetylene flame AAS method, to achieve a hotter and more stable flame. The Lowry method [10] was used to measure protein in cell samples.

RESULTS

Ho^{3+} at 1 or 5 mM (not if lower) raised hepatocyte calcium, expressed either as % of total in cells and supernatant or on a protein basis (Table 1); that Ca displacement by Ho from glassware was not the cause was shown by the near-constancy of total Ca. Assuming a cytocrit for the resuspended cells of 5%, one gets [Ca] = 0.6 mM in the untreated cells, which agrees quite closely with reported values of 1-2 mM total Ca in mammalian cells [11].

The viability values were 85.6% and 88.9% in 2 experiments when assessed by Trypan Blue just after isolation in Ca^{2+}-free incubation medium, but much lower (Table 1) as assessed by LDH leakage after incubating for 30 min. The rapid fall in viability during incubation was most likely due to the 'poor' medium (phosphate-free, initially Ca^{2+}-free) required for work with the rare earths. Whereas LDH leakage was 26.6 ±0.3% (±S.D., n = 3) in Ca^{2+}-free basic medium, and 18.0 ±0.9

with 2.5 mM Ca^{2+} present, it was only 11.4 ±0.5% in a cell-culture medium (Leibovitz L15, Gibco).

Holmium treatment surprisingly reduced LDH leakage by ~25% at 1 or 5 mM (Table 1). A separate experiment using lysed cell suspensions showed that holmium inhibited LDH activity by 3% at 0.2 mM and 12% at 10 mM. However, this slight inhibition cannot account for the 25% reduction of leakage, signifying cell protection, at higher concentrations of holmium.

DISCUSSION

The increase in total cell calcium following exposure to 1 or 5 mM Ho^{3+} was an interesting and somewhat surprising result in view of the ability of lanthanides to replace Ca^{2+} at Ca^{2+}-binding sites. While the foregoing preliminary studies with hepatocytes do not explain the mechanisms by which Ho^{3+} interferes with cellular Ca, our results suggest that the influx of exogenous Ca^{2+} is not affected but that the intracellular increase is due to inhibition of the active efflux of Ca^{2+}. The increase is consistent with the reported effects of lanthanides on the inhibition of the p.m. Ca^{2+}-ATPase pump [7] and the accumulation of Ca^{2+} in mitochondria [12]. In contrast to our findings, Langer & Frank [13] reported that 0.5 mM La^{3+} initially displaced Ca^{2+} from rat heart myocytes but then inhibited further Ca^{2+} influx or efflux. Their results, obtained under rather different experimental conditions, probably reflect the effects of rare earths on free intracellular Ca^{2+} and not total Ca. In order to understand the mode of action of lanthanides on cellular calcium levels we clearly need to distinguish between free and bound pools of calcium.

Since the hepatocyte viabilities are the same with and without Ho^{3+} and indeed LDH leakage is reduced at high Ho^{3+} concentrations, the increase in total cellular calcium cannot be explained by non-specific binding of Ca^{2+} to ligands in permeant dead cells. We also consider that the increase in Ca with Ho^{3+} treatment cannot be attributed to the lack of phosphate in the incubation medium, since all control and experimental cell suspensions were incubated in this medium.

In cell suspensions treated with 5 mM Ho^{3+}, 90% of the holmium was found in the supernatant and 10% was associated with the cells. Although we have at present no data for the amount of bound holmium at lower Ho^{3+} concentrations, from our results it appears that 1 mM Ho^{3+} may saturate the available binding sites on the hepatocytes. Concentrations

<1 mM therefore may not have affected the cellular calcium because there was insufficient Ho^{3+} to saturate a significant number of the binding sites.

References

1. Haley, T.J. (1965) *J. Pharm. Sci. 54*, 663–670.
2. Haley, T.J. (1979) in *Handbook of the Physics and Chemistry of the Rare Earths* (Gschneider, K.A. & Eyring, L., eds.), North-Holland, Amsterdam, pp. 553–580.
3. Arvela, P. (1979) *Prog. Pharmacol. 2*, 69–114.
4. Evans, C.H. (1983) *Trends Biochem. Sci. 8*, 445–449.
5. Evans, C.H. (1988) in *Calcium in Drug Actions: Handbook of Experimental Pharmacology*, Vol. 83 (Baker, P.F., ed.), Springer-Verlag, Berlin, pp. 527–546.
6. Martin, R.B. & Richardson, F.S. (1979) *Quart. Rev. Biophys. 12*, 181–209.
7. Schatzmann, H.J., Luterbacher, S., Stieger, J. & Wutrich, A. (1986) *J. Cardiovasc. Pharmacol. 8*, 533–537.
8. Reese, J.A. & Byard, J.L. (1981) *In Vitro 17*, 935–940.
9. Benford, D.J. & Hubbard, S.A. (1985) in *Biochemical Toxicology: a Practical Approach* (Snell, K. & Mullock, B., eds.), IRL Press, Oxford, pp. 57–79.
10. Lowry, O.H. Rosenborough, N.J., Farr, A.L. & Randall, R.J. (1951) *J. Biol. Chem. 193*, 265–275.
11. Alberts, B., Bray, D., Lewis, J., Raff, M., Roberts, K. & Watson, J.D. (1983)   *Molecular Biology of the Cell*, Garland Publishing Co., New York: see p. 286.
12. Kolbeck, R.C. & Speir, W.A., jr. (1986) *J. Mol. Cell Cardiol. 18*, 733–738.
13. Langer, G.A. & Frank, J.S. (1972) *J. Cell Biol. 54*, 441–455.

#ncA.4

A Note on

ACTIVATORS AND INACTIVATORS OF Ca^{2+} CHANNELS[@]

M. Spedding

Department of Pharmacology, Syntex Research Centre,
Research Park, Riccarton, Edinburgh EH14 4AP, U.K.

Although the predominant control mechanism for the activity of voltage-dependent Ca^{2+} channels (VOC's, L channels) is membrane depolarization, there are high affinity binding sites on the channels for a variety of different chemicals (dihydropyridines (DHP's; *abbreviation by Ed.)*, phenylalkyl-amines, etc.), occupation of which can change the channel state, favouring increased channel opening or closure, dependent on membrane potential. Study of the effects of these drugs and of their sites of action is critically dependent on the experimental conditions, which influence the effects of the drugs. Drug binding sites are allosterically linked on the Ca^{2+}-channel protein, which can lead to complex interactions in ligand binding studies although the functional consequences may not be so marked. Ligand binding studies, with membrane homogenates, evidently do not measure effects under physiological conditions, but the high affinity of drugs such as the DHP's (Kd's frequently <1 nM) indicates that these drugs label the inactivated, high-affinity state of the channel; affinity corresponds to the affinity shown for the inactivated state of the channel seen in electrophysiological experiments [1]. The drugs therefore label the channel (according to the simplest form of kinetics):

closed channel $\rightleftharpoons$ open $\rightleftharpoons$ inactivated
$$\updownarrow$$
dihydropyridines

Thus simple membrane homogenates in rather non-physiological buffers (HEPES or Tris) may be used to quite satisfactorily measure the potency of related DHP's to displace ^{3}H-DHP's from their binding sites, assuming that appropriate precautions are taken to reach equilibrium, and that the concentration of divalent cations is not too low. DHP binding requires >1 µM Ca^{2+} to maintain the integrity of the binding sites; this concentration is present in most buffers which do not contain chelating agents. It would therefore appear appropriate to control the concentration of Ca^{2+} during the assays; but few workers normally do so. Yet the affinity for the channel for DHP's in brain and smooth muscle homogenates

[@] with an **ADDENDUM** focussed on **NEURAL TISSUE**

resembles the potency seen in functional experiments in smooth muscle for inhibitory effects [review: 2], assuming that the functional experiments are set up so as to allow assessment of high-affinity pharmacology, i.e. there is also a high proportion of the inactivated state. This is normally performed by depolarizing tissues with K^+ over a long time period, so as to avoid intracellular Ca^{2+} release and to allow equilibrium of the drugs with the inactivated state of the channels; apparently competitive kinetics can be obtained if these conditions are optimized [3]:

$$\text{closed channel} \overset{K^+}{\rightleftharpoons} \text{open} \rightleftharpoons \text{inactivated}$$
$$\mathclose{\updownarrow}$$
$$\textit{dihydropyridines}$$

Bean [1] estimated that >70% of the channels may be inactivated under constant depolarizing conditions, so that, as the open state is very short (msec), the equations approximate to those used to describe receptor occupation [3]. [Arts. in Vol. 13, this series, are pertinent.- Ed.] This means that DHP's may have very high affinity for tissues which are relatively depolarized and where there is a high proportion of inactivated channels (viz. smooth muscle). Thus under the correct experimental conditions radioligand binding experiments and functional experiments may yield similar results, as regards simple affinity for channels.

However, this may not hold true if the ability of drugs to activate channels is measured, as affinity only for the inactivated state is measured. Sanguinetti et al. [4] consider that activator DHP's, e.g. Bay K 8644, which increase Ca^{2+} entry do so by binding to the Ca^{2+} channel, but slowing the rate of closure of the channels. As the drugs bind to the inactivated state of the channel, which is the high-affinity state for DHP's, ligand binding studies cannot address the problem of Ca^{2+} channel activation as yet.

Furthermore, drugs may interact with other sites on Ca^{2+} channels which are allosterically linked to the DHP site [2, 5, 6].-

Classes of drugs interacting with Ca^{2+} channels

I. DIHYDROPYRIDINES: nifedipine, nicardipine, nimodipine, nitrendipine, nisoldipine, felodipine, isradipine, darodipine, nivaldipine, niguldipine, amlodipine, ryodipine, mesudipine, etc.

II. PHENYLALKYLAMINES (verapamil, gallopamil, diclofurime, etc.) and DILTIAZEM analogues.

III. DIPHENYLALKYLAMINES: fluspirilene, flunarizine, cinnarizine, fendiline, prenylamine, pimozide.

Again, experimental conditions are critical to defining these interactions. Diltiazem, rather than inhibiting DHP binding to membrane fragments, increases it, presumably by an allosteric interaction with the channel, which slows dissociation of the DHP's in the presence of diltiazem [see 5]. As verapamil reduces DHP binding and enhances DHP dissociation from the channel, verapamil and diltiazem have been thought to bind to different sites and hence have been classed differently. However, the effect of diltiazem is highly temperature-dependent and diltiazem reduces binding of DHP's at 0°. Mir & Spedding [6] have shown that one of the most potent Ca^{2+}-antagonists known, diclofurime, binds to the diltiazem site and at 25° has little effect on DHP binding because at this temperature, as commonly used to screen drugs for affinity for the Ca^{2+} channel, the drug has nil effect with stimulant and inhibitory effects cancelling out. When the assays were performed at 0°, diclofurime was a very potent inhibitor of DHP binding. Thus binding assays should be run at several temperatures, and preferably a range of tritiated ligands should be used to define binding sites. [See Laduron in Vol. 13.- *Ed.*]

Functional effects may be used to verify binding assays. The interactions of different Ca^{2+} antagonists may be differentiated by interactions with Ca^{2+}-channel activators, such as Bay K 8644, which reverse the inhibitory effects of some but not all Ca^{2+}-antagonists depending on their site of action. Thus, DHP Ca^{2+}-antagonists are competitively reversed by Bay K 8644 under constant depolarizing conditions, whereas Bay K 8644 non-competitively reverses the effects of the group II antagonists [3, 7-9] but it does not reverse at all the effects of the group III drugs [7].

Are there endogenous ligands for these binding sites? Lipid metabolites, *e.g.* acylcarnitines, accumulate in the myocardial sarcolemma, particularly during ischaemia. Low concentrations (~1 μM) of palmitoylcarnitine cause similar effects to Bay K 8644 in heart preparations [10, 11], implying that these substances may be important in modulating ischaemic damage; here there is involvement of Ca^{2+} channels. These findings indicate that experimental conditions, particularly as regards substrate, may be critical in defining the effects of Ca^{2+}-channel ligands when assessing effects in *in vivo* techniques.

References

1. Bean, B.P. (1984) *Proc. Nat. Acad. Sci. 81*, 6388-6392.
2. Janis, R.A., Silver, P.J. & Triggle, D.J. (1987) *Adv. Drug. Res. 16*, 309-591.
3. Spedding, M. (1985) *J. Pharmacol. (Paris) 16*, 319-343.
4. Sanguinetti, M.C., Krafte, D.S. & Kass, R.S. (1986) *J. Gen. Physiol. 88*, 369-392.

5. Glossmann, H., Ferry, D.R., Goll, A., Striesnig, J. &
    Zernig, G. (1985) *Arzneim.-Forsch. 35*, 1917-1935.
6. Mir, A.K. & Spedding, M. (1987) *J. Cardiovasc. Pharmacol.
    9*, 469-477.
7. Spedding, M. & Berg, C. (1984) *Naunyn-Schmiedeberg's Arch.
    Pharmacol. 328*, 69-75.
8. Spedding, M. (1985) *Naunyn-Schmiedeberg's Arch. Pharmacol.
    328*, 464-466.
9. Su, C.M., Swamy, V.C. & Triggle, D.J. (1984) *Canad. J.
    Physiol. Pharmacol. 62*, 1401-1410.
10. Patmore, L., Duncan, G. & Spedding, M. (1989) *Br. J.
    Pharmacol.*, in press.
11. Knabb, M.T., Saffitz, J.E., Corr, P.B. & Sobel, B.E.
    (1986) *Circ. Res. 58*, 230-240.

ADDENDUM: BIOCHEMICAL CORRELATES OF Ca^{2+} CHANNEL ACTIVATION IN NEURAL TISSUE

Neurotransmitter release is normally Ca^{2+}-dependent, but although voltage-operated Ca^{2+} channels (VOC's, L channels) exist in the CNS their role in modulating transmitter release is still unclear, and ligands with high affinity for VOC's (e.g. Ca^{2+}-antagonists) have only slight effects on transmitter release under physiological conditions. In contrast, activators of VOC's, e.g. BAY K 8644, tend to augment transmitter release and this augmented release is susceptible to Ca^{2+}-antagonists. Several explanations have been advanced to account for these findings.-

1. Ca^{2+} entry may be via other activation pathways (Na^+/Ca^{2+} exchange, etc.) at nerve endings.

2. Different types of VOC's (N or T channels) may be present at nerve endings, which are resistant to Ca^{2+}-antagonists; L-channels may predominate in the cell bodies.

3. Neural action potentials may be too short to allow association of the drugs, the effects of which are highly use- or depolarization-dependent: inhibitory effects would become apparent only with high-frequency stimulation or with tissue depolarization.

If the latter situation holds, Ca^{2+}-antagonists might be effective in depolarized or ischaemic situations; the experimental conditions would be critical for demonstration of effects. *In vitro* and animal models show that there are marked differences amongst Ca^{2+}-antagonists in their protective effects against ischaemic damage. Furthermore, Ca^{2+} channel number, as assessed by binding studies, changes after ischaemic insults in both neural and myocardial tissue. Certain Ca^{2+}-antagonists may therefore be effective in pathological conditions, yet be without deleterious effects on transmitter release under normal conditions.

#ncA.5

A Note on

L-TYPE CALCIUM CHANNEL IN SMOOTH MUSCLE CELLS AS STUDIED WITH $^{45}Ca^{2+}$-UPTAKE TECHNIQUES

Urs T. Rüegg, Cynthia Cauvin, Jean-François Zuber and Robert Hof

Preclinical Research, Sandoz Ltd.,
4002 Basle, Switzerland

Smooth muscle cells appear to have two major pathways for Ca^{2+} entry: POC's and ROC's[*] [1, 2]. While little is known about the regulation of the activity of ROC's, POC's can be subdivided into at least two, perhaps three, types [3, 4]: a channel that is activated at a relatively low membrane potential and that inactivates rapidly (transient channel, 'T-type', perhaps 'N-type') and a channel that requires a larger decrease in membrane potential but is more slowly inactivated (slow Ca^{2+} channel, 'L-type').

We have studied the regulation of the L-type Ca^{2+} channel in two cellular systems [5, 6]: the A_7r_5 cell line which originated from foetal rat aorta, and a primary culture of rat aortic smooth muscle cells subsequently passaged up to 40 times. Both cell types show the typical morphology of vascular smooth muscle cells but do not contract upon vasoconstrictor addition.

Even though smooth muscle cells do contain channels with T-type properties which can be seen by electrophysiological techniques, they cannot be detected with tracer flux methods, probably because they are too 'short-lived'. L-type channels can be studied quite well by flux measurements. To achieve a good signal/noise ratio, the cells have to undergo a pre-treatment leading to an emptying of intracellular Ca^{2+} pools. Using $^{45}Ca^{2+}$ efflux techniques we have noted earlier [7] that the release of Ca^{2+} from these pools occurs with a $t_\frac{1}{2}$ of ~4 min (at 37°). Accordingly, only 10% of the Ca content of the pools remains if the cells are pre-incubated in Ca^{2+}-free EGTA solution for 15 min. When Ca^{2+} is added back together with tracer, influx occurs predominantly <u>via</u> L-channels.

[*]*Abbreviations*.- POC, potential-operated channel; ROC, receptor-operated channel; DHP, dihydropyridine; PSS, physiological salt solution; SDS, sodium dodecyl sulphate.

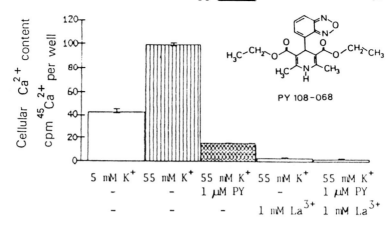

Fig. 1. $^{45}Ca^{2+}$ uptake into smooth muscle cells (A$_7$r$_5$ cell line) as affected by depolarization or by Ca^{2+}-entry blockers - organic (**PY** 108-068; darodipine) or inorganic (LaCl$_3$; **La^{3+}**). A confluent monolayer of cells in 16 mm culture dishes was pre-incubated for 15 min with PSS buffer containing no CaCl$_2$ but 0.1 mM EGTA to remove extracellular Ca^{2+} and to empty the intracellular pools. Uptake measurements were performed in PSS with 0.1 mM Ca^{2+} (0.2 ml per well) containing 0.25 µCi of $^{45}CaCl_2$. After 30 min at 37° the cells were washed (×4) with 1 ml of ice-cold PSS and were solubilized with 0.5 ml 0.1% SDS. The blocker was added together with the $^{45}CaCl_2$.

Upon exposure to depolarizing (55 mM KCl) solution, a 2- to 3-fold increase in $^{45}Ca^{2+}$ uptake can now be observed (Fig. 1). This increase is inhibited by Ca^{2+}-entry blockers (DHP's) in concentrations equivalent to those able to relax K$^+$-induced contractures of aortic rings [6]. The IC$_{50}$ values are typically: verapamil, 200 nM; nitrendipine, 3 nM; PN 200-100 (isradipine), 0.8 nM.

As can be seen from Fig. 1, Ca^{2+}-entry blockers can reduce the $^{45}Ca^{2+}$ uptake to below that in controls (5 mM K$^+$). This is most likely due to the fact that the A$_7$r$_5$ cells show spontaneous electrical activity [8]. Early (5-10) passages of primary cultures of rat aortic muscle cells do not show this. Note that the inorganic blocker La^{3+} is a more complete inhibitor than the organic one (Fig. 1). Presumably other pathways including the less specific ones are also blocked by this ion.

Ca^{2+}-entry activators can also be studied with these cells, as exemplified by a DHP, SDZ 202-791. Its optical isomers, known to have opposing activities with respect to their ability to modulate smooth muscle contraction [9],

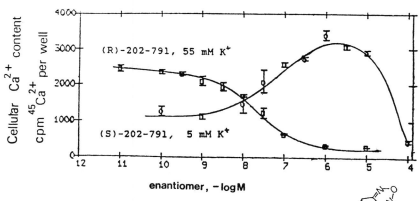

Fig. 2. Uptake into cultured cells, as in Fig. 1: concentration-dependent effects of the *R*- and *S*-enantiomers of SDZ 202-791. After the pre-treatment, the cells were exposed to PSS (5 mM K^+) or to an isotonic PSS (in which Na^+ had been partially replaced with 55 mM K^+) to depolarize the cells. The compounds were added together with these solutions which also contained 0.1 mM $^{40}Ca^{2+}/^{45}Ca^{2+}$. Care was taken to exclude short-wavelength UV light (<400 nm) because DHP's are generally light-sensitive.

(+)-(S)-202-791

(−)-(R)-202-791

also have opposing effects on cellular Ca^{2+} uptake (Fig. 2). In depolarizing solution (55 mM KCl), the (*R*)-enantiomer displayed purely antagonistic activity (IC_{50} 20 nM) whereas the (*S*)-enantiomer stimulated $^{45}Ca^{2+}$ uptake in normal PSS (EC_{50} 80 nM). The blocking effect of this activator at concentrations $>10^{-5}$ M (Fig. 2) is probably ascribable to non-specific membrane interactions. It is observed with many compounds including other activators (<u>e.g.</u> Bay K 8644) or DHP-like molecules that are inactive as Ca^{2+} entry modulators.

We have used this technique for studying the concentration-dependent effects on the L-channel of hundreds of different molecules, including well-known Ca^{2+}-entry blockers [6] of which many are clinically useful.

The correlation of inhibition of Ca^{2+} uptake into the smooth muscle cells and inhibition of contraction of rabbit aorta was good for all compounds acting solely at the L-channel. We conclude therefore that the type of L-channel seen in cultured aortic cells resembles, at least with respect to its pharmacology, the one that is functionally important in regulating smooth muscle contraction.

References

1. Bolton, T.B. (1979) *Physiol. Rev. 59*, 606-718.
2. van Breemen, C., Aaronson, P. & Loutzenhiser, R. (1979) *Pharmacol. Rev. 30*, 167-208.
3. Reuter, H. (1985) *Nature 316*, 391.
4. Nowycky, M.C., Fox, A.P. & Tsien, R.W. (1985) *Nature 316*, 440-443.
5. Rüegg, U.T., Doyle, V.M., Zuber, J-F. & Hof, R.P. (1985) *Biochem. Biophys. Res. Comm. 130*, 447-453.
6. Rüegg, U.T., Doyle, V.M. & Hof, R.P. (1985) *J. Hypertens. 3 (Suppl. 3)*, 557-559.
7. Doyle, V.M. & Rüegg, U.T. (1985) *Biochem. Biophys. Res. Comm. 131*, 469-476.
8. Van Renterghem, C., Romey, G. & Lazdunski, M. (1988) *Proc. Nat. Acad. Sci. 85*, 9365-9369.
9. Hof, R.P., Rüegg, U.T., Hof, A. & Vogel, A. (1985) *J. Cardiovasc. Pharmacol. 7*, 689-693.

Discussion at the Forum (& see p. 152)

W.E.J.M. Ghijsen asked whether it is because of an influence of $[Ca^{2+}]_i$ on membrane polarity that L-channel pick-up in smooth muscle cells needs pre-depletion of Ca^{2+} by extracellular EGTA. **Reply.-** Of the several possibilities we can consider two, both conjectural. (1) Our cultured cells have very large Ca^{2+} pools (mM $[Ca^{2+}]$ values). The cytoplasm, although larger in volume, has $[Ca^{2+}]_i$ ~10^{-7} M. One can argue that a large (e.g. 10-fold) elevation of cytosolic $[Ca^{2+}]_i$ induced by K-depolarization isn't enough to be seen as a signal when the Ca^{2+} content of the total cells is measured. If the pools are empty, they fill up more quickly in depolarizing solution because the cells use the L-channel, the major plasma-membrane Ca^{2+} transport system for this process. (2) On the other hand, we and others can't detect any cytosolic $[Ca^{2+}]_i$ rise when we suspend fura-2-loaded cells in a cuvette and add 55 mM KCl.- Conceivably even resting $[Ca^{2+}]_i$ is high enough to inactivate the L-type channel. **O.H. Petersen remarked** (cf. p. 153) on the non-utility of a POC/ROC distinction in some situations (e.g. where L-channels are coupled via a G-protein to a receptor) [Hescheler et al. (1988) *EMBO J. 7*, 619-624]. **Rüegg:** indeed an adrenal cell-line has such a channel, but it is still a POC. I guess that the ROC is quite different. Benham & Tsien [(1987) *Nat. 328*, 275-278] have patch-clamped an ATP-operated channel which is not dependent on membrane potential, like the blocker-insensitive ROC we detect with Ca^{2+} fluxes (*Biochem. Biophys. Res. Comm.*, 1988).

#ncA.6

A Note on

PROTEIN KINASE C-MEDIATED EFFECTS ON ^{45}Ca^{2+} FLUXES AND ON [Ca^{2+}]$_i$ IN SMOOTH MUSCLE CELLS AND ON CONTRACTION OF MESENTERIC RESISTANCE VESSELS

Urs T. Rüegg, Andreas Wallnöfer, Jürg C. Peter and Cynthia Cauvin

Preclinical Research, Sandoz Ltd., 4002 Basle, Switzerland

One of our salient investigative interests is measuring Ca^{2+} fluxes and their correlation to tissue contraction. Here we outline a study which yielded only partial information when flux measurements were the only source of data. Using phorbol esters to activate PKC*, which may modulate Ca^{2+} entry and/or Ca^{2+} sensitivity in smooth muscle [1,2], we have carried out parallel studies on cultured smooth-muscle cells[t] and on MRV's, measuring ^{45}Ca^{2+} uptake and contraction, respectively. After depletion of the Ca^{2+} pools in the cultures, phorbol esters increased ^{45}Ca^{2+} uptake [3, 4] (Fig. 1), to a maximal extent similar to that obtained by depolarization with KCl or exposure to a Ca^{2+}-entry activator -- Bay K 8644 (Fig. 1) or SDZ (+)202-791. With 4 known PKC activators, stimulation potency for PKC closely parallelled that for Ca^{2+} uptake (Fig. 2); PMA had the highest activity (EC$_{50}$ 0.5 nM). This indicated that PKC was the target enzyme causing increased Ca^{2+} uptake. At higher concentrations, however, both phorbol esters inhibited the stimulated ^{45}Ca^{2+} uptake such that a bell-shaped curve resulted (Fig. 2). It is unclear whether or not this effect is PKC-mediated.

PMA-induced ^{45}Ca^{2+} uptake was inhibited by L-type Ca^{2+} channel blockers (verapamil, DHP's) with essentially the same potencies as when cells were depolarized with 55 mM KCl. As PMA did not modify ^{86}Rb$^+$ efflux, it was unlikely that inhibition of K$^+$ channels was causing depolarization leading to increased Ca^{2+} uptake. However, when the [Ca^{2+}]$_i$ was not depleted before applying test compounds, PMA and other activators had hardly any effect. So the influx-triggering needs depleted pools.

[Ca^{2+}]$_i$ measured with fura-2 was unchanged by PMA with extracellular Ca^{2+} present. The response to vasoconstrictor hormones was, however, down-regulated by PMA [2][t] — whereas

Abbreviations.- DHP, dihydropyridine; MRV's, isolated mesenteric resistance vessels; PKC, protein kinase C; & see Fig.1 legend.
[t]See Fig. 2 legend concerning cultures <u>etc.</u>; 'down-regulation' observations are in press (Rüegg <u>et al.</u>, *J. Cardiovasc. Pharmacol.*).

Fig. 1. Stimulation of ^{45}Ca uptake
in primary (9 passages) cultures
(cell-line cultures gave similar
results), by depolarizing solution,
a DHP Ca^{2+}-entry activator used at
its maximally stimulating concentration, or phorbol 12-myristate
13-acetate (PMA). Control: physiological salt solution (PSS) with
5 mM KCl. See preceding article,
p. 121, for amplification (cultures
- from rat **aorta** - and procedures
used, including Ca^{2+} depletion).

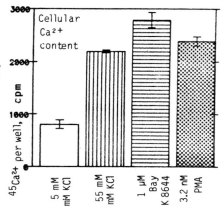

Fig. 2.
Concentration-
dependent
effects of known
PKC activators
on ^{45}Ca^{2+} uptake
into smooth
muscle cells.
The % values
represent the
difference
between PSS
(5 mM KCl) and
55 mM KCl as
shown in Fig. 1.

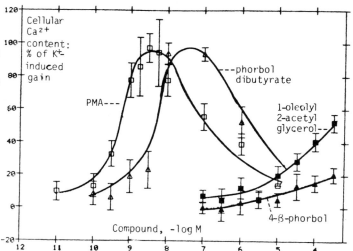

MRV's exposed to PMA (1-1000 nM) showed a slow (hours) contraction,
<5% of that produced within seconds by, e.g., vasopressin
or noradrenaline (perhaps an effect on myofilament sensitivity
[1]). PMA did not influence the tension response to subsequently
applied agonists - as observed likewise for angiotensin II-induced
contractions of rat aortic rings, unaffected by PMA.

Seemingly, then, phorbol esters cannot be regarded as
general vasoconstrictors. Possibly they inhibit Ca^{2+} entry
through ROC's by down-regulation and, on the other hand, stimulate
the refilling of depleted intracellular pools, in an unknown way.

References

1. Rasmussen, H., Takuwa, Y. & Park, S. (1987) *FASEB J. 1*, 177-185.
2. Aiyar, N., Nambi, P., Whitman, M., Stassen, F. & Crooke, S.
   (1987) *Molec. Pharmacol. 31*, 180-184.
3. Rüegg, U.T. (1986) *J. Hypertens. 4 (suppl. 6)*, S711.
4. Fish, D.R., Sperti, G., Colucci, W.S. & Clapham, D.E.
   (1988) *Circulation Res. 62*, 1049-1054.

#ncA.7

A Note on

Na$^+$/Ca^{2+} EXCHANGE AS A TARGET FOR INOTROPIC AND ANTIHYPERTENSIVE DRUGS

G. Cargnelli, S. Bova, P. Debetto and S. Luciani

Department of Pharmacology, University of Padova, Largo E. Meneghetti 2, 35131 Padova, Italy

The intracellular Ca^{2+} concentration is regulated by plasma-membrane and cytoplasmic mechanisms. The former include Ca^{2+}-channels for Ca^{2+} entry, and Ca^{2+}-ATPase and Na$^+$/Ca^{2+} exchanger for Ca^{2+} extrusion. This exchanger can also operate as a Ca^{2+}-entry system, depending upon the direction of the Na$^+$ gradient across the membrane and the membrane potential. It has been shown recently that drugs (amiloride and amiloride derivatives) that inhibit the activity of the Na$^+$/Ca^{2+} antiport can induce modifications of the cardiac [1, 2] and vascular smooth muscle [3] contractility.

Materials and methods.- In guinea-pig left atria, the positive inotropic effect of amiloride (0.5 mM) was characterized by driving the atria at various rates and at different [Ca^{2+}] and [Na$^+$] levels in the medium maintained at 35°. In guinea-pig aortic strips, the effect of amiloride (10-500 µM) on the contractions induced by omitting Na$^+$ or K$^+$, or by adding ouabain to the medium, was studied [4].

RESULTS

In the atria the appearance of the positive inotropic effect of the drug was strictly dependent on the frequency of stimulation and on the extracellular [Ca^{2+}] and [Na$^+$]. A positive inotropic effect (denoted '+') was not manifest below a certain Ca^{2+} concentration for a given stimulation rate, <u>e.g.</u> below 1.8 mM Ca^{2+} for 0.5 Hz:

- 3.6 mM: 0.1 Hz, +; 0.5 Hz, +; 1 Hz, +.
- 2.7 mM: -;   +;   +.
- 1.8 mM: -;   +;   +.
- 0.9 mM: -;   -;   +.
- 0.45 mM: -;   -;   -.

Since contractility of isolated cardiac preparation is strongly influenced also by extracellular [Na$^+$], and since the intracellular [Ca^{2+}] and force of contraction have the approximate relationship [Ca^{2+}]$_o$/[Na$^+$]$_o^2$, the pattern of the cardiac action of amiloride with respect to this ratio at 1 and 0.1 Hz was examined. A positive inotropic response was obtained at 0.1 Hz when the ratio was increased from 8×10^{-5} to 16×10^{-5} and at 1 Hz from 2×10^{-5} to 8×10^{-5}, but only when the ratio was raised by increasing $[Ca^{2+}]_o$ and not by decreasing $[Na^+]_o$. Thus whether or not the drug elicited a positive inotropic response at a given ratio depended on how the value was reached.

In guinea-pig aortic strips, amiloride inhibited the contraction induced by Na^+-free or KCl-free medium, or by ouabain, when it was added before applying the contraction stimulus, but did not relax the contracted strips (Table 1). Moreover, the drug reduced the rate of relaxation of contracted aortic strips (KCl-free medium) induced by re-introducing KCl into the medium.

Table 1. Inhibition (%) of the contraction induced in aortic strips by Na^+-free or KCl-free medium, or by ouabain. (± = S.E.M.)

Amiloride, μM	Na^+-free	KCl-free	Ouabain, 25 μM
10	36 ±4.5	38.3 ±5	11.5 ±3
50	51 ±5.0	83.8 ±6	35.5 ±9
500	89 ±2.5	90 0 ±9	92.2 ±2.9

Since the above experimental conditions make the function of the Na^+/Ca^{2+} antiport predominant, they are particularly suitable for studying the functional modifications induced by drugs acting on this exchange system. Our findings suggest that the Na^+/Ca^{2+} exchanger may be a physiological system susceptible to pharmacological modulation for the positive inotropic and antihypertensive effects.

References

1. Floreani, M. & Luciani, S. (1984) *Eur. J. Pharmacol.* *105*, 317-322.
2. Floreani, M., Tessari, M., Debetto, P., Luciani, S. & Carpenedo, F. (1987) *Naunyn Schmiedeberg's Arch. Pharmacol. 336*, 661-669.
3. Bova, S., Cargnelli, G. & Luciani, S. (1986) *Ann. N.Y. Acad. Sci. 488*, 543-545.
4. Bova, S., Cargnelli, G. & Luciani, S. (1988) *Br. J. Pharmacol. 93*, 601-608.

#ncA.8

A Note on

STUDIES ON THE CARDIOVASCULAR ROLE OF Na$^+$/Ca^{2+} EXCHANGE USING ISOLATED PREPARATIONS AND LIPOSOMES

S. Luciani, P. Debetto, G. Cargnelli, M. Tessari, M. Antolini and F. Cusinato

Department of Pharmacology, University of Padova, Largo E. Meneghetti 2, 35131 Padova, Italy

In cardiac and vascular smooth muscle, the Na$^+$/Ca^{2+} exchange plays a key role in the regulation of intracellular Ca^{2+} levels and hence in contractile control. The exchanger is a large-capacity system which catalyzes electrogenic counter-transport of Na$^+$ ions on either side of the membrane for Ca^{2+} on the opposite side. The electrochemical gradients for Na$^+$ and Ca^{2+}, and the transmembrane potential, determine the direction of net Ca^{2+} flux.

Our experimental approaches deal both with protein-structure study and with defining the physiological functions of this antiporter in isolated atria and aortic strips, in cardiac myocytes and sarcolemmal vesicles. To determine the relative importance of Na$^+$/Ca^{2+} exchange in the mechanism of muscle contractility, we investigated the pharmacological effects on cardiac and vascular smooth muscle contraction elicited by drugs (amiloride and amiloride derivatives) that are active as inhibitors of Na$^+$/Ca^{2+} exchange activity.

Protein structure study entailed isolation and purification of the Na$^+$/Ca^{2+} exchanger extracted from bovine cardiac sarcolemma, then its reconstitution into proteoliposomes of known composition (phosphatidylcholine : phosphatidylserine, 9:1) [1]. The exchange activity was affected by modifications of the redox status both in the proteoliposomes and in isolated cardiac myocytes, where 200 μM diamide (a sulphydryl oxidizing agent) caused a 40% inhibition of the exchanger.

In electrically driven guinea-pig left atria and right ventricular papillary muscle, amiloride and some derivatives (*o*-chlorobenzamil and 3',4'-dichlorobenzamil) produced a positive inotropic effect [2]. In guinea-pig aortic strips amiloride inhibited the contraction induced by Na$^+$-free medium, KCl-free medium and ouabain addition. The drug also reduced the rate of relaxation of contracted aortic strips induced by re-introduction of the physiological salt solution [3].

Table 1. Intravesicular drug levels attained during preincubation, and inhibition (%) of Na^+/Ca^{2+} exchange activity. After 20 min at 37°, vesicle-associated drug was measured fluorimetrically (363/410 nm). From the $\mu g/\mu g$-protein levels for the drug, its concentration inside the vesicles (amiloride$_i$, mM) was calculated assuming an internal vol. of 8.5 $\mu l/mg$ protein.

[Amiloride$_o$] as set	[Amiloride$_i$], observed	Exchange inhibition
0.54	0.00249 = *1.11 mM*	15
1.04	0.00374 = *1.66 mM*	31
1.70	0.00872 = *3.87 mM*	65
2.15	0.01370 = *6.08 mM*	70
3.10	0.02573 = *11.42 mM*	83
4.20	0.04109 = *18.25 mM*	91

A highly purified preparation of sarcolemmal vesicles from bovine ventricular tissue [1] was used to study the kinetics of amiloride inhibition on Na^+/Ca^{2+} exchange. The drug could act from either the outside or the inside of the vesicles [4]. During pre-incubation, amiloride entered the vesicles, wherein it reached levels in the mM range (Table 1).

Amiloride inhibited the Na^+/Ca^{2+} exchange activity at the cis and trans sides of the sarcolemmal membrane with similar affinities and estimated K_i's [5]. The study of the pattern of interaction between amiloride and the Na^+/Ca^{2+} exchange in cardiac sarcolemmal vesicles also revealed functional asymmetries of this antiporter. In fact, while Na^+ protected against inhibition by amiloride when acting on the same side of the vesicle membrane as the drug, it synergistically interacted with amiloride to inhibit exchange activity when acting on the opposite side to the drug [5].

In conclusion, comparison of the functional characteristics of the Na^+/Ca^{2+} exchanger in systems differing in functional complexity may help illuminate the relative contribution of this antiporter in the mechanism of cardiac and vascular smooth muscle contractility.

References

1. Luciani, S. (1984) *Biochim. Biophys. Acta 772*, 127-134.
2. Floreani, M., Tessari, M., Debetto, P., Luciani, S. & Carpenedo, F. (1987) *Naunyn Sch'berg's Arch. Pharmacol 336*, 661-669.   [608.
3. Bova, S., Cargnelli, G. & Luciani, S. (1988) *Br. J. Pharmacol. 93*, 601-
4. Debetto, P., Floreani, M., Carpenedo, F. & Luciani, S. (1987) *Life Sci. 40*, 1523-1530.
5. Debetto, P., Tessari, M., Floreani, M., Luciani S. & Carpenedo, F. (1988) *Pharmacol. Res. Comm. 20*, 619-620.

#ncA.9

A Note on

THIN FILAMENT ACTIVATION BY PHOTOLYSIS OF CAGED-CALCIUM IN SKINNED MUSCLE FIBRES

*C.C. Ashley, [†]R.J. Barsotti, [†]M.A. Ferenczi,
*T.J. Lea and *[†]I.P. Mulligan

[†]National Institute for Medical Research,
London NW7 1AA, U.K.

and *University Laboratory of Physiology,
Oxford OX1 3PT, U.K.

Recent work [1] with intact muscle has shown in detail how thin-filament activation precedes cross-bridge attachment and force generation. We have attempted to determine the Ca^{2+}-dependence of this activation process. For this purpose we used the photolabile Ca^{2+} chelator nitr-5 [2] ($k_{off} = 3000$ sec^{-1}) and chemically or mechanically skinned semitendinosus fibres of *Rana temporaria*. A single pulse of laser light (75 mJ in <200 nsec at 347 nm) photolysed 50% of the initial nitr-5. When the fibres were incubated in 3 mM nitr-5 with sufficient Ca^{2+} to give pCa 6.5 in the presence of 5 mM Mg-ATP, single light pulses resulted in a fast increase in Ca^{2+} concentration to pCa 5.6.

Tension signals recorded from these initially relaxed fibres showed that the light pulse induced a contraction which reached the same force level as that obtained by immersing fibres in a solution at pCa 4.5 with 5 mM Mg-ATP present. At pulse intensities of 75 mJ and above, the time-course of tension rise was approximately exponential with a half-time of 40 ±2 msec (S.E.M., n = 14) at 12°. The normalized rate of tension rise observed in these fibres is close to that observed by Kress <u>et al.</u> [1] for tetanically stimulated frog sartorius ($t_{\frac{1}{2}}$ = 32.5 msec at 14°). It is therefore likely that *in vivo* the rate of tension rise is not limited by the availability of Ca^{2+} in the vicinity of the thin filaments. As the laser pulse intensity was decreased, the rate of tension development remained constant while less than the full isometric force was developed. However, at still lower laser energies which resulted in <50% of full isometric force, both the level and the rate of tension development decreased as a function of pulse intensity.

The tension rise following photolysis of 2 mM caged-ATP in the presence of pCa 4.5 had $t_{\frac{1}{2}}$ 22 ±2 msec under similar conditions. These fibres were initially in rigor, with thin filaments activated by the presence of Ca^{2+} and rigor cross-bridges. Under these conditions the rate of tension rise is limited not by activation of the thin filament but by cross-bridge detachment/attachment and force-generation steps which appear to be faster than the step controlling thin-filament activation as observed in the nitr-5 experiments.

When fibres were incubated in solutions containing 2 mM caged-ATP and 3 mM nitr-5 at pCa 6.5, the observed tension rise at a high pulse intensity (75 mJ) had $t_{\frac{1}{2}}$ 19 ±1 msec (n = 5). The rate of tension rise was equal to that found in fibres with fully activated thin filaments, and faster than that when thin filaments were not pre-activated by rigor cross-bridges. Evidently Ca^{2+} release by caged-calcium photolysis does not limit the rate of force development at pCa's <6.0.

ADDENDUM (Ashley, Mulligan & Lea; *abridged by Ed.*).-Our chemical destruction of sarcolemmal and s.r. membranes allowed control of the filament environment but caused loss of soluble components, altered ionic composition of the filament space, increased lattice spacing, and increased series compliance (the cut ends being damaged when mounted with aluminium clips). Yet the main limitations to force-development rate seem to be the same as for intact fibres ([1]; initial propagation of depolarization and Ca^{2+} release from s.r. are negligibly short). If $[Ca^{2+}]$ is high we find a plateau ($t_{\frac{1}{2}}$ = 37 msec), suggesting that Ca^{2+} binding to the regulatory sites on troponin C is not then rate-limiting. Force-development kinetics do not match a 1- or 2-step model, assuming independent constant-affinity binding of Ca^{2+} to the 2 Ca^{2+}-specific sites and and proportionality of force to the number of troponin complexes with Ca^{2+} bound to both sites [3, 4]. Besides the ion-binding kinetics, there may well be a contribution to overall activation from cooperative interactions along the thin filament between Ca^{2+} binding and force generation (cf. the steady-state relation between pCa and force). This cooperativity may be due to envisaged nearest-neighbour interactions [5], where the states of adjacent regulatory units govern between-state transition probabilities.

Acknowledgements.- We thank Dr Roger Tsien for early samples of nitr-5 and Dr David Trentham for facilities at Mill Hill.

References

1. Kress, M., Huxley, H.E., Faruqi, A.R. & Hendrix, J. (1986) *J. Mol. Biol. 188*, 325-342.
2. Adams, S.R., Kao, J.P.Y., Grynkiewicz, G., Minta, A. & Tsien, R.Y. (1988) *J. Am. Chem. Soc. 110*, 3212-3220.
3. Moisescu, D.G. (1976) *Nature 262*, 610-613 [cf. (1972) *237*, 208-211].
4. Hill, T.L. (1983) *Biophys. J. 44*, 383-396.
5. Shiner, J.S. & Solaro, R.J. (1982) *Proc. Nat. Acad. Sci. 79*, 4637-4641.

#ncA.10

A Note on

CALCIUM UPTAKE BY SYNAPTOSOMES WITH LOW AND HIGH Na+ CONTENT, AND EFFECT OF Ca²⁺ ANTAGONISTS

C.A.M. Carvalho, C.B. Duarte, D.L. Santos,
E.J. Cragoe jr., and A.P. Carvalho

Center for Cell Biology, Department of Zoology,
University of Coimbra, 3049 Coimbra, Portugal

We have used synaptosomes isolated from rat brain to study Ca^{2+} influx through Ca^{2+} channels and by Na^+/Ca^{2+} exchange. With our experimental design these two mechanisms of Ca^{2+} uptake can be distinguished under K^+ depolarization, and we have found that the contribution of each mechanism for Ca^{2+} uptake depends strongly on the Na^+ content of synaptosomes [1-3]. We have used two synaptosomal preparations, one being a described preparation [4] in which sucrose medium is used throughout. These synaptosomes have ~20 mM Na^+ and ~120 mM K^+, whereas the second preparation [5] is much richer in Na^+ (~80 mM) with 80-90 mM K^+. The two preparations are now compared in respect of Ca^{2+} uptake through Ca^{2+} channels and Na^+/Ca^{2+} exchange, and IC_{50} values, *vs.* k_p's*, are given for various Ca^{2+} antagonists which act on both mechanisms of Ca^{2+} uptake.

Ca²⁺ UPTAKE BY LOW-Na+ and Na-RICH SYNAPTOSOMES

In Fig. 1A we show the $^{45}Ca^{2+}$ uptake by low-Na^+ synaptosomes isolated in sucrose medium [4]. KCl addition to give K_o^+ = 66, with lowering of Na_o^+ from 133 to 67, causes an increase in Ca^{2+} uptake from 8 to 11 nmol/mg protein per 30 sec. If instead of KCl we add choline chloride (Ch_o^+ = 66, Na_o^+ = 67), no Ca^{2+}-uptake difference is discerned relative to the control (Na_o^+ = 133). We therefore conclude that the Ca^{2+}-uptake increase due to K^+ addition (Na_o^+ = 67) is due to depolarization and probably occurs through voltage-dependent Ca^{2+} channels, which open upon depolarization.

From Fig. 1B, showing similar experiments carried out with Na^+-rich synaptosomes [5], it is evident that the Ca^{2+} uptake is about doubled. In addition, K^+-depolarization

*Abbreviations.- Ch, choline; CBZ-DMB, chlorobenzyl-dimethyl-benzamil; k_p, partition coefficient; K_o^+ (e.g.) signifies the added cation's concentration (mM) in the medium.

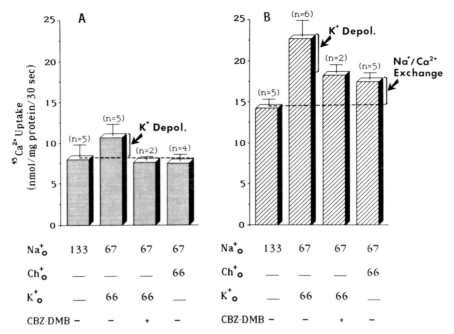

Fig. 1. Ca^{2+} influx – **via** Ca^{2+} channels (due to K^+ depolarization) **vs.** by Na^+/Ca^{2+} exchange (Na^+ gradient altered) – in
rat brain cortex synaptosomes of two types: **A**, low-Na^+ (isolated in non-ionic sucrose media [4]); **B**, Na^+-rich (isolated
in 145 mM NaCl [5]). Each incubate received 0.6 mg protein/ml.
Media for $^{45}Ca^{2+}$ uptake had the mM compositions shown, plus
1 mM $MgCl_2$, 10 mM glucose, 10 mM HEPES-Tris pH 7.4, and 1 mM
$^{45}CaCl_2$ (2.5 µCi/µmol). CBZ-DMB, where added, was 30 µM.
After 30 sec at 30° the reactions were stopped by adding, per
ml, 0.12 ml 1.0 mM $LaCl_3$ in choline chloride (133 mM) medium at
0-4°. After vacuum filtration through Whatman GF/B fiberglass
filters, the medium for pre-soaking and washing being 0.32 M
sucrose/10 mM HEPES-Tris pH 7.4/0.1 mM $LaCl_3$, liquid scintillation counting was performed on the filters and media [2, 7].
The S.D. is shown above each block.

(K_0^+ 66, Na_0^+ 67), or Ch without K^+ (Ch_0^+ 66, Na_0^+ 67), causes
Ca^{2+} uptake to increase by ~8 and ~3 nmol/mg protein per
30 sec respectively. We therefore conclude that in Na^+-rich
synaptosomes Ca^{2+} uptake occurs through Ca^{2+} channels opened
by K^+-depolarization and by Na^+/Ca^{2+} exchange due to decreasing
Na_0^+ and substituting either K^+ or Ch^+ (Fig. 1B).

   CBZ-DMB, an amiloride derivative [6], inhibits Ca^{2+} uptake
in both preparations (Fig. 1, A & B). In the low-Na^+ synaptosomes
(Fig. 1A), whose uptake was increased if external Na^+ were
replaced by K^+ but not by Ch, the effect of CBZ-DMB must

be on Ca^{2+} influx through the voltage-dependent Ca^{2+} channels, since this preparation shows no Na^+/Ca^{2+} exchange (the last column in Fig. 1A being equal to the 133 mM Na^+ control). However, in Na^+-rich synaptosomes we detect a Na^+/Ca^{2+} exchange component (Fig. 1B, last column); also (second column), when the synaptosomes are depolarized by K^+ and, besides, the Na^+ gradient in→out is increased by decreasing Na_0^+, the Ca^{2+} uptake is further increased by ~8 nmol/mg protein per 30 sec. This increase is composed of a Ca^{2+} fraction which enters through the Ca^{2+} channels (K^+ depolarization) and of another fraction due to Na^+/Ca^{2+} exchange.

The effect of CBZ-DMB on Na^+-rich synaptosomes, under conditions in which Ca^{2+} enters through both Ca^{2+} channels and Na^+/Ca^{2+} exchange, seems to be on Ca^{2+} entry through the Ca^{2+} channels since, in the presence of the drug, the K^+-depolarized synaptosomes (Fig. 1B, column 3) take up as much Ca^{2+} as the synaptosomes in which only Na^+/Ca^{2+} exchange takes place (column 4).

These results on the effect of CBZ-DMB on Ca^{2+} uptake on both synaptosomal preparations, at the concentration of CBZ-DMB utilized (30 μM), suggests that this amiloride derivative inhibits predominantly Ca^{2+} fluxes through Ca^{2+} channels rather than the Ca^{2+} uptake through Na^+/Ca^{2+} exchange as has been more conventionally reported for amiloride derivatives [6].

INHIBITION OF Ca^{2+} UPTAKE BY Ca^{2+} ANTAGONISTS

In Table 1 we summarize the effect of various Ca^{2+} antagonists on Ca^{2+} uptake by K^+ depolarization and by Na^+/Ca^{2+} exchange in synaptosomes [2, 3, 7, 8]. It is observed that for nifedipine, verapamil and *d-cis*-diltiazem the IC_{50} values range between 4×10^{-5} and 4×10^{-4} for Ca^{2+} uptake by both mechanisms. As reported previously [2, 7], these concentrations are much higher than those needed for the drugs to saturate their specific binding sites in synaptosomal membranes.

However, pimozide and flunarizine inhibit Ca^{2+} uptake at concentrations 100-fold lower than those found for the other listed Ca^{2+} antagonists, and they are more specific for Ca^{2+} fluxes through Ca^{2+} channels than for Ca^{2+} uptake by Na^+/Ca^{2+} exchange (Table 1) [8]. It is interesting that these two drugs are more lipophilic than the other Ca^{2+} antagonists, as suggested by their higher k_p's in synaptosomal membranes [8]. As Table 1 shows, these k_p values in isolated synaptic plasma membranes are ~6.5×10^3 and ~19×10^3 respectively, as compared with ~464 and ~361 for nitrendipine and verapamil respectively.

Table 1. Efficacy of Ca^{2+} antagonists, $vs.$ k_p's in isolated synaptic plasma membranes, in inhibiting synaptosomal Ca^{2+} uptake due to K^+ depolarization ('depolar.') or Na^+/Ca^{2+} exchange ('exch.'). Methods as in [7, 8]. The ± values are S.D.'s.

Antagonist	IC_{50} (M), depolar.	IC_{50} (M), exch.	k_p (& n)
Nifedipine[1]	1.3×10^{-4}	1.3×10^{-4}	–
Nitrendipine[1]	–	–	464 ±75 (4)
Verapamil	4.0×10^{-5}	2.5×10^{-5}	361 ±40 (4)
d-cis-Diltiazem[2]	3.3×10^{-4}	1.0×10^{-4}	–
Pimozide[3]	1.9×10^{-7}	3.2×10^{-6}	6.5×10^3 (2)
Flunarizine[3]	1.0×10^{-6}	6.0×10^{-6}	19.0×10^3 (2)

[1]from Sandoz, Portugal.   [2]from BAYER AG, Japan.
[3]from Dr. J. Leysen, Janssen Research Foundation, Belgium.

Conclusions.- Low-Na^+ synaptosomes when depolarized by K^+ take up Ca^{2+} through Ca^{2+} channels only, whereas Na-rich synaptosomes take up Ca^{2+} by Na^+/Ca^{2+} exchange also. CBZ–DMB appears to inhibit only the former process. The classical Ca^{2+}-channel blockers inhibit Ca^{2+} influx through Ca^{2+} channels in synaptosomes with IC_{50}'s of 10^{-5}-10^{-4} M and, in this concentration range, they also inhibit the Na^+/Ca^{2+} exchanger. However, pimozide and flunarizine inhibit Ca^{2+} influx through synaptosomal Ca^{2+} channels and the Na^+/Ca^{2+} exchanger with lower IC_{50}'s (10^{-7}-10^{-6} M), and they are ~5- to 10-fold more specific for Ca^{2+} channels (Table 1); their greater inhibitory effects on Ca^{2+} fluxes in synaptosomes may be related to their higher lipophilicity in the membranes.

Acknowledgments.- This work was supported by INIC, JNICT and the Calouste Gulbenkian Foundation.

References

1. Coutinho, O.P., Carvalho, C.A.M. & Carvalho, A.P. (1984) *Brain Res. 290*, 261-271.
2. Carvalho, C.A.M., Coutinho, O.P. & Carvalho, A.P. (1986) *J. Neurochem. 47*, 1774-1784.
3. Carvalho, A.P., Santos, M.S., Henriques, A.O., Tavares, P. & Carvalho, C.M. (1988) in *Cellular and Molecular Basis of Synaptic Transmission*, NATO ASI Series, Vol. H21 (Zimmermann, H., ed.), Springer-Verlag, Berlin, pp. 263–284.
4. Hajós, F. (1975) *Brain Res. 93*, 485-489.
5. Krueger, B.K., Ratzlaff, R.W., Strichartz, G.R. & Blaustein, M.P. (1979) *J. Membr. Biol. 50*, 287-310.
6. Zaczorowski, G.J., Barros, F., Dethmers, J.K., Trumble, M.J. & Cragoe, E.J., jr. (1985) *Biochemistry 24*, 1394-1403.
7. Carvalho, C.M., Santos, S.V. & Carvalho. A.P. (1986) *Eur. J. Pharmacol. 131*, 1-12.
8. Carvalho, C.A.M. & Santos, D.L. (1989) *Annu. N.Y. Acad. Sci.*, in press.

#ncA.11

A Note on

THE IMPORTANCE OF Na^+/Ca^{2+} EXCHANGE IN REGULATING SYNAPTOSOMAL Ca^{2+} CONCENTRATION

C.A.M. Carvalho, [†]O.P. Coutinho[*] and A.P. Carvalho

Center for Cell Biology, University [†]University of Braga,
of Coimbra, 30 49 Coimbra, Portugal   4719 Braga, Portugal

Neurotransmitter release at nerve terminals is normally triggered by a rise in $[Ca^{2+}]_i$ in response to depolarization of the plasma membrane. The rapid termination of release may be due to a reduction of $[Ca^{2+}]_i$ to resting levels. In the present study we determined whether Na^+/Ca^{2+} exchange is responsible for extruding the Ca^{2+} load which accumulates in synaptosomes during depolarization.

We created conditions of Ca^{2+} influx parallel to those necessary for GABA release [1] by applying a pulse of K^+-depolarizing medium containing $^{45}Ca^{2+}$ and subsequently replacing the superfusion medium with basal Na^+ medium. We then followed the efflux of $^{45}Ca^{2+}$ from rat brain synaptosomes and observed that K^+-depolarization or K^+-depolarization with simultaneous Na^+/Ca^{2+} exchange permits different levels of Ca^{2+} loading, as we previously demonstrated [2]. The superfusion with Na^+ medium of the synaptosomes loaded with $^{45}Ca^{2+}$ caused the release of $^{45}Ca^{2+}$ by Na^+/Ca^{2+} exchange however $^{45}Ca^{2+}$ loading was brought about.

The rate of $^{45}Ca^{2+}$ efflux induced by Na^+, determined by rapid filtration, indicates a fast phase which is completed in <10 sec and a slower phase which continues for at least 2 min. The fast phase of Ca^{2+} efflux has a Km value for Na^+ of ~40 mM and is completely blocked by La^{3+} in the mM range.

In conclusion, these results indicate that the Na^+/Ca^{2+} exchanger is responsible for rapidly reducing $[Ca^{2+}]_i$ inside synaptosomes after different Ca^{2+} loading levels attained in physiological resting conditions or after activation of isolated nerve terminals.

---

[*]addressee for any correspondence

References

1. Carvalho, C.A.M., Santos, S.V. & Carvalho, A.P. (1986)
   Eur. J. Pharmacol. *131*, 1-12.
2. Carvalho, C.A.M., Coutinho, O.P. & Carvalho, A.P. (1986)
   J. Neurochem. *47*, 1774-1784.

#ncA.12

A Note on

STUDIES ON Ca²⁺-DEPENDENT EXCITATION-SECRETION COUPLING IN PURIFIED NERVE TERMINALS

W.E.J.M. Ghijsen and M. Verhage

Department of Experimental Zoology,
University of Amsterdam,
Kruislaan 320, 1098-SM Amsterdam, The Netherlands

One of the most important regulators of presynaptic signal transduction is Ca^{2+}. Upon arrival of an action potential, voltage-dependent Ca^{2+}-channels are activated in nerve terminals, mediating rapid Ca^{2+}-influx from the extracellular compartment. The subsequent rise in intrasynaptic $[Ca^{2+}]_i$ triggers NT* release to neighbouring (post)synaptic elements (receptors). **Sy**'s, readily purified from brain homogenates by density-gradient centrifugation, constitute sealed structures surrounded by polarized p.m.'s [1]; they still contain intracellular organelles such as synaptic vesicles, mitochondria and e.r. [2]. In addition, (auto)receptors are still present on the outer **Sy** membrane surface, and depolarization-induced release of diverse NT's can occur.

For use of **Sy**'s as a biochemical model system to study neurotransmission, the purified fraction must retain optimal function of the regulatory mechanisms, <u>e.g.</u> for Ca^{2+} homeostasis that keeps intrasynaptic $[Ca^{2+}]_i$ at a sub-μM level. **Sy** p.m.'s contain voltage-operated Ca^{2+}-channels as well as ATP-dependent Ca^{2+}-extrusion and Na^+/Ca^{2+}-exchange systems. Moreover, Ca^{2+}-cycling across membranes of intrasynaptic mitochondria [3] and e.r. [4] is intact. In order to couple calcium-regulation to NT release, these components must be studied in concert. Here we outline our methodology, with rat-brain purified **Sy**'s, in studying the quantitative correlation between calcium-entry, $[Ca^{2+}]_i$ and NT release.

Measurement of resting $[Ca^{2+}]_i$ and changes after stimulation (a rise being essential for NT release) is informative. As **Sy**'s are of diam. <1 μm, direct $[Ca^{2+}]_i$ assessment by

---

Abbreviations (mostly by Ed.).- $[Ca^{2+}]_i$, intracellular (usually intrasynaptic) Ca^{2+} concentration; NT, neurotransmitter; **Sy**, synaptosome/synaptosomal; e.r., endoplasmic reticulum; p.m., plasma membrane; AM, acetoxymethyl ester (<u>e.g.</u> of the probe fura-2); Glu, glutamate; LDH, lactate dehydrogenase; MP, membrane potential.

Fig. 1. Cytosolic [Ca²⁺] in different **Sy** preparations loaded with fura-2 (5 µM) and incubated at 30° or 37° for 30 min. Fluorescence excitation was alternately at 336 nm and 380 nm; emission was measured at 510 nm. For details see [7].

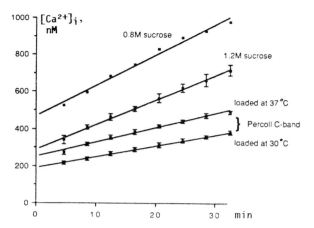

impaled Ca²⁺-selective electrodes is impossible; but it is now attainable with dyes. Thereby widely diverging **Sy** [Ca²⁺]$_i$'s have been reported since 1984 in different preparations and with different dyes, ranging from 100 to 400 nM (refs. in [5]). The fluorescent fura-2 and quin-2 are favoured, being transportable into intact **Sy**'s as lipophilic esters which within **Sy** yield the free acids that have a high affinity for Ca²⁺. With **Sy**'s as distinct from neuronal cell cultures, however, quin-2 did not readily give absolute [Ca²⁺]$_i$ values [6] (cf. #B-7, this vol.-*Ed.*). We too have concluded that quin-2 results are questionable, being very dependent on quin-2 concentration; with fura-2, applied at a 10-fold lower concentration, there was no such dependence [5]. [Ca²⁺]$_i$ values and the rate of the linear rise with time during measurement depended on the **Sy** gradient-purification procedure, being higher in hyperosmotic sucrose preparations [7] than in isoosmotic Percoll preparations [8] (Fig. 1). [Ca²⁺]$_i$ was lower if fura-2 loading was at 30° rather than 37° (Fig. 1); at 37° there appeared to be poorer **Sy** viability (suggested by reduced cytosolic LDH latency). Leakage of fura-2 out of the **Sy**'s only partly accounted for the [Ca²⁺]$_i$ increase.

The between-preparation differences in [Ca²⁺]$_i$ did not reflect differences in how well the **Sy**'s were sealed, the LDH latencies being similar (near 90%; Table 1). **Sy** mitochondrial content closely paralleled [Ca²⁺]$_i$ (Table 1, *vs.* Fig. 1). The tabulated mitochondrial levels are not absolute (no serial sections analyzed), but show relative differences relatable to better Ca²⁺-handling by **Sy**'s collected from heavier density-gradient fractions, being highest with Percoll. Such a higher Ca²⁺-buffering capacity could be explained by direct mitochondrial Ca²⁺-accumulation or (indirectly) by higher ATP delivery to extrude Ca²⁺ from the cytosolic compartment [9].

Fig. 2. Sy $[Ca^{2+}]_i$ as
affected by low extra-
cellular Ca^{2+} concentrations. $[Ca^{2+}]_e$ was
initially ~100 nM after
equilibration with only
20 μM $[Ca^{2+}]_e$. ■: $[Ca^{2+}]_e$
changed to 38 μM, then
(by EGTA) to 200 nM.
□: lowering to <1 nM by
excess EGTA; then Ca^{2+} added.
Sy's were Percoll preparations (in Figs. 3 & 4 also).

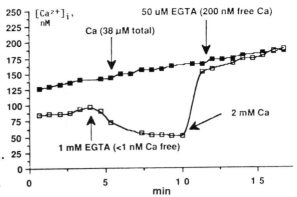

Table 1. Biochemical and morphological characteristics of
different **Sy** preparations. Each mean ±S.E.M. is followed () by
no. of expts. (LDH) or no. of electron micrographs as taken
randomly from 2 expts. Mitochondria (within **Sy**'s) denoted **mit**.

Preparation	LDH, % latency	% of **Sy**'s with **mit**	**mit** per **Sy**
0.8 M sucrose	86.0 ±5.0 (4)	33.0 ±3.8 (83)	0.38 ±0.05 (83)
1.2 M sucrose	89.1 ±3.8 (3)	56.1 ±2.6 (119)	0.67 ±0.04 (119)
Percoll	89.2 ±2.5 (5)	64.4 ±4.3 (73)	0.86 ±0.08 (73)

In summary, these results indicate optimal $[Ca^{2+}]_i$ regulation in isoosmotic Percoll-purified **Sy**'s. A steady-state
$[Ca^{2+}]_i$ of 200-250 nM seem atypically high *vs.* values reported
[10] for other secretory cell types, but is probably an
over-estimate due to the fraction of **Sy**'s with limited Ca^{2+}-
buffering capacity still present in the preparation.

Ca^{2+}-influx into Sy's.- The Ca^{2+}-channels through which
Ca^{2+} is thought to enter prior to NT release from terminals
are not measurable by patch-clamping in individual **Sy**'s. For
MP-dependent Ca^{2+}-influx, $^{45}Ca^{2+}$ accumulation in **Sy**'s under
polarized (low $[K^+]_e$) and depolarized (high $[K^+]_e$) conditions
has been a routine approach, commonly with **Sy**'s stored in
Ca^{2+}-deficient (μM) or even Ca^{2+}-free media such that the
initial Ca^{2+} uptake might be drastically affected since
intra-**Sy** pools of Ca (free and bound) are far below normal
resting levels. There seems to be a significant rapidly
mobilizable pool within intact **Sy**'s [11].

Fig. 2 illustrates the effect of low $[Ca^{2+}]_e$ on **Sy**
$[Ca^{2+}]_i$. $[Ca^{2+}]_i$ was unaffected by lowering $[Ca^{2+}]_e$ to 200 nM,
but rapidly fell at $[Ca^{2+}]_e$ <1 nM. We therefore studied K^+-depolari-
zation dependency of $^{45}Ca^{2+}$ uptake after pre-equilibration

Fig. 3. Effect of K$^+$-induced depolarization on rapid Ca^{2+} uptake in **Sy**'s: ^{45}Ca accumulation after 1 sec at 37° as a function of membrane polarity, at 1-2 mM [Ca^{2+}]$_e$. See text for amplification.

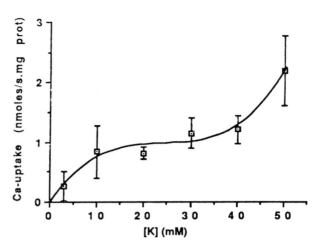

with high [Ca^{2+}]$_e$ (1-2 mM), _i.e._ under steady-state conditions. If measured after incubating for 1, 2 and 5 sec, K$^+$-evoked uptake deviated progressively from linearity, possibly due to inactivation of Ca^{2+}-influx after prolonged depolarization [12] and/or back-flux of ^{45}Ca^{2+} from the **Sy**'s after >1 sec of incubation. With conditions that measured primarily Ca^{2+}-influx, the threshold for its stimulation was already reached as [K$^+$]$_e$ approached 10 mM, and a plateau at 20-40 mM. At 50 mM there tended to be a further rise, but not significant (n = 3, P>0.2)

Ca^{2+}-dependent glutamate release. - Purified **Sy**'s still contain the NT-releasing machinery; but in defining the actual exocytotic NT component in the total release, the technique applied is crucial. Most NT release studies with **Sy**'s entail pre-loading with radiolabelled compounds or precursors. We studied the **Sy** regulation of Glu release in hippocampal tissue, where Glu plays an important role in adaptive changes in neurotransmission (neuronal plasticity). After pre-loading **Sy**'s with ^{3}H-L-Glu, no significant Ca^{2+}-dependent ^{3}H release could be measured upon K$^+$-induced depolarization. Recent studies by Nicholls & Sihra [13] showed predominant labelling of the cytosolic Glu pool under these conditions, which could explain the mainly Ca^{2+}-independent release we observed.

We developed a method to analyze, after reverse-phase HPLC separation, endogenous release of amino acid NT's from **Sy**'s superfused in small (1 ml) chambers. Thereby a broad spectrum of endogenous amino acids could be measured very sensitively (pmol range): Asp, Glu and GABA were specifically increased by K$^+$-induced depolarization during 1 min. In the representative experiment shown (Fig. 4, left) on Glu release,

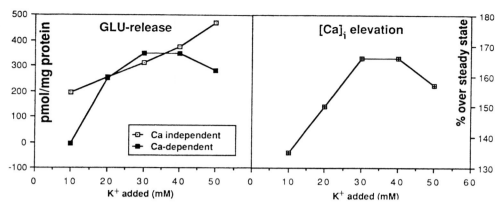

Fig. 4. Effect of K^+-induced depolarization on **Sy** Glu (GLU) release and $[Ca^{2+}]_i$. $[Ca^{2+}]_e$ was 2 mM for the Ca^{2+}-dependent component and 200 nM (see Fig. 2) for the Ca^{2+}-independent component.

Ca^{2+}-dependent and Ca^{2+}-independent components are distinguished. The former, which probably reflects the exocytotic pathway, appeared already maximal at moderate depolarizing stimulations with 20-30 mM extracellular K^+. Changes in $[Ca^{2+}]_i$ as a function of $[K^+]_e$ (Fig. 4, right) closely followed Ca^{2+}-dependent Glu release. In contrast, Ca^{2+}-independent Glu release showed a continuous increase with rising $[K^+]_e$ above 20 mM, becoming a major component in total Glu release at 50 mM (>60%). This release seemed to change as a function of the MP, and probably reflects carrier-mediated efflux of Glu across the **Sy** p.m. [14].

Stimulus-secretion coupling in synaptosomes.- An integrated picture of coupling between nerve-terminal excitation and Glu release can be created by relating Ca^{2+}-influx, $[Ca^{2+}]_i$ and Ca^{2+}-dependent release as functions of K^+-induced membrane polarization. Comparison of the results presented above shows a close relationship between extent of depolarization and optimal stimulation of these parameters. In all three cases a plateau was reached on raising $[K^+]$ to 30 mM, <u>i.e.</u> after depolarization from -70 to -40 mV [15]. Maximal stimulation of Glu release was already obtained at a $[Ca^{2+}]_i$ value of ~400 nM. The threshold for stimulation lies between 10 and 20 mM K^+. Such a high sensitivity was achieved by using **Sy**'s with proper Ca^{2+}-buffering capacity (Percoll) and with fura-2 as Ca^{2+}-indicator. Moreover, the different Ca^{2+}-dependent pathways studied were measured under steady-state conditions, without severe depletion of **Sy** Ca^{2+}_i. Finally, measurement of endogenous NT release allows quantitative discrimination between the different intra-**Sy** pools involved in the total release.

Acknowledgements

We thank Dr. J. Odink (Inst. of Applied Research, Zeist) for providing HPLC facilities and Prof. F.H. Lopes da Silva for his encouraging interest. We gratefully acknowledge the technical assistance of Mrs. E. Besselsen, Mrs. G. Scholten and Mr. H. Sandman.

References

1. Bradford, H.F. (1986) in *Chemical Neurobiology* (Bradford, H.F., ed.), Freeman Press, New York, pp. 311-352.
2. McGraw, C.F., Somlyo, A.V. & Blaustein, M.P. (1980) *J. Cell Biol. 85*, 228-241.
3. Akerman, K.E.O. & Nicholls, D.G. (1983) *Trends Biochem. Sci. 8*, 63-64.
4. Rasgado-Flores, H. & Blaustein, M.P. (1987) *Am. J. Physiol. 252*, C588-C594.
5. Verhage, M., Besselsen, E., Lopes da Silva, F.H. & Ghijsen, W.E.J.M. (1988) *J. Neurochem. 91*, 1667-1674.
6. Meldolesi, J., Huttner, W.B., Tsien, R.Y. & Pozzan, T. (1984) *Proc. Nat. Acad. Sci. 81*, 620-624.
7. Gray, E.G. & Whittaker, V.P. (1962) *J. Anat. 96*, 79-88.[*]
8. Nagy, A. & Delgado-Escueta, A.V. (1984) *J. Neurochem. 43*, 1114-1123.
9. Nachshen, D.A. (1985) *J. Physiol. 363*, 87-101.
10. Meldolesi, J., Malgaroli, A., Wollheim, C.B. & Pozzan, T. (1987) *In Vitro Methods for Studying Secretion* (Poisner, A. & Trifaro, J.M., eds.), Elsevier, New York, pp. 289-313.
11. Ghijsen, W.E.J.M., Besselsen, E., Verhage, M., Melchers, B.P.C., Pennartz, C.M.A. & Lopes da Silva, F.H. (1987) in *Cellular Calcium and Phosphate Transport in Health and Disease* (Bronner, F. & Peterlik, M., eds.), Alan Liss, New York, pp. 209-214.
12. Nachshen, D.A. (1985) *J. Physiol. 361*, 251-268.
13. Nicholls, D.G. & Sihra, T.S. (1986) *Nature 321*, 772-773.
14. Erecinska, M. & Nelson, D. (1987) *FEBS Lett. 213*, 61-66.
15. Adam-Vizi, V. & Ligeti, E. (1986) *J. Physiol. 372*, 363-377.

----

[*]cf. art. #C-4 by Whittaker et al. in Vol. 15, this series (1985), *Investigation of Antibody Combining Sites* (Reid, E., et al., eds.), Plenum, New York.

#ncA.13

A Note on

MECHANISM OF Ca²⁺ TRANSPORT ACROSS THE PLASMA MEMBRANE OF YEAST CELLS

Y. Eilam⊗, D. Chernichovsky and M. Othman

Department of Bacteriology,
The Hebrew University-Hadassah Medical School,
Jerusalem, Israel

The mechanism of Ca²⁺ influx into yeast cells is not yet understood. Ca²⁺ influx proceeds only in the presence of metabolic substrate, but the driving force is not known [1]. Since in yeast cells cellular Ca²⁺ is mostly sequestered within the vacuole [2], it is difficult to distinguish in intact cells between Ca²⁺ transport across the p.m.* and the uptake into the vacuole by the vacuolar Ca²⁺/H⁺ antiport [3]. A method is presented which enables the initial rates of Ca²⁺ influx across the p.m. of *Saccharomyces cerevisiae* to be determined in intact cells without interference by the vacuolar transport system. In addition we discuss a method for preparing cells with different values of membrane potential [4]. Experiments based on these methods led to the suggestion that Ca²⁺ influx across *S. cerevisiae* p.m. is mediated by channels which open at membrane potentials (MP's) below a threshold value.

MATERIALS AND METHODS

S. cerevisiae strain N123 (genotype MATα/αhis 1) was maintained at 4° on YPD-agar slopes and grown at 30° in YPD-broth (Bacto yeast extract 10 g/1, Bacto peptone 20 g/1, glucose 20 g/1). Cells were collected from an overnight culture by centrifugation, washed 3 times by resuspension in distilled water and finally resuspended in the indicated medium. Cells were pre-incubated for 90 min at 30°, with shaking, in two different media: (a) 'Ca²⁺ medium': glucose 10 mM, buffer*, CaCl₂ 1 μM; (b) 'Ca²⁺-free medium', identical but omitting CaCl₂. Following the pre-incubation the cells were centrifuged and suspended in MES/Tris buffer for transport measurement.

⊗ to whom any correspondence should be sent

*p.m. denotes plasma membrane(s); TFP = trifluoperazine; MP = membrane potential (Δψ); 'buffer' = 10 mM MES/Tris pH 6.0 (MES = 2-[*N*-morpholino]ethanesulphonic acid); TPP⁺ - see overleaf.

Ca^{2+} influx was initiated by adding 0.5 ml of cells suspended in buffer to 0.5 ml transport medium containing buffer, glucose 20 mM (when indicated) and CaCl$_2$ (2 μM) with ^{45}Ca^{2+} (1 μCi/ml). Transport after different times at 30° was terminated by adding 1 ml washing solution: MgCl$_2$ 20 mM, LaCl$_2$ 0.1 mM. The cells were filtered immediately on glass-fibre filters or 0.45 μm membrane filters and washed 5 times with 2 ml portions of washing solution. The filters were dried and the radioactivity determined by liquid scintillation counting (toluene-based fluid). Ca^{2+} binding was determined by 'zero time measurements', obtained by adding 0.5 ml cell suspension to 0.5 ml of buffer solution containing Ca^{2+} (2 μM) labelled with ^{45}Ca^{2+} (1 μCi/ml) and LaCl$_2$ (0.2 mM). The cells were immediately filtered as above.

Cells with different MP values were prepared by pre-incubation for 30 min at 30° in media containing glucose (100 mM), buffer and different concentrations (0 up to 30 μM) of TFP. It was previously found that TFP at low concentrations hyper-polarizes the cells [5, 6]. Following the pre-incubation the cells were centrifuged, washed once with buffer and finally resuspended in buffer (10^8 cells/ml) for transport measurements.

MP measurements were based on the steady-state distribution between cells and medium of ^{3}H-tetraphenylphosphonium ion (TPP$^+$): it was added (1 μM, 0.05 μCi/ml) to all pre-incubation media and to the washing buffer so that the cells were exposed to it for at least 90 min. Then, after adding 2 ml of washing solution containing MgCl$_2$ (20 mM), the cells were collected on glass-fibre filters and washed 5 times with 2 ml portions of washing solution. The filters were dried and the radioactivity determined. The results were corrected for binding of ^{3}H-TPP$^+$ and the MP values were calculated [6]. In studies with ^{45}Ca^{2+}, its differential extraction from cytosol and vacuole after influx for 0, 2 or 180 sec was done as described previously [2, 4].

RESULTS AND DISCUSSION

After pre-incubation in 'Ca^{2+} medium' the time course of ^{45}Ca^{2+} uptake was near-linear up to the end of the experiment at 180 sec. Under these conditions ^{45}Ca^{2+} influx observed represents ^{45}Ca^{2+}/Ca^{2+} exchange. The time course of ^{45}Ca^{2+} uptake into cells incubated in 'Ca^{2+}-free medium' comprised a saturable component and, from 60–90 sec after the start of Ca^{2+} influx, a linear component. This transport represents Ca^{2+} influx under non-steady state conditions [4]. Binding of Ca^{2+} to the cells was determined by 'zero-time measurement' (see METHODS) and was similar for both types of transport.

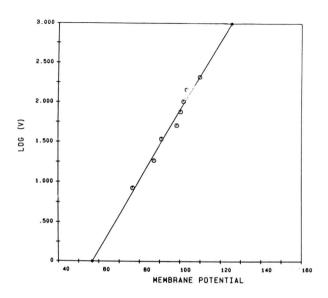

Fig. 1. Initial rate of $^{45}Ca^{2+}$ influx as a function of MP as varied by varying TFP (see text, for other conditions also; influx was measured during 20 sec). Calculation of V ($10^{-14} \times$ mol $Ca^{2+}/10^8$ cells per sec) was from 20 sec and zero-time data. Units for MP: mV.

In cells (pre-incubated in Ca^{2+}-free medium) differentially extracted as above [4], 99% of $^{45}Ca^{2+}$ was found to be bound at zero-time. After 20 sec of influx, 56.1% of total radioactivity was in the cytosol, 0.05% in the vacuole and 43.8% was bound Ca^{2+}; after 3 min, the respective values were 3.7%, 68.5% and 27.7% [4]. It is concluded that after pre-incubation for 90 min in Ca^{2+}-free medium $^{45}Ca^{2+}$ influx into yeast cells during the first 20 sec represents transport across p.m. (after subtracting the zero-time value which represents binding).

The mechanism of Ca^{2+} influx across p.m. was studied (Fig. 1) under non-steady state conditions during the first 20 sec of influx. The role of MP was studied in cells with different MP values obtained as described above using TFP (which induces K^+ efflux and hence the hyperpolarization) [5, 6]. After removing TFP by the washing, Ca^{2+} influx was measured in buffer-containing glucose-free medium. It was found that MP drove Ca^{2+} influx into these cells in the absence of glucose, and a linear log-relation was seen between initial influx rates and MP values (Fig. 1). Similar results were obtained in energy-depleted cells, in the presence of metabolic inhibitors [4].

It is suggested that Ca^{2+} influx across the yeast-cell p.m. is mediated by channels which open when the negative membrane potential is below a threshold value. Evidently the influx was driven electrophoretically by MP.

References

1. Borst Pauwels, G.W.F. (1981) *Biochim. Biophys. Acta 650*, 88–127.
2. Eilam, Y., Lavi, H. & Grossowicz, N. (1985) *J. Gen. Microbiol. 131*, 623–629.
3. Ohsumi, Y. & Anraku, Y. (1983) *J. Biol. Chem. 258*, 5614–5617.
4. Eilam, Y. & Chernichovsky, D. (1987) *J. Gen. Microbiol. 133*, 1641–1649.
5. Eilam, Y. (1984) *Biochim. Biophys. Acta 769*, 601–610.
6. Eilam, Y., Lavi, H. & Grossowicz, N. (1985) *J. Gen. Microbiol. 131*, 2555–2564.

#ncA

COMMENTS related to

PLASMA MEMBRANE; Ca^{2+} MOVEMENT; MUSCLE AND NERVE

Comments on #**A-1**: E. Carafoli *et al.* - PLASMA MEMBRANE Ca^{2+} PUMP
#**A-2**: A. Vieyra - KIDNEY PROXIMAL TUBULE Ca^{2+} PUMP*

Question to E. Carafoli by W.E.J.M. Ghijsen.- Is there regulation of the Ca^{2+} pump by membrane potential besides the molecular regulation by CaM and cAMP-dependent protein kinase? **Reply** (unverified).- There is some debate about electroneutrality of the Ca^{2+} pump (if a $Ca^{2+}/2H^+$ exchanger is assumed). Membrane potential is observable with purified reconstituted pump protein. Other groups including Rasmussen at Yale claim electrogeneity of the pump in *intact* tissue. **Remark by R. Huch.**- In relation to the assumption that the Ca^{2+} pump of the p.m. contains two Ca^{2+}-binding loops, a Ca^{2+}-binding assay such as equilibrium or flow dialysis would be worth performing on the purified enzyme (**reply**: not yet tried); I reckon that such a study is easily done, and could give a reasonable indication whether there indeed exist 'EF-hand'-like domains within the primary structure, more convincingly than by making deductions from a constructed, average 'EF-hand' based on sequences of other Ca^{2+}-binding proteins.

F.R. Maxfield, to A. Vieyra.- Does the reversal of the pump occur to a physiologically significant degree in intact tissue? **Reply.**- I don't know. We demonstrated that ligands found inside the cell and/or in the peritubular space may promote the reversal of the enzyme. It is of interest that in *intact* hepatocytes (not in kidney cells!) the exchange reaction was demonstrable, and was stimulated by low ATP/ADP as in kidney p.m. vesicles.

Comments on #**A-3**: T.J. Rink - RECEPTOR-MEDIATED Ca ENTRY
#**A-4**: C. Enrich & W.H. Evans - LIVER BINDING PROTEINS

T.J. Rink, answering C.J. Kirk.- Depletion of internal Ca^{2+} could not account for the latency of the Ca^{2+} response to EGTA or Ni^{2+}: these are added only a few seconds before stimulation. **Question by I.S. Watts.**- In your studies with human platelets and fura-2 you have demonstrated that aggregatory agonists induce an influx of extracellular Ca^{2+}. Since platelets can undergo functional changes (*e.g.* shape, charge) independently of extracellular Ca^{2+} or indeed, as you have

---
* See p. 156 for a further Forum discussion item on #**A-2**

shown, intracellular Ca^{2+}, have you any evidence of a functional role in platelet aggregation or secretion for transmembrane Ca^{2+} influx? **Reply.-** This is a very difficult question to answer. However, the rapidity of the influx observed with agonists such as ADP would be consistent with the time course observed for aggregation induced by this agonist.

Rink, replying to C.W. Taylor.- As yet we have had no success with experiments to indicate whether, if Ca^{2+} entry occurs through e.r., the e.r. pool can be refilled with Mn^{2+} and then mobilized in response to further addition of agonist. **Rink, answering T.J. Brown** concerning the fura-2 response of parotid cells in suspension to carbachol (initial mobilization of $[Ca^{2+}]_i$, then Ca^{2+} influx).— We have not observed an oscillatory Ca^{2+} response to carbachol in single parotid cells, so comparison with the 2-phase response of suspensions is precluded.

U.T. Rüegg asked whether pharmacological tools had been used to differentiate between an ADP-stimulated influx into pools as seen with carbachol-stimulated platelets and a direct influx as seen with parotid cells. **Rink, in reply:** Benham et al. have found that Mn^{2+} ions easily enter platelets but not parotid cells when they are stimulated. **W.H. Evans, answering E. Carafoli.-** It was the conditions used in your laboratory that we used for detecting CaM-binding activity, so they should be appropriate. I agree that we are dealing merely with minor proteins.

Comments on #**A-5:** S.J. Smith & P.J. England - MYOFIBRIL STUDIES
        #**A-6:** B. Wilffert et al. - Ca^{2+} IN ISOLATED AORTA
        #**A-7:** J.P. Huggins & co-authors - PHOSPHOLAMBAN

Remark by U.T. Rüegg to P.J. England.- Presumably your compounds affect the dissociation constant for Ca^{2+}, with detrimental prolonged contraction in consequence. **Response.-** Prolongation of contraction will occur only if the rate-limiting step in relaxation is the dissociation of Ca^{2+} from Tn. Present evidence suggests that this may occur in some (but not all) circumstances. This can be tested only by the discovery of compounds which *selectively* increase the Ca^{2+}-sensitivity of cardiac myofibrils. **Remark by U.T. Rüegg.-** I understand that there are species differences for the compounds made by Boehringer, some being more active in human tissue - implying that the sites on the proteins must be different.

B. Wilffert, replying to S. Luciani.- We have no measurements to show whether there is significant uptake of R 56865 into the cytosol; concerning the possibility that it acts on Na^+/Ca^{2+} exchange, electrophysiological studies by a colleague gave no evidence of inhibition. **Replies to U.T. Rüegg.**- When using noradrenaline as a contractile agent we did not add a β-blocker; but we did find the concentration-response curve for noradrenaline to be unaffected by propranolol pre-treatment. The curves relating contraction to ^{45}Ca uptake are not really different in slope (as the legends indicate, Fig. 5 and Fig. 6 differ in the basis for calculating contraction). **Reply to Ghijsen.**- It is indeed conceivable that inactivation of the Ca^{2+} channel by increased $[Ca^{2+}]_i$ could account for the loss of correlation between Ca^{2+} uptake and muscle contraction at high $[K^+]$; but what I want to stress is the usefulness of the methodology. **J.P. Huggins, answering M.B. Vallotton.**- We have not considered the interesting possibility that, in our studies, Ca^{2+} was not acting directly on the fragmentation and reassembly of the e.r., but was affecting e.r. indirectly by acting on the cytoskeleton, which serves as a scaffold for the e.r. and, when disrupted, can no longer support the e.r. which then fragments.

Comments on **#ncA.1:** H.J. Hilderson & co-authors – EXCIMER STUDIES
  #ncA.2: S. Lotersztajn et al. – GLUCAGON & Ca^{2+} PUMP
  #ncA.3: M. Dobrota et al. – RARE–EARTH CATION EFFECTS

H.J. Hilderson, replying to A.P.R. Theuvenet.- The effects we measure are indeed dependent on the phospholipid composition of the artificial membrane; it is a major goal of our research to investigate the interplay of all different phospholipid components. **S. Lotersztajn, answering Rüegg:** we don't yet know whether the peptide circulates; one of its actions is to increase phosphorylation of liver phosphorylase. **N. Crawford asked** whether the modulation of the Ca^{2+} pump by GTPγs had been examined after pre-treatment of the vesicles with GDPβs. **Reply.**- We have some confusing results with GDPβs since it must be in high concentration to block the effect of GTPγs; at 1 mM it inhibits the Ca^{2+} pump by itself.

Comment by C.A. Pasternak to M. Dobrota.- A 'biphasic' response like you described is also seen with divalent cations such as Zn^{2+} and Cd^{2+}.

Comments on **#ncA.4:** M. Spedding – CHANNEL (IN)ACTIVATORS
  #ncA.6: U.T. Rüegg – FLUXES etc.; *likewise* #ncA.5 *

Spedding, answering L. Ver Donck *re* fluspirilene levels used (cf. palmitoylcarnitine studies): 10 nM and 0.2 μM in smooth-muscle and myocyte experiments respectively. **Reply to M. Caulfield.**- Little is known about effects of acylcarnitines

* partly on p. 124

on Ca channels in neurones; but they may be released from endothelium of cerebral vasculature in ischaemia and may contribute to neurotoxic Ca influx. **Reply to E. Johnson.-** In studies with Ca antagonists we usually administered them i.p., but at least some are also effective orally; we have not tried intracerebroventricular administration. Concerning model systems: (1) (**response to B.F. Trump**) in preliminary trials, *in vitro* effects were less striking than those in the animal models, but brain slices seemed promising (nimodipine seemed the best agent); (2) (**comparison with work by Pauwels;** cf. #E-3) results for the protective effects of Ca-channel antagonists in ischaemic gerbil hippocampus need to be reconciled with those in KCN-treated hippocampal slices; (3) (**reply to M. Caulfield**) in the gerbil model, Bay K 8644 (which causes 5-HT release) does cause some damage to the hippocampus, but less than with ischaemia, and as we gave it i.p. there may not have been adequate access to the CNS. **P.H. Cobbold remarked** that, in the ischaemic/reperfused hippocampus, maybe the protective effect of Ca-antagonists might in part be due to their ATP-sparing action and consequent depression of spontaneous electrical activity.

T.J. Brown asked Rüegg (reply: No!) whether with high K^+ $[Ca^{2+}]_i$ rose in Ca^{2+}-replete A_7r_5 cells. **Rüegg replied** to a question about pre-incubation with PMA: none necessary to see effects on K^+-induced influx, but 1-2 min needed if influx is receptor-activated; with <10 min as adopted, no desensitization to the PMA effect was seen. **Reply to M.B. Vallotton,** who mentioned reports of PMA-induced contraction in vascular myocytes.- We saw only slow and slight contraction. **Question by L. Ver Donck,** *re* the ROC $^{45}Ca^{2+}$ uptake protocol where the key feature was DHP-insensitivity which some VOC's/POC's (e.g.'T-type' channels) also show: did AVP depolarize the cultured cells? **Reply:** No! It hyperpolarizes them (due to Ca^{2+}-activated K^+ channels?); hence no POC involvement.

Comments on #**ncA.12:** W.E.J.M. Ghijsen - SYNAPTOSOMAL PHENOMENA

O.H. Petersen, to Ghijsen.- Your secretion data seem not to match the Ca^{2+} influx curve which, rather than manifesting a plateau even at ~10 mM K^+, seemed to show a very considerable increase in Ca^{2+}-dependent amino acid secretion in the 10-30 mM K^+ range. **Reply.-** This apparent discrepancy is probably explained by experimental difficulties appearing around the stimulation threshold for both Ca^{2+} uptake and Ca^{2+}-dependent amino acid release (between 10 mM and 20 mM K^+). My point was the close correlation between the optimal activation of both phenomena at moderate, physiologically relevant, depolarization levels. **P.W. Pauwels commented** on the possible true physiological role of the Ca^{2+}-independent Glu uptake, in the light of the Na^+-dependent Glu uptake which may function under pathophysiological conditions.

SOME LITERATURE PERTINENT TO #A THEMES, *noted by Senior Editor*
- with **bold type** for test material or other 'keyword'

Not particularly muscle or nerve

'ATP-driven **pumps** and related transport: calcium, proton
and potassium pumps'.- Fleischer, S. & Fleischer, B. (1988)
Meth. Enzymol. 157 [Biomembranes, Pt. Q], 678 pp.

'Calcium channels'. - Petersen, O.H. (1988) *Nature 336*, 528.-
Concerning 'real' receptor-operated calcium channels (**ROCC's**)
as discussed by Rink [in *Nature*; cf. #A-3], data taken from
a cited example do not conclusively demonstrate a ROCC. Direct
or G-protein-linked Ca^{2+}-channel opening should be distinguished
from second-messenger-operated calcium channels (**SMOCC's**). The
term ROCC should be reserved for the latter category. A
'proper' voltage-gated channel may, because of requisite modulation, operate as an SMOCC. It is as yet an unproven
hypothesis that Ca^{2+} entry needs $InsP_3$ and $InsP_4$ conjointly.
Both voltage-sensitive and -insensitive Ca^{2+} channels exist,
each type controllable by receptor-activation directly (via
G-protein) or indirectly (via, e.g., InsP's or diacylglycerol).
A particular cell type might lack one of the possible Ca^{2+}-gating
mechanisms.

'Calcium **antagonists** — Pharmacological and clinical research'
(1988) *Ann. N.Y. Acad. Sci. 522*, 802 pp.; also *ibid.:* 'Calcium
channels: Structure and function' (1989) in press.

'Calcium channel **ligands**'.- Triggle, D.J. & Janis, R.A.
(1987) *Annu. Rev. Pharmacol. Toxicol. 27*, 247-269.

'The role of calcium in **drug action**'.- Denborough, M.,
ed. (1987), Pergamon, Oxford, 200 pp.

'Lack of effect by prostaglandin $F_{2\alpha}$ and verapamil on
calcium uptake by isolated **corpora lutea** from pseudopregnant
rats': voltage-dependent channels evidently lacking; procedures
validated, e.g. A 23187 stimulates.- Lahav, M., Shariki-Sambag, K.
& Rennert, H. (1989) *Biochem. Pharmacol. 38*, 546-548.

'A calmodulin-dependent Ca^{2+}-activated K^+ channel in the
adipocyte plasma membrane'.- Pershadsingh, H.A., Gale, R.D.,
Delfert, D.M. & McDonald, J.M. (1986) *Biochem. Biophys. Res.
Comm. 135*, 934-9411

'Effects of exogenous fatty acids on calcium uptake by
brush-border membrane vesicles from rabbit **small intestine**'.-
Merrill, A.R., Proulx, P. & Szabo, A. (1986) *Biochim. Biophys.
Acta 855*, 337-344.

'Intraluminal calcium modulates lipid dynamics of rat
intestinal brush-border vesicles'.- Dudeja, P.K., Brasitus, T.A.,
Dahiya, R., Brown, M.D., Thomas, D. & Lau, K. (1987) *Am. J.
Physiol. 252*, G398-G403.

The elevating effect of glucagon on [cAMP]$_i$ in **hepatocytes** was augmented or, if cAMP degradation was blocked, was lowered by A 23187 - whose desensitization-like action was interpreted in terms of a functional uncoupling of glucagon-stimulated adenylate cyclase.- Irvine, F.J. & Houslay, M.D. (1988) *Biochem. Pharmacol. 37*, 2773-2780.

'Cellular and paracellular calcium transport in the rat **ileum......**': Vitamin D and dexamethasone investigated.-Karbach, U. & Rummel, W. (1987) *Naunyn-Schmiedeberg's Arch. Pharmacol. 336*, 117-124.

Calcium transport by rat **duodenal** villus and crypt baso-lateral membranes'.- Walters, J.R.F. & Weiser, M.M. (1987) *Am. J. Physiol. 252*, G170-G177.

'Sulfone analogues of taurine as modifiers of calcium uptake and protein phosphorylation in rat **retina**'.- Liebowitz, S.M., Lombardini, J.B. & Allen, C.I. (1989) *Biochem. Pharmacol. 38*, 399-406.

'**EF-hand** structure-domain of calcium-activated neutral proteinase (CANP) can bind Ca^{2+} ions'.- Minami, Y., Emori, Y., Kawasaki, H. & Suzuki, K. (1987) *J. Biochem. (Tokyo) 101*, 889-895.

'Effect of ruthenium red upon Ca^{2+} and Mg^{2+} uptake in ***Saccharomyces cerevisiae.*** Comparison with the effect of La^{3+}.'-van der Pal, R.H.M., Belde, P.J.M., Theuvenet, A.P.R., Peters, P.H.J. & Borst-Pauwels, G.W.F.H. (1987) *Biochim. Biophys. Acta 902*, 19-23.

'**Electron probe** analysis, **X-ray** mapping and **electron energy-loss** spectroscopy of calcium, magnesium and monovalent ions in log-phase and in dividing *E. coli* B cells'.- Chang, C.F., Shuman, H. & Somlyo, A.P. (1986) *J. Bact. 167*, 935-939.

Muscle and nerve (Some cardiovascular refs. in #ncE also)

'Voltage-dependent calcium **channels** of excitable membranes'.-Stanfield, P.R. (1986) *Br. Med. Bull. 42*, 359-367.

'**Toxins** that affect voltage-dependent calcium channels'.-Hamilton, S.L. & Perez, M. (1987) *Biochem. Pharmacol. 36*, 3325-3329.

'Calcium-permeable **channel** in sarcoplasmic reticulum of rabbit **skeletal muscle**'.- Sekiguchi, T., Kawahara, S. & Shimizu, H. (1987) *J. Biochem. (Tokyo) 102*, 307-312.

'**Purification** and reconstitution of the calcium release channel from **skeletal muscle**'.- Lai, F.A., Erickson, H.P., Rousseau, E., Liu, Q-Y. & Meissner, G. (1988) *Nature 331*, 315-319. POINTS FROM A FORUM ABSTRACT (Lai & Meissner; **cardiac** as well as skeletal muscle).- Using ryanodine as a channel-specific ligand, its receptor was purified on sucrose density gradients. The resulting complex (polypeptides of M_r ~400,000) was shown by e.m. etc. to be identical with 'feet' structures and with the channel for Ca^{2+} release as induced by T-system depolarization.

'Characterization of Ca^{2+} uptake and release by vesicles of **skeletal-muscle** sarcoplasmic reticulum'.- McWhirter, J.M., Gould, G.W., East, J.M. & Lee, A.G. (1987) *Biochem. J. 245*, 731-738 (also 723-730 & 739-749).

Evidence that in **skeletal muscle** "InsP$_3$ can modulate Ca^{2+} channels of transverse tubules from plasma membrane", rather than affecting exclusively the s.r. membrane.- Vilven, J. & Coronado, R. (1988) *Nature 336*, 587-589.

'Increase in calcium sensitivity of **cardiac myofibrils** contributes to the cardiotonic action of sulmazole'; stereo-isomers compared, and **smooth muscle** studied too.- van Meel, J.C.A., Zimmermann, R., Diederen, W., Erdman, E. & Mrwa, U. (1988) *Biochem. Pharmacol. 37*, 213-220.

With a DHP affinity column, **binding** of CaM and of troponins (**skeletal** and **cardiac**) was investigated, and the site found to be in the amino-terminal domain.- Bostrom, S.L., Westerlund, C., Rochester, S. & Vogel, H.J. (1988) *Biochem. Pharmacol. 37*,3723-3728.

Amiloride effects (pH-dependent and non-specific, but can be informative) on the Na^+/Ca^{2+} exchanger of **cardiac** sarco-lemmal vesicles: evidence for 'sided-ness' of the drug action. - Debetto, P., Luciani, S., Tessari, M., Floreani, F. & Carpenedo, F. (1989) *Biochem. Pharmacol. 38*, 1137-1145 (cf. #ncA.8, this vol.).

'Regulation of **blood flow**' - (1988) *Biochem. Soc. Trans. 16*, 479-505: incl. 'Histamine and a guanyl nucleotide modulate calcium accumulation in a microsomal fraction from pig aortic smooth muscle'(Blayney, L. & Newby, A.C., 479-480) and 'Calcium channels in smooth muscle cells' (Bolton, T.B., et al., 492-493), and 'Ca^{2+} regulation of the thin filaments' (Marston et al., 494-497).

'Effect of **anaesthetics** on intracellular calcium ions in **muscle** (Mini-review)' - Wali, F. (1986) *Life Sci. 38*, 1441-1443.

'Intracellular calcium and **smooth muscle** contraction'. - Sommerville, I.E & Hartshorne, D.J. (1986) *Cell Calcium 7*, 353-364.

Anti-CaM drugs and ^{45}Ca transport in plasmalemmal vesicles from gastric **smooth muscle** - Lucchesi, P.A. & Scheid, C.R. (1988) *Cell Calcium 9*, 87-94.

*REVIEW by M. Spedding on **drugs and Ca^{2+} transport**:* ref. 2, p. 120.

'Beta-adrenergic agonists and cyclic AMP decrease intracellular resting free-calcium in ileum **smooth muscle**'.- Parker, I., Ito, Y., Kuriyama, H. & Miledi, R. (1987) *Proc. Roy. Soc. B 230*, 207-214.

In s.r. vesicles from **vascular smooth muscle** (not cardiac or skeletal) an $InsP_3$-gated Ca^{2+} channel is demonstrable, very different from the Ca^{2+}-gated channel of striated muscle s.r. in single-channel conductance and pharmacology.- Ehrlich, R.E. & Watras, J. (1988) *Nature 336*, 583-586.

'Morphine inhibition of calcium fluxes, neurotransmitter release and protein and lipid phosphorylation in brain **slices** and **synaptosomes**'.- Crowder, J.M., Norris, D.K. & Bradford, H.F. (1986) *Biochem. Pharmacol. 35*, 2501-2507.

'Adenosine-A_1 receptors are not coupled to Ca^{2+} release in rat brain **synaptosomes**'.- Garritsen, A., Ijzerman, A.P. & Soudun, W. (1989) *Biochem. Pharmacol. 38*, 693-695.

'Existence of a Ca^{2+}-dependent K^+ channel in **synaptic** membrane and postsynaptic density fractions isolated from canine cerebral cortex and cerebellum, as determined by apamin binding'.- Wu, R., Carlin, P., Sachs, L. & Siekevitz, P. (1985) *Brain Res. 360*, 183-194.

Intra-**synaptosomal** $[Ca^{2+}]$ in guinea pig: resting level 0.4 μM by arsenazo III but much lower by quin-2.- Akerman, K.E.O., Heinonen, E., Kaila, K. & Scott, I.G. (1986) *Biochim. Biophys. Acta 858*, 275-284.

'Ruthenium red inhibits the voltage-dependent increase in cytosolic free calcium in cortical **synaptosomes** from guinea pig'.- Taipale, H.T., Kauppinen, R.A. & Komulainen, H. (1989) *Biochem. Pharmacol. 38*, 1109-1113.

'Immunological identification of the **synaptic** plasma membrane Na^+-Ca^{2+} exchanger'.- Barzilai, A., Spanier, R. & Rahamimoff, H. (1987) *J. Biol. Chem. 262*, 10315-10320.

Regulation of $[Ca^{2+}]_i$ by Mg^{2+} features in a Commentary, 'Magnesium ions in **cardiac** function. Regulator of ion channels and second messengers'.- White, R.E. & Hartzell, H.C. (1989) *Biochem. Pharmacol. 38*, 859-867.

———————————

#*Further comment on* #**A-2** (A. Vieyra - KIDNEY).- **W.E.J.M. Ghijsen asked**, concerning the Ca^{2+}-pump recycling with osmolytes present, where Na^+ also plays a role, whether the Na^+/Ca^{2+} exchange that is also occurring is involved too. **Vieyra's reply**.- Osmolytes appear without effect on this exchange, at least at pH 7.4. The reduction in Ca^{2+} uptake when Na^+ is added (efflux due to exchange) is unaffected by the presence of urea or TMA→O (or both), or sucrose. Maybe there would be an effect at pH 8.5 (not investigated).

———————————

Section #B

CYTOSOLIC AND SEQUESTERED Ca^{2+}; INOSITOL PHOSPHATES

#B-1

THE MAJOR 55 kDa CALCIUM-BINDING STAINS-ALL-POSITIVE PROTEIN IN MAMMALIAN CELLS IS IN THE RETICULOPLASM OF THE ENDOPLASMIC RETICULUM

[1]G.L.E. Koch[†], [1]D.R.J. Macer and [2]F.B.P. Wooding

[1]MRC Laboratory of Molecular Biology,
Hills Road, Cambridge CB2 2QH, U.K.

and [2]AFRC Institute of Animal Physiology,
Babraham, Cambridge CB2 4AT, U.K.

The major 55 kDa[*] Ca^{2+}- and Stains-All-binding protein in mammalian cells (CRP 55) was purified from reticuloplasm isolated from a plasmacytoma cell line. Immuno-e.m. showed that the protein is present throughout the e.r. and not in discrete sites or organelles. N-terminal analyses showed that the protein is similar or identical to the protein called calregulin. Analysis of fragments derived form the protein show that it has an unusually asymmetric sequence with a high density of acidic residues, Stains-All binding and Ca^{2+} binding being associated with the CRP 55-terminal region of the protein. Thus the major 'calsequestrin-like' protein in mammalian cells is not calsequestrin itself and is a component of the general reticuloplasm rather than a specific organelle.

There is growing support for the hypothesis that the intracellular calcium stores play an important role in signalling processes involving Ca^{2+} [1, 2]. The exact identity of these stores remains obscure. Originally a mitochondrial locus was proposed, but the currently accepted view is that the e.r. or an associated site is the most likely source of Ca^{2+} for intracellular signalling [3]. Recently it has been proposed that the direct source of Ca^{2+} for signalling is a novel organelle called the calciosome [4] which appears

[†] to whom any correspondence should be addressed
[*]*Terms & abbreviations.*- The authors' unit kDa has been kept for mol. wts., understood to be "apparent" (term used in the MS.), *i.e.* really M_r. Generally their term calcium has been altered to Ca^{2+}. For Stains-All consult Sigma catalogue; 'Coomassie Blue ' is the Brilliant Blue R; for CRP 55 see above; e.m., electron microscopy; e.r., endoplasmic reticulum.

to be a separate structure from the e.r. The main evidence for this is that the putative calciosome contains a protein which is calsequestrin-like on the basis of its staining with Stains-All and its apparent cross-reaction with antibody to calsequestrin. During our own studies on the luminal proteins of the e.r. [5, 6] we discovered a protein (CRP 55) which also showed a close resemblance to calsequestrin with respect to size, staining properties, acidic residue content and Ca^{2+} binding. Furthermore it was shown that the expression of this protein is regulated by agents which affect the Ca^{2+} level in cells [6].

In this study we have addressed the question of the possible identity of CRP 55 and calsequestrin, and show unequivocally that they are different proteins. CRP 55 is probably the same as the protein previously called CAB 63 or calregulin [7]. Immuno-e.m. shows that the protein is distributed throughout the general reticulum and not localized to specific regions or organelles.

MATERIALS AND METHODS

CRP 55 was purified from reticuloplasm isolated from MOP C315 cells [6]. The sample of isolated reticuloplasm was applied to a column of DEAE-Sephadex equilibrated with 10 mM Tris-HCl pH 7.5, and eluted with a linear gradient (10 to 500 mM) of the same buffer. Polyacrylamide gel electrophoresis showed that the protein eluted at ~350 mM Tris-HCl. The fractions were pooled, concentrated and applied to a column of Sephadex G150 equilibrated with 100 mM Tris-HCl pH 7.5. CRP 55 eluted at a volume corresponding to mol. wt. ~50 kDa, confirming that it is a monomer. The eluted protein is >95% pure by gel electrophoersis.

Affinity-purified antibodies to CRP 55 were prepared as described before [6] and tested by immunoblotting on whole cell extracts to confirm specificity towards the protein. Immuno-e.m. was carried out as described before [8].

RESULTS AND DISCUSSION

Analysis of the Stains-All-positive proteins in reticuloplasm isolated from the plasmacytoma cell line shows that such preparations contain two major proteins, 100 and 55 kDa (Fig. 1). These two proteins co-migrate with endoplasmin and CRP 55 respectively and are consistent with the previous demonstration that these two proteins are the major Ca^{2+}-binding proteins in such preparations. There was no evidence for another Stains-All-binding protein in the 55 kDa mol. wt. range. Fig. 1 also confirms that the 55 kDa species is CRP 55 since the purified protein clearly binds Stains-All.

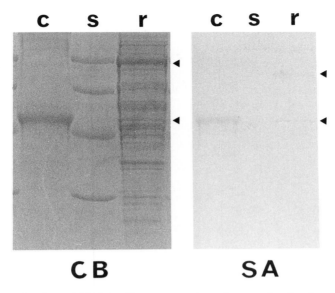

Fig. 1. Stains-All-binding proteins in reticuloplasm (**r**), purified [6] and fractionated on a SDA-polyacrylamide gel [9] which was stained successively with Stains-All (SA)[10], then Coomassie Blue (CB). The two major Stains-All-positive species co-migrate with the *arrowed* proteins on the CB-stained track. The differential swelling properties of the gel in the two staining media prevent superimposition of the two staining patterns; hence migration identity is based on their positions relative to the protein markers (**s**; **c** = cells).

The N-terminal sequence of purified CRP 55 was determined and found to be G-P-A-I-Y-F-K-E-Q-F. This does not show any homology with the sequence of calsequestrin but is clearly similar to that of the protein called calregulin [11], suggesting that CRP 55 and calregulin are the same.

Previous studies have indicated that CRP 55/calregulin is associated with the e.r. However, both these studies used immunofluorescence which lacks the resolution to distinguish between conventional e.r. and an e.r.-associated system such as the calciosome. Hence immuno-e.m. was used to examine the location of the protein more precisely. The results (Fig. 2) show that CRP 55 is distributed throughout the reticulum and that its distribution is indistinguishable from that of endoplasmin which is located in the reticuloplasm. Therefore these studies show that CRP 55 is not located in a special organelle such as the calciosome but is a component of the e.r. generally.

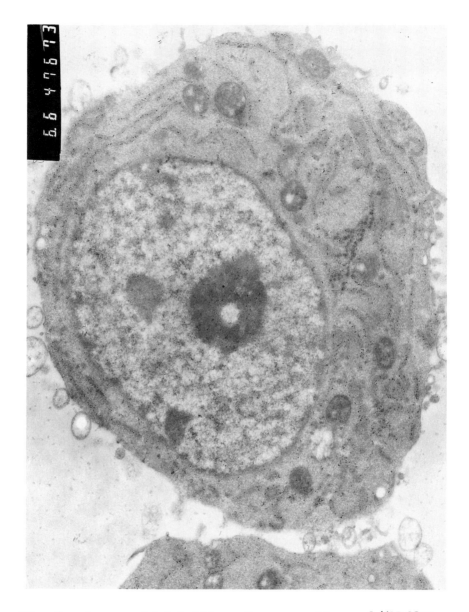

Fig. 2. Immunogold labelling of e.m. sections of MOP C315 cells for CRP 55. Note that the gold labelling is distributed throughout the e.r. and does not appear to be associated with any specific region or organelle within the e.r. Staining of parallel sections for endoplasmin yielded a very similar pattern of staining. × 12,500.

Fig. 3. Stains-All binding by cyanogen
bromide fragments of CRP 55, fractionated
on an SDS-polyacrylamide gel in duplicate.
Staining was with Stains-All (SA), then
with Coomassie Blue (CB). The strongest
SA-staining band, ~10 kDa, does not stain
with CB *(lowest pair of arrows)*. The
upper band in the doublet at 21 kDa
appears to be a partial cleavage product
since its yield is low. Thus the major
fragments appear to be 21, 14 and 10 kDa
respectively. The 14 kDa fragment has
the same N-terminal sequence as the
whole protein.

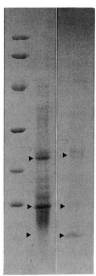

CB SA

The Stains-All staining pattern was likewise determined
for the fragments of CRP 55 produced after cyanogen bromide
cleavage. Coomassie Blue staining after gel separation showed
two major fragments, respectively 21 and 14 kDa (Fig. 3).
Stains-All staining revealed that the 14 kDa fragment does
not stain at all, the 21 kDa fragment stains relatively
weakly; it stained strongly a ~10 kDa fragment which is
not detected at all with Coomassie Blue.

These studies showed that the Stains-All binding regions
of the protein are not randomly distributed throughout the
sequence. Evidence that these regions correspond to the
C-terminus of the protein was obtained. An endogenous proteolytic fragment of 28 kDa was also isolated and purified
from reticuloplasm. This fragment and the 14 kDa CNBr
fragment had the same N-terminal amino acid sequence as
the intact protein (G-P-A-I-Y-F-K-E-Q-F), showing that they
were derived from the N-terminal region. The 28 kDa N-terminal
proteolytic fragment stained very weakly with Stains-All.
Thus most of the Stains-All binding capacity seems to be
associated with the C-terminal part of the protein. This
is consistent with the amino acid analyses of the protein
and the fragments. The protein contains a total of 35%
ASX + GLX*. In contrast, the N-terminal 28 kDa fragment

*X signifies acid + amide

region is estimated to have altogether 54% ASX + GLX. Since
the presence of negatively charged residues is the basis
of Stains-All staining, both lines of evidence show that
the C-terminus of the protein is more negatively charged
than the N-terminus. This is also consistent with the
fact that the N-terminal proteolytic fragment does not bind
significant amounts of Ca^{2+} in the overlay assay which has
been used to demonstrate Ca^{2+} binding by the intact protein
[6].

These studies clearly confirm that CRP 55 and calsequestrin
are different proteins and show that CRP 55 and the protein
called calregulin are probably the same. Furthermore they
formally establish that the protein is a general e.r. protein
and not a component of a specialized region of the e.r.
This distinguishes it from the proposed calsequestrin-like
protein described in other studies. Interestingly, we have
not obtained evidence for another 55 kDa Stains-All- and
Ca^{2+}-binding protein in the cells we have examined systematically. Thus if a true calsequestrin homologue exists in
these cells it must be present in very small quantities.

The demonstration that CRP 55 and calregulin are probably
the same raises the question of the physiological relevance
of the Ca^{2+} binding of these proteins. Calregulin has
been studied extensively with respect to its Ca^{2+}-binding
properties in the high-affinity range and found to contain
a single such site [11]. In the studies on CRP 55 emphasis
was placed on examining the low-affinity sites [6] since
these might be the relevant ones in terms of calcium storage.
Ultimately the crucial issue is the concentration of calcium
within the e.r. This is still an unknown quantity, but
estimates for the rough e.r. by electron microprobe analysis
indicate a value of ~3 mM within the luminal space [12].
Therefore the demonstration that calregulin/CRP 55 is within
this space suggests that it is the low-affinity sites that
are physiologically relevant.

These studies have also revealed an interesting property
in the stucture of CRP 55, namely that it appears to be
made up of two distinct regions: an N-terminal 'half' which
is neutral or even somewhat positively charged, and a C-terminal
'half' which is rich in negatively charged residues. The
isolation of a proteolytic fragment corresponding to the
N-terminal 27 kDa suggests that the two halves might actually
be folded into separate domains. One consequence of this
is that the Ca^{2+} binding activity is located in the C-terminal
domain. It also raises the possibility that the two domains
interact with each other in such a manner as to modulate

the Ca^{2+}-binding activity of the C-terminal domain. Such a protein could prove particularly suited to the release of its Ca^{2+} as a rapid burst through cooperative interactions within the protein molecule itself. This in turn would render the protein suitable for participation in Ca^{2+} signalling. We are therefore examining whether CRP 55 plays any special role in the release of Ca^{2+} from intracellular stores by agents such as $InsP_3$.

References

1.　Berridge, M.J. (1984) *Nature 312*, 315-321.
2.　Berridge, M.J. (1987) *Ann. Rev. Biochem. 56*, 159-193.
3.　Carafoli, E. (1987) *Ann. Rev. Biochem. 56*, 395-433.
4.　Volpe, P., Krause, K.H., Hashimoto, S., Zorzato, F., Pozzan, T., Meldolesi, J. & Lew, D.P. (1988) *Proc. Nat. Acad. Sci. 85*, 1091-1095.
5.　Koch, G.L.E. (1987) *J. Cell Sci. 87*, 491-492.
6.　Macer, D.R.J. & Koch, G.L.E. (1987) *J. Cell Sci. 87*, 61-70.
7.　Waisman, D.M., Salimath, B.P. & Anderson, M.J. (1985) *J. Biol. Chem. 260*, 1652-1660.
8.　Koch, G.L.E., Macer, D.R.J. & Wooding, F.B.P. (1988) *J. Cell Sci. 90*, 485-491.
9.　Laemmli, U.K. (1970) *Nature 227*, 680-685.
10.　Campbell, K.P., Maclennan, D.H. & Jorgensen, A.O. (1983) *J. Biol. Chem. 258*, 11267-11273.
11.　Somlyo, A.P., Bond, M. & Somlyo, A.V. (1985) *Nature 314*, 622-625.
12.　Khanna, N.C., Tokuda, M. & Waisman, D.M. (1986) *J. Biol. Chem. 261*, 8883-8887.

#B-2

CALCIUM-ION UPTAKE AND RELEASE BY
RAT-LIVER ENDOPLASMIC RETICULUM

A.P. Dawson, J.G. Comerford, P.J. Cullen
and D.V. Fulton

School of Biological Sciences,
University of East Anglia,
Norwich NR4 7TJ, U.K.

Microsomes derived from rat-liver e.r.*, in common with those from many other cell types, have an ATP-driven Ca^{2+} accumulation system associated with a Mg^{2+}-dependent, Ca^{2+}-stimulated ATPase. At 37° their initial Ca^{2+}-accumulation rate is 16 nmol Ca^{2+}/min per mg protein, with K_m 0.2 μM Ca^{2+} (determined with EGTA buffers present). The steady-state level of accumulation (oxalate absent) is ~20 nmol Ca^{2+}/mg at pH 7.0. Loading is decreased at more alkaline pH values, due to an increase in Ca^{2+} permeability.

A '36,000 g fraction' of microsomes releases very little Ca^{2+} in response to Ins(1,4,5)P_3, whereas microsomes sedimenting in the mitochondrial fraction respond markedly. Both fractions become very much more responsive after incubation with GTP/5% PEG - a treatment which itself causes some Ca^{2+} release. The GTP effects can be explained by GTP-dependent membrane fusion, furnishing much larger microsomal vesicles. Microsomes from other cell types show much more marked GTP-dependent Ca^{2+} release, without enhancement of the effects of InsP_3. This may reflect differences in the distribution of InsP_3 receptors on the e.r.

The InsP_3-responsive vesicles produced by GTP treatment of rat-liver microsomes provide a very convenient test system for investigating the effects of InsP's on Ca^{2+} movements.

The e.r., or perhaps a specialized part of it, is now generally recognized as an important intracellular store for Ca^{2+} [1]. Some or all of this store is mobilizable in response to polyphosphoinositide breakdown and the

---
Abbreviations.*- e.r., endoplasmic reticulum; p.m., plasma membrane(s); InsP's, inositol phosphates, <u>e.g.</u> InsP_3, usually signifying Ins(1,4,5)P_3; PEG, polyethylene glycol. **Mt: see text.

generation of InsP$_3$ following hormone-mediated stimulation of phospholipase C [2]. Many of the properties of this intracellular compartment have been very successfully studied using a wide variety of cell preparations where the p.m. has been permeabilized (by electric shock or digitonin) to expose the e.r. to the experimental medium. Such studies have great value, because the structure of the e.r. is probably largely maintained. They have, however, several limitations. Isolation of cells from whole tissues such as liver is rather time-consuming and expensive. Growth of cultured cells *in vitro* is also rather expensive, in some cases gives rather small quantities of experimental material and is, of course, limited to immortal cell lines rather than normal differentiated tissues.

The use of microsomes, isolated from tissue homogenates, as an experimental system suffers from the major problem that the gross structure of the e.r. is destroyed. However, it has the advantages of speed, economy and, for washed preparations, the absence of soluble enzymes which may metabolize InsP's. The latter can considerably add to the complications of interpreting data from permeabilized cells. Furthermore, a source such as a rat-liver homogenate produces very large quantities of material. As described below, the use of GTP to produce a fused microsome preparation from rat liver microsomes allows the latter to be used as a cheap, readily available and highly reproducible test system for compounds which modulate e.r. Ca^{2+} transport.

MICROSOME PREPARATIONS FROM RAT LIVER

Moore et al. [3] isolated Ca^{2+}-transporting microsomes from rat liver by homogenizing it in 0.25 M sucrose, followed by centrifugation of a post-mitochondrial supernatant at 105,000 **g** for 60 min. High-speed centrifugation (86,000 **g**, 30 min) is also used in the preparation method of Brattin et al. [4], their homogenizing medium being based on 80 mM KCl. In our hands, the quickest and most reproducible method of microsome preparation is based on that of Reinhart & Bygrave [5], in which rat liver is first homogenized in 250 mM sucrose/5 mM HEPES-KOH pH 7.0/1 mM dithiothreitol/ 0.5 mM EGTA, using a Dounce homogenizer. Following centrifugation at 1000 **g** for 5 min, the supernatant is centrifuged at 8000 **g** for 10 min (yielding the 'mitochondrial fraction' - **Mt**; *Ed.'s abbreviation*). The supernatant is then centrifuged at 36,000 **g** for 20 min (giving the '36,000 g fraction'). For use in Ca^{2+}-transport studies, the **Mt** and/or the 36,000 **g** fraction are washed by resuspension in pH 7.0 medium as above but with 10 mM KCl in place of EGTA. This wash removes EGTA, which can otherwise alter the free [Ca^{2+}]

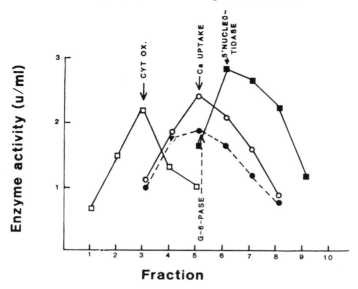

Fig. 1. Density-gradient centrifugation of 36,000 **g** fraction: 2.5 ml (75 mg protein), mixed with 10 ml 25% (w/v) Percoll in the medium given in the text (10 mM KCl, no EGTA). After centrifugation at 39,000 **g** for 60 min at 4°, ~1 ml fractions were collected ('1' being at the bottom of the gradient). For marker enzyme assays see [5, 7]. Ca^{2+} uptake was measured with ruthenium red (see legend to Fig. 2).

in experimental incubations; the KCl seems to aid reproducible vesicle formation.

Both the **Mt** and the 36,000 **g** fraction contain substantial ruthenium red-insensitive, ATP-driven Ca^{2+}-accumulating activity [5-7] and glucose-6-phosphatase (G-6-Pase) activity. While the **Mt** is, of course, substantially enriched in mitochondrial marker enzymes, it also contains the heavy microsomal fraction, which has properties differing somewhat from those of the 36,000 **g** fraction ([7] and see below). The 36,000 **g** fraction is substantially enriched in G-6-Pase activity and contains rather little cytochrome oxidase [5, 7]. Percoll gradient centrifugation of the 36,000 **g** fraction (Fig. 1) shows that ruthenium red-insensitive Ca^{2+}-accumulation parallels the G-6-Pase (microsomal) marker and is clearly separated from cytochrome oxidase (mitochondrial) and 5'-nucleotidase (p.m.) markers. However, Volpe _et al_. [8] have observed, using HL 60 cell microsomal fractions, that InsP₃-stimulated Ca^{2+} release does not correlate with G-6-Pase activity. Ca^{2+}-accumulating activity is largely lost on storage overnight at 0°, or on freezing.

Fig. 2. Ca^{2+} uptake
(at 37°) by 36,000 g
fraction (1.0 mg pro-
tein), in 1.0 ml con-
taining 100 mM KCl,
5 mM $MgCl_2$, 20 mM
HEPES-KOH pH 7.0, 1 mM
dithiothreitol, 3 μM
ruthenium red, 60 μM
Ca^{2+} (total) & 0.2 μCi
$^{45}Ca^{2+}$. After 5-min
pre-incubation, uptake
started by adding ATP
to 2.5 mM. Aliquots
(0.1 ml) filtered (filters
0.45 μm, pre-wetted with

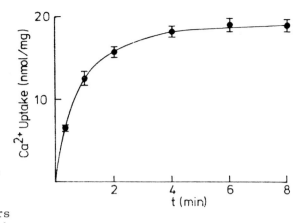

$KCl/MgCl_2$/HEPES-KOH buffer - <u>cf</u>. above), and washed on the filter
with 2 × 1 ml of this buffer at 0°. Data are the means ±S.E.M.
(bars) for 4 preparations from individual male rats.

MEASUREMENT OF Ca^{2+} ACCUMULATION

Routinely, two techniques are used for measuring ATP-
driven Ca^{2+}-accumulation by microsomes:- (a) $^{45}Ca^{2+}$ accumulation, followed by filtration (details: [6]); (b) Ca^{2+}-
sensitive electrode to measure changes in extra-vesicular
free Ca^{2+} [9]. $^{45}Ca^{2+}$ and filtration, coupled with measurement
of total Ca by atomic absorption, gives a precise measure
of intra-vesicular Ca^{2+} and, since strong Ca^{2+} buffers can
be used, closely controlled free $[Ca^{2+}]$ conditions. The
Ca^{2+}-sensitive electrode, on the other hand, does not give
a very precise value for intra-vesicular Ca^{2+}, since it
does not measure Ca^{2+} present in the vesicles initially,
does not allow very precise control of extra-vesicular free
$[Ca^{2+}]$ (since changes in the latter parameter are what are
observed) but does give a quick, easy and continuous read-out
of the progress curve, with good time-resolution.

Fig. 2 exemplifies a progress curve for Ca^{2+} accumulation,
measured using $^{45}Ca^{2+}$. In the absence of precipitating
anions such as oxalate, there is an initial, rapid Ca^{2+}
accumulation (reaching a plateau after ~5 min at 37°), half-
maximal at 0.2 μM free Ca^{2+} and having $\underline{V}_{max}$ 16 nmol Ca^{2+}/min
per mg protein. Measurement of $^{45}Ca^{2+}$ exchange at steady
state [10] or of Ca^{2+} efflux following removal of ATP by
washing [6] strongly suggests that the plateau level of
Ca^{2+} accumulation is a true steady state, where the accumulation
rate equals the rate of passive outward Ca^{2+} leakage. The
latter is strongly pH-dependent [6], so that at pH's >7.6

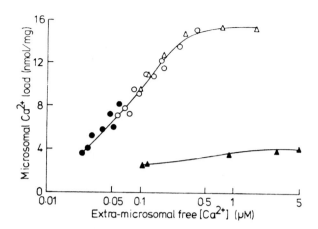

Fig. 3. Steady-state Ca^{2+} loading as a function of external free $[Ca^{2+}]$. Microsomes (1 mg protein/ml) were suspended in a medium containing 100 mM KCl, 20 mM HEPES-KOH – pH 6.8 (o, •, Δ) or 7.6 (▲), 5 mM $MgCl_2$, 0.15 µCi $^{45}Ca^{2+}$/ml, 0.5 µCi ^3H_2O/ml, 1 mM dithiothreitol, 1.5 µM ruthenium red, and EGTA – 18 µM (Δ, ▲), 36 µM (o) or 136 µM (•). Total $[Ca^{2+}]$ was adjusted over the range 10-30 µM for each EGTA concentration. Uptake was started after 5 min by adding ATP (to 2.5 mM); 1 ml aliquots removed after 5 and 8 min were immediately centrifuged for 2.5 min at 15,000 **g**. Pellet and supernatant were counted for ^{45}Ca and ^3H, and ^{45}Ca in the pellet was corrected for supernatant contamination using ^3H as an aqueous space marker. From ^{45}Ca distribution and total Ca^{2+} measurements, free and microsomal $[Ca^{2+}]$ were calculated. *From [6], by permission of the Biochemical Society©.*

rather little Ca^{2+} is accumulated, due to the fast outward Ca^{2+} leakage. The steady-state loading observed is a function of extra-vesicular free $[Ca^{2+}]$ (Fig. 3). At pH 7.0, it is half-maximal with ~0.1 µM external free $[Ca^{2+}]$.

Ca^{2+}-STIMULATED ATPase ACTIVITY

As might be expected, associated with the ATPase-driven Ca^{2+}-uptake, there is a Ca^{2+}-stimulated, Mg^{2+}-dependent ATPase activity [3, 10]. This is difficult to detect when measured under the precise conditions used for Ca^{2+} accumulation (Fig. 2), since intra-vesicular Ca^{2+} is inhibitory [11]. When A 23187 is present, to prevent a Ca^{2+} build-up in the vesicles, there is a readily detectable enhancement of ATPase activity on substituting Ca^{2+} for EGTA in the presence of Mg^{2+}

Table 1. ATPase activities of 36,000 g microsomes. Activities were measured in 1 ml containing 100 mM KCl, 20 mM HEPES–KOH pH 7.0, 5 mM $MgCl_2$, 1 µg oligomycin, 1 µg A 23187, and 0.5–1.0 mg protein. In addition, there was **either** 1 mM EGTA ([Ca^{2+}] <1 nM) or 62 µM Ca^{2+} + 60 µM EGTA ([Ca^{2+}] = 5 µM), to give values for Mg^{2+}-ATPase and Mg^{2+}/Ca^{2+}-ATPase activities. After pre-incubaton at 37° for 5 min, the reaction was started by adding ATP to 1 mM, and stopped after 10 min with trichloroacetic acid; P_i was measured by the Fiske-SubbaRow method. Ca^{2+}-stimulated ATPase activity is the paired difference between that in the presence of Mg^{2+} and that with both Mg^{2+} and Ca^{2+} present, for a given microsomal sample. Values are µmol P_i/mg per 10 min.

Condition (& n)	Mg^{2+} ATPase	Ca^{2+}-stimulated ATPase
Control (8)	0.29 ±0.013	0.16 ±0.01
36-h starved (8)	0.61 ±0.03**	0.10 ±0.016*

** $P < 0.001$;   * $P < 0.01$

([10] & Table 1). This stimulation is half-maximal at 0.19 µM free Ca^{2+}, and V_{max} is 160 nmol P_i produced/10 min per mg protein at 37°. The Ca^{2+}-stimulated ATPase is inhibited by vanadate [10] and has a phosphorylated intermediate of mol. wt. ~100,000 on SDS polyacrylamide gel electrophoresis [12, 13].

EFFECTS OF HORMONAL STATE

The pre-history of the rats used to prepare microsomes has a considerable impact on the Ca^{2+}-transporting activity of the vesicles produced. An early observation [3] was that microsomes from livers of female rats showed very much lower Ca^{2+}-accumulating activity than those from male rats. Glucagon injection [5, 14] and glucocorticoids [15] enhance Ca^{2+} uptake of subsequently isolated microsomes, while starvation significantly decreases it [16, 17]. Starvation also significantly decreases the Ca^{2+}-stimulated ATPase activity (Table 1), while increasing the Mg^{2+}-dependent activity measured in the absence of Ca^{2+}.

MEASUREMENT OF Ca^{2+} RELEASE FROM RAT–LIVER MICROSOMES

While it was shown at an early stage that $InsP_3$ could release Ca^{2+} from intracellular stores in permeabilized cell preparations from rat liver [18], attempts to measure Ca^{2+} release from rat-liver microsomal fractions were initially unsuccessful [18]. Finally, however, an effect of $InsP_3$ was found using **Mt** microsomes [7, 18]. Traces shown in Fig. 4, obtained with a Ca^{2+}-sensitive electrode, are typical

Fig. 4. InsP₃-stimulated
Ca²⁺ release (addition at **I**;
1 µM) from microsomal frac-
tions: (**a**) **Mt**, (**b**) 36,000 g
fraction; respectively 2.5
and 0.7 mg protein/ml medium,
added as marked (**Mc**, **Mi**).
Traces obtained by Ca²⁺-sensi-
tive electrode. Medium (30°):
150 mM sucrose, 50 mM KCl,
10 mM HEPES pH 7.0, 5% (w/v)
PEG, 5 mM ATP, 2 mM MgCl₂,
10 mM creatine phosphate,
0.1 mg/ml creatine kinase,
1 µg/ml oligomycin, 1 mM dithiothreitol & 2 µM ruthenium red.

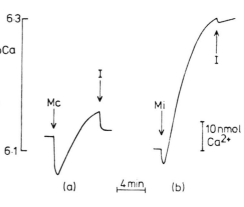

of those obtained from **Mt** and 36,000 g fraction microsomes.
The effect of InsP₃ on the latter is just about detectable,
given a degree of optimism, but a quite readily measurable
amount of Ca²⁺ is released (0.6-1.0 nmol/mg protein) on
adding InsP₃ to **Mt** microsomes. Even so, the amount of
Ca²⁺ released is a rather small percentage (<10%) of the
total intravesicular Ca²⁺, making detection of the effect
using ⁴⁵Ca and the filtration assay very difficult. In
the latter case, sampling, washing and counting inevitably
introduce errors of ~±3% at best.

Using the **Mt**, it was possible to show that the effect
of InsP₃ was half-maximal at ~0.5 µM, with maximal release
at ≤1 µM.

EFFECTS OF GTP ON Ca²⁺ RELEASE

We found that GTP, at quite low concentrations, had
very dramatic effects on Ca²⁺ release from microsomal fractions
[9], with the characteristics now summarized.-
(1) GTP addition (20 µM) to microsomes after they have
accumulated Ca²⁺ causes a slow release of Ca²⁺.
(2) After GTP treatment the amount of Ca²⁺ releasable by
InsP₃ is greatly enhanced, in both **Mt** and 36,000 g microsomes.
(3) The effect of GTP on release is time-dependent and
cannot be produced by non-hydrolyzable GTP analogues; the
latter (GTP-γ-[S], GPPPNHP, GPPCP) are inhibitors of the
effect of GTP.
(4) The effect of GTP on isolated microsomes requires the
presence of 5% PEG (mol. wt. 8000) in the assay medium.
[Our assay medium already contained this (Fig. 4), since
we found that it stabilized the effect of InsP₃ to some
extent.]

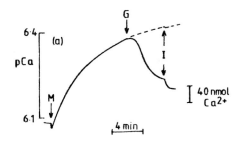

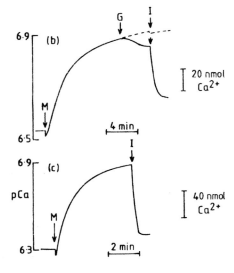

Fig. 5. Effect of GTP (20 µM, added at **G**), and of InsP$_3$ (1 µM, at **I**), on Ca^{2+} release from microsomes (**M**; the 36,000 **g** fraction) – 1.0 or (**c**) 1.5 mg protein/ml medium, which was otherwise as in Fig. 4. In (**a**) extra Ca^{2+} was added to give the desired pCa. Shown thus, ----: progress curves with no GTP addition. In (**c**) the GTP was present prior to adding the microsomes.

These effects of GTP on Ca^{2+} release are illustrated for relatively high (Fig. 5a) and much lower (Fig. 5b) free [Ca^{2+}]. When the ambient free [Ca^{2+}] (and the intravesicular Ca^{2+} loading) is high, GTP addition causes release of a major part of the accumulated Ca^{2+}. InsP$_3$ addition then causes release of a relatively small proportion of the remaining Ca^{2+} (although substantially higher than without GTP treatment). At low free [Ca^{2+}], the Ca^{2+} release caused by GTP is smaller (Fig. 5b), and the effect of InsP$_3$ is very large, up to 80% of the remaining intravesicular Ca^{2+} being released. (Fig. 5c is referred to later.)

Interpretation of the GTP effect

There is still debate on the nature of the effect of GTP on Ca^{2+} mobilization, which can occur with microsomes and permeabilized cells from many different sources (refs.: e.g. in [19]). In general effects on microsomes require PEG, while with permeabilized cells GTP is in many cases quite effective in the absence of PEG. Liver microsomes (and liver cells [20]) are at the moment unique, in that GTP, as well as causing a slow and limited Ca^{2+} release, dramatically enhances the effect of InsP$_3$. In other microsomal and cell systems investigated, GTP causes a very rapid and massive Ca^{2+} release, without significantly increasing the amount of Ca^{2+} mobilizable by InsP$_3$.

Our view of the effect of GTP is that it is a structural one. We have presented evidence from electron microscopy [21] and fluorescence resonance energy transfer [22] that GTP causes liver microsomal vesicles to fuse together, forming much larger structures. If the original e.r. contained many Ca^{2+}-ATPase molecules and very few $InsP_3$ receptor sites (or they were localized in particular regions), then on fragmentation during homogenization one could envisage that rather few of the vesicles produced would have the competence to both accumulate and release Ca^{2+} (Fig. 6). Fusion of small vesicles together to form larger structures would: (a) increase the Ca^{2+} pool size available to a given $InsP_3$ receptor; (b) possibly increase the passive Ca^{2+} permeability if leaky vesicles, or vesicles with open Ca^{2+} channels, were fused into the structure. Gill's group, working on N1E 115 neuroblastoma cells, have suggested that GTP causes connections to form between different Ca^{2+} pools [19]. In contrast to the fusion mechanism, these connections could be quite transient and might not result in any gross structural change, although the result in respect of Ca^{2+} movements would be similar. In any case, the requirement for PEG in the microsomal system appears to be due to the ability of PEG to cause aggregation of microsomes, allowing membranes to come into relatively close contact. Such a requirement would not be present in permeabilized cells, where the membranes will already be close together. Physiologically, it seems quite likely that the size of the intracellular hormone-sensitive Ca^{2+} pool could be regulated by this GTP-dependent mechanism.

TEST SYSTEM FOR INOSITOL PHOSPHATES (InsP's)

The normal protocol for testing InsP analogues is shown in Fig. 5c, where GTP is present initially in the incubation medium with low free $[Ca^{2+}]$. Following establishment of the steady state, an $InsP_3$ [e.g. $Ins(1,4,5)P_3$, the normal agent] is added. Thereby we find that $Ins(1,3,4)P_3$ is totally inactive in mobilizing Ca^{2+} and $Ins(1,2 \text{ cyclic } 4,5)P_3$ has ~3% of the activity of $Ins(1,4,5)P_3$. $Ins(2,4,5)P_3$ is ~10-fold less effective than $Ins(1,4,5)P_3$.

Acknowledgements

This work is supported by a grant from the Wellcome trust. P.J.C. thanks the SERC for a research studentship, and we are very grateful to Dr. R.F. Irvine (Inst. of Animal Physiology, Babraham, Cambridge) for generous gifts of inositol phosphates.

Fig. 6.
*Legend
opposite.*

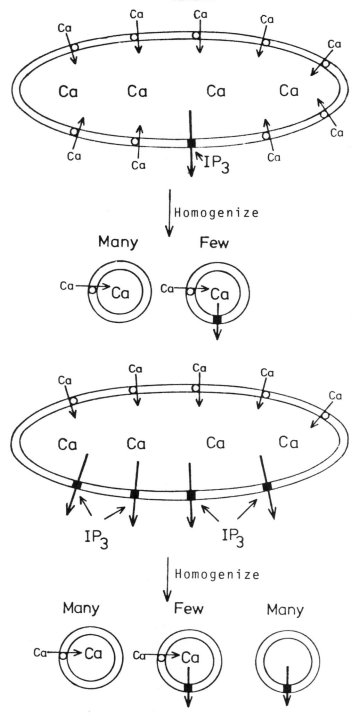

Fig. 6, *opposite.* Schemes for fragmentation of the e.r. during homogenization. In (**a**), it is envisaged that e.r. contains many Ca^{2+}-ATPase molecules (o) and very few $Ins(1,4,5)P_3$-binding sites (■), resulting in the production of a small minority of vesicles which can both accumulate and release Ca^{2+}. (**b**) A similar result can be obtained if ATPase molecules and the binding sites are in equal numbers but are spatially separated on the e.r. cisternae.

References

1. Somlyo, A.P., Bond, M. & Somlyo, A.V. (1985) *Nature 314*, 622-625.
2. Berridge, M.J. & Irvine, R.F. (1984) *Nature 312*, 315-321.
3. Moore, L., Chen, T., Knapp, H.R. & Landon, E.J. (1975) *J. Biol. Chem. 250*, 4562-4568.
4. Brattin, W.J., Waller, R.J. & Recknagel, R.O. (1982) *J. Biol. Chem. 257*, 10044-10051.
5. Reinhart, P.H. & Bygrave, F.L. (1981) *Biochem. J. 194*, 541-549.
6. Dawson, A.P. (1982) *Biochem. J. 206*, 73-79.
7. Dawson, A.P. & Irvine, R.F. (1984) *Biochem. Biophys. Res. Comm. 120*, 858-864.
8. Volpe, P., Krause, K.H., Hashimoto, S., Zorzato, F., Pozzan, T., Meldolesi, J. & Lew, D.P. (1988) *Proc. Nat. Acad. Sci. 85*, 1091-1095.
9. Dawson, A.P. (1985) *FEBS Lett. 185*, 147-150.
10. Dawson, A.P. & Fulton, D.V. (1983) *Biochem. J. 210*, 405-410.
11. Brattin, W.J. & Waller, R.L. (1983) *J. Biol. Chem. 258*, 6724-6729.
12. Heilman, C., Spamer, C. & Gerok, W. (1985) *J. Biol. Chem. 260*, 788-794.
13. Fleschner, C.R., Kraus-Friedmann, N. & Wibert, G.J. (1985) *Biochem. J. 226*, 839-845.
14. Andia-Waltenbaugh, A.M., Lam, A., Hummel, L. & Friedmann, N. (1980) *Biochim. Biophys. Acta 630*, 165-175.
15. Friedmann, N. & Johnson, F.D. (1980) *Life Sci. 27*, 837-842.
16. Dawson, A.P. (1984) *3rd Eur. Bioenerg. Congr. Short Rep.*, 529-530.
17. Osuna, C., Galvan, A. & Lucas, M. (1987) *FEBS Lett. 211*, 41-43.
18. Joseph, S.K., Thomas, A.P., Williams, R.J., Irvine, R.F. & Williamson, J.R. (1984) *J. Biol. Chem. 259*, 3077-3081.
19. Mullaney, J.M., Yu, M., Ghosh, T.K. & Gill, D.L. (1988) *Proc. Nat. Acad. Sci. 85*, 2499-2503.
20. Thomas, A.P. (1988) *J. Biol. Chem. 263*, 2704-2711.

21. Dawson, A.P., Hills, G. & Comerford, J.G. (1987) *Biochem. J.* *244*, 87-92.
22. Comerford, J.G. & Dawson, A.P. (1988) *Biochem. J. 249*, 89-93.

#B-3

USE OF MONOCLONAL ANTIBODIES WITH PERMEABILIZED PLATELETS AND ISOLATED MEMBRANES TO STUDY Ca^{2+} SEQUESTRATION AND ITS RELEASE

[1]N. Crawford, [2]N. Hack and [1]K.S. Authi

[1]Department of Biochemistry
and Cell Biology,
Hunterian Institute, Royal College
of Surgeons of England,
Lincoln's Inn Fields,
London WC2A 3PN, U.K.

[2]Department of Medicine,
University of Toronto,
King's College Street,
Toronto,
Canada, M58 1A8

Increases in platelet [Ca^{2+}]$_i$[*] *can occur during agonist-induced activation sequences which can trigger various Ca^{2+}-dependent intracellular processes. The [Ca^{2+}]$_i$ rise is due largely to release from intracellular storage sites, mainly e.r. but also ligand-gated Ca^{2+} influx channels in the platelet p.m. A panel of platelet-directed MAb's has been produced using as immunogen highly purified platelet membrane fractions. One MAb specifically immunoblots a 100 kDa e.r. polypeptide, and inhibits ⁴⁵Ca^{2+} uptake into saponin-treated cells or isolated membranes by action against an e.r. Ca^{2+} channel associated with the Ca^{2+}/Mg^{2+}-ATPase. This channel differs from that associated with the putative receptor for InsP$_3$. The various findings are considered in the context of the overall platelet Ca^{2+}-mobilizing scenario and the various mechanisms for maintaining Ca^{2+} homeostasis in the quiescent platelet.*

Calcium plays a key role in many important platelet functions. During agonist-induced activation sequences [Ca^{2+}]$_i$ may reach thresholds for triggering a variety of Ca^{2+}-dependent intracellular processes - some regulated by relatively low

---

[*][Ca^{2+}]$_i$, free intracellular (cytosolic) Ca^{2+} concentration.
Abbreviations.- Ab, antibody (MAb if monoclonal); CaM, calmodulin; InsP$_3$, inositol 1,4,5-trisphosphate; PAGE, polyacrylamide gel electrophoresis; e.r. (s.r.), endoplasmic (sarcoplasmic) reticulum; p.m., plasma membrane; LDH, lactate dehydrogenase. GPIIb and GPIIIa are the most prominent glycoproteins in platelet membranes; formation of the heterodimer GPIIb/IIIa is Ca^{2+}-dependent.

$[Ca^{2+}]_i$ through CaM action, whilst others may be CaM-insensitive and operational at higher $[Ca^{2+}]_i$ thresholds. Ca^{2+}-regulated intracellular enzymes and functions (roles as indicated) include the following.-

1. Myosin light-chain kinase.- Actin/myosin interactions.
2. Protein kinase-C.- Secretion, feedback regulation.
3. Ca^{2+}-dependent proteases.- Cytoskeletal modifications.
4. Microtubule assembly/disassembly.- Cell shape maintenance.
5. Phospholipase-A_2.- Eicosanoid synthesis.
6. Cyclic nucleotide phosphodiesterase.- cAMP (?cGMP) regulation.
7. Glycogen phosphorylase.- Energy supply.
8. Ca^{2+}/Mg^{2+}-ATPase.- Control of $[Ca^{2+}]_i$.

Some of these processes are manifest immediately following the action of an agonist at the p.m. Others may appear later in the response sequence. In the resting or quiescent platelet $[Ca^{2+}]_i$ is low (reckoned to be ~50-100 nM). Elevations can vary with different activators, but values of ~1-2 μM have been recorded with fluorescent dye probes, which allow a continuous record of $[Ca^{2+}]_i$ status [see opening art. by A.K. Campbell - *Ed.*].

CALCIUM STORAGE, MOVEMENT AND HOMEOSTASIS IN PLATELETS

Amongst membrane-bound Ca^{2+} storage pools within the platelet are the serotonin storage granules (or dense bodies), the mitochondria and the e.r. The e.r. is the most likely contributor to $[Ca^{2+}]_i$ changes: it can both release Ca^{2+} into, and evacuate it from, the cytosolic compartment. The serotonin storage granules liberate their Ca^{2+} directly into the extracellular environment during exocytosis, and the mitochondrial mechanisms for mobilizing Ca^{2+} generally require higher concentrations than have been recorded during agonist-induced responses. - This suggests that the role of mitochondria in Ca^{2+} sequestration and release is normally less important, but they may operate in events involving membrane damage. An additional pool of mobilizable Ca^{2+} may, however, lie in the p.m.: a proportion of the Ca^{2+} may be surface-oriented at saturable sites in equilibrium with the extracellular environment, but the remainder could be available at the cytoplasmic face of the membrane. The contribution (if any) of the latter component to changes in $[Ca^{2+}]_i$ is as yet unknown. During activation some surface-associated Ca^{2+} is involved in GPIIb/IIIa complex expression which appears to be a requirement for optimum fibrinogen binding and aggregation.

For $[Ca^{2+}]_i$ homeostasis the platelet must be protected against external Ca^{2+}, which is ~10,000-fold higher. Whilst the red cell can deal with any adventitious leak of Ca^{2+} into its intracellular space by the action of an active Ca^{2+}/Mg^{2+}-ATPase-linked Ca^{2+}-extrusion pump in the p.m., there is at present some controversy about the existence of such an efflux pump in the platelet p.m. We and some others have failed to identify any components or functions of such a pump [1-3], whilst others have claimed its existence [4, 5].

Platelet activation by certain agonists (ADP and thrombin) is associated with Ca^{2+} entry into the cell, and there is now considerable evidence for the presence at the surface-membrane level of Ca^{2+}-influx channels which are in some way operationally linked to receptors for activating ligands [6, 7]. In the resting platelet, Ca^{2+}-influx appears to be restricted, but at least part of the $[Ca^{2+}]_i$ rise that occurs during agonist-induced responses is now believed to be attributable to receptor-operated Ca^{2+}-influx channels, not voltage-dependent. These channels may be analogous to the voltage-insensitive and mitogen-regulated movement of Ca^{2+} into lymphocytes in which internally generated second-messenger molecules appear to have a mediating role [8]. In the platelet these influx channels, though initiated by agonist-receptor interactions, do not appear to require continued receptor occupancy to maintain a transmembrane Ca^{2+} flux [9]. Recently voltage-insensitive channels selective for divalent cations have been demonstrated in isolated platelet membrane vesicles fused to a planar phospholipid bilayer [10]. These channels are activated by thrombin at the whole-cell level before membrane isolation, and appear to survive the subcellular fractionation.

A role for Na^+/Ca^{2+} exchange mechanisms in $[Ca^{2+}]_i$ regulation in platelets has not yet been clearly established. Such mechanisms in other cell types are electrogenic with the flux imbalance between the two cations providing the driving force. Increased Na^+ influx has been reported in platelets stimulated by agonists [11], but since $[Ca^{2+}]_i$ is unaffected by ouabain the exchange of extracellular Na^+ for intracellular Ca^{2+} is unlikely to be a significant mechanism in maintaining Ca^{2+} homeostasis.

Our continuing interest in platelet Ca^{2+} homeostasis has focussed on the role of the dense tubular membrane system (DTS) or platelet e.r., both as a site for sequestration of Ca^{2+} and as the principal storage organelle responsive to second-messenger-sensitive release mechanisms. The use of high-voltage continuous-flow electrophoresis to separate a mixed membrane fraction, from platelet homogenates, into well separated and highly purified subfractions of different cellular origin has allowed us to study Ca^{2+}-mobilizing processes in p.m. and intracellular membranes without cross-contaminant artefacts.

These intracellular membranes have an active uptake mechanism for Ca^{2+} which is associated with a Ca^{2+}/Mg^{2+}-ATPase activity. Like muscle s.r. the platelet pump activity can be enhanced by oxalate or phosphate acting as an intravesicular sink for Ca^{2+}. Although others have shown differences in Ca^{2+}-pumping enzymes between platelet membranes and muscle s.r. based upon structural distinctions after tryptic digestion [12], two features appear to be analogous: (1) the vesicle sequestration of Ca^{2+} is itself Ca^{2+}-dependent, and (2) the Ca^{2+} uptake appears to be associated with the phosphorylation of a 100 kDa polypeptide present both in the platelet intracellular membrane and in muscle s.r.

After pre-loading platelet intracellular membrane vesicles with $^{45}Ca^{2+}$, ~40-50% of the A23187-releasable Ca^{2+} pool can be liberated rapidly (in 30-60 sec) by $InsP_3$ at μM levels [13]. Thereafter Ca^{2+} is taken up again to restore the original steady-state levels. This re-uptake process is believed to be associated with the presence of an $InsP_3$-hydrolyzing 5-phosphatase having Km 5 μM for $InsP_3$ [13]. A similar release pattern of pre-stored $^{45}Ca^{2+}$ has been demonstrated when $InsP_3$ is introduced into the cytosol of saponin-permeabilized platelets [14].

Our use of monoclonal antibodies (MAb's).- In order to explore the structure/function relationship of this Ca^{2+}-sequestering ATPase and that of other important proteins and enzymes known to be present in platelet intracellular membranes, we have prepared a panel of MAb's, using highly purified intracellular membrane vesicles as immunogen. The media from successful fusions were screened for Ab production against intracellular membranes using an ELISA procedure, and the hybridomas generating positive supernatants were then screened by immunoblotting _vs._ membranes separated by SDS-PAGE. Hybridomas positive towards intracellular membranes were cloned by limiting dilution and injected into primed mice for ascites production. Within this Ab panel, one (PL/IM 430) strongly and consistently reacted in blots with a 100 kDa polypeptide present in gel bands from whole platelets, mixed membrane fractions and isolated intracellular membranes. In functional screening of this Ab against activities already known to reside in human platelet intracellular membranes, it inhibited markedly (~70%) the uptake of $^{45}Ca^{2+}$ into freshly isolated intracellular membrane vesicles [15]. The remainder of this article concerns the use of this Ab in further characterization of the Ca^{2+}-uptake property of these membranes and with some investigations of its action on Ca^{2+} mobilization at the whole-platelet level when the cell's surface membranes have been permeabilized by saponin.

Fig, 1. PAGE performed
on 1% SDS extracts of whole
platelets (lanes **a, c**) or
intracellular membranes (**b, d**).
The polypeptide bands were
stained with Coomassie Blue
(**a, b**) or transferred to nitrocellulose for immunostaining
with PL/IM 430 (**c, d**). See
[15] for procedures. The
100 kDa component is the
prominent band common to
c and **d**. A non-platelet-
directed IgG_1 was used in
the controls.

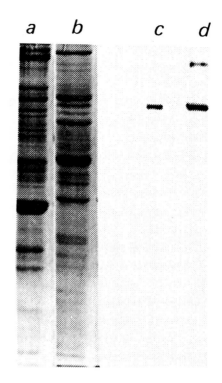

RESULTS

The PL/IM 430 IgG was isolated from ascites fluid or
concentrated culture supernatants by affinity chromatography
on a protein-A Sepharose column. The IgG isotype was characterized by double diffusion against a range of rabbit anti-mouse
IgG's and shown to be of the IgG class. Human platelet
mixed membranes were separated by sorbitol density-gradient
sedimentation from sonicates of fresh human platelets [16].
These mixed membranes were separated into subfractions representing surface and intracellular membranes by continuous-flow
electrophoresis [details: 17, 18].

Fig. 1 shows patterns for gel separation and for immunoblotting with the MAb supernatant. A prominent 100 kDa band
for both whole platelets and membrane extracts can be seen.

Possible effects of the MAb (20 µg/ml of membrane suspension) were investigated with intracellular membrane vesicles
from 5 different human platelet preparations. There was
no effect on Ca^{2+}/Mg^{2+}-ATPase activity, for which the means
(±S.D.; nmol/min per mg protein) were 101.4 ±6.4 with MAb
$vs.$ 101.8 ±12.3 in the controls. There was, however, a
strong inhibition (62%; $P < 0.001$) of $^{45}Ca^{2+}$ uptake, expressed

Fig. 2. Time course of $^{45}Ca^{2+}$ uptake by human platelet intra-cellular membrane vesicles incubated with 3 μM external $[Ca^{2+}]$ and 1 mM ATP in the presence of MAb PL/IM 430 (o, □) or non-immune IgG_1 (•, ■), measured in the presence (o, •) or absence (□, ■) of 2.5 mM potassium oxalate. See [13] for details of uptake buffer and measurement of sequestered $^{45}Ca^{2+}$.

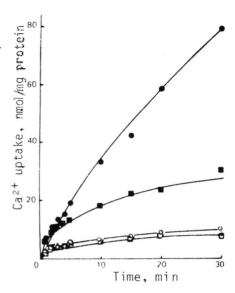

as nmol/min sequestered per mg protein at steady-state levels (12-min incubation): the value with MAb was 5.4 ±0.3, $vs.$ 14.1 ±0.7 in the controls. A similar investigation was made (Fig. 2) of the time course of Ca^{2+} uptake, with or without uptake enhancement by added oxalate. The inhibitions by the MAb were similar.

In further studies the phosphorylation of the 100 kDa polypeptide which had been shown to parallel Ca^{2+} uptake was measured during an uptake assay in the presence of the MAb. The MAb had no significant effect upon the rate or extent of formation of the phosphoprotein: the amount of ^{32}P incorporated into the platelet membrane protein (as nmol/mg) was 3.8 in the no-IgG control, 3.0 with non-immune IgG present, and 3.3 with the MAb present. Moreover, when $^{45}Ca^{2+}$ was measured in assays with ATP present in varying amount (0.1-2.0 mM), the MAb produced essentially the same inhibition (60-70%)

Using a protocol that we had earlier developed to demons-trate the role of $InsP_3$ as a Ca^{2+}-releasing agent operating on internal membrane-stored calcium, we investigated this mechanism in the presence of MAb PL/IM 430 (Fig. 3). After intracellular membrane vesicles had reached steady-state Ca^{2+} levels through incubation with $^{45}Ca^{2+}$ and ATP in the presence of either control MAb or PL/IM 430 (both at 20 μg/ml) $InsP_3$ was added and the ^{45}Ca content of the vesicles measured after Millipore membrane filtration at intervals thereafter. Since the MAb substantially inhibited Ca^{2+} sequestration,

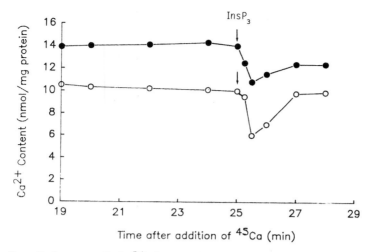

Fig. 3. Release of Ca^{2+} from platelet intracellular membrane vesicles by $InsP_3$. After $^{45}Ca^{2+}$ sequestration to steady-state levels (15 min), Ca^{2+} content was measured, and re-measured at intervals after addition of MAb PL/IM 430 (o) or a control MAb (●) to 20 µg/ml and, 10 min later, of $InsP_3$ (to 5 µM).

the steady-state Ca^{2+} levels reached in its presence are lower than with the control Ab. However, the magnitude of the $InsP_3$-induced Ca^{2+} release was similar in both Ab-containing preparations (Fig. 3).

Studies with saponin-permeabilized platelets

The effect of MAb PL/IM 430 was also investigated with whole platelets rendered permeable by pre-treatment with saponin. Using only ~10-12 µg/ml in the suspension, the membrane pores created by the detergent allow only low mol. wt. compounds to diffuse, and little LDH is lost from the cell. In fact we had earlier shown that under such permeabilizing conditions, $InsP_3$ (normally not membrane-penetrating) can be introduced directly into the platelet cytosol and that intracellular (non-mitochondrial) stored $^{45}Ca^{2+}$ is liberated (40-50% in 30-60 sec [19]). This release by $InsP_3$ is believed to be due to the presence of a receptor in the platelet e.r. controlling a Ca^{2+} channel.

With higher concentrations of saponin the pores become more permeable to high mol. wt. material, as evidenced by substantial losses of LDH. MAb's will then diffuse into the cell from the external medium. Accordingly, to study the effect of PL/IM 430 on Ca^{2+} sequestration and $InsP_3$-induced release at the whole-cell level, platelets were incubated

Fig. 4. Relationship between $^{45}Ca^{2+}$ uptake inhibition by 20 μg/ml MAb PL/IM 430 and the specific binding of ^{125}I-labelled PL/IM 430 (1 h, room temp.) in platelets permeabilized by the presence of different levels of saponin. Non-specific binding was measured at each level by adding 100-fold excess of unlabelled PL/IM 430 and subtracted. Permeabilization was monitored by cytosolic LDH release measured on separate aliquots taken after incubation with saponin for 6 min.

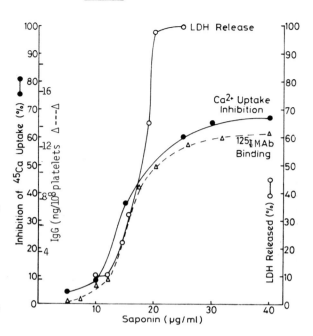

with $^{45}Ca^{2+}$ in the presence of saponin, at levels increasing from 5 to 40 μg/ml, and either the MAb (20 μg/ml) or control IgG (likewise 20 μg/ml). After incubating 20 min to ensure steady-state concentrations, aliquots of the cells were rapidly filtered through Millipore membranes, washed with ice-cold buffer and the $^{45}Ca^{2+}$ content determined. In other experiments similarly performed, aliquots were taken after 6 min of incubation and centrifuged (1200 **g**, 2 min) to prepare supernatants for determination of released LDH, which was expressed as % of the amount present in untreated cells solubilized in 0.2% Triton X-100. Fig. 4 shows that LDH release is slight with <12 μg/ml saponin but rises steeply and at ~20-25 μg saponin/ml is effectively total. The inhibition of Ca^{2+} uptake into intracellular stores rises with increasing saponin concentrations to a maximum inhibition at ~30 μg/ml.

To confirm that this inhibition relates to the entry of MAb into permeabilized cells, binding studies were performed in parallel using ^{125}I-labelled IgG (MAb PL/IM 430) throughout the saponin concentration range. Corrections were made for non-specific binding with controls containing 100-fold excess of unlabelled PL/IM 430. With non-permeabilized platelets specific binding was negligible; at the lower saponin concentrations. Specific binding with the permeabilized platelets increased to attain a plateau at ~20-25 μg saponin/ml. Fig. 4 includes the binding curve, which is seen to virtually follow that for the progressive inhibition of Ca^{2+} uptake.

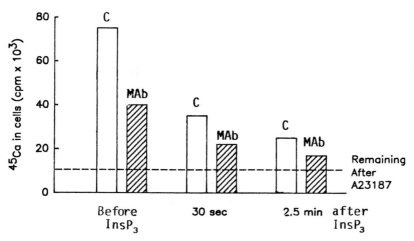

Fig. 5. $InsP_3$-induced Ca^{2+} release from intracellular sites in saponin-permeabilized platelets: effect of MAb PL/IM 430. Saponin-treated (25 µg/ml) platelets were allowed to sequester $^{45}Ca^{2+}$ for 25 min in the presence (20 µg/ml) of the MAb or of control (**C**) Ab. The reactions were terminated at 30 sec and 2.5 min by rapid filtration and washing, and the $^{45}Ca^{2+}$ was measured. Sequestered Ca^{2+} not released by 5 µM A23187 shown---.

To investigate the effect of $InsP_3$ on these saponin-treated platelets, they were allowed to sequester $^{45}Ca^{2+}$ in the presence of saponin and either PL/IM 430 or a control Ab. After 25 min $InsP_3$ was added (to 15 µM) and at timed intervals the incubations were terminated by rapid filtration (Millipore) and washing with ice-cold buffer. With control Ab the $InsP_3$ addition rapidly decreased the Ca^{2+} content, within 60 sec and maximally at 2-3 min; then there was slow re-accumulation (Fig. 5). With PL/IM 430 present the Ca^{2+} uptake into intracellular stores is inhibited, the steady-state levels attaining only 55% of the controls, but $InsP_3$ addition induces a similar release event, its time course resembling that with control Ab. With either Ab present, the $InsP_3$-induced release is ~60-70% of the intravesicular A23187-releasable Ca^{2+}. The half-maximal Ca^{2+} release values with $InsP_3$ were respectively 1.4, 1.8 and 0.5 µM measured in the presence of the MAb, the control IgG, and no Ab.

Investigation of the 100 kDa polypeptide present in solubilized human platelet intracellular membranes was undertaken in a further study after immunoaffinity purification on a Sepharose 4B-PL/IM 430 column. The purified protein was then subjected to digestion with bovine pancreatic trypsin (5 min, 30°). Fig. 6(a) shows the SDS gel pattern of the

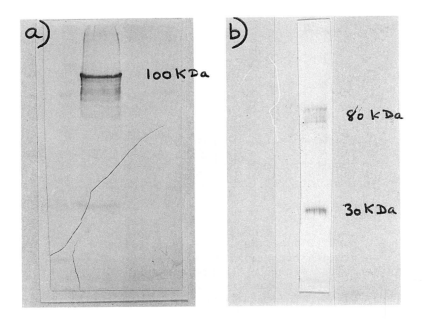

Fig. 6. SDS-PAGE of (**a**) affinity-purified 100 kDa polypeptide immunoblotted with PL/IM 430 after transfer to nitrocellulose, and (**b**) the product after trypsin, similarly immunoblotted.

purified protein immunoblotted with PL/IM 430, and Fig. 6(b) the pattern after trypsin treatment and similar blotting. As shown, a major component of the digests which could be immunoblotted with the PL/IM 430 Ab was a low mol. wt. digestion product of ~30 kDa. This gave a strong immunoperoxidase staining, as also did a component, ~80 kDa, which migrated ahead of the 100 kDa band.

DISCUSSION

The opening survey dealt with the role of the platelet e.r. in Ca^{2+} mobilization, with vesicles prepared therefrom which show ATP-dependent Ca^{2+} pumping, and with $InsP_3$-induced release of the stored Ca^{2+}. PL/IM 430, one of our panel of MAb's to intracellular membrane antigens, inhibits Ca^{2+} uptake into the vesicles by as much as 70%. It appears not to affect the Ca^{2+}/Mg^{2+}-ATPase of the membrane, nor the phosphorylation of a 100 kDa polypeptide previously shown to be phosphorylated during Ca^{2+} uptake. The MAb strongly immunoblots this polypeptide in solubilized preparations of mixed membranes and intracellular membranes. Cross-reactivity with a similar

100 kDa polypeptide has also been demonstrated, by immunoblotting with solubilized preparations   of U937, WT46 and
endothelial cells (data not presented here), and some cross-
reactivity has been observed, although not consistently, with
a 100 kDa component present in rabbit muscle s.r.

The inhibition of $^{45}Ca^{2+}$ sequestration into an intracellular storage site has also been demonstrated, using platelets
permeabilized by saponin, which with increasing concentration
gives an inhibition curve that parallels the specific intracellular binding of ^{125}I-labelled Ab.   The Ab had no effect
on InsP$_3$-induced Ca^{2+} release by either intracellular membrane
vesicles or permeabilized whole cells which had been prelabelled
with $^{45}Ca^{2+}$ by an ATP-dependent uptake.   Clearly the action
of the Ab on the Ca^{2+} channel associated with ATP-dependent
Ca^{2+} uptake in no way interferes with the InsP$_3$-inducible
Ca^{2+} release channel present in the same intracellular membrane.

When the 100 kDa polypeptide was isolated and purified
from solubilized intracellular membranes by immunoaffinity
chromatography using Sepharose-linked PL/IM 430, tryptic
digestion revealed two main digestion products immunodetectable
in Western blots.   These proteolytic fragments, ~80 and
30 kDa, appear to closely resemble the fragments observed
by Fischer et al. [12] after tryptic digestion of a platelet
microsomal preparation.

In the latter report the 78 kDa and 25 kDa tryptic
fragments had no Ca^{2+}-transporting activity, and loss of
this activity correlated closely with the loss of the 100 kDa
pump protein during tryptic digestion.   In contrast, however,
the tryptic digestion fragments still retained some Ca^{2+}/Mg^{2+}-
ATPase activity.   From the immunoblotting patterns seen in
the present study with platelets after tryptic digestion
of the purified 100 kDa protein, one might suggest that
one or other of the tryptic fragments includes the Ca^{2+}-
translocating domain of the pump protein with a functional
requirement for high-affinity Ca^{2+} binding.   In the panel
of Ab's raised to platelet intracellular antigens, a number
of others gave good immunoblotting of the 100 kDa polypeptide
with no apparent effect on either Ca^{2+} transport or enzymic
functions.   Studies with these are in progress to investigate
further the structure-function relationships of this important
intracellular Ca^{2+}-mobilizing protein of the platelet e.r.

Acknowledgements

We thank the British Heart Foundation for support, Mr.
Andrew Sankar for technical assistance with some of the antibody
characterization, and Miss Heather Watson for her skilled
assistance with the preparation of this manuscript.

References

1. Hack, N., Croset, M. & Crawford, N. (1986) *Biochem. J.* *233*, 661-668.
2. Steiner, B. & Luscher, E.F. (1985) *Biochim. Biophys. Acta* *818*, 299-309.
3. Fauvel, J., Chap, H., Roques, V., Levy-Toledano, S. & Douste-Blazy, L. (1986) *Biochim. Biophys. Acta* *856*, 155-164.
4. Enouf, J., Bredoux, R., Bordeau, N. & Levy-Toledano, S. (1987) *J. Biol. Chem.* *262*, 9293-9297.
5. Enyedi, A., Sarkadi, B., Foldes-Papp, Z., Monostory, S. & Gardos, G. (1986) *J. Biol. Chem.* *261*, 9558-9563.
6. Rink, T.J. & Hallam, T.J. (1984) *Trends Biochem. Sci.* *9*, 215-219.
7. Sage, S.O. & Rink, T.J. (1986) *Biochem. Biophys. Res. Comm.* *136*, 1124-1129.
8. Kuno, M. & Gardner, P. (1987) *Nature* *326*, 301-304.
9. Moffat, K.J. & MacIntyre, E. (1987) *Thrombos. Haemostas.* *58*, 509.
10. Zschauer, A., van Breemen, C., Bahler, F.R. & Nelson, M.T. (1988) *Nature* *334*, 703-705.
11. Rengasamy, A., Soudabeh, S. & Feinberg, H. (1987) *Thrombos. Haemostas.* *57*, 337-340.
12. Fischer, T.H., Campbell, K.P. & White, G.C. (1985) *J. Biol. Chem.* *260*, 8996-9001.
13. Authi, K.S. & Crawford, N. (1985) *Biochem. J.* *230*, 247-253.
14. Authi, K.S., Hornby, E., Evenden, B.J. & Crawford, N. (1987) *FEBS Lett.* *213*, 95-101.
15. Hack, N., Wilkinson, M.J. & Crawford, N. (1988) *Biochem. J.* *250*, 355-361.
16. Menashi, S., Weintroub, H. & Crawford, N. (1981) *J. Biol. Chem.* *256*, 4095-4101.
17. Carey, F., Menashi, S. & Crawford, N. (1982) *Biochem. J.* *204*, 847-851.
18. Hack, N. & Crawford, N. (1984) *Biochem. J.* *222*, 235-246.
19. Authi, K.S., Evenden, B.J. & Crawford, N. (1986) *Biochem. J.* *233*, 709-718.

#B-4

THE CHARACTERIZATION OF INOSITOL TETRAKISPHOSPHATES IN WRK 1 MAMMARY TUMOUR CELLS

C.J. Kirk, R.H. Michell, A.J. Morris and C.J. Barker

Department of Biochemistry, University of Birmingham,
P.O. Box 363, Birmingham B15 2TT, U.K.

Receptor-mediated hydrolysis of PtdIns(4,5)P$_2$* and the consequent accumulation of the intracellular messengers Ins(1,-4,5)P$_3$ and diacylglycerol is a widespread mechanism of signal transduction in stimulated cells. Ins(1,4,5)P$_3$, which mobilizes Ca^{2+} from intracellular stores in such cells, may be metabolized to a variety of other inositol phosphates inside the cell. These include at least two inositol tetrakisphosphates. Many cells also contain inositol pentakisphosphate and inositol hexakisphosphate (phytate). This article describes the characterization of the inositol tetrakisphosphates which accumulate in hormone-stimulated WRK 1 rat mammary tumour cells.

Since Ins(1,4,5)P$_3$ was first recognized as an intracellular second messenger in 1983 [1], considerable effort has been directed towards elucidation of the further metabolism of this compound [e.g. 2-8]. Interest in this area derives partly from the need to characterize the mechanisms responsible for the removal of the Ins(1,4,5)P$_3$ message from the cell and partly from speculation that other phosphorylated inositol derivatives may act as messengers [9-11].

In addition to being dephosphorylated to free inositol, it is now clear that Ins(1,4,5)P$_3$ can be further phosphorylated to Ins(1,3,4,5)P$_4$ in stimulated cells [3, 4] by an enzyme that is regulated by Ca^{2+}/calmodulin [12]. In cellular homogenates, Ins(1,3,4,5)P$_4$ is hydrolyzed to Ins(1,4,5)P$_3$ by a 5-phosphatase [5] and this compound may be either degraded to free inositol or further phosphorylated to yield Ins(1,3,4,-6)P$_4$ [5, 6]. These pathways are summarized in Fig. 1.

**Abbreviations.*- Ins, InsP$_2$, InsP$_3$ and InsP$_4$ are *myo*inositol and its mono-, bis-, tris- and tetrakis-phosphates, with D-locants designated where appropriate. PtdIns(4,5)P$_2$, phosphatidylinositol 4,5-bisphosphate. Ins4P (e.g.) is inositol 4-phosphate.

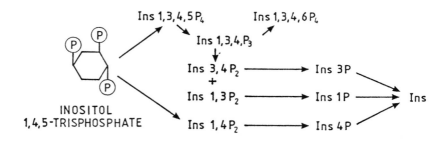

Fig. 1. Known pathways of Ins(1,4,5)P$_3$ metabolism. For abbreviations, here without parentheses (), see foot of first page. Evidence for the existence of these pathways is derived from several tissues (further details: text & [2-6]).

Since many of the pathways in Fig. 1 have been delineated in cellular homogenates, we have also sought to characterize the inositol phosphates in intact cells. This brief contribution describes the techniques used to identify the inositol tetrakisphosphates which accumulate in WRK 1 rat mammary tumour cells that have been stimulated with the hormone vasopressin.

INOSITOL TETRAKISPHOSPHATES IN WRK 1 CELLS

If WRK 1 cells are labelled to equilibrium with ^{3}H-inositol and stimulated with vasopressin as described previously [13], their accumulated inositol phosphates may be separated by HPLC (column: Whatman Partisphere WAX) [8]. Fig. 2 shows the InsP$_4$ regions from the chromatograms of extracts of WRK 1 cells which have been stimulated with vasopressin for various times. In control cells, prior to hormonal stimulation, two InsP$_4$ peaks (I & III were evident, neither of which co-chromatographed with a [^{32}P]Ins(1,-3,4,5)P$_4$ standard. After stimulation for 30 sec with vasopressin, a new peak (II) appeared which was coincident with this standard, but there were no other changes in the elution pattern of the inositol tetrakisphosphates. After 10 min of stimulation, all three inositol tetrakisphosphates were present in greater abundance than in control cells (Fig. 2).

The identification of inositol polyphosphates which have been separated by HPLC necessitates the use of several complementary analytical techniques. Co-chromatography with a particular inositol phosphate standard is not sufficiently unambiguous to identify a particular peak. We have used the methods of Stephens et al. [7] to investigate the identity of the three InsP$_4$ peaks shown in Fig. 2. These methods rely upon the ability of periodate, in appropriate conditions, to oxidize the inositol ring between adjacent hydroxylated C

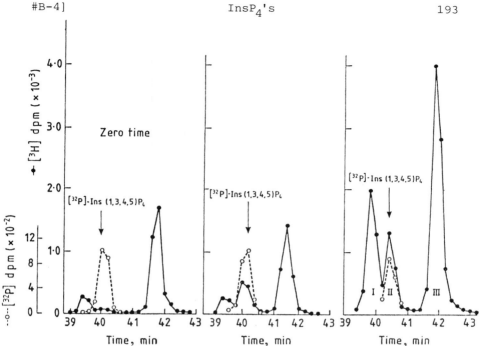

Fig. 2. InsP$_4$'s in WRK 1 cells, cultured with [^{3}H]inositol
(10 μCi/ml) for 5 days and stimulated with vasopressin
(400 nM) for the times shown [13]. Incubations (at 37°) were
quenched with a neutral, phenol-based extraction buffer [13].
InsP's were spiked with a [^{32}P-4,5]Ins(1,3,4,5)P$_4$ standard
and separated by HPLC on a Partisphere WAX column (125×45 mm,
Whatman) pre-equilibrated with deionized water. Elution was
at 1.0 ml/min with a gradient (A = 1.0 M ammonium formate brought
to pH 3.7 with phosphoric acid) and collection of 0.2 ml fractions for liquid scintillation counting. The Figure shows the
region of InsP$_4$'s (for **I**, **II** & **III**, see text). Elution gradient:
0-5 min, 0% A; 5-12 min, A increased linearly to 17%; 37-43 min,
40% **A**; 43-67 min, A→ 80%; 67-72 min, A→100%; 72-73 min, **A** decreased
linearly → 0%.

atoms. If inositol tetrakisphosphates are subjected to
periodate oxidation, followed by reduction with sodium boro-
hydride and dephosphorylation with alkaline phosphatase, those
susceptible to periodate attack will yield a variety of
open-chain polyols which may be separated by HPLC on a
Brownlee Polypore Pb$^+$ column [7] (Fig. 3).

When the [^{3}H]InsP$_4$ fractions were periodate-oxidized,
reduced and dephosphorylated as described above, peaks **I**
and **II** yielded [^{3}H]inositol (*i.e.* they were not oxidized
by periodate) and **III** yielded [^{3}H]iditol. As shown in
Fig. 3 and Table 1, 9 of the 16 possible InsP$_4$ isomers are

InsP$_4$ ISOMER	ENANTIOMERIC PAIR	RESULT OF PERIODATE OXIDATION
1,2,4,5	2,3,5,6	Resistant
1,3,4,6	—	
1,3,4,5	1,3,5,6	Resistant, but yield a pentitol if solo phosphate is first removed by alkaline phosphatase
1,2,4,6	2,3,4,6	
1,2,3,5	—	
2,4,5,6	—	
1,4,5,6	3,4,5,6	D-Iditol / L-Iditol
1,2,5,6	2,3,4,5	D-Glucitol / L-Glucitol
1,2,3,4	1,2,3,6	L-Altritol / D-Altritol

eg)

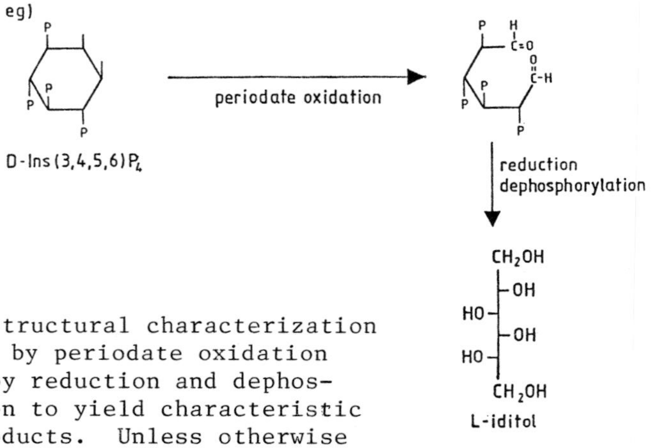

D-Ins(3,4,5,6)P$_4$

periodate oxidation →

reduction
dephosphorylation
↓

CH$_2$OH
—OH
HO—
—OH
HO—
CH$_2$OH

L-iditol

Fig. 3. Structural characterization of InsP$_4$'s by periodate oxidation followed by reduction and dephosphorylation to yield characteristic polyol products. Unless otherwise stated, inositol phosphate isomers are numbered according to the D-numbering system.

insensitive to periodate attack because they do not carry adjacent hydroxyl groups. Each one of these will yield inositol on periodate oxidation followed by reduction and dephosphorylation. Furthermore, since *myo*inositol has a plane of symmetry between C-2 and C-5, some of these InsP$_4$ isomers are enantiomers (Fig. 3). Hence additional approaches are required in order to identify the inositol tetrakisphosphates shown in Fig. 2.

Some periodate-resistant isomers of InsP$_4$ may be rendered periodate-sensitive by the removal of lone phosphate groups with alkaline phosphatase [7]. However, we have adopted an alternative approach in which samples of peaks **I** and **II** are subjected to non-selective ammoniacal dephosphorylation by heating with 10 M NH$_4$OH in a sealed tube at 110° for 72 h. This treatment results in an elimination of inositol

Table 1. Polyol products of periodate oxidation, reduction and dephosphorylation of *myo*-inositol bis-, tris- and tetrakis-phosphates. All isomers are numbered according to the D-numbering system.

Polyols formed		InsP$_4$	InsP$_3$	InsP$_2$
Hexitols:	Inositol	1,3,4,5	1,3,5	
	(not oxidized)	1,2,4,5	2,4,6	
		1,2,4,6		
		1,3,5,6		
		2,3,5,6		
		2,3,4,6		
		1,3,4,6		
		1,2,3,5		
		2,4,5,6		
	D-Iditol	1,4,5,6	1,4,5	
			1,4,6	
	L-Iditol	3,4,5,6	3,5,6	
			3,4,6	
	D-Glucitol	2,3,4,5	2,4,5	
			2,3,5	
	L-Glucitol	1,2,5,6	2,5,6	
			1,2,5	
	D-Altritol	1,2,3,6	1,3,6	
			1,2,4	
	L-Altritol	1,2,3,4	1,3,4	
			2,3,6	
Pentitols:	Xylitol		3,4,5	3,5
			4,5,6	4,6
			1,5,6	1,5
	Adonitol		1,2,3	1,3
	D-Arabitol		1,2,6	2,6
	L-Arabitol		2,3,4	2,4
Tetritols:	Erythritol			1,2
				2,3
	D-Threitol			4,5
				1,6
	L-Threitol			5,6
				3,4
Polyols destroyed:				*1,4; 2,5; 3,6*

monoester phosphate groups without rearrangement of the remaining, esterified phosphates [8]. In appropriate conditions, therefore, this technique yields a mixture of inositol monophosphate isomers which retain the configuration of their parent inositol polyphosphates. The products from the ammoniacal hydrolysis of **I** and **II** were analyzed by an HPLC system which resolves Ins2P, Ins5P and the enantiomeric pairs Ins1P/Ins3P and Ins4P/Ins6P [8]. The ^{3}H-labelled material co-eluting with the latter 3 peaks was observed in the ammoniacal hydrolysates from both peaks **I** and **II**, but neither contained radioactivity that co-eluted with Ins2P. Hence the InsP$_4$ isomers represented by peaks **I** and **II** cannot contain a monoesterified phosphate in the D-2 position.

As shown in Fig. 3 and Table 1, only 3 periodate-resistant InsP$_4$ isomers have no phosphate in the 2-position, *viz.* Ins(1,3,4,6)P$_4$, Ins(1,3,4,5)P$_4$ and Ins(1,3,5,6)P$_4$. The latter two isomers are an enantiomeric pair and are therefore inseparable on the non-chiral column used in these experiments. Consequently, peaks **I** and **II** must represent Ins(1,3,4,6)P$_4$ and either Ins(1,3,4,5)P$_4$ or Ins(1,3,5,6)P$_4$.

Peak II, as noted above and shown in Fig. 2, co-eluted with a [^{32}P]Ins(1,3,4,5)P$_4$ standard. The identity of this peak was confirmed by exposing it to the Ins(1,4,5)P$_3$/Ins(1,3,4,5)P$_4$ 5-phosphatase of human erythrocyte membranes in conditions which achieve complete dephosphorylation of an internal standard of [^{32}P]Ins(1,3,4,5)P$_4$ [5]. We previously showed that this enzyme will not attack authentic [^{3}H]Ins(1,3,4,6)P$_4$ prepared from rat liver [5]. After 5-phosphatase treatment **II** was entirely converted to an [^{3}H]InsP$_3$ which co-eluted on HPLC with an Ins(1,3,[^{32}P]4)P$_3$ standard. When this [^{3}H]InsP$_3$ was subjected to periodate oxidation, reduction and dephosphorylation as above, it yielded a compound which co-eluted with altritol on a Polypore Pb$^+$ HPLC column. Reference to Table 1 reveals that four InsP$_3$ isomers yield either D- or L-altritol following periodate treatment. The stereo-isomeric configuration of the altritol was determined using the L-polyol dehydrogenase of *Candida utilis* [7] which has an absolute specificity for the L-isomers of altritol and iditol: it catalyzed the oxidation of the [^{3}H]altritol derived from peak **II** to [^{3}H]psicose. Only two isomers of InsP$_3$ yield L-altritol after periodate attack, *viz.* Ins(1,3,4)P$_3$ and Ins(2,3,6)P$_3$. We have already shown that the InsP$_4$ in peak **II** has no phosphate in the 2-position, so the [^{3}H]InsP$_3$ arising by dephosphorylation of this material must be Ins(1,3,-4)P$_3$. Hence the InsP$_4$ in **II** may be unambiguously identified as Ins(1,3,4,5)P$_4$.

Peaks I and III.- As explained above, **I** is another periodate-insensitive isomer of InsP$_4$ without a phosphate group on the 2-carbon. Since it elutes separately it cannot be an enantiomer of Ins(1,3,4,5)P$_4$; hence it must be Ins(1,3,4,-6)P$_4$ (see Fig. 3 & Table 1). The InsP$_4$ in **III** (Fig. 2) yielded iditol following periodate oxidation, reduction and dephosphorylation as above. On treatment with the L-polyol dehydrogenase this was oxidized to sorbose at the same rate as authentic L-iditol. Thus the product of periodate treatment of **III** is L-iditol and this peak must be Ins(3,4,5,6)P$_4$ (see Fig. 3 & Table 1).

Thus we have identified the three InsP$_4$ isomers present in WRK1 cells as Ins(1,3,4,6)P$_4$ (peak **I**), Ins(1,3,4,5)P$_4$ (**II**) and Ins(3,4,5,6)P$_4$ (**III**). This latter isomer is the most abundant InsP$_4$ in both quiescent and vasopressin-stimulated cells. However, Ins(1,3,4,5)P$_4$ accumulates most rapidly following hormone stimulation, showing a 10-fold increase in concentration within 30 sec. After 10 min of stimulation, all 3 InsP$_4$ isomers accumulate in these cells.

Ins(1,3,4,5)P$_4$ accumulates rapidly in many stimulated cells as a result of 3-kinase action on the Ins(1,4,5)P$_3$ product of receptor-mediated PtdIns(4,5)P$_2$ hydrolysis [3, 4]. We have shown that Ins(1,3,4,6)P$_4$ can be formed by the action of a 6-kinase upon Ins(1,3,4)P$_3$ which arises as a hydrolysis product of Ins(1,3,4,5)P$_4$ [5, 6]. Ins(3,4,5,6)P$_4$ may be an intermediate in the synthesis of InsP$_5$ and InsP$_6$ which are present at high concentrations in a variety of cells [8, 11, 14]. However, its metabolic origin and likewise the significance of accumulation of this compound in stimulated cells remain obscure.

CONCLUSIONS

This article has described a variety of experimental approaches which we have used to identify the InsP$_4$'s which accumulate in WRK1 cells in culture. Although we have focused on the tetrakisphosphates, similar techniques may be used to characterize other inositol phosphates. In view of the wide range of such molecules which are now being found in living cells[*], it is vital that rigorous methods be adopted before structures are assigned to these compounds.

References

1. Streb, H., Irvine, R.F., Berridge, M.J. & Schulz, I. (1983) *Nature 306*, 67-69.
2. Shears, S.B., Storey, D.J., Morris, A.J., Cubitt, A.B., Michell, R.H. & Kirk, C.J. (1987) *Biochem. J. 242*, 393-402.
3. Batty, I.R., Nahorski, S.R. & Irvine, R.F. (1985) *Biochem. J. 232*, 211-215.

---

[*]Art. #B-7 (Fasolato et al.) is pertinent.- *Ed.*

4. Irvine, R.F., Letcher, A.J., Heslop, J.P. & Berridge, M.J.
    (1986) *Nature 320*, 631-634.
5. Shears, S.B., Parry, J.B., Tang, E.K.Y., Irvine, R.F.,
    Michell, R.H. & Kirk, C.J. (1987) *Biochem. J. 246*, 139-147.
6. Balla, T., Guillemette, G., Baukal, A.J. & Catt, K.J.
    (1987) *J. Biol. Chem. 262*, 9952-9955.
7. Stephens, L., Hawkins, P.T., Carter, N., Chahwala, S.B.,
    Morris, A.J., Whetton, A.D. & Downes, C.P. (1988)
    Biochem. J. 249, 271-282.
8. Stephens, L., Hawkins, P.T., Barker, C.J. & Downes, C.P.
    (1988) *Biochem. J. 253*, 721-733.
9. Irvine, R.F. & Moor, R.A. (1986) *Biochem. J. 240*, 917-920.
10. Morris, A.P., Gallacher, D.V., Irvine, R.F. &
    Petersen, O.H. (1987) *Nature 330*, 653-655.
11. Vallejo, M., Jackson, T., Lightman, S. & Hanley, M.R.
    (1987) *Nature 330*, 656-658.
12. Biden, T.J., Comte, M., Cox, J.A., & Wollheim, C.B.
    (1987) *J. Biol. Chem. 262*, 9437-9440.
13. Wong, N.S., Barker, C.J., Shears, S.B., Kirk, C.J. &
    Michell, R.H. (1988) *Biochem. J. 252*, 1-5.
14. Stephens, L.R., Hawkins, P.T., Morris, A.J. & Downes, C.P.
    (1988) *Biochem. J. 249*, 283-292.

#B-5

ACTIONS AND METABOLISM OF SYNTHETIC
INOSITOL TRISPHOSPHATE ANALOGUES

Colin W. Taylor and **Barry V.L. Potter**

Department of Zoology,   Department of Chemistry,
University of Cambridge, University of Leicester,
Downing Street,   Leicester LE1 7RH, U.K.
Cambridge CB2 3EJ, U.K.

Activation of many cell-surface receptors evokes both hydrolysis of phosphatidylinositol 4,5-bisphosphate to give $Ins(1,4,5)P_3$[†] and an increase in intracellular Ca^{2+} concentration. While $Ins(1,4,5)P_3$ is recognized as the intracellular messenger that stimulates release of Ca^{2+} from intracellular stores, its complex and rapid metabolism has confused interpretation of its possible roles in oscillatory changes in intracellular Ca^{2+} concentration and in controlling Ca^{2+} entry at the p.m.[†] $DL\text{-}Ins(1,4,5)P_3S_3$, in which each phosphate group of $DL\text{-}Ins(1,4,5)P_3$ is replaced by a phosphorothioate group, was synthesized as a potentially phosphatase-resistant analogue of $DL\text{-}Ins(1,4,5)P_3$. In a variety of cell types $DL\text{-}Ins(1,4,5)P_3S_3$ elicits the same maximal release of Ca^{2+} as $DL\text{-}Ins(1,4,5)P_3$, and half-maximal release occurs at a ~3-fold higher concentration, implying that $DL\text{-}Ins(1,4,5)P_3S_3$ is a full agonist and only 3-fold less potent than $DL\text{-}Ins(1,4,5)P_3$. There is a similar 3-fold difference in the affinities of $DL\text{-}Ins(1,4,5)P_3$ and $DL\text{-}Ins(1,4,5)P_3S_3$ for high-affinity $D\text{-}Ins(1,4,5)P_3$-binding sites of permeabilized hepatocytes. While the ability of $DL\text{-}Ins(1,4,5)P_3$ to mobilize Ca^{2+} is completely inactivated by either dephosphorylation to $Ins(1,4)P_2$ or phosphorylation to $Ins(1,3,4,5)P_4$, $DL\text{-}Ins(1,4,5)P_3S_3$ is resistant to these inactivation pathways. We suggest that $DL\text{-}Ins(1,4,5)P_3S_3$ will be useful as both an active stable analogue of the natural intracellular messenger and in allowing the direct actions of $Ins(1,4,5)P_3$ to be distinguished from those that require its metabolism.

Recognition that $D\text{-}Ins(1,4,5)P_3$ is a cytosolic messenger that mobilizes intracellular Ca^{2+} stores [1, 2] has provoked renewed interest in the chemistry of inositol derivatives.

[†]*Abbreviations:- see over; authors' system,* IP, *altered to* InsP.
N.B. Ring numbering by biochemists typically clockwise, by chemists (cf. Fig.1) typically anticlockwise. - *Ed.*

Two books [3, 4] provide useful introductions to the chemistry of these compounds, and an excellent review [5] describes their stereochemistry and nomenclature. The biological functions of inositol phosphates are the subjects of many reviews; the book edited by Putney [6] and a recent conference proceedings [7] provide a comprehensive introduction.

Recent synthetic work on inositol phosphates and analogues has addressed three distinct problems: (1) the preparation of suitably protected inositol derivatives which allow selective phosphorylation of the required hydroxyl groups; (2) improvements in the phosphorylation methodologies traditionally used for chemical phosphorylation, and especially the avoidance of the formation of 5-membered cyclic phosphates that is prevalent when phosphorylation of vicinal diols is attempted using the $P(V)^\dagger$ phosphorylation procedures; (3) the final deprotection of the phosphorylated synthetic intermediate.

Although inositol phosphates had been successfully synthesized before the recent revival of interest, the procedures used – especially the phosphorylation steps – were not entirely satisfactory. Most interest is now focused on D-Ins(1,4,5)P$_3$, but before 1987 no chemical synthesis was available for this compound: the first was reported by Ozaki et al. [8], followed by other groups [9-13] including ours [14]. The complex cellular metabolism of D-Ins(1,4,5)P$_3$, which is a substrate for both a specific 3-kinase [15] and a 5-phosphatase [16], has provoked interest in the chemical synthesis of the products of each of these reactions: Ins(1,3,4,5)P$_4$ [11, 17-19], its dephosphorylation product Ins(1,3,4)P$_3$ [18], and inositol mono- and bis-phosphates [17, 20]. Thus new methods of protection have been developed [8-14, 17-23], and the success of each of these syntheses has depended upon the development of two phosphorylation strategies: the use of tetrabenzyl pyrophosphate [10, 17, 18, 20, 21] and more recently of phosphite or $P(III)^\dagger$ reagents [9, 11-14].

We have used N,N-diisopropylamino-(2-cyanoethyl)chlorophosphine [24] to rapidly phosphitylate the vicinal diol [25] in the synthesis of Ins(4,5)P$_2$, and subsequently applied the method to the synthesis of DL-Ins(1,4,5)P$_3$ [14] and Ins(1,4)P$_2$ [26]. Perhaps more important is the application of this method to the synthesis of an analogue of DL-Ins(1,4,5)P$_3$, DL-Ins(1,4,5)P$_3$S$_3$ (Fig. 1: **2**) [27], wherein each phosphate group has been replaced by a phosphorothioate group. Such phosphoro-

---

†*Abbreviations (& see Figs.).*- Ins(1,4,5)P$_3$, *myo*-inositol 1,4,5-trisphosphate; Ins(1,4,5)P$_3$S$_3$, inositol 1,4,5-trisphosphorothioate; InsP$_2$, an inositol bisphosphate, typically Ins(4,5)P$_2$; p.m., plasma membrane; P(III) = tri-, P(V) = penta-valent P.

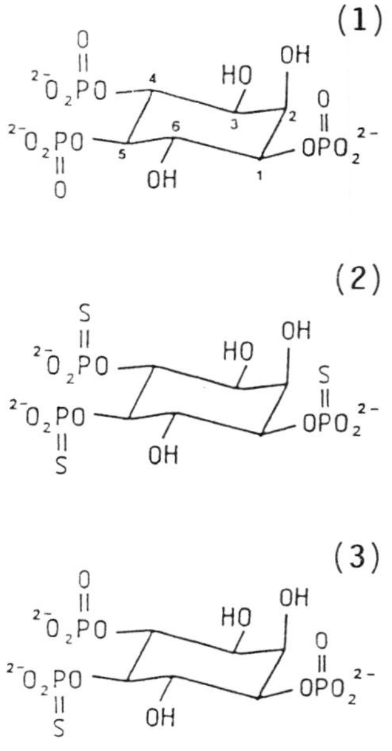

Fig. 1. Structures of InsP₃ analogues. Single enantiomers are shown. (1) D-Ins(1,4,5)P₃; (2) D-Ins(1,4,5)P₃S₃; (3), D-Ins(1,4)P₂(5)PS, _i.e._ -5-phosphorothioate. *Footnote, opposite, gives abbreviations.*

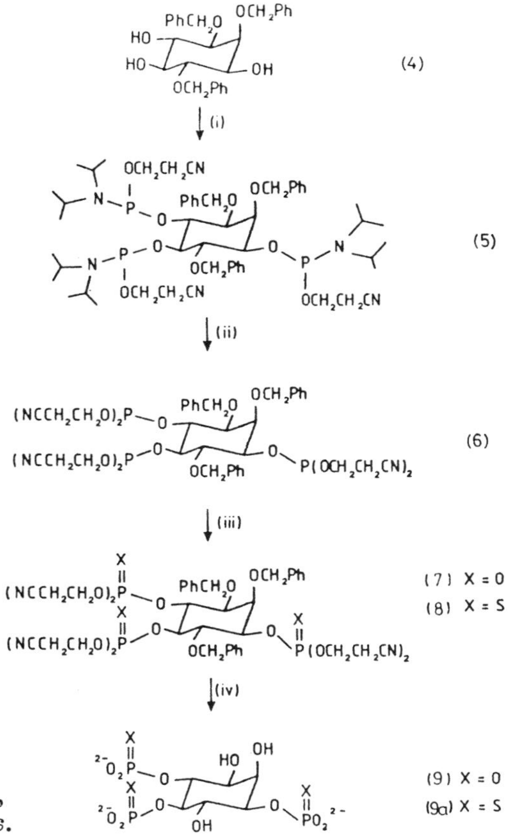

Fig. 2 *(right).* Synthesis of DL-Ins(1,4,5)P₃ and DL-Ins(1,4,5)P₃S₃ (phosphorothioate). Reagents: (**i**) ClP(OCH₂CH₂CN)NPrⁱ₂, EtNPrⁱ₂; (**ii**) NCCH₂CH₂OH-tetrazole; (**iii**) t-BuOOH (X=O) or sulphur-pyridine (X=S); (**iv**) Na-liq. NH₃. *Amplified in text.*

thioate analogues, _e.g._ of ATP, GTP and cyclic AMP, have found widespread use as phosphatase-resistant analogues of the naturally occurring materials [28]. In this article we summarize our recent studies of the biological activity and metabolic stability of DL-Ins(1,4,5)P₃S₃.

METHODS

Synthesis of InsP analogues.- DL-Ins(1,4,5)P₃ and DL-Ins-(1,4,5)P₃S₃ were synthesized by a phosphite chemistry approach, from a tri-_O_-benzyl-_myo_inositol (**4**, Fig. 2; -OH protection by isopropylidene groups also) [22] using the reagent _N,N_-diisopropylamino(2-cyanoethyl)chlorophosphine (Fig. 2, reaction **i**) [24]. The intermediate was phosphitylated to the trisphosphoramidite

Fig. 3. Synthesis of
DL-Ins(1,4)P$_2$(5)PS, *i.e.*
the 1,4-bisphosphate-
5-phosphorothioate.

Reagents: (**i**) & (**ii**)
(CCl$_3$CH$_2$O)$_2$POCl-pyridine,
then, for (**i**), H$_3$O$^+$.
(**iii**) t-BuOOH (X=O) or
sulphur-pyridine (X=S).
(**iv**) Na-liq. NH$_3$.

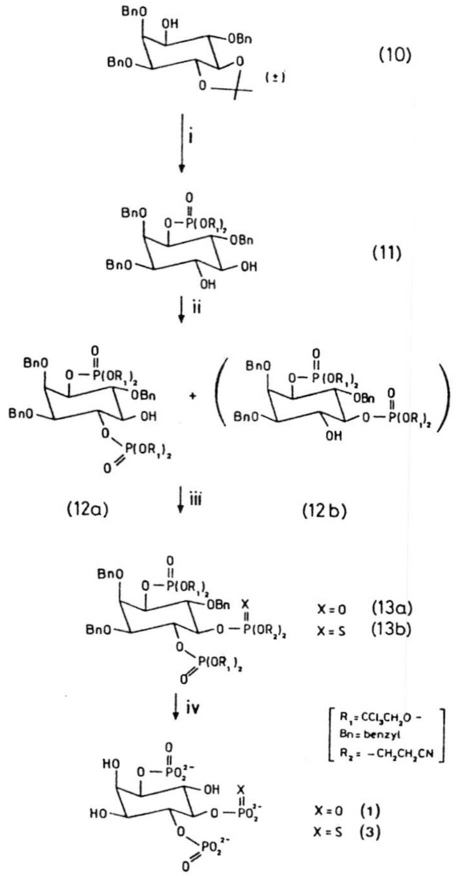

(**5**) and converted to (**6**), the hexacyanoethyl trisphosphite
ester. Then oxidation was carried out with alternative reagents
(**iii**): t-butyl hydroperoxide to generate the hexacyanoethyl
trisphosphate triester (**7**) or sulphur in pyridine to form
the corresponding phosphorothioate triester (**8**). The blocking
groups were removed using sodium in liquid ammonia (**iv**),
and the products (**9** or **9a**) purified by ion-exchange chromatography.

Synthesis of DL-Ins(1,4)P$_2$(5)PS (**3** in Fig. 1) employed
a mixed P(V) and P(III) approach (Fig. 3) [29]. The synthesis
depends on formation of the novel 1,4-substituted bisphosphate
triester (**12a**) which selectively crystallized from a mixture
of isomers prepared by phosphorylation of **11**, itself prepared
from the protected precursor **10**. Phosphorylation of **12a**
using our P(III) approach was followed by oxidation of the
product either with t-BuOOH to give **13a** or with sulphur

to give **13b**. The protecting groups were then removed with
sodium in liquid ammonia to yield **3**, DL-Ins(1,4)P$_2$(5)PS.

**Actions of synthetic InsP's on intracellular Ca^{2+} stores
of permeabilized hepatocytes.-** Hepatocytes, prepared by
collagenase digestion of livers from male Wistar rats, were
suspended in a Ca^{2+}-free, cytosol-like medium containing (mM)
KCl 140, NaCl 20, MgCl$_2$ 2, EGTA 1, PIPES, 20, pH 6.8; albumin,
2%. They were permeabilized at 37° by saponin (75 µg/ml),
and after 10 min the cells were washed and resuspended in
the same medium but containing ~120 nM free [Ca^{2+}] [30].
Then [final concentrations stated] ^{45}CaCl$_2$ (1 µCi/ml),
oligomycin (10 µM) and antimycin (10 µM) were added, and
1 min later ^{45}Ca^{2+} uptake into non-mitochondrial pools was
initiated by adding ATP (1.5 mM), creatine phosphate (5 mM)
and creatine phosphokinase (1 u./ml). After 20 min InsP$_3$
was added, and 5 min later 200 µl samples of cells were
diluted into 2 ml cold iso-osmotic sucrose (310 mM) with
EGTA (4 mM) and [^{3}H]mannose (0.3 µCi/ml), and rapidly filtered
through Whatman GF/C filters [31]. After correction for
trapped volume, the cell ^{45}Ca content was calculated and
the Ins(1,4,5)P$_3$-induced release expressed as a fraction of
ATP-dependent uptake.

Metabolism of InsP$_3$ and analogues.- The metabolism of
D-Ins(1,4,5)P$_3$ was followed by including tracer amounts of
[^{3}H]D-Ins(1,4,5)P$_3$ (Radiochemical Centre, Amersham; 1 Ci/mmol)
with the incubations and then separating the products on
a Partisil 10-SAX anion-exchange HPLC column [31].

The susceptibilities of InsP's to phosphorylation by
a crude preparation of rat-brain Ins(1,4,5)P$_3$-3-kinase were
determined by incubating them with the crude enzyme, stopping
the reactions and then determining the remaining activity
of the InsP's in a Ca^{2+}-release assay. The medium for
incubation (37°) included 310 mM sucrose, 500 mM tris-maleate
pH 7.5, 20 mM MgCl$_2$ and 2% rat-brain supernatant [32]. Samples
(20 µl) were transferred to tubes, stopped by incubation
at 70° for 2 min and then stored frozen. Earlier experiments
established that this protocol caused no loss of activity
of either Ins(1,4,5)P$_3$ or Ins(1,4,5)P$_3$S$_3$. After taking initial
samples of medium and adding DL-Ins(1,4,5)P$_3$ or DL-Ins(1,4,5)P$_3$S$_3$
(to 100 µM), further samples were taken after 30 sec. Then
ATP (10 mM), creatine phosphate (10 mM) and creatine phosphokinase (5 u./ml) were added and the incubation continued for
a further 60 min. The activities of the InsP samples were
subsequently determined by adding 200 µl of permeabilized
hepatocytes pre-labelled to steady-state with ^{45}Ca and terminating the incubations 5 min later. Results, corrected for
the small ^{45}Ca release stimulated by incubation medium alone,
are expressed as a fraction of ATP-dependent ^{45}Ca uptake.

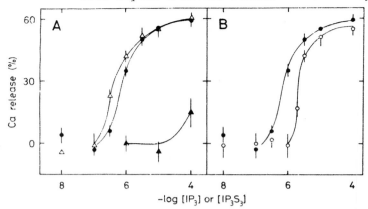

Fig. 4. Actions of Ins(1,4,5)P$_3$ and Ins(1,4,5)P$_3$S$_3$ on Ca^{2+}
pools of permeabilized hepatocytes. Ca^{2+} release, as % of
ATP-dependent uptake, is shown as mean ±S.E.M. of duplicate
determinations (4-9 experiments). ▲, L-Ins(1,4,5)P$_3$;
△, D-Ins(1,4,5)P$_3$; ●, DL-Ins(1,4,5)P$_3$; o, DL-Ins(1,4,5)P$_3$S$_3$.

RESULTS

Addition of D-Ins(1,4,5)P$_3$, DL-Ins(1,4,5)P$_3$ or DL-Ins(1,4,5)-
P$_3$S$_3$ to permeabilized hepatocytes labelled to steady-state
with ^{45}Ca^{2+} stimulated rapid release of up to 60% of the
accumulated ^{45}Ca^{2+}. Maximal concentrations of each compound
were equally effective, although the µM concentrations that
elicited a half-maximal response differed: 0.4 (△), 0.75 (●)
and 2.1 (o) respectively (Fig. 4). L-Ins(1,4,5)P$_3$, the synthetic
enantiomer of D-Ins(1,4,5)P$_3$, failed to stimulate ^{45}Ca^{2+} release
at concentrations up to 10 µM, and even at 100 µM it stimulated
release of only 15% of the ATP-dependent ^{45}Ca pool (Fig. 4).
The ^{45}Ca^{2+} releases elicited by near-maximal (10 µM) or submaximal (0.3 µM) concentrations of D-Ins(1,4,5)P$_3$ were
unaffected by the simultaneous application of L-Ins(1,4,5)P$_3$
(10 µM). Under the conditions of these experiments, where
the cells were washed after permeabilization and then incubated
at low cell density (only 0.2-0.4 mg protein/ml), there was
negligible metabolism of added [^{3}H]D-Ins(1,4,5)P$_3$ [final concentration of D-Ins(1,4,5)P$_3$ = 0.2 µM].

When permeabilized hepatocytes were incubated at much
higher cell density (2-3 mg protein/ml) with DL-Ins(1,4,5)P$_3$
(10 µM) labelled with [^{3}H]D-Ins(1,4,5)P$_3$, it was rapidly degraded
(half-time ~10 min) to products characteristic of the 5-phos-
phatase pathway - Ins(1,4)P$_2$, InsP$_1$ and inositol - with
only minor (~8%) formation of products characteristic of
the 3-kinase pathway, <u>viz</u>. Ins(1,3,4,5,)P$_4$ and Ins(1,3,4)P$_3$.
We therefore examined the effects of DL-Ins(1,4,5)P$_3$ and DL-Ins-
(1,4,5)P$_3$S$_3$ on the intracellular Ca^{2+} pools of permeabilized

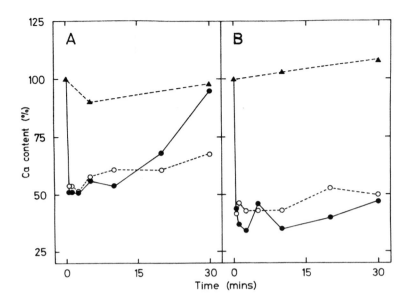

Fig. 5. Resistance of DL-Ins(1,4,5)P$_3$S$_3$ to inactivation by 5-phosphatase action. After permeabilized cells were labelled to steady-state with ^{45}Ca, DL-Ins(1,4,5)P$_3$ ($\bullet$; 10 µM), DL-Ins-(1,4,5)P$_3$S$_3$ (o; 10 µM) or dist. water ($\blacktriangle$) were added. Results are means of duplicate determinations from a single experiment with different cell densities: 2.82 (**A**) or 0.33 (**B**) mg protein per ml. Two further experiments gave similar results.

cells, incubated at both low and high cell densities, so as to determine whether the two compounds differed in their suscepti-bility to inactivation by the 5-phosphatase pathway.   When applied to cells incubated at a low density, both compounds stimulated rapid release of ~60% of accumulated ^{45}Ca^{2+}, and the intracellular pools then remained depleted for the remaining 30 min of the experiment.   In contrast, when applied to cells at high cell density, both compounds again stimulated a rapid release of ^{45}Ca^{2+} - falling to 52.9 ±9% of control for DL-Ins(1,4,5)P$_3$ and 61.9 ±9% for DL-Ins(1,4,5)P$_3$S$_3$ (n = 3); but whereas the pools soon refilled after stimulation by DL-Ins(1,4,5)P$_3$ (to 99 ±4% of control), they remained empty (67 ±2%) after stimulation by DL-Ins(1,4,5)P$_3$S$_3$ (Fig. 5).

To determine whether DL-Ins(1,4,5)P$_3$ and DL-Ins(1,4,5)P$_3$S$_3$ differ in their susceptibilities to inactivation by the 3-kinase pathway, samples of each were incubated with a crude preparation

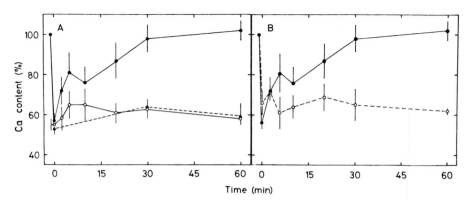

Fig. 6. Resistance of DL-Ins(1,4,5)P$_3$S$_3$ to inactivation by 3-kinase action. **A:** DL-Ins(1,4,5)P$_3$ was incubated for the times shown with a brain supernatant preparation alone (o), together with ATP (●) or after heat-inactivation of the enzyme preparation (▲), and the Ca^{2+}-releasing activity of the remaining DL-Ins(1,4,5)P$_3$ was determined (see text). **B:** Relative stabilities of DL-Ins(1,4,5)P$_3$ (●) and DL-Ins-(1,4,5)P$_3$S$_3$ (o) to inactivation when incubated with active kinase preparation in the presence of ATP. All results are means ±S.E.M. of duplicate assays in 3 independent experiments.

of the enzyme, and the reaction products assayed for their abilities to mobilize intracellular Ca^{2+} pools. Earlier experiments had established that Ins(1,3,4,5)P$_4$, the product of phosphorylation of D-Ins(1,4,5)P$_3$, did not stimulate Ca^{2+} release at concentrations up to 5 µM (the maximal concentration attainable if all the added DL-Ins(1,4,5)P$_3$ were phosphorylated). Phosphorylation of Ins(1,4,5)P$_3$ is therefore accompanied by a loss of its ability to mobilize intracellular Ca^{2+} pools, and hence a bioassay can be used to assess the progress of the reaction. The results (Fig. 6) demonstrate that incubation of DL-Ins(1,4,5)P$_3$ with the crude kinase in the presence, though not in the absence, of ATP rapidly inactivated its ability to release intracellular Ca^{2+} stores. Subsequent analysis by HPLC of the products of these incubations indicated that the inactivation was entirely attributable to phosphorylation of D-Ins(1,4,5)P$_3$ to D-Ins(1,3,4,5)P$_4$. In parallel incubations where DL-Ins(1,4,5)P$_3$S$_3$ was incubated with ATP and the crude kinase preparation, there was no loss of the ability of the DL-Ins(1,4,5)P$_3$S$_3$ to release Ca^{2+} even if the incubations were prolonged for 60 min (Fig. 6).

DISCUSSION

The inability of L-Ins(1,4,5)P$_3$ to either mobilize intra-cellular Ca^{2+} pools or antagonize the actions of D-Ins(1,4,5)P$_3$ in permeabilized hepatocytes has since been confirmed in *Xenopus*

oocytes [33] and in GH$_3$ and Swiss 3T3 cells [34]. A similar
strict stereoselectivity has been observed for specific high-
affinity D-Ins(1,4,5)P$_3$ binding sites in cerebellar membranes
[35] and in permeabilized hepatocytes (D.L. Nunn, B.V.L. Potter
& C.W. Taylor, to be published).

We have shown DL-Ins(1,4,5)P$_3$S$_3$ to be a full agonist for
intracellular Ca^{2+} mobilization in hepatocytes, only ~3-fold
less potent than DL-Ins(1,4,5)P$_3$. Similar results have been
reported for several other cell types including *Xenopus* oocytes
[33], Swiss 3T3 cells [33, 34] and GH$_3$ cells [34]. Recently
(with D.L. Nunn) we have demonstrated that DL-Ins(1,4,5)P$_3$
and DL-Ins(1,4,5)P$_3$S$_3$ also differ by a factor of ~3 in their
affinities for a high-affinity D-Ins(1,4,5)P$_3$ binding site
in permeabilized hepatocytes, whereas in cerebellar membranes
a 6-fold difference in affinities was observed [36]. Together
these results provide convincing evidence that in a wide
variety of cell types DL-Ins(1,4,5)P$_3$S$_3$ - presumably only
the D-isomer is active - is a full agonist for mobilization
of intracellular Ca^{2+} pools and only ~3-fold less potent
than the natural messenger. In preliminary experiments DL-Ins-
(1,4)P$_2$(5)PS, in which only the 5-phosphate group of DL-Ins(1,4,5)P$_3$
has been substituted by a phosphorothioate group, was near
equipotent with DL-Ins(1,4,5)P$_3$S$_3$ both in displacing D-Ins(1,4,5)P$_3$
from specific binding sites in cerebellar membranes and in
stimulating Ca^{2+} release from permeabilized Swiss 3T3 and
GH$_3$ cells [29].

Under conditions where there is minimal degradation of
D-Ins(1,4,5)P$_3$, both DL-Ins(1,4,5)P$_3$ and DL-Ins(1,4,5)P$_3$S$_3$ can
discharge and keep empty the responsive intracellular Ca^{2+}
stores. These results confirm an earlier suggestion that
the receptor that mediates Ins(1,4,5)P$_3$-induced Ca^{2+} release
does not desensitize during prolonged stimulation [37]. The
prolonged actions of DL-Ins(1,4,5)P$_3$S$_3$ under conditions where
DL-Ins(1,4,5)P$_3$ is rapidly dephosphorylated and thereby inactivated must therefore result from the resistance of the former
to dephosphorylation by the 5-phosphatase pathway. This
result has been confirmed by the demonstration that treatment
of DL-Ins(1,4,5)P$_3$ with human erythrocyte 5-phosphatase
abolished its ability to displace [^{3}H]D-Ins(1,4,5)P$_3$ from specific
binding sites in cerebellar membranes, whereas the activity
of DL-Ins(1,4,5)P$_3$S$_3$ was unaffected [36]. Although we have
not yet determined whether DL-Ins(1,4,5)P$_3$S$_3$ competes with
the natural substrate for the liver 5-phosphatase, it does
appear to be a potent inhibitor (K$_i$ = 6 μM for the racemic
mixture) of the 5-phosphatase of human erythrocytes [38].

Although DL-Ins(1,4,5)P$_3$ was rapidly phosphorylated to
Ins(1,3,4,5)P$_4$ and was thereby inactivated when incubated

with ATP and a crude preparation of Ins(1,4,5)P$_3$ 3-kinase, there was no loss of activity of DL-Ins(1,4,5)P$_3$S$_3$. This result confirms that DL-Ins(1,4,5)P$_3$S$_3$ resists inactivation not merely by the 5-phosphatase pathway but also by the only other route by which D-Ins(1,4,5)P$_3$ is known to be metabolized - the 3-kinase pathway. Some additional supporting evidence has emerged from the following recent finding: DL-Ins-(1,4,5)P$_3$S$_3$ (10 μM) did not inhibit the phosphorylation of D-Ins(1,4,5)P$_3$ during a prolonged incubation (60 min) with the crude 3-kinase in the course of which the concentration of D-Ins(1,4,5)P$_3$ declined from 1 μM to ~200 nM (well below the K$_m$, 0.6 μM (15]) [31]. These results suggest that DL-Ins(1,4,5)P$_3$-S$_3$ is neither a substrate of nor a competitor for D-Ins(1,4,5)P$_3$ 3-kinase.

We have shown that in a wide variety of cells DL-Ins(1,4,5)P$_3$-S$_3$ is a potent analogue of the natural intracellular messenger D-Ins(1,4,5)P$_3$, being only ~3-fold less potent. However, whereas D-Ins(1,4,5)P$_3$ is rapidly metabolized in intact cells to Ins(1,3,4,5)P$_4$ and Ins(1,4)P$_2$, DL-Ins(1,4,5)P$_3$S$_3$ is not a substrate for either of these pathways. While both the phosphorylation and dephosphorylation pathways abolish the effects of D-Ins(1,4,5)P$_3$ on intracellular Ca^{2+} pools, it has been suggested that at least one of the products, D-Ins(1,3,4,-5)P$_4$, may itself be an intracellular messenger and regulate Ca^{2+} transport at the p.m. [39, 40]. The rapid turnover of D-Ins(1,4,5)P$_3$ and the possibility that it may be the source of other active intracellular messengers has caused considerable experimental problems. Radioligand binding experiments and concentration-effect relationships, for example, may often be confused by metabolism of D-Ins(1,4,5)P$_3$. Furthermore, the direct actions of D-Ins(1,4,5)P$_3$ have not been readily distinguishable from those that require its metabolism. One model that seeks to explain hormone- and D-Ins(1,4,5)P$_3$-stimulated oscillations in cytosolic [Ca^{2+}] suggests that it reflects oscillatory changes in intracellular [D-Ins(1,4,5)P$_3$], while another suggests that the oscillations result from oscillatory release of Ca^{2+} from an overloaded intracellular store that remains overloaded only as long as Ins(1,4,5)P$_3$ is able to keep the InsP$_3$-sensitive store empty [41]. We suggest that the metabolic stability of DL-Ins(1,4,5)P$_3$S$_3$ will make it the ligand of choice in studying the actions of the natural messenger; in addition, it may allow the proposed mechanisms underlying oscillations in cytosolic [Ca^{2+}] to be distinguished and the relative roles of Ins(1,4,5)P$_3$ and InsP(1,3,4,5)P$_4$ in regulating Ca^{2+} entry at the p.m. to be determined.

Acknowledgements

The work was supported by grants from the SERC, The Research Corporation Trust, Merck, Sharp & Dohme, and the Erna and Victor Hasselblad Foundation. B.V.L.P. is a Lister Institute Research Fellow. C.W.T. is a Royal Society Locke Research Fellow.

References

1. Streb, H., Irvine, R.F., Berridge, M.J. & Schulz, I. (1983) *Nature 306*, 67-69.
2. Berridge, M.J. & Irvine, R.F. (1984) *Nature 312*, 315-321.
3. Posternak, T. (1965) *The Cyclitols*, Holden-Day, San Francisco, 431 pp.
4. Cosgrove, D.J. (1980) *Inositol Phosphates, their Chemistry, Biochemistry and Physiology*, Elsevier, Amsterdam, 191 pp.
5. Parthasarathy, R. & Eisenberg, F. jr. (1986) *Biochem. J. 235*, 313-322.
6. Putney, J.W. jr. (1986) *Phosphoinositides and Receptor Mechanisms*, Alan R. Liss, New York, 400 pp.
7. Berridge, M.J. & Michell, R.H., eds. (1988)* *Inositol Lipids and Transmembrane Signalling*, The Royal Society, London, 200 pp.
8. Ozaki, S., Watanabe, Y., Ogasawara, T., Kondo, Y., Shiotani, N., Nishii, H. & Matsuki, T. (1986) *Tetrahedron Lett. 27*, 3157-3160.
9. Reese, C.W. & Ward, J.G. (1987) *Tetrahedron Lett. 28*, 2309-2312.
10. Vacca, J.P., de Solms, S.J. & Huff, J.R. (1987) *J. Am. Chem. Soc. 109*, 3478-3479.
11. Dreef, C.A., Tuinman, R.J., Elie, C.J.J., van der Marel, G.A. & van Boom, J.H. (1988) *Rec. Trav. Chim. Pays-Bas 107*, 395-397.
12. Yu, K-L.& Fraser-Reid, B. (1988) *Tetrahedron Lett. 29*, 979-982.
13. Meek, J.L., Davidson, F. & Hobbs, F.W. (1988) *J. Am. Chem. Soc. 110*, 2317-2318.
14. Cooke, A.M., Gigg, R. & Potter, B.V.L. (1987) *Tetrahedron Lett. 28*, 2305-2308.
15. Irvine, R.F., Letcher, A.J., Heslop, J.P. & Berridge, M.J. (1986) *Nature 320*, 631-634.
16. Storey, D.J., Shears, S.B., Kirk, C.J. & Michell, R.H. (1984) *Nature 312*, 374-376.
17. Billington, D.C. & Baker, R. (1987) *J. Chem. Soc. Chem. Comm.*, 1011-1013.
18. de Solms, S.J., Vacca, J.P. & Huff, J.R. (1987) *Tetrahedron Lett. 28*, 4503-4506.
19. Ozaki, S., Kondo, Y., Nakahiura, H., Yamaoka, S. & Watanabe, Y. (1987) *Tetrahedron Lett. 28*, 4691-4694.

*appears also in *Phil. Trans. Roy. Soc. B 320*, 237-436.

20. Billington, D.C., Baker, R., Kulagowski, J.J. &
    Mawer, I.M. (1987) *J. Chem. Soc. Chem. Comm.*, 314-316.
21. Watanabe, H., Nakahira, H., Buyna, M. & Ozaki, S. (1987)
    Tetrahedron Lett. 28, 4179-4180.
22. Gigg, J., Gigg, R., Payne, S. & Conant, R. (1987) *J. Chem.
    Soc. Perkin Trans. 1*, 423-429.
23. Gigg, J., Gigg, R., Payne, S. & Conant, R. (1987) *J. Chem.
    Soc. Perkin Trans. 1*, 2411-2414.
24. Sinha, N.D., Biernat, J., McManus, J. & Koester, H. (1984)
    Nucl. Acids Res. 12, 4539-4557.
25. Hamblin, M.R., Gigg, R. & Potter, B.V.L. (1987) *J. Chem.
    Soc. Chem. Comm.*, 626-627.
26. Hamblin, M.R., Flora, J.S. & Potter, B.V.L. (1987) *Biochem.
    J. 246*, 771-774.
27. Cooke, A.M., Gigg, R. & Potter, B.V.L. (1987) *J. Chem. Soc.
    Chem. Comm.*, 1515-1526.
28. Eckstein, F. (1985) *Annu. Rev. Biochem. 54*, 367-402.
29. Cooke, A.M., Noble, N.J., Payne, S., Gigg, R.,
    Strupish, J. Nahorski, S.R. & Potter, B.V.L. (1988)
    Biochem. Soc. Trans. 16, 992-993.
30. Burgess, G.M., McKinney, J.S., Fabiato, A., Leslie, B.A.
    & Putney, J.W. jr. (1983) *J. Biol. Chem. 258*, 2716-2725.
31. Taylor, C.W., Berridge, M.J., Cooke, A.M. & Potter, B.V.L.
    (1989) *Biochem. J.*, in press.
32. Tennes, K.A., McKinney, J.S. & Putney, J.W. jr. (1987)
    Biochem. J. 242, 797-802.
33. Taylor, C.W., Berridge, M.J., Cooke, A.M. & Potter, B.V.L.
    (1988) *Biochem. Biophys. Res. Comm. 150*, 626-632.
34. Strupish, J., Cooke, A.M., Potter, B.V.L., Gigg, R. &
    Nahorski, S.R. (1987) *Biochem. J. 253*, 901-905.
35. Willcocks, A.L., Cooke, A.M., Potter, B.V.L. &
    Nahorski, S.R. (1987) *Biochem. Biophys. Res. Comm. 146*,
    1071-1078.
36. Willcocks, A.L., Cooke, A.M., Potter, B.V.L. &
    Nahorski, S.R. (1988) *Eur. J. Pharmacol. 155*, 181-183.
37. Prentki, M., Corkey, B.E. & Matschinsky, F.M. (1985) *J.
    Biol. Chem. 260*, 9185-9190.
38. Cooke, A.M., Nahorski, S.R. & Potter, B.V.L. (1989) *FEBS
    Lett. 242*, 373-377.
39. Irvine, R.F. & Moor, R.M. (1986) *Biochem. J. 240*, 917-920.
40. Morris, A.P., Gallacher, D.V., Irvine, R.F. &
    Petersen, O.H. (1987) *Nature 330*, 653-655.
41. Berridge, M.J., Cobbold, P.H. & Cuthbertson, K.S.R. (1988)
    Phil. Trans. Roy. Soc. B 320, 325-343.

#B-6

AEQUORIN MEASUREMENTS OF REPETITIVE CALCIUM TRANSIENTS IN SINGLE MAMMALIAN CELLS

P.H. Cobbold, N.M. Woods, J. Dixon and K.S.R. Cuthbertson

Department of Human Anatomy and Cell Biology,
University of Liverpool,
P.O. Box 147, Liverpool L69 3BX, U.K.

The concept of cytoplasmic Ca^{2+} being modulated as repetitive transients in hormone-stimulated cells is proving to be applicable to a wide range of cell types. Our aequorin studies in single rat hepatocytes revealed repetitive transients that were frequency-modulated according to the hormone concentration. Since the time course of individual transients was different for different agonists, we have proposed a model mechanism involving repetitive generation of $Ins(1,4,5)P_3^*$. The fluorescent indicator fura-2 shows a suggestion of $[Ca^{2+}]$ oscillations in hormone-stimulated liver cells, but the transients are not as clear-cut as with aequorin. However, single epithelial cells loaded with fura-2 respond to histamine in a manner highly reminiscent of hepatocytes. Whatever the reasons for different responses with different indicators, all the single-cell data point to a complexity of behaviour of $[Ca^{2+}]$ that is not hinted at in population measurements. Understanding how $[Ca^{2+}]$ is controlled demands studies using single cells.

FREE CALCIUM TRANSIENTS IN SINGLE CELLS

Recently an increasing number of laboratories have measured $[Ca^{2+}]$ in single cells and found responses to hormones that involve oscillations in $[Ca^{2+}]$ [1, 2]. Such complexity of behaviour of $[Ca^{2+}]$ was masked in measurements on populations of millions of cells, presumably because of cell asynchrony and/or heterogeneity, so an understanding of how $[Ca^{2+}]$ is controlled in cells will require that measurements be made in single cells by either photoproteins or fluorescent indicator [3]. (Art. #ncE.1 illustrates use of the authors' approaches.)

---

*Abbreviations.- InsP, an inositol phosphate; PVP, polyvinyl-pyrrolidone. The authors' term 'free Ca' has been replaced by Ca^{2+}, often implying its concentration, $[Ca^{2+}]$. Use of the probe fura-2 is amplified in (e.g.) A.K. Campbell's article.

AEQUORIN TECHNIQUE

The properties of aequorin that make it so suitable for reporting [Ca^{2+}] have been reviewed recently [3], with citations of our original descriptions of the technique and later modifications. [In this series, aequorin features especially in Vol. 6.-*Ed.*] Here we describe some recent innovations that further improve the technique.

Microdialysis.- Freeze-dried samples of aequorin are dissolved at ~150 mg/ml in microdroplets of EDTA buffer, held under liquid paraffin in glass tubes and stored frozen at -70°. Every few days ~100-200 nl of this stock solution is dialyzed in a 5 mm length of Biorad Biofiber 50A microdialysis tubing. The aequorin column fills ~$\frac{2}{3}$rds of the length of the tubule which is plugged at each end with short columns of liquid paraffin. The tubule, supported on two hooks formed out of teflon-coated Pt wire glued to a small piece of plastic, can then be immersed in dialysis buffer (for composition see [3]). Dithiothreitol (1 mM) may be added to this buffer; we find this to improve the ease of filling of micropipettes and the flow of aequorin into the cell, possibly by reducing denaturation of the protein during dialysis and storage. The column of aequorin tends to slowly swell, and should be shrunk by evaporation every 10-30 min for the 4 h to 6 h dialysis period. This can be avoided by having PVP (mol. wt. 4×10^4) present in the medium (25 mg/ml), allowing uninterrupted dialysis to be carried out overnight in a refrigerator. We do not know how much PVP enters the tubule, nor if it is truly inert within the cell. Indeed, the inertness of any injected material is difficult to establish. The golden rule is to keep additions to a minimum, and minimize the volume injected. We typically aim for an injection volume of <2% of the cell's volume.

Filling micropipettes.- At ~100 mg/ml aequorin forms a viscous solution and the small volume we dialyze is far too small to filter, and would suffice to back-fill only one micropipette at best. The aequorin droplet can be stored on the surface of a plastic petri dish covered with liquid paraffin (BDH Chemicals, light grade). On dipping the tip of a freshly-pulled micropipette into the oil for a few sec, a column of oil will enter, followed spontaneously by a short column (~50 µm) of aequorin solution. This amount suffices for injecting a single cell, the aequorin droplet volume never being limiting, and such 'tip-filled' pipettes are disposable. Tips that are reluctant to fill can be helped by gently scraping across the surface of the petri dish, or across small scratches made in the plastic.

Scanning electron-microscopy reveals the bore of the micro-pipettes to be ~0.1-0.3 μm; they have a resistance of ~5-10 MΩ when filled with 3 M KCl.

Injection.- The short column of aequorin can be forced in a second or so out of the pipette tip with gas (N_2) pressure of ~10-70 psi (0.6-5 atm.). A backing pressure of ~7 psi (0.5 atm.) is applied while the pipette is in the extracellular medium, to deter inflow of the Ca^{2+}-rich solution. We use a custom-built pressure-controlling apparatus, but a commercially available system provides similar facilities (Digitimer/Medical Systems Corp., 'Picoinjector'). The volume injected into the cell is estimated from the tip-to-meniscus length and the profile of the pipette bore.

Cell impalement.- From our experience with micro-injecting a range of cell types, we find that the biggest procedural unknown is how easily the plasmalemma is penetrable by the micropipette tip. This is not a question of cell size, some of the biggest mammalian cells being either very difficult (e.g. oocytes) or very easy (e.g. heart ventricle myocytes) to impale. Liver cells are easy, the smaller chromaffin cells not impossible, while HL60 leukaemic cells – the same size as chromaffin cells – seem to be impossible to inject. Isolated neonatal rat hippocampal CA1 neurones have proved very difficult even when attached to a glass surface, while primary cultured vascular smooth-muscle cells from rat aorta, when spread on a glass surface, are easily injected, as are fibroblasts. We imagine the cell-to-cell differences reside in the elasticity and thickness of the plasmalemma, glyco-calyx and sub-plasmalemmal cortical network.

Recently one of us (K.S.R.C.) devised an electrostatic method to achieve an oscillating voltage across the micropipette tip to mimic 'buzzing' technique used by neurophysiologists to puncture neuronal cells. By connecting the bore of a tip-filled micropipette to a wire connected to the centre electrode of the snout of a domestic piezo-electric gas lighter (actually a Bosch model), and insulating the connection from the outer circular electrode with a short sleeve of rubber tubing, the electrostatic discharge can be made to occur across the micropipette tip. A gentle squeeze of the gas-lighter's trigger will be found to allow the tip, already dimpling the cell surface, to enter the cell. An earth return wire to the metal microscope stage improves consistency. We have no idea of the size of the voltage near the pipette tip. Presumably the voltage pulse is capacitatively coupled across the 50 μm-long oil column and air bubbles near the tip. The procedure is well tolerated

by heart and liver cells and mouse oocytes, and we have
the impression that it is more benign than mechanical prodding
of the cell, when the tip tends to penetrate deeper into
the cytoplasm.

INTERPRETATION OF AEQUORIN SIGNALS

A microcomputer-driven system for recording photon counts
with time at, typically, 50 msec intervals allows off-line
normalization of the signal (counts/sec) as a fraction of
the total remaining aequorin counts. Cobbold & Rink [3]
have shown two calibration curves appropriate for mammalian
cells. Between 10^{-7} and 10^{-5} M Ca^{2+} the fractional rate
of consumption has a cube-law relationship with $[Ca^{2+}]$, such
that a doubling of $[Ca^{2+}]$ leads to an 8-fold increase in the
signal. This non-linearity will tend to emphasize idealized
regions of high $[Ca^{2+}]$, and this fact has sometimes been
used to criticize the technique. However, many Ca^{2+}-triggered
enzymes, including calmodulin, may favour the detection of
a physiologically significant Ca^{2+} rise where linear indications would not. Our signals are calibrated assuming continual
homogeneity of Ca^{2+}.   But for a homogeneous Ca^{2+} level
of 600 nM (the peak level reached in a transient in a hormone-
stimulated liver cell - see below), the same signal could
arise from 1200 nM in one-eighth (12%) of the cell, or
from 2400 nM in 1.5%, or 4800 nM in 0.2%. As % of cell
volume falls to low values, the size of the putative Ca^{2+}
zone falls to sub-μm levels, which is unlikely to persist
for more than a fraction of a second due to the speed
of diffusion of Ca^{2+} out of the zone.

The actual extent and concentration of the $[Ca^{2+}]$ changes
may also alter with time, and would need to be imaged (which
is not feasible using aequorin in mammalian cells: see [3]).
In those large cells in which imaging is feasible, the
loss of light in the microscope lenses usually dictates
that the sensitivity of the system is orders of magnitude
worse than with a photomultiplier arrangement designed to
collect the isotropically radiated light, causing a loss
of detectable signal at Ca^{2+} levels well above the resting
level. This may be the reason why Miyazaki et al. [4]
reported, through imaging of an oocyte at fertilization,
discrete Ca^{2+} transients while our own more sensitive photomultiplier system recorded a sustained Ca^{2+} rise with a superimposed
oscillation similar to the frequency of Miyazaki's transients
[5]; the imaging system had 'cut off' the base of the oscillations because of low collection efficiency of the lens
and the noisier detector. A sensitive aequorin system with
a good, low-noise detector and good light-collection efficiency
will generate signals that are impossible to display faithfully

as linear recordings. Linear plots contain only a fraction of the information that should be present; some form of logarithmic normalization of the signals is highly desirable.

FUTURE PROSPECTS FOR AEQUORIN AND OBELIN

The instrumentation needed for making fluorescence measurements of Ca^{2+} using fura-2 in single cells is ~10 times as expensive and much more complicated than the simple photon-counting, computer-driven photomultiplier system needed to record photoprotein signals. Rather, the popularity of fura-2 lies in the ease with which the probe can be loaded into cells. No micro-injection is needed. The recent cloning of a gene for aequorin [6, 7] and isolation of mRNA for aequorin [7] opens up the possibility of eliminating micro-injection by transfecting cell lines, or of producing transgenic animals. Campbell et al. [8, & final art., this book] have shown that mRNA for obelin may be introduced into neutrophils in immunoliposomes to produce functional apo-protein, which can be converted into the Ca^{2+}-sensitive photoprotein obelin by addition of coelenterazine, a chemically synthesized lumino-phore, to the re-sealed cells' medium. This approach, when extended to cells transfected with the gene, is likely to lead to the simplest single-cell Ca^{2+} measurements yet devised.

REPETITIVE Ca^{2+} TRANSIENTS IN HEPATOCYTES

Single isolated rat hepatocytes injected with aequorin respond to several Ca^{2+}-mobilizing hormones by generating repetitive Ca^{2+} transients that are frequency-modulated by the hormone concentration [9, 10]. A key observation, which is particularly meaningful when made in the same individual cell, is that different agonists induce transients differing subtly in time-course, the rate of fall of Ca^{2+} from the peak being slower for vasoactive peptides and ATP compared with phenylephrine or ADP [10-12]. On the basis of these observations we have proposed a model in which the phospho-inositol cycle is also activated transiently with a time course for $InsP_3$ production superimposable on the Ca^{2+} transient [2, 12]. The hypothesis predicts a ~1 sec lifetime for $InsP_3$, which might be tested by recording Ca^{2+} levels while injecting $InsP_3$ into a cell. Other experimental tests will require modification of $InsP_3$ metabolism by injecting purified $InsP_3$-metabolizing enzymes (for which the techniques developed for obelin will be ideal), or using injected antibodies to modify the activity of enzymes in the cycle, e.g. phospho-lipase-C or protein kinase C. We see the ability to micro-inject single cells as being central to such single-cell studies.

Another implication of the agonist-specific time courses of the transient concerns the marked difference between ATP

and ADP [11].   The simplest interpretation would be that
ATP and ADP act via different receptors with different dynamics
of inactivation following the Ca^{2+} peak.   The liver-cell
receptor is of the $P_{2\gamma}$ class of purinoceptors, identified
solely by rank potency studies of a series of analogues.
So the possibility of ATP and ADP having discrete receptors
is open.   Another possibility could be that the falling
phase of the transient depends upon the speed with which
an agonist dissociates from an inactivated receptor.
Conceivably ADP might dissociate more rapidly than ATP, to
explain the faster switch-off of Ca^{2+}.   But experiments
comparing arg^8-vasopressin with oxytocin, which has a ~300-fold
lower affinity for the hepatic vasopressin receptor, fail
to show any marked difference in the transient's time course
(J. Dixon, unpublished), arguing against agonist dissociation
being an important factor.   At present we favour the idea
that ATP and ADP act via different receptors, or that ATP
also acts on a separate ATP-selective receptor in addition
to a shared $P_{2\gamma}$ receptor.

AEQUORIN VERSUS FURA-2 IN HEPATOCYTES

Fura-2 has been used by Monck et al. [13] in single
rat hepatocytes with conditions and experimental protocols
closely resembling those used in the aequorin studies. However,
clear-cut repetitive Ca^{2+} transients were not recorded,
although one cell showed an oscillatory sequence in response
to phenylphrenine. Since the peak height of the Ca^{2+} responses
was considerably less than recorded with aequorin (~200-300 nM
compared with 600-1000 nM), the disparity could be caused
by buffering of the Ca^{2+} changes by aequorin. The suggestion
by Monck et al. [13] that aequorin emphasizes locally elevated
Ca^{2+}, whereas the linear response of fura-2 is less biased
by local Ca^{2+} changes, would not explain the very different
time-courses of the transients, which should remain more
or less the same despite different calibrated peak heights.
The failure of Ca^{2+} in fura-2-loaded cells to return to
rising levels after the first upstroke of $[Ca^{2+}]$ is particularly
suggestive of buffering.

Other problems with fura-2 such as subcellular localization and leakage have been reviewed recently [2]. However,
Jacob et al. [14] have reported repetitive Ca^{2+} transients,
frequency-modulated by the histamine concentration, in single
fura-2-loaded endothelial cells; the parallels with the
aequorin data in liver cells are dramatic.   We conclude
that Ca^{2+} indicators may work differently in different cells
and that many more single-cell observations will be required
to enable us to confidently distinguish artefact from reality.

References

1.   Berridge, M.J., Cuthbertson, K.S.R. & Cobbold, P.H. (1988)
     Phil. Trans. Roy. Soc. B 320, 313–328.
2.   Cuthbertson, K.S.R. (1988) in *Neuronal and Cellular
     Oscillators* (Jacklet, J.W., ed.), Marcel Dekker, New
     York, pp. 437–460.
3.   Cobbold, P.H. & Rink, T.J. (1987) *Biochem. J. 248*, 313–328.
4.   Miyazaki, S., Hashimoto, N., Yoshimoto, Y., Kishimoto, T.,
     Igusa, Y. & Hiramoto, Y. (1986) *Devel. Biol. 118*, 259–267.
5.   Cuthbertson, K.S.R. & Cobbold, P.H. (1985) *Nature 316*,
     541–542.
6.   Inouye, S., Noguchi, M., Sakaki, Y., Takagi, Y.,
     Miyata, T., Iwanagu, S., Miyata, T. & Tsuji, K.J. (1985)
     Proc. Nat. Acad. Sci. 82, 3154–3158.
7.   Prahser, D., McCann, R.O. & Cormier, M.J. (1985) *Biochem.
     Biophys. Res. Comm. 126*, 1259–1268.
8.   Campbell, A.K., Patel, A.K., Ragavi, Z.S. & McCapra, F.
     (1988) *Biochem. J. 252*, 143–149.
9.   Woods, N.M., Cuthbertson, K.S.R. & Cobbold, P.H. (1986)
     Nature 319, 600–602.
10. Woods, N.M., Cuthbertson, K.S.R. & Cobbold, P.H. (1987)
     Cell Calcium 8, 79–100.
11. Cobbold, P.H., Woods, N.M., Wainwright, J. &
     Cuthbertson, K.S.R. (1988) *J. Receptor Res. 8*, 481–491.
12. Cobbold, P.H., Cuthbertson, K.S.R. & Woods, N.M. (1988)
     in *Hormones and Cell Regulation*, 12th Symp. (Nunez, J. &
     Dumont, J.E., eds.), INSERM/Libbey, London/Paris, pp. 135–146.
13. Monck, J., Reynolds, E.E., Thomas, A.P. & Williamson, J.R.
     (1988) *J. Biol. Chem. 263*, 4569–4575.
14. Jacob, R., Merrith, J.E., Hallan, T.J. & Rink, T.J.
     (1988) *Nature 335*, 40–45.

#B-7

MODULATION OF $[Ca^{2+}]_i$ HOMEOSTASIS BY BRADYKININ IN PC 12 CELLS: ROLE OF INOSITOL PHOSPHATES AND PROTEIN KINASE-C

[1]Cristina Fasolato, [2]Atanasio Pandiella,
[2]Jacopo Meldolesi & [1,3]Tullio Pozzan

[1]Institute of General Pathology, University of Padova,
C.N.R. Center of Biomembranes, Padova, Italy[†]

[2]Department of Pharmacology, University of Milano;
C.N.R. Center of Pharmacology and Scientific Institute
of S. Raffaele, Milano, Italy

[3]Institute of General Pathology, University of Ferrara,
Ferrara, Italy

BK* added to fura-2-loaded PC 12 cells induces a dose-dependent increase in $[Ca^{2+}]_i$ which depends initially on Ca^{2+} redistribution from intracellular stores and later depends on a prolonged Ca^{2+} influx from the extracellular medium. Concomitantly, outward and inward currents and inositol phosphate formation are triggered by BK. Data are presented indicating that $InsP_3$ generation, $[Ca^{2+}]_i$ redistribution and activation of the outward current (due to Ca^{2+}-activated K^+ channels) are causally related. Ca^{2+} influx and the inward current remained constantly associated under a variety of experimental conditions, suggesting that they are due to the same process, i.e. the activation of an unspecific cation channel. The outward current and Ca^{2+} influx could on the other hand be dissociated from the kinetics of generation of $InsP_3$ isomers (1,4,5; 1,3,4), suggesting that inositol phosphates are not primary regulators of Ca^{2+} influx triggered by BK. Protein kinase-C activation on the other hand appears to play a feedback inhibitory role.

[†]Via Loredan 16, 35131 Padova (Prof. Pozzan's mailing address)
*Abbreviations.- BK, bradykinin; Ptd, phosphatidyl; $InsP_2$/ $InsP_3$/$InsP_4$: inositol phosphates – the 4,5/1,4,5/1,3,4,5 isomers unless otherwise indicated; AM, acetoxymethyl ester (of the probe fura-2); Ap, apamin; $I-B_1$ & $I-B_2$: see text; KRH, standard medium as specified in text; NGF, nerve growth factor; p.m., plasma membrane; PMA, phorbol 12-myristate,13-acetate; TCA, trichloroacetic acid; TEA, tetra-ethylammonium (chloride used).

A variety of experimental evidence demonstrates that following activation of receptors coupled to the hydrolysis of phosphoinositides, the transient release of intracellularly stored Ca^{2+} (due to $InsP_3$ production) is accompanied by a more persistent increase of Ca^{2+} influx across the p.m. This latter process, variable in extent and duration depending on the cell type and the receptor involved, has been the subject of extensive investigation in the last three years, but its mechanism is still obscure. Various hypotheses have been proposed: opening of a true receptor-operated channel [1]; Ca^{2+}-dependent activation [2]; modulation by, respectively, $Ins(1,4,5)P_3$ [3] or $Ins(1,3,4,5)P_4$ [4][⊗] or these two in combination [5, 6].

Utilizing the nonapeptide BK as a stimulus, we have addressed this problem in a cell line (PC 12) derived from a rat phaeochromocytoma and widely used as a convenient model for studying neuroendocrine cell physiology *in vitro* [7].

BK was found to cause phosphoinositide hydrolysis, $[Ca^2]_i$ rises and p.m. potential changes <u>via</u> the activation of a receptor of the B_2 type. Various observations suggest that BK-induced depolarization and Ca^{2+} influx are both due to the opening of a non-selective cation channel. This article considers the role of inositol phosphates and protein kinase-C in the regulation of such a channel.

MATERIALS AND METHODS

Composition (mM) of media.- KRH medium [with variations in some experiments]: NaCl [or choline-HCl] 125; KCl, 5; $MgSO_4$, 1; Na_2HPO_4 [or K_2HPO_4], 1; glucose, 5.5; $CaCl_2$ [or EGTA], 1; HEPES [or tris] (pH 7.4 at 37°), 20. Sucrose-based medium: sucrose, 250; $KHCO_3$, 5; $MgSO_4$, 1; glucose, 5.5; HEPES, 20; pH adjusted to 7.4 with tris at 37°.

Measurement of $[Ca^{2+}]_i$.- Loading with fura-2 [8] was performed esssentially as described [9]. The medium was routinely supplemented with sulphynpirazone (200 µM) to block fura-2 leakage.

Membrane potential was measured with bis-oxonol essentially as described by Di Virgilio <u>et al</u>. [10].

Patch clamping.- Measurement of p.m. conductance was performed in the whole-cell mode of the patch clamp technique [11] as described by Wanke <u>et al</u>. [12].

[⊗] $InsP_4$'s feature in art. #B-4 (Kirk <u>et al</u>.). - *Ed.*

Inositol extraction and separation.- PC 12 cells were incubated for 24-48 h in RPMI 1640 inositol-free medium, supplemented with foetal calf serum (to 1%) and 5 μCi/ml of ^{3}H-*myo*inositol. The cells were then washed and resuspended in one of the above saline media. The reaction was stopped by adding an equal vol. of 15% (w/v) ice-cold TCA. In order to increase the recovery of InsP$_4$, a phytic acid hydrolysate was included in the extractant (0.6 mg P$_i$/ml) [13]. The different inositol phosphates were eluted by HPLC according to Batty et al. [14].

Materials.- The B$_2$ specific inhibitor was kindly gifted by Dr. D. Regoli (Sherbrooke, Canada). All other materials were analytical or highest available grade.

RESULTS

Effects of BK on [Ca^{2+}]$_i$

[Ca^{2+}]$_i$ was measured with fura-2 in the presence of the organic anion transporter inhibitor sulphynpirazone. This drug was routinely used because in cell suspensions fura-2 slowly leaks out from intact cells [9]. Recently Di Virgilio et al. [15] found that this drug blocks fura-2 leakage without major adverse effect on PC 12 cell functions; its use makes calibration of fluorescent signal in terms of [Ca^{2+}]$_i$ more accurate, and it is particularly helpful when following long-lasting increases in [Ca^{2+}]$_i$.

BK caused a dose-dependent increase in [Ca^{2+}]$_i$: the threshold level was ~0.1 nM, and >100 nM gave maximal increases (not shown; see [16]). The kinetics of the [Ca^{2+}]$_i$ transients were biphasic: an abrupt, sudden increase peaking at ~5-10 sec (at BK concentrations >10 nM) and a lower sustained increase whose duration depended on BK concentration (>10 min at high [BK]; see Fig. 1a). In EGTA-containing medium the [Ca^{2+}]$_i$ spike was largely maintained, but Ca^{2+} addition after exposing cells to BK in a Ca^{2+}-free medium caused [Ca^{2+}]$_i$ to rise again to levels approaching the plateau seen in the Ca^{2+}-containing medium. The latter rise was completely dependent on BK since it was not seen in cells merely switched from a Ca^{2+}-free to a Ca^{2+}-containing medium (Fig. 1c). These observations suggest that, as in other experimental systems, the stimulation of BK receptors causes both Ca^{2+} redistribution and increased Ca^{2+} influx.

So as to characterize the type of BK receptor triggering [Ca^{2+}]$_i$ rises in PC 12 cells, three types of BK competitive inhibitor were used: desArg^9[Leu^8] BK (I-B$_1$) specific for the B$_1$ subtype, [Thi^{5,8},D-Phe^7] BK partially specific for B$_2$, and (Fig. 2; I-B$_2$) Arg^0[Hyp^3,Thi^{5,8},D-Phe^7] BK specific for B$_2$. I-B$_1$

Fig. 1. Effect
of BK on $[Ca^{2+}]_i$
in fura-2-
loaded cells.
a: Cells resus-
pended in KRH
containing Ca^{2+};
BK and (to 1 mM)
EGTA added as
indicated.
b & **c**: Ca^{2+}-free
KRH with 1 mM
EGTA; after
different pre-
exposures to BK
(**c** = control),
$CaCl_2$ added to
2 mM *(arrows)*.
Calibration:
see below.

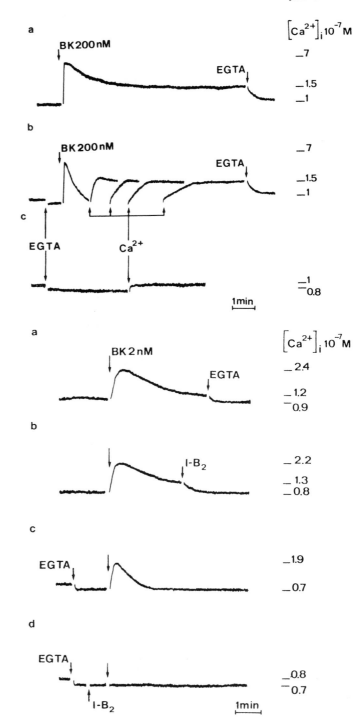

Fig. 2. Effect
of $I-B_2$ (B_2
receptor inhib-
itor; to 2 µM)
on $[Ca^{2+}]_i$ rises
induced by BK
(2 nM).
a & **b**: KRH
containing Ca^{2+}.
c & **d**: KRH
without Ca^{2+}.
Arrows denote
additions (to
1 mM in case
of EGTA).
Incubations
were at 37°.
In both Figs.
the calibration
of the fluores-
cent signal in
terms of $[Ca^{2+}]_i$
is shown.

Fig. 3. Effects of pre-incubation with PMA (100 nM, 5 min) on fura-2-loaded cells resuspended in Ca^{2+}-free KRH (+ 1 mM EGTA); BK added at different concentrations.
a: Effect on Ca^{2+} redistribution.
b: Effect on Ca^{2+} influx. $CaCl_2$ was added (2 mM) 90 sec after BK, and the initial rate of influx was measured.

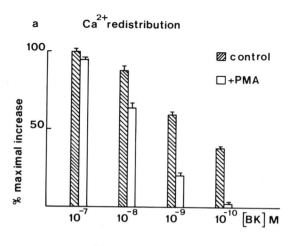

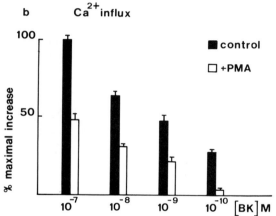

had no effect up to 0.1 mM, while $I-B_2$ appears to effectively block both the sustained (Fig. 2a,b) and the transient phase (Fig. 2c).

Another agent that induced a clear inhibition of the BK-induced $[Ca^{2+}]_i$ transients was PMA. Its effects were complex, depending on BK concentrations, and were distinctly different on redistribution and influx (Fig. 3). As has been described for other receptors [17], the inhibition required PMA concentrations in the 10^{-9}–10^{-7} M range and became appreciable within a few sec. after PMA addition (not shown; see [16]). However, unlike behaviour with other receptors, the inhibition of $[Ca^{2+}]_i$ redistribution disappeared above 10^{-8} M BK (Fig. 3a). On the contrary, influx was appreciably reduced by PMA at any concentration of BK employed, the % inhibition being remarkably similar over the range 10^{-9}–10^{-6} M BK (Fig 3b).

Fig. 4. Effect
of BK on memb-
rane potential,
in KRH medium
with (**a-c**) or
without (**d**)
Ca^{2+}.
a: fura-2
measurement of
$[Ca^{2+}]_i$.
b-d: bis-oxonol
measurement of
membrane poten-
tial. Indicated
additions
(arrows) besides
BK: EGTA (1 mM),
Ap (0.3 μM), TEA
(10 mM), ionomy-
cin (0.2 μM),
$I-B_2$ (30 μM) &
KCl (15 mM).

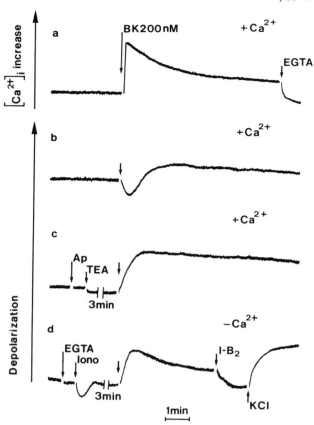

The sensitivity of BK-induced Ca^{2+} influx and redistribution to a number of pharmacological treatments was also
investigated. Neither process was affected by pertussis toxin,
cAMP analogues, cyclo-oxygenase inhibitors or blockers of
voltage-gated Ca^{2+} channels. Furthermore, 7-10 days of
culturing of PC 12 cells with NGF did not alter the response
of PC 12 to BK, in contrast with other stimuli (<u>e.g.</u> muscarinic
agonists).

Membrane potential

The effect of BK on membrane potential of PC 12 cells
was investigated with the potential-sensitive dye bis-oxonol
[10]. Under standard conditions (KRH medium with Ca^{2+}) BK
caused first a hyperpolarization with a return to and above
the resting level in 30-60 sec (Fig. 4b). The hyperpolarization
was blocked by inhibitors of Ca^{2+}-activated K^+ channels,
Ap and TEA (Fig. 4c), as well as by buffering $[Ca^{2+}]_i$ rises
with a high concentration of quin-2 (not shown; see [16]);
it coincided with the $[Ca^{2+}]_i$ peak caused by BK (Fig. 4a).

A hyperpolarization similar to that caused by the peptide was induced by the Ca^{2+} ionophore ionomycin, added in Ca^{2+}-free KRH medium containing EGTA.   In cells treated first with ionomycin in that medium, and then challenged with BK, the latter's hyperpolarizing effect was abolished.   The protein kinase-C activator PMA (100 nM) had no detectable effect on membrane potential, but induced a partial inhibition of hyperpolarization at low BK concentrations (<10 nM), i.e. under conditions in which the phorbol ester severely curtailed the BK-induced Ca^{2+} release from stores (not shown; see [16]). On the contrary, PMA inhibited BK-induced depolarization at all the nonapeptide levels tested (not shown; see [16]).

When either blockers of Ca^{2+}-activated K^+ channels or treatments that inhibit Ca^{2+} redistribution were employed, BK induced - with no lag phase - a net and persistent depolarization similar to that caused by 15 mM KCl (Fig. 4, c & d).   Depolarization was slightly reduced in extent and was shorter-lived in Ca^{2+}-free medium and was inhibited by PMA at all BK concentrations tested [16]. Pre-depolarization of the cells by increasing $[KCl]_o$ inhibited BK-induced responses [16]. In the presence of K^+ channel inhibitors no depolarization was triggered when BK was applied in a medium based on sucrose (250 mM).   However, BK-induced depolarization reappeared upon addition of Na^+ or Ca^{2+} to the sucrose-based medium.   The BK effect was found to be ~3-fold greater with 1 mM $CaCl_2$ than with 30 mM NaCl (see Fasolato et al. [16]).

In order to study the role of $[Ca^{2+}]_i$ elevations in the BK-induced depolarization, PC 12 cells were pre-treated with ionomycin (to deplete intracellular Ca^{2+} stores) in the Ca^{2+}-free, EGTA-containing medium. Under these conditions, $[Ca^{2+}]_i$ is unaffected by BK; however, the depolarization induced by BK was still largely preserved (Fig. 4d).   In contrast, when $[Ca^{2+}]_i$ was reduced well below the resting level by loading the cell with quin-2 in the Ca^{2+}-free, EGTA-containing medium, depolarization was markedly (50-100%) inhibited [see 16].

The bis-oxonol studies of membrane potential were complemented by preliminary electrophysiological studies carried out in the whole-cell mode of the patch-clamp technique [11]. When the peptide (5 nM) was applied to cells internally dialyzed with standard solution (potassium aspartate, 150 mM; Ca^{2+} buffered with EGTA at 200 nM), it induced the appearance of a persistent, inwardly directed current. When the calcium concentration of the intracellular perfusion fluid was decreased below 10^{-9} M, the BK-induced inward current failed

Fig. 5. Effect of BK on generation of (**a**) InsP$_3$, and (**b**) InsP$_4$. Cells were labelled with ^{3}H-*myo*-inositol for 48 h [16] and stimulated by 200 nM BK in KRH; *open bars* refer to cells pre-loaded with 50 μM quin-2/AM (and BK-stimulated) in Ca^{2+}-free KRH containing EGTA. Under these latter conditions resting [Ca^{2+}]$_i$ fell to ~10 nM and it did not increase in response to BK. Reaction stoppage at the different times was by ice-cold TCA (see text – also for the HPLC, with a Partisil SAX column).

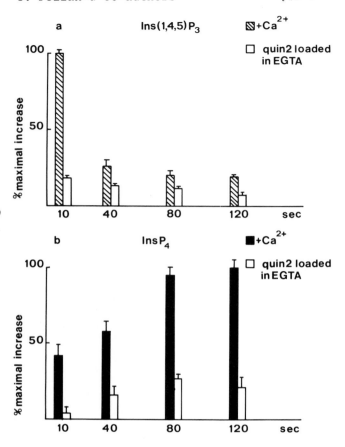

to appear in 3 cells out of 4 examined. **In summary:-**

CHARACTERISTICS OF BK-INDUCED DEPOLARIZATION

Cation selectivity: Ca^{2+} > Na$^+$ > choline$^+$
Ca^{2+} sensitivity: 10^{-8} M

Inhibited by:
1. Phorbol esters
2. Pre-depolarization
3. Specific receptor antagonists

Unaffected by:
1. Buffering [Ca^{2+}]$_i$
2. Dihydropyridines
3. Pertussis toxin
4. [cAMP]$_i$
5. Cyclo-oxygenase inhibitors
6. NGF

Inositol phosphate generation

Addition of BK (from 10^{-10} to 10^{-6} M) induces, in PC 12 cells, a rapid accumulation of various InsP's, in particular InsP$_3$ (1,3,4 and 1,4,5 isomers) and InsP$_4$, as revealed by standard HPLC analytical procedures [14]. Fig. 5 shows that

at early times the Ins(1,4,5)P$_3$ peak manifests the largest
increase. At later times (>30 sec) while this peak declined
towards basal levels, InsP$_4$ continued to accumulate. The
InsP$_4$ peak separated under standard HPLC conditions is known
to contain at least 3 isomers, co-eluting with ^{32}P standards
of Ins(D-1,3,4,6)P$_4$, Ins(D-1,3,4,5)P$_4$ and Ins(L-1,4,5,6,)P$_4$
[18]. Preliminary studies were carried out with a new
HPLC procedure which gives good resolution of InsP$_4$ isomers
[18].

By this procedure it was possible to demonstrate that
Ins(1,3,4,5)P$_4$ was massively (>7-fold) increased at early
times after BK, while the other two isomers increased more
slowly and to a lesser extent [see 16]. The new procedure
is, however, very laborious, and although the changes of
the total InsP$_4$ pool were systematically measured, the contribution of the various isomers in the different experimental
conditions has not been evaluated.

The rises in Ins(1,4,5)P$_3$ and InsP$_4$ were hardly affected
by removing Ca^{2+} from the incubation medium and by pre-treatment
with ionomycin (2×10^{-7} M), 3 min before BK, to deplete intracellular Ca^{2+} stores (not shown). On the other hand, Fig. 5
shows that lowering [Ca^{2+}]$_i$ by quin-2 in Ca^{2+}-free EGTA-
containing medium almost entirely abolished Ins(1,4,5)P$_3$ and
InsP$_4$ generation induced by BK.

A clear dissociation between the Ins(1,4,5)P$_3$ level on
the one hand, and Ca^{2+} influx and depolarization on the
other hand, is demonstrated by comparing the kinetics of
these three parameters.

In Ca^{2+}-containing medium depolarization and Ca^{2+} influx
are still well above basal at 6 min after addition of BK
(to 200 nM) while the level of InsP$_3$ (both 1,4,5 and 1,3,4
isomers) is indistinguishable from that of resting cells
(Fig. 6a). A dissociation between InsP$_3$ and depolarization
is also observed in cells pre-treated with ionomycin in
Ca^{2+}-free, EGTA-containing medium (Fig. 6b). Under these
conditions depolarization remained elevated (~40%) when the
levels of both InsP$_3$ isomers were indistinguishable from
basal. A rapid return of membrane potential to its resting
value is obtained by adding the specific B$_2$ antagonist
(Fig. 4d). On the other hand, depolarization and Ca^{2+} influx
remained closely associated, in respect not only of time
course but also of dose dependence (Fig. 7). We did not
attempt to correlate InsP$_4$ and Ca^{2+} influx-depolarization
because of the uncertainties in the relative isomer concentration in the HPLC peak, particularly at later incubation times.

Fig. 6. Time-course comparison of InsP's, Ca^{2+} influx and depolarization induced by BK (200 nM); values expressed as % of maximal BK increase. KRH medium: **a**, with Ca^{2+}; **b**, Ca^{2+}-free (+ EGTA, 1 mM) and with ionomycin added 3 min prior to adding BK. Ca^{2+} influx measured by adding EGTA at various times after BK [16].

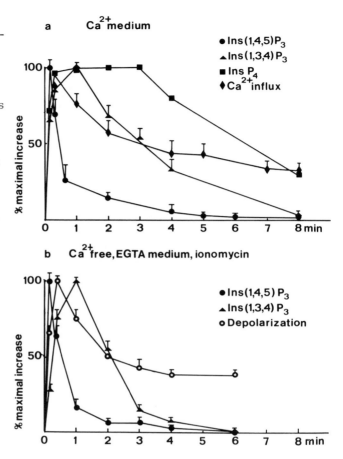

DISCUSSION

The natural nonapeptide BK elicits, in PC 12 cells, a number of phenomena which, depending on their time course, can be distinguished as transient or prolonged. The transient phenomena, viz. $Ins(1,4,5)P_3$ generation, Ca^{2+} redistribution and p.m. hyperpolarization, reached their peak in a few seconds (at 37°) and declined to basal levels in a few minutes. In contrast, the more persistent events (accumulation of various InsP's, Ca^{2+} influx and depolarization), although not showing an appreciable lag phase, reached their maximum in 20-60 sec and remained elevated for several minutes thereafter. A B_2 subtype of BK receptor appears responsible for all these phenomena: a B_1 antagonist was completely without effect, while a potent specific B_2 inhibitor competitively blocked both transient and persistent events.

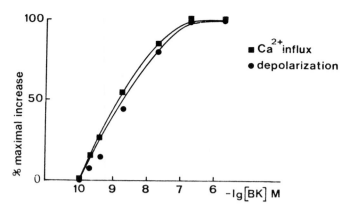

Fig. 7. Dependence of Ca^{2+} influx and depolarization on BK concentration (log scale), taking the maximal rate of each (at high BK concentrations) as 100%. Influx measured as in Fig. 6a, and depolarization as initial rate of BK-induced increase in bis-oxonol fluorescence signal.

As in other experimental systems, Ins(1,4,5)P_3 generation, Ca^{2+} redistribution and hyperpolarization appear causally related. However, there is no simple correlation between the peak $[Ca^{2+}]_i$ rise and the maximal extent of InsP$_3$ generation since, depending on the experimental conditions (PMA pre-treatment and BK concentrations), the same increase in InsP$_3$ is accompanied by different rises in $[Ca^{2+}]_i$ (Fasolato et al., to be published). The simplest interpretation of this result is that the amplitude of the $[Ca^{2+}]_i$ rise depends not only on the amount of InsP$_3$ eventually formed, but also on the speed at which this second messenger is generated.

The correlation between rises in InsP$_3$ and $[Ca^{2+}]_i$ is further complicated by the heterogeneity of cellular responses to agonists. On the one hand the correlation can be made only in populations of cells; on the other hand, measurement of $[Ca^{2+}]_i$ in single cells has revealed that, particularly at low concentrations of agonists, Ca^{2+} redistribution is asynchronous and often $[Ca^{2+}]_i$ oscillations are observed [19].

Interpretations, in relation to receptors

As for other receptors coupled to PtdIns(4,5)P_2 hydrolysis, the BK receptor in PC 12 cells appears to be modulated in a negative way by pre-treatment with active phorbol esters, and thus presumably by protein kinase-C-dependent phosphorylation(s). It is not clear, however, whether protein kinase-C affects the affinity of the receptors for the agonist or the coupling mechanism or both.

As mentioned above, while transient effects were subsiding, more persistent phenomena became appreciable and, at later times, predominated. In particular, Ca^{2+} influx and p.m. depolarization persisted for >10 min after BK addition. A prolonged increase of Ca^{2+} influx, not attributable to voltage-gated Ca^{2+} channels, is not a peculiarity of the response of PC12 cells to BK, since it has been observed also for muscarinic receptors in this cell type [20] and in many other experimental models with various agonists [7].

The prolonged increase in p.m. permeability to Ca^{2+} is a general consequence of the stimulation of receptors linked to $PtdInsP_2$ hydrolysis; but its mechanism is still obscure. In most systems it is not even clear whether a channel is involved or whether Ca^{2+} influx is attributable to an antiporter or to some other process. If a cation channel is opened as a consequence of BK receptor stimulation it is predicted that an inward current should be activated with consequent depolarization of the p.m. Both these phenomena were observed. The whole-cell patch-clamping studies, although preliminary, revealed that BK activates an inward current, with doubling of p.m. conductance. This result excludes a main role for the M current (whose inhibition would be expected to cause a decrease of membrane conductance) as has been suggested by Brown & Higashida [21] in astroglioma-neuroblastoma hybrid cells. We could not exclude, however, a Cl^- channel as the cause of this inward current. The data on membrane potential obtained with bis-oxonol strongly argue against the latter possibility and suggest the involvement of a BK-activated cation channel. In fact in a sucrose-based medium, devoid of permeant cations, the activation of a Cl^- current would be expected to cause depolarization; yet no effect was observed under these conditions. On the contrary, when Na^+ or, better, Ca^{2+} was added, a net depolarization was immediately observed.

Taken together, these observations suggest the involvement of a cation channel of low specificity, which might be similar to that described for ATP in smooth muscle cells. Since this BK-activated channel appears also permeable to Ca^{2+}, it is the logical candidate to account for the Ca^{2+} influx elicited by the peptide. The association of depolarization and Ca^{2+} influx in a variety of experimental conditions also supports this conclusion. Both phenomena were in fact persistent, had a similar concentration-dependence for BK, and were inhibited by PMA over the whole range of peptide concentration.

Recently receptor-activated cation channels of variable selectivity for divalent cations have been described in a number of cell types [1-3, 22-25].

Several hypotheses have been proposed very recently to explain Ca^{2+}-influx activation triggered by receptors coupled to InsP generation: (a) $[Ca^{2+}]_i$-regulated cation channel [2]; (b) InsP-modulated channel [3, 5, 6]; and (c) tight coupling between receptor and channel independent of second messenger [1]. Our data appear incompatible with (a), since increased Ca^{2+} influx was observed upon Ca^{2+} re-addition, at basal $[Ca^{2+}]_i$, to BK-treated cells and, more importantly, depolarization was seen after ionomycin depletion of intracellular Ca^{2+} stores. $[Ca^{2+}]_i$, however, appears to have a permissive role in BK activation of the inward current since the latter was inhibited by clamping $[Ca^{2+}]_i$ below resting level.

Role of inositol phosphates (InsP's)

A role for $Ins(1,3,4)P_3$ seems unlikely because this metabolite varies widely amongst various batches of PC 12 cells, as well as because its concentration decreased to basal well before Ca^{2+} influx/depolarization returned to the pre-stimulatory level. A similar time-course discrepancy was observed between Ca^{2+} influx, which peaked 30-60 sec after BK addition, and $Ins(1,4,5)P_3$ which peaked in 10-20 sec and, more importantly, was indistinguishable from basal when influx was still 40% of its maximal value. The latter observation also argues against models in which $Ins(1,4,5)P_3$ and $InsP_4$ act in tandem, the former opening a channel in the intracellular store and the latter activating a pathway directly from the extracellular space into the microsomal store [5, 6].

Because $InsP_4$ isomers are hard to separate we cannot exclude a direct effect of one of them in Ca^{2+}-influx activation. However, all electrophysiological studies performed so far have failed to reveal an effect of $Ins(1,3,4,5)P_4$ alone on p.m. conductance [25]. Indirect arguments against its having a primary role emerge from study of the kinetics of $Ins(1,3,4)P_3$ - a slowly metabolizable compound, which is generally assumed to be generated from $Ins(1,3,4,5,)P_4$ - whose return to basal level could reasonably be expected to parallel or follow that of $Ins(1,3,4,5)P_4$.

The experiment shown in Fig. 6b demonstrates that the concentration of $Ins(1,3,4)P_3$ is very similar to that of unstimulated cells when p.m. potential is still depolarized. Thus, although we have no direct data in support of the third mechanism [(c), above], a tight coupling between the channel and the receptor seems to us the most appealing mechanism. Whether, similarly to other channels (K^+ channels activated by muscarinic receptors [26] and voltage-gated Ca^{2+} channels [27]), the coupling is mediated by a G protein or whether the inositol phosphates play an ancillary, modulatory role, must remain speculative.

Acknowledgements.- We thank, besides Dr. Regoli (see **Materials**),
Dr. E. Wanke for the patch-clamp studies, Drs. C.P. Downes
and L. Stephens (SK&F, U.K.) for help in InsP isomer separation,
and Ms. L. DiGiorgio for secretarial assistance. A.P. was
a Fellow of the S. Romanello Foundation, Milan. Part-supported
by grants to T.P. from the Italian CNR 'Special Project Oncology'.

References

1. Benham, C.D. & Tsien, R.W. (1987) *Nature 328*, 275-278.
2. Tscharner, V., Prod'hom, B., Baggiolini, M. & Reuter, H. (1986) *Nature 324*, 369-372.
3. Kuno, M. & Gardner, P. (1987) *Nature 326*, 301-303.
4. Tertoolen, L.G.J., Tilly, B.C., Irvine, R.F. & Moolenaar, W.H. (1987) *FEBS Lett. 214*, 365-369.
5. Irvine, R.F. & Moor, R.M. (1986) *Biochem. J. 240*, 917-920.
6. Morris, A.P., Gallacher, D.V., Irvine, R.F. & Petersen, O.H. (1987) *Nature 330*, 653-656.
7. Meldolesi, J. & Pozzan, T. (1987) *Exp. Cell Res. 171*, 271-283.
8. Grynkiewicz, G., Poenie, M. & Tsien, R.Y. (1985) *J. Biol. Chem. 260*, 3440-3445.
9. Malgaroli, A., Milani, D., Meldolesi, J. & Pozzan, T. (1987) *J. Cell Biol. 105*, 2145-2153.
10. Di Virgilio, F., Milani, D., Leon, A., Meldolesi, J. & Pozzan, T. (1987) *J. Biol. Chem. 262*, 9189-9195.
11. Hamill, O.P., Marty, A., Neher, E., Sakmann, B. & Sigworth, F.J. (1981) *Pfluger's Arch. 391*, 85-100.
12. Wanke, E., Ferroni, A., Malgaroli, J., Ambrosini, A., Pozzan, T. & Meldolesi, J. (1987) *Proc. Nat. Acad. Sci. 84*, 4313-4317.
13. Wregget, K.A., Howe, L.R., Moore, J.P. & Irvine, R.F. (1987) *Biochem. J. 245*, 933-934.
14. Batty, I.R., Nahorski, S.R. & Irvine, R.F. (1985) *Biochem. J. 232*, 211-215.
15. Di Virgilio, F., Fasolato, C. & Steinberg, T.H. (1988) *Biochem. J. 256*, 959-963.
16. Fasolato, C., Pandiella, P.A., Meldolesi, J. & Pozzan, T. (1988) *J. Biol. Chem. 263*, 17350-17359.
17. Vincentini, L.M., Di Virgilio, F., Ambrosini, A., Pozzan, T. & Meldolesi, J. (1985) *Biochem. Biophys. Res. Comm. 127*, 310-317.
18. Stephens, L., Hawkins, P.T., Barker, C.J. & Downes, C.P. (1988) *Biochem. J. 253*, 721-733. [<u>Cf.</u> Kirk et al., art.#B-4.]
19. Jacob, R., Merritt, J.E., Hallam, T.J. & Rink, T.J. (1988) *Nature 335*, 40-45.
20. Pozzan, T., Di Virgilio, F., Vicentini, L.M. & Meldolesi, J. (1986) *Biochem. J. 234*, 547-553.
21. Brown, D.A. & Higashida, H. (1988) *J. Physiol. 397*, 185-207.
22. Mohr, F.C. & Fewtrell, C. (1987) *J. Immunol. 138*, 1564-1570.
23. McDougall, S.L., Grinstein, S. & Gelfand, E.W. (1988) *Cell 54*, 229-234.
24. Zschauer, A., Van Breemen, C., Buhler, F.R. & Nelson, M.T. (1988) *Nature 334*, 703-705.
25. Penner, R., Mattwer, G. & Neher, E. (1988) *Nature 334*, 499-504.
26. Brown, A.M. & Birnbaumer, L. (1988) *Am. J. Physiol. 254*, H401-H410.
27. Codina, J., Yatani, A., Grenet, D., Brown, A.M. & Birnbaumer, L. (1987) *Science 236*, 442-445.

#ncB

NOTES and COMMENTS relating to

CYTOSOLIC AND SEQUESTERED Ca^{2+}; INOSITOL PHOSPHATES

'COMMENTS' are on pp. 249-254, starting with Forum discussions
on the preceding main articles, then on the 'Notes'

#ncB.1

A Note on

CELLULAR CONTROL OF CALCIUM IONS IN RELATION TO MESSENGER EFFECTIVENESS

Ernesto Carafoli

Laboratory of Biochemistry,
Swiss Federal Institute of Technology (ETH),
CH-8092 Zurich, Switzerland

The intracellular messenger function of Ca^{2+} requires its rigorous control at very sub-μM concentrations. The mechanism adopted by evolution for the control process is the reversible complexation by special proteins, which are intrinsic to membranes, soluble in the cytoplasm or organized in non-membranous structures. The non-membranous proteins consist of two α-helices flanking a non-helical loop where Ca^{2+} coordinates to 6-8 oxygen atoms of carboxyl side-chains, and to carbonyl oxygen atoms in the peptide backbone. The helices are oriented perpendicularly to each other. This structural arrangement has been perfected by evolution to a point where only limited variability in the binding loop is now tolerated. The buffering of cell Ca^{2+} by these proteins is quantitatively limited by their total cellular amount. Their main function, rather than the buffering of Ca^{2+}, is thus the processing of the Ca^{2+} signal: they express hydrophobic surfaces upon complexing Ca^{2+}, and transmit the Ca^{2+} information to enzyme targets. CaM* is the best known example of this class of Ca^{2+}-binding proteins.

The buffering of cell Ca^{2+} is performed essentially by membranous proteins, which transport Ca^{2+} reversibly amongst membrane boundaries. They are found in the p.m., in the membrane of (endo)sarcoplasmic reticulum, in the mitochondrial inner membrane and in the membranes of lysosomes and of the Golgi body. They transport Ca^{2+} by four basic modes: ATPases, exchangers, channels and electrophoretic uniporters, only the first mode being capable of high-affinity interaction with Ca^{2+}. The existence of low- and high-affinity membrane transporters is demanded by the functional cycle of cells, which requires both the fine-tuning of Ca^{2+} and its less precise regulation. Structural studies on the Ca^{2+}-binding proteins in membranes are sparse: however, the primary structure

---

*Abbreviations.- CaM, calmodulin; p.m., plasma membrane.

of the Ca^{2+}-ATPase of sarcoplasmic reticulum of skeletal and heart muscle is known, and does not contain domains resembling the helix-loop-helix model of CaM. On the other hand, the Ca^{2+} pump of the p.m., whose sequence is also known, apparently contains domains resembling the Ca^{2+}-binding sites of CaM.

The membrane organelles contribute differently to the homeostasis of cell Ca^{2+}. While the concerted operation of the importing and exporting systems of the p.m. maintains the 10^4-fold gradient of Ca^{2+} between cells and ambient fluid, the total amount of Ca^{2+} exchanged with the intracellular medium is a minor fraction of the total Ca^{2+} used by cells. The Ca^{2+} penetration from the external spaces, however, triggers essential processes, e.g. the liberation of massive amounts of Ca^{2+} from intracellular stores.

The endo(sarco)plasmic reticulum is responsible for the rapid and precise regulation of Ca^{2+} at the sub-µM level. The mitochondrion is a low-affinity system which transports Ca^{2+} at a slow rate to regulate some Ca-sensitive matrix dehydrogenases. The mitochondrion also responds to abnormal increases of cell Ca^{2+} with the accumulation of Ca^{2+} and P_i, and with the deposition of calcium phosphate granules in the matrix. Since the free Ca^{2+} concentration is not altered significantly the matrix dehydrogenases are not deranged. When cell Ca^{2+} has returned to normal, mitochondria release the excess Ca^{2+} at a slow rate, compatible with the Ca^{2+}-exporting p.m. systems. This Ca^{2+}-buffering function of mitochondria plays an important defence role in cell injury, when the cytosol frequently experiences abnormally high Ca^{2+} concentrations.

#ncB.2

A Note on

METHODOLOGICAL PROBLEMS IN THE MEASUREMENT OF INTRACELLULAR FREE CALCIUM IN ENDOTHELIAL CELLS BY MEANS OF FLUORESCENT PROBES

Andreas Lückhoff, Roland Zeh and Rudi Busse

Department of Applied Physiology,
University of Freiburg,
Hermann-Herder-Str. 7, D-7800 Freiburg i. Br., F.R.G.

The fluorescent-dye approach devised in R.Y. Tsien's laboratory for measuring $[Ca^{2+}]_i$* is now very widely practised. To calibrate the signal in terms of $[Ca^{2+}]$ some major assumptions must be valid, notably: (1) only one Ca^{2+}-sensitive fluorochrome is present, in either Ca^{2+}-bound or free form; (2) the dye is strictly confined to the cytosol; (3) the entire $[Ca^{2+}]_i$-independent background signal can be considered a term additive to the $[Ca^{2+}]_i$-sensitive fluorescence; (4) intracellular probe behaviour is identical with that in calibration solutions.

Seldom in published work on $[Ca^{2+}]_i$ are these assumptions experimentally checked. Here we report some methodological problems observed during work with cultured endothelial cells and indo-1. One or more of the above assumptions were not necessarily true. Great care therefore has to be applied in the quantitative evaluation of some experimental conditions.

Methods (details in [1,2]).- Bovine-aorta endothelial cells were grown in subculture (2-4 stages) on quartz cover-slips. After incubation for 30-90 min with 1-2 µM indo-1/AM, [indo-1]$_i$ attained 20-100 µM (calculated assuming 1 pl intracellular volume).

Sequestration of indo-1/AM

We previously reported that a considerable fraction of indo-1/AM is incorporated within the endothelial cells but

*Abbreviations.- $[Ca^{2+}]$ or (intracellular/cytosolic) $[Ca^{2+}]_i$, free calcium concentration (similarly [indo-1]$_i$); AM, acetoxymethyl ester (of the probe indo-1); CaM, calmodulin; Cz, calmidazolium; Fn, fendiline; Kd, dissociation constant; LDH, lactate dehydrogenase; PGI$_2$, a prostaglandin; TMB-8, 8-(N,N-diethylamino)octyl-3,4,5-trimethoxybenzoate; FI, fluorescence intensity.

not metabolized. This indo-1/AM is not completely released by digitonin but almost completely with Triton X-100 (0.05% w/v). It is therefore assumed that this indo-1/AM is confined to intracellular membranes, probably by pinocytotic or phago-cytotic processes. Incorporation of particles without diges-tion is a prominent feature of cultured endothelial cells [3]. We are at present attempting to improve indo-1 loading by using the detergent pluronic F-127 [4]. Uncleaved ester disturbs little the emission of Ca^{2+}-bound indo-1 (recorded at 400 nm; excitation 330-355 nm) but seriously disturbs that of Ca^{2+}-free indo-1 (recorded at 480-500 nm; Fig. 1). The reason is that fluorescence of indo-1/AM is enhanced by light emitted at 400 nm by indo-1 (Fig. 1b). $[Ca^{2+}]_i$ must therefore be calculated from fluorescence emitted at 400 nm (FI_{400}).

For cells exposed to ionomycin (0.1 µM) the spectrum is very similar to that of Ca^{2+}-saturated indo-1 in the range 390-430 nm. As a test of whether the FI_{400} of indo-1 is as sensitive to changes in $[Ca^{2+}]_i$ as to $[Ca^{2+}]$ changes in a cell-free system, the following experimental protocol was devised.- A buffer containing 120 mM NaCl, 25 mM KCl, 10 mM HEPES and 0.3 nM indo-1 (pH 8.2, 20°) was titrated with EGTA until the indo-1-derived FI_{400} did not decrease any further, indicating <10 nM $[Ca^{2+}]$; this needed 2-4 µM EGTA. Cells incubated with indo-1/AM in a $CaCl_2$-free buffer were put into this buffer and ionomycin was added (to 10 µM). The total amount of cellular indo-1 exceeded that of the buffer by a factor of 10, whereas the extracellular volume exceeded intracellular volume by a factor of 10^4. Ca^{2+} (as $CaCl_2$) was cumulatively added, so increasing buffer $[Ca^{2+}]$ in 0.5 µM steps. The results (Fig. 2) are quite concordant with the assumptions that indo-1 binds Ca^{2+} with Kd = 250 nM, and that $[Ca^{2+}]_i$ may be calculated from FI_{400} as*

$$[Ca^{2+}]_i = Kd \cdot (FI_{400} - FI_{min}) / (FI_{max} - FI_{400})$$

$[Ca^{2+}]_i$ measurements in the presence of CaM antagonists

CaM plays a pivotal role in many $[Ca^{2+}]_i$-mediated cellular reactions [5]. Surprisingly, the CaM antagonists Cz and Fn enhance the ATP-stimulated release of PGI_2 from endothelial cells [6], although activation of one key enzyme of PGI_2 synthesis, phospholipase A_2, is thought to be under control of Ca^{2+}-CaM [7]. We studied the effects of these two CaM antagonists on endothelial $[Ca^{2+}]_i$. Fig. 3 shows the changes in FI_{400} evoked by addition of Cz to indo-1-loaded cells

*FI_{max}: FI of Ca^{2+}-saturated indo-1; F_{min}: FI of free indo-1

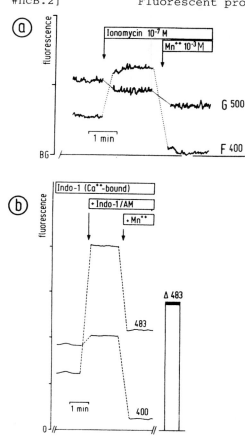

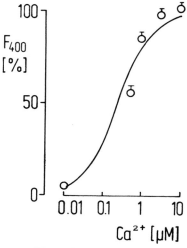

Fig. 2. Changes of
indo-1-derived fluor-
escence. To cells
exposed to ionomycin
in a Ca^{2+}-free solu-
tion, Ca^{2+} was added
as titrant. Each
circle is a mean
FI_{400} (±S.E.M.; n = 4)
as % of FI_{400} in the
presence of 2 mM Ca^{2+}.
The line indicates
the theoretical FI_{400}
of pure indo-1 assum-
ing a Kd of 250 nM.

Fig. 1. (a) Cellular indo-1-derived fluorescence as affected
by ionomycin (BG = background), which gave a FI_{400} ('F400') rise
that indicates an increase in $[Ca^{2+}]_i$ from 130 nM to a concentration that saturates indo-1 with Ca^{2+}. The fall in 500 nm
emission ('G500') was, however, less than expected from spectral
analysis of indo-1 in calibration solutions. G500 partly arises
from uncleaved indo-1 and is influenced by light emitted from
Ca^{2+}-saturated indo-1 at 400 nm - an effect demonstrated in b.
(b) In a cell-free system, indo-1/AM was added (to 0.1 μM) to
Ca^{2+}-bound indo-1 (0.1 μM). FI_{483} increased by Δ483 *(right column)*.
Then indo-1 fluorescence was quenched by Mn^{2+}. The remaining
fluorescence at 483 nm was markedly smaller than Δ483.
....

in the presence of 50 μM extracellular Mn^{2+}. Mn^{2+} binds
avidly to indo-1 and thereby totally quenches all indo-1-derived
fluorescence. At low concentrations, Mn^{2+} does not normally
disturb $[Ca^{2+}]_i$ measurements, since it enters the cytosol
only slowly. However, there was an immediate quenching

after Cz addition (Fig. 3), indicating that indo-1 and Mn^{2+} had come into contact with each other. This may be caused by influx of Mn^{2+} and/or by efflux of indo-1. In further experiments, Cz was directly demonstrated to release indo-1 as well as the cytosolic enzyme LDH from endothelial cells.

Fig. 4 shows effects of Fn.[*] It increased FI_{400}, indicating an increase in $[Ca^{2+}]_i$ since the increase in FI_{400} was not inhibited by Mn^{2+}. The effects were strictly dependent on extracellular Ca^{2+}, suggesting that Fn evoked a transmembrane Ca^{2+} influx. When ATP was added to Fn-treated cells, the cell membrane became permeable to Mn^{2+}, as evidenced by the decrease in FI_{400}. However, the amount of indo-1 released by the combined application of Fn and ATP was only barely detectable. No release of LDH was evident. It is concluded that the two lipophilic substances Cz and Fn penetrate the cell membrane, thereby interfering with the membrane integrity and enhancing permeability towards small molecules. This may lead to a Ca^{2+} influx and an increase in $[Ca^{2+}]_i$, and may explain the mechanism of action whereby the two substances stimulate endothelial production of PGI_2. However, there is no evidence that these effects are related to the CaM-antagonistic properties of Cz or **Fn**.

Measurements of $[Ca^{2+}]_i$ in the presence of TMB-8

TMB-8 is an intracellular Ca^{2+} antagonist that interferes with the mobilization of Ca^{2+} from intracellular stores [8]. In endothelial cells it severely inhibits the bradykinin-induced production of PGI_2. It is a strong fluorochrome: at concentrations required to inhibit PGI_2 production (30 µM) it almost doubled FI_{400} of indo-1-loaded endothelial cells. Bradykinin (10 nM, normally sufficient to increase $[Ca^{2+}]_i$ ~5-fold) only marginally enhanced FI_{400} in TMB-8-treated cells. A marked increase in FI was induced by ionomycin (3 µM) together with $CaCl_2$ (2.5 mM). Spectral analysis was consistent with indo-1 being saturated with Ca^{2+} under these conditions.

When the background signal was determined by addition of Mn^{2+} (to 3 mM), FI_{400} before exposure to ionomycin and $CaCl_2$ was calculated to be 62 ±4% (±S.E.M., n = 5) of F_{max}, which would correspond to an $[Ca^{2+}]_i$ of 300-500 nM. Similar experiments were performed with fura-2. Calculated $[Ca^{2+}]_i$ values in resting and bradykinin-stimulated endothelial cells were not significantly different from those obtained with indo-1, but the calculated $[Ca^{2+}]_i$ values in the TMB-8-treated cells were significantly lower (220 ±40 nM; n = 8).

---

[*]In Fig. 4, the purpose of ionomycin as used in (**a**) was to saturate indo-1 with Ca^{2+}; Mn^{2+} served to quench FI derived from indo-1.

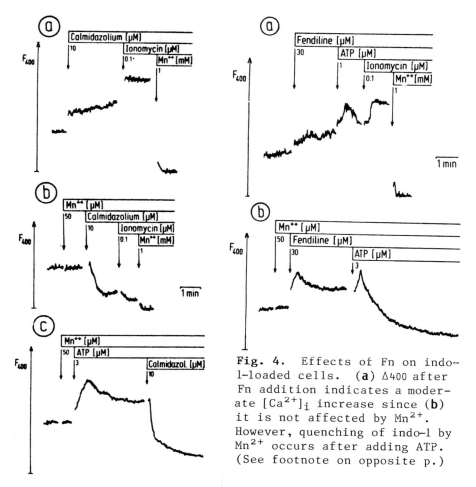

Fig. 4. Effects of Fn on indo-1-loaded cells. (**a**) Δ400 after Fn addition indicates a moderate $[Ca^{2+}]_i$ increase since (**b**) it is not affected by Mn^{2+}. However, quenching of indo-1 by Mn^{2+} occurs after adding ATP. (See footnote on opposite p.)

Fig. 3. Effects of Cz on indo-1 loaded cells. Δ400 after adding Cz (**a**) is reversed to a decrease in the presence of Mn^{2+} (**b**), a quencher of extracellular indo-1. The rate of quenching is enhanced after ATP pre-treatment (**c**).

These results may be interpreted as suggesting that TMB-8 induces some increase in $[Ca^{2+}]_i$. However, there is no biological reason to assume that TMB-8 raises rather than lowers $[Ca^{2+}]_i$. A more likely explanation is that the experiments were biased by methodological artefacts. As described early in this article, complex interactions may occur between different intracellular fluorochromes. Accordingly, measurements of $[Ca^{2+}]_i$ in the presence of a fluorescent compound such as TMB-8 should be interpreted with caution. Furthermore, we found by spectral analysis of cells loaded with TMB-8 but no $[Ca^{2+}]_i$ indicator that combined application of ionomycin and Mn^{2+} significantly lowered the recorded fluorescence. This was more pronounced at 400 nm than at 500 nm, and may have contributed to the surprising effects of TMB-8 on the calculated $[Ca^{2+}]_i$ values. In a cell-free system, Mn^{2+} did not interfere with the fluorescence of TMB-8. We therefore suggest that intracellular TMB-8 metabolism occurs, probably favoured by the reducing properties of the cytosol.

Concluding comments

It is essential to validate experimentally the major assumptions, listed at the start of this article, in measuring $[Ca^{2+}]_i$ with fluorescent indicators. With additional fluorochromes present, $[Ca^{2+}]_i$ calculations may yield incorrect values. Particular care should be applied in modern computer-based systems which perform on-line calculation of $[Ca^{2+}]_i$ from fluorescence ratios.

SUMMARY

Endothelial cells incubated with indo-1/AM do not completely hydrolyze the incorporated ester to the $[Ca^{2+}]_i$-sensitive dye. Uncleaved ester severely interferes with the fluorescence of the Ca^{2+}-free but not of the Ca^{2+}-bound form of indo-1. Calmidazolium (Cz; a CaM-antagonist that surprisingly enhances endothelial PGI_2 production) permeabilizes the cell membrane and thus causes indo-1 efflux and Ca^{2+} influx, precluding measurement of $[Ca^{2+}]_i$ changes. Pre-incubation with TMB-8, a strong fluorochrome, leads to suspiciously high $[Ca^{2+}]_i$ values, maybe due to complex interactions.

Acknowledgment.- Support came from Stiftung Volkswagenwerk (Grant I/62689).

References

1. Lückhoff, A. (1986) *Cell Calcium 7*, 233-248.
2. Lückhoff, A., Busse, R. & Bassenge, E. (1987) *Hypertension 9*, 295-303.
3. Ryan, U.S. (1988) *News Physiol. Sci. 3*, 93-96.
4. Owen, C.S. (1988) *Cell Calcium 9*, 141-147.
5. Cheung, W.Y. (1980) *Science 207*, 19-27.
6. Busse, R., Lückhoff, A., Winter, I., Mülsch, A. & Pohl, U. (1988) *Nauyn-Schmiedebergs Arch. Pharmacol. 337*, 79-84.
7. Craven, P.A. & De Rubertis, F.R. (1983) *J. Biol. Chem. 258*, 4814-4823.
8. Chiou, C.Y. & Malagodi, M.H. (1975) *Br. J. Pharmacol. 53*, 279-285.

#ncB.3

A Note on

SIGNALLING ROLE OF CONCENTRATIONS OF FREE CYTOSOLIC Ca^{2+} IN CHONDROCYTES

[1]A. Beit-Or[⊗], [1]Z. Nevo and [2]Y. Eilam[T]

[1]Department of Chemical Pathology,
The Sackler School of Medicine,
Tel Aviv University, Ramat Aviv, Israel

and [2]Department of Bacteriology,
The Hebrew University-Hadassah Medical School,
Jerusalem, Israel

Cartilaginous tissues are characterized by well defined age-related changes, viz. reduction in the rate of cell proliferation early in life [1], and typical alterations in the composition and structure of the PrG, occurring gradually over the years [2]. Cultured chondrocytes undergo similar changes but over a much shorter time-scale [3, 4]. During 30 days in culture the rate of cell division markedly decreases and that of PrG synthesis increases. The composition of the PrG formed is gradually altered to a typical mature matrix, characterized by a reduced content of chondroitin sulphate and an increased content of keratan sulphate. Accordingly, cultured chondrocytes may serve as a useful model to study the process of differentiation and ageing in cartilage.

Previous studies gave us some indications that the concentration of free cytosolic Ca^{2+} ($[Ca^{2+}]_i$) may signal the transition from proliferation to differentiation, which is manifested by enhanced PrG synthesis [5-7]. Under a variety of conditions low $[Ca^{2+}]_i$ was associated with enhanced PrG synthesis and a low rate of cell division, and *vice versa*. Here we review the results and discuss the possible signalling role of $[Ca^{2+}]_i$ in chondrocytes. The method for measuring $[Ca^{2+}]_i$ in cells surrounded by extracellular matrix is discussed.

MATERIALS AND METHODS

Cell culture.- Chick epiphyseal chondrocytes were isolated from 11 day-old embryos by mechanical disintegration and trypsin-EDTA digestion [8]. The single-cell suspension

---

[⊗]This work in part served towards Ph.D. requirements (Dept. as in [1]).
[T]to whom any correspondence should be sent. *Abbreviation by Ed.:*
PrG, proteoglycan; FI, fluorescence intensity. *Others:* BSA,
bovine serum albumin; GAG glycosaminoglycan; *see over for* AM.

obtained was applied to plates covered with soft agar. The cultures were grown in F-12 medium with 10% foetal calf serum, glutamine and antibiotics, under air/CO$_2$ (90:10) in a 37° incubator. Besides this 'regular' atmosphere (giving 18% O$_2$ tension), a different atmosphere was also used [7] - O$_2$/CO$_2$/N$_2$, 5:10:85 (~8% O$_2$ tension) - in examining the effect of O$_2$ tension on the ageing process. Cells were counted and sized in a Coulter Counter Model D Industrial with a 100 μm orifice. Generation times of chondrocytes were calculated by following cell counts.

   Rates of PrG GAG synthesis were measured in triplicate cultures of chondrocytes grown in the presence of carrier-free ^{35}S-sulphate (10 μCi/ml medium) for 24 h at 37° whilst the cultures were maintained on a rotary shaker in the chosen gas atmosphere. The incubation was terminated by boiling cells and medium for 3 min. GAG was isolated as previously described [8]. ^{35}SO$_4$ (Radiochem. Cent., Amersham) was 1100 mCi/mmol.

   Growth factor sources [5] were embryonic chick epiphyseal cartilage (C-CDF) and swarm rat chondrosarcoma (SRC-CDF).

   [Ca^{2+}]$_i$ determination.- Cells (~15 × 10^6) were transferred to 0.6 ml of MEM-HEPES-BSA medium, which contained 20 mM HEPES and 0.5 % BSA dissolved in bicarbonate-free MEM and brought to pH 7.4 at 37° by NaOH. After addition of quin-2 (to 75 μM) or fura-2 (3 μM) as the 2-acetoxymethyl (AM) esters, and incubation for 45 min at 37°, 3 ml of the medium was added to quin-2-loaded cells for a further 45 min of incubation (unnecessary with fura-2-loaded cells). To remove probe adsorbed onto the extracellular matrix, the spun-down cells were thoroughly washed with medium (× 4) to obviate masking of intracellular by extracellular fluorescence. Prior to measurement the cells, suspended in the same medium, were quickly centrifuged and suspended in salt solution containing (mM) NaCl, 145; KCl, 5; Na$_2$HPO$_4$, 1; Ca^{2+}, 1; MgSO$_4$, 0.5; glucose, 5; HEPES, 10; taken with NaOH to pH 7.4 at 37°.

   The settings for fluorescence measurements on loaded cells (37°; 10 nm slit) were 339/492 nm (quin-2) or 340/500 nm (fura-2). Calibration was by a procedure [9, 10] entailing initial addition of EGTA in Tris base (pH 8.1) to assess extracellular indicator (<5% of total). The fluorescence after cell lysis by Triton X-100 (0.4%) represented FI$_{min}$; CaCl$_2$ addition (to 4 mM) then gave FI$_{max}$. Calculation [9] then gave the [Ca^{2+}]$_i$ value.

RESULTS AND DISCUSSION

   Age-dependent changes in [Ca^{2+}]$_i$ were detected in fura-2-loaded chondrocytes: it attained 186 nM after 5 days, then decreased over 25 days to 72 nM (Table 1). During this

Table 1. Changes with age in fura-2-assessed $[Ca^{2+}]_i$, rates of PrG synthesis (^{35}S cpm/10,000 cells) and generation time.

Days in culture	$[Ca^{2+}]_i$, nM	PrG syn. rate	Generation time, h
5	186	717	40-45
14	151	1143	72
30	72	3143	>180

Table 2. Effect of external stimuli on quin-2-assessed $[Ca^{2+}]_i$ just after effector addition and, 24 h afterwards, on rates of PrG synthesis (as % of control) and on cell counts (±S.D.) as measure of proliferation rate. Culture, for 14 days, was under 'regular' atmosphere except for line 5.

Effector	$[Ca^{2+}]_i$, nM	PrG syn.	Cells × 10^{-4}	Gener^n. time, h	Ref.
- (control)	93.4	100	4.2 ±0.3	72 ±2	5, 6
C-CDF (50 µl)	70.2	250.3	4.0 ±0.4		5, 6
SRC-CDF (50 µl)	58	451.7	5.7 ±0.1		5, 6
A-23187 (0.3 µM)	150	28.4	11.3 ±0.5		6
Low O_2 tension	178	40.0	-	45 ±7	7

time the PrG synthesis rate increased >4-fold and the generation time, which inversely correlates with the cell-division rate, markedly increased.

Previously [5] we had found that at 12-14 days external factors which affect the PrG synthesis rate also affect $[Ca^{2+}]_i$. $[Ca^{2+}]_i$ fell immediately and PrG synthesis rose after adding C-CDF - with no effect on the low cell-division rate - or SRC-CDF, which slightly raised cell counts, contrary to the general inverse relation between proliferation and PrG synthesis and maybe reflecting the factor's chondrosarcoma origin. Ca^{2+} ionophore decreased PrG synthesis and enhanced prolifation [5, 6] (Table 2). In these experiments $[Ca^{2+}]_i$ was measured with quin-2, which gave rather lower absolute values for $[Ca^{2+}]_i$ than fura-2.

A signalling role of $[Ca^{2+}]_i$ in cell proliferation has been suggested for several cell types [11-13], wherein stimuli that induce proliferation induced a rapid and transient increase in $[Ca^{2+}]_i$. Agents which increase $[Ca^{2+}]_i$, e.g. Ca^{2+}-ionophore, accelerated cell division [14]. We suggest that in chondrocytes the cell-division rate is dependent on the steady-state $[Ca^{2+}]_i$ level, a decrease being associated with slower cell division and faster PrG synthesis. The latter may be due directly to the decrease in cell division. However, our experiments indicate that division is undiminished on adding growth factors

such as C–CDF or SRC–CDF; they merely enhance PrG synthesis
and decrease $[Ca^{2+}]_i$. These results indicate that the primary
signal for 'differentiation', _i.e._ enhanced PrG synthesis,
may be the decrease in $[Ca^{2+}]_i$.

In contrast to other systems, in which a transient
increase in $[Ca^{2+}]_i$ represents the $[Ca^{2+}]_i$ signal [15], in
chondrocytes the steady-state level of $[Ca^{2+}]_i$ seems to be
the factor which determines the rates of proliferation and
PrG synthesis. Thus gradual changes in $[Ca^{2+}]_i$ may be the
factor determining the time-scale of ageing in cartilagenous
tissue.

References

1. Stockwell, R.A. (1979) in _Biology of Cartilage Cells,
    Biological Structure and Function_ (Harrison, R.J. &
    McMinn, R.M.H., eds.), Cambridge Univ. Press, Cambridge,
    pp. 213–240.
2. Roughley, P.J. & Mort, J.S. (1986) _Clin. Sci. 71_, 337–344.
3. Hunter, S.J. & Caplan, A.I. (1983) in _Cartilage_ (Hall, B.K.,
    ed.), Vol. 2, Academic Press, New York, pp. 87–120.
4. Schuckett, R. & Malemud, C.J. (1986) _Mech. Ageing Develop.
    34_, 73–90.
5. Eilam, Y., Beit-Or, A. & Nevo, Z. (1985) _Biochem.
    Biophys. Res. Comm. 32_, 770–778.
6. Eilam, Y., Beit-Or, A. & Nevo, Z. (1987) in _Current
    Advances in Skeletogenesis III_ (Hurwitz, S. & Sela, J., eds.),
    Heiliger Publ. Co., Jerusalem, pp. 132–139.
7. Nevo, Z., Beit-Or, A. & Eilam, Y. (1988) _Mech. Ageing
    Develop. 45_, 157–165.
8. Nevo, Z., Horwitz, S. & Dorfman, A. (1972) _Develop. Biol.
    28_, 219–228.
9. Tsien, R.Y., Pozzan, T. & Rink, T.J. (1982) _J. Cell Biol.
    94_, 325–335.
10. Wollheim, C.B. & Pozzan, T. (1984) _J. Biol. Chem. 259_,
    2262–2267.
11. Tsien, R.Y., Pozzan, T. & Rink, T.J. (1982) _Nature 295_,
    68–71.
12. Busa, W.B. & Nuccitelle, R. (1985) _J. Cell Biol. 100_,
    1325–1329.
13. Hesketh, T.R., Moore, J.P., Morris, J.D.H., Taylor, M.V.,
    Rogers, J., Smith, G.A. & Metcalfe, J.C. (1985) _Nature
    313_, 481–484.
14. Resch, K., Bouillon, D. & Gemsa, D. (1978) _J. Immunol.
    120_, 1514–1520.
15. Rasmussen, H. (1986) _N. Engl. J. Med. 17_, 1094–1169. and

#ncB.4

A Note on

MODULATION OF PHAGOCYTE CHEMILUMINESCENCE BY VARYING CONCENTRATIONS OF CALCIUM AND SODIUM IN THE EXTERNAL MEDIUM

S. Janah, [†]I. Das and A. Fleck

Departments of Chemical Pathology and [†]Psychiatry, Charing Cross & Westminster Medical School, Fulham Palace Road, London W6 8RF, U.K.

An earlier series of experiments [1] using f-MLP[*] as the stimulating agent revealed a difference between the luminol-enhanced CL response of human PMN's treated with heparin and EDTA. This was followed up by the present series where $[Ca^{2+}]$ was varied in the external medium (pH 7.4), this being either Na^+-fortified Krebs-Ringer (KR; 151 mM Na^+) or KR buffer containing 10 mM HEPES also (KRH; 164 mM Na^+).

The response to f-MLP induced in PMN's has a well established pattern. It is initiated by binding of the ligand to the membrane receptor and the appropriate G-protein, which in turn stimulates the hydrolysis of phosphatidyl-$InsP_2$ to $InsP_3$ and diacylglycerol. The former releases internal stores of Ca^{2+} and the latter activates protein kinase-C which phosphorylates various effector proteins [2]. This leads to the initiation of many cellular functions, of which superoxide generation and degranulation are linked with the CL response of these cells. The amplification and maintenance of these functions coincide with influx of Ca^{2+} from the external medium [3, 4].

There was an amplification of the CL response in parallel with the progressively increasing levels of Ca^{2+} in the medium (Table 1). The KRH buffer appeared to enhance the CL response more than the KR buffer. This could possibly be attributed to the higher Na^+ content of the KRH buffer. The results of further experiments to establish the regulatory role of the external Na^+ gradient in influencing the CL

---

[*]*Abbreviations, besides buffers* (KR, KRH).- CL, chemiluminescence; InsP, an inositol phosphate - 1,4,5-substituted if $InsP_3$; f-MLP, formyl-methionyl-leucyl-phenylalanine; PMN, polymorphonuclear leucocyte.

Table 1. Height (as mV) of second CL peak for PMN cells
(4 × 10^5) treated with 0.5 μM f-MLP in the presence of 50 μM
luminol. For the buffers (KR, KRH), see text.

Ca^{2+} in medium, mM	Peak with KR	Peak with KRH
0.2	300	425
0.5	940	1000
1.0	1200	1440

response are being published elsewhere. Evidence of similar
enhancement has been reported in studies on chemotaxis [5]
and platelet function [6].

References

1. Janah, S. & Das, I. (1988) _Biochem. Soc. Trans. 16_, 289-291.
2. Pozzan, T., Lew, D.P., Wollheim, C.B. & Tsien, R.Y. (1983)
   Science 221, 1413-1415.
3. Synderman, R., Smith, C.D. & Verghese, M.W.(1986) _J.
   Leuk. Biol.40_, 785-800.
4. Korchak, H.M., Vienne, K., Rutherford, L.E.,
   Wilkenfield, C., Finkelstein, M.C. & Weissmann, G. (1984)
   J. Biol. Chem. 259, 4076-4082.
5. Simchonitz, L. (1985) _J. Biol. Chem. 260_, 13248-13255.
6. Siffert, W. & Akkerman, J.W.N. (1988) _Trends Biochem. Sci.
   13_, 148-151.

#ncB

COMMENTS related to

CYTOSOLIC AND SEQUESTERED Ca²⁺; INOSITOL PHOSPHATES

Comments on #B-1: G.L.E. Koch et al. - 55 kDa BINDING PROTEIN

SUMMARY OF THE BROADER TALK given at the Forum by
G.L.E. Koch:
A simple method has been developed for the isolation and
fractionation of reticuloplasm, the luminal content of the
e.r., from plasma cells. Five major proteins make up the
bulk of the material. Using a nitrocellulose-based Ca^{2+}-binding
assay, it was found that several of the e.r. proteins bind
Ca^{2+} in the mM range. The expression of these proteins
is also increased specifically by agents such as Ca^{2+} ionophores, which perturb cellular calcium, suggesting a possible
role in calcium homeostasis/storage. Sequencing and protein
chemistry suggest that one of the proteins called CRP 55
might be best suited to perform a calsequestrin-type calcium
signalling function. Studies with Ca^{2+} ionophores also suggest
that Ca^{2+} ions are involved in several aspects of e.r.
structure and function.

Koch was asked whether ionophore action is demonstrable
in Ca^{2+}-free medium - which would distinguish loss of e.r.
Ca with elevated cytosolic Ca^{2+} from loss of e.r. Ca with
lowered cytosolic Ca^{2+}. **Reply:** not investigated; ionophores
were always added in Ca^{2+}-containing medium. But Ca^{2+} removal
has somewhat similar effects to ionophore; this would reflect
e.r. Ca depletion. **Reply to E.M.Bevers.-** A23187 was the
ionophore (removed by harvesting the cells), at μM levels;
we cannot express levels relative to phospholipid (cf. the
membrane-perturbing action of ionophores) since this was not
determined.

Comments on #B-2: A.P. Dawson et al.- Ca^{2+} MICROSOMAL STUDIES
#B-3: N. Crawford et al. - PLATELET STUDIES

M.B. Vallotton, to A.P. Dawson.- It would be informative
to check whether the density of the e.r. markers is different
for the fused vesicles compared with the rest of the
e.r.: conceivably in that respect the e.r. might not be
homogeneous and consequently the fusion might not be random.
Reply.- This is indeed an interesting suggestion which we
plan to follow up. **N. Crawford** asked whether (1) aggregated
vesicles in close association would, in the energy transfer

experiments, be revealed as fused by transfer of fluorescence energy, and (2) the GTP-enhancing effect on InsP$_3$ release is itself Ca^{2+}-dependent. **Replies.-** (1) Aggregation of vesicles induced by PEG addition is not sufficient in itself to lead to fluorescence energy transfer; but it is possible that the sort of transient junctions envisaged by Gill's group might suffice, because it would behave as a transient fusion. (2) It is difficult to answer the question at the moment. Certainly the fusion rate, measured by fluorescence energy transfer, is independent of [Ca^{2+}]. However, there is a much larger InsP$_3$-stimulated Ca^{2+} release with low rather than high [Ca^{2+}]; either this is something to do with the InsP$_3$-dependent release mechanism or it might reflect the formation of different-sized vesicles at the different Ca^{2+} concentrations: we do not know as yet. **F.R. Maxfield asked,** concerning the GTP effect, whether (1) it is specific for Ca^{2+} rather than due to leaky fusion, and (2) similar effects might occur *in vivo*. **Replies.-** (1) There may be some increased leakiness: for example, the latency of mannose-6-phosphate is reduced. (2) Probably this is the case, since similar effects of GTP are demonstrable with intact-cell preparations. **Dawson, answering F.A. Lai.-** No one has looked at the effect of guanidine on e.r. vesicles, but we plan to do so.

N. Crawford, answering A.P. Dawson.- We have indeed used immuno-gold labelling to see where the MAb goes in permeabilized platelets: it appears to label most intensely the apical regions of the cell, which is the main location for the dense tubular membrane system and for microtubules. **Remarks by T.J. Rink.-** (1) Up to 1 mmol Ca^{2+} per litre of cell water can be discharged into the cytosol of stimulated platelets. (2) Sakardi *et al*. and Rink & Sage have quite convincing evidence for Ca^{2+} pumping in platelet p.m. (3) The intrinsic Ca^{2+}-buffering sites of cytosol are not well defined but include Ca^{2+}-binding proteins and organic polyphosphates (although Mg^{2+} will be occupying many such sites, since up to 10-15 mmol Mg^{2+}/litre cell water is at least 90% bound to these sites).

Comment on #**B-4**: C.J. Kirk *et al*. - CHARACTERIZING InsP$_4$'s IN WRK 1 CELLS

C.J. Kirk, answering A.K. Campbell.- InsP$_3$ concentrations appear from various measurements to be in the low-μM range; similarly for InsP$_4$, using equilibrium-labelled cells, but InsP$_5$ and InsP$_6$ may be much higher, maybe >100 μM in some cells. We indeed see cyclic InsP's in stimulated WRK 1 cells: cyclic Ins(1,5)P$_2$ accumulates to ~0.1% of the levels of the non-cyclic compound. I suspect that it accumulates as a by-product of the protein kinase C mechanism; the kinetics of its accumulation do not obviously correlate with any physiological changes in the cells.

Comments on #**B-5**: C.W. Taylor - Studies with InsP$_3$ analogues
#**B-6**: P.H.   Cobbold et al. - Ca^{2+} transients

M.B. Vallotton asked C.W. Taylor whether he believed that the Ca stores remain empty exclusively as a consequence of the continuous presence of non-metabolizable InsP$_3$S$_3$ — or whether he also considered the possibility that its inability to be phosphorylated to InsP$_4$ prevents refilling of the Ca stores.   **Reply.-** In our system InsP$_4$ is devoid of any effect on Ca^{2+} transport.   **Taylor, in reply to** queries by **T.J. Rink, P.H. Cobbold and F.R. Maxfield,** centered on whether InsP$_3$S$_3$-induced oscillations in Xenopus oocytes argue against a role of InsP$_3$ (maybe remaining high) in causing Ca^{2+} oscillations.— When we inject InsP$_3$S$_3$ into the oocytes, we still see oscillations of membrane potential (reflecting Ca^{2+} oscillations), but they die down, perhaps because the analogues may be diluted by diffusion into these large cells – the analogues may also trigger oscillations in the natural InsP's.   Altogether Xenopus oocytes are not the best system in this connection; InsP$_3$ induces oscillations in many cells, but may initiate a feedback loop leading to further oscillatory functions of InsP$_3$; in Xenopus InsP$_3$S$_3$ is effectively inactivated, despite its metabolic stability, by dilution into a large cytoplasmic volume.

T. Pozzan, to Cobbold concerning his 'model'.- I can see two difficulties: (1) there are receptors which are insensitive to protein kinase C feedback inhibition; (2) with several thousands of receptors you would expect asynchronous individual activations.   **Cobbold, answering E. Johnson** concerning hepatocyte stimulation by 'AVP' (Arg-vasopressin).- At low concentrations a 4-min interval is suffient for re-stimulation to induce the response (and induce oscillations); but with high concentrations the cell desensitizes very quickly: only one large transient peak is observed (and no oscillations), and the cell remains refractory to re-stimulation.   But as soon as oscillations are induced, a second response is always achievable.   **A.P. Dawson asked** whether the spikes are discernible all over the cell, or whether individual spikes can arise from particular regions.   **Cobbold's reply.-** Aequorin does not produce enough counts to look at particular regions of the cell; fura-2 should be better for imaging oscillations. **A. Malgaroli** wondered about the importance of opening of SMOC's during receptor activation in a cell's oscillatory behaviour.   ——————————

Comment on #**ncC** .5: O.H. Petersen (more pertinent to #**B** than to #**C**)

P.J. England asked Petersen whether the synergistic effect of InsP$_4$ and InsP$_3$ is mediated by conjoint direct action or by InsP$_4$ plus the Ca^{2+} released by InsP$_3$. **Reply.**- The effect of InsP$_4$ is entirely dependent on the presence of InsP$_3$. The simplest model concept is that InsP$_4$ allows Ca^{2+} entry into a pool from which InsP$_3$ can release Ca^{2+}.

Comments on #**B-7**: T. Pozzan & co-authors - [Ca^{2+}]$_i$ MODULATION
          #**ncB.2**: A. Lückhoff <u>et al.</u> - [Ca^{2+}]$_i$ MEASUREMENT

T. Pozzan, replying to F.R. Maxfield.- PMA treatment inhibits InsP$_3$ production in response to BK, but there is not a straightforward good correlation between [InsP$_3$]$_i$ and the maximal rise in [Ca^{2+}]$_i$, which rather seems related to the net rate of production of InsP$_3$ as distinct from its level. **W.E.J.M Ghijsen, to T. Pozzan.**- PC 12 cells also contain L-type Ca^{2+} channels which are linked to catecholamine release. BK apparently affects another type of Ca^{2+} channel which is less specific and which does not induce secretion. Conceivably this phenomenon could, for certain agonists, be a discriminatory route to uncoupled Ca^{2+}-entry and secretion. **Pozzan's response.**- Indeed dihydropyridines did not affect BK-activated Ca^{2+} influx. **M. Caulfield asked** whether the possibility of catecholamine release serving as the mediator of BK-induced hyperpolarization in PC 12 cells had been eliminated. **Reply.**- Yes, because hyperpolarization is still seen in Ca^{2+}-free medium. **Remarks by O.H. Petersen.**- Non-selective Ca^{2+} channels are in general unlikely sources of regulated Ca^{2+} entry, which even if of slight extent needs ingress of large amounts of Na$^+$. Activation of non-selective channels has, however, been demonstrated directly under certain circumstances [Maruyama, Y. & Petersen, O.H. (1982) *Nature* 299, 159-161 & *300*, 61-63] and may certainly operate in the circumstances you describe. **Reply to T.J. Brown:** in PC 12 cells the observed changes in membrane potential do not influence the generation of InsP's.

Caroline E.M. Jones asked A. Lückhoff, who had stated that Pluronic was toxic, whether he doesn't use it in his system or whether he has an alternative. **Reply.**- We now use single-cell suspensions, so don't need Pluronic. But I feel that although it produces good data it imperils cell integrity; alternatives tend to be specific to particular cell systems. **F.R. Maxfield commented** that some workers use as alternatives carrier proteins, <u>e.g.</u> albumin, or sonicate the solution; sometimes they centrifuge the solution to remove the microcrystals, to prevent them being phagocytosed.

============

SOME LITERATURE PERTINENT to #B *THEMES, noted by Senior Editor*
- with **bold type** for test material or other 'keyword'

Purification of **hepatocalcin-55**.- Gultekin, H., Horitsu, K.
& Domagk, G.F. (1987) *Int. J. Biochem. 19*, 671-677.

CaM assay by competitive binding.- LeVine, H., Sahyoun, N.
& Cuatrecasas, P. (1986) *Anal. Biochem. 152*, 183-188.

Purification (from myocardium) of **calpastatin** - inhibitor
of calpains (intracellular Ca^{2+}-dependent proteinases).-
Mellgren, R.L., Nettey, M.S., Mericle, M.T., Renno, W. &
Lane, R.D. (1988) *Prep. Biochem. 18*, 183-197.

'A spatial-temporal model of **cell activation**', involving
Ca^{2+}, the inositol lipid system and protein kinase C: it
"can account for sustained cellular responses during sustained
application of diverse physiological stimuli that mediate
endocrine secretion, neurotransmission, or cell proliferation".-
Alkon, D.L. & Rasmussen, H. (1988) *Science 239*, 998-1005.

'Perturbation of liver microsomal calcium **homeostasis** by
ochratoxin A': it apparently impairs the e.r. membrane probably
<u>via</u> enhanced lipid peroxidation.- Khan, S., Martin, M.,
Bartsch, H. & Rahimtula, A.D. (1989) *Biochem. Pharmacol. 38*,
67-72.

Neutrophil activation: $[Ca^{2+}]_i$ and arachidonic acid lipoxygenation metabolite(s) seem to mediate **granule exocytosis**
elicited with IL-1.- Smith, R.J., Epps, D.E., Justen, J.M.,
Sam, L.M., Wynalda, M.A., Fitzpatrick, F.A. & Yein, F.S. (1987)
Biochem. Pharmacol. 36, 3851-3858.

Amongst agents compared for inhibition of $InsP_3$-induced
Ca^{2+} release from **platelet membrane vesicles**, cinnarizine
and flunarizine were notably potent and could serve as probes
in studying the channels concerned.- Seiler, S.M., Arnold, A.J.
& Stanton, H.C. (1987) *Biochem. Pharmacol. 36*, 3331-3337.

$[Ca^{2+}]_i$ role in **exocytosis** from 3 subcellular compartments
in human neutrophils.- Lew, P.D., Monod, A., Waldvogel, F.A.,
Dewald, B., Baggiolini, M. & Pozzan, T. (1986) *J. Cell Biol.
102*, 2197-2204.

Non-mitochondrial Ca^{2+} **pumping sites** in human neutrophils:
microsomal, but not in e.r. (gradient centrifugation; maybe
endosomes or Golgi.- Kraus, K-H. & Lew, P.D. (1987) *J. Clin.
Invest. 80*, 107-116.

A23187 treatment of hepatocytes in presence of Ca^{2+} perturbs
both synthesis and degradation of **cAMP** (functional uncoupling
which mimics desensitization).- Irvine, F.J. & Houslay, M.D.
(1988) *Biochem. Pharmacol. 37*, 2773-2780.

'The Ca signal from fura-2 loaded **mast cells** depends strongly on the method of dye loading'.- Almers, W. & Neher, E. (1985) *FEBS Lett. 192*,   13-18. With fura-2/AM, degranulation results in ~50% loss of fluorescence, although $[Ca^{2+}]_i$ stays near-constant. After micro-injection of the impermeant K salt of fura-2, there is no loss of fluorescence; transient fluor- escence changes occur, indicating a transient $[Ca^{2+}]_i$ increase. Conclusion: the AM ester accumulates in the secretory granules and is lost from the cell by exocytosis.

Heterogeneity of $[Ca^{2+}]_i$ responses among mitogen-stimulated (PHA, anti-CD3) **T-cells**: indo-1 and immunofluorescence used simultaneously with flow cytometry.- Rabinovitch, P.S., June, C.H., Grossman, A. & Ledbetter, J.A. (1986) *J. Immunol. 137*,   952-961.

Cancer cells with **multiple drug resistance** have not given clear-cut results for a role of Ca^{2+}.- Beck, W.T. (1987) *Biochem. Pharmacol. 36*, 2879-2887.

The $[Ca^{2+}]_i$ rise in stimulated **taste cells** is mediated, in the case of a bitter agent, not electrogenically but by a surface receptor.- Akabas, M., Dodd, J. & Al-Awqati, Q. (1988) *Science 242*,   issue of 18 Nov.

Synthesis and characterization of **19F-NMR** chelants for measuring $[Ca^{2+}]_i$.- Levy, L.A., Murphy, E. & London, R.E. (1987) *Am. J. Physiol. 252*, C441-C449.

Ca²⁺ TRANSIENT MEASUREMENTS IN THE CARDIAC CYCLE BY ¹⁹F NMR

- Ed.'s summary of a Forum abstract by J. METCALFE & co-authors, who cited their papers in *Cell Calcium 6*,183-195 (1985) and *Proc. Nat. Acad. Sci.: 80*, 7178-7182 (1983) *& 85*, 9017-9021 (1988).

The $[Ca^{2+}]_i$ transient, pH_i and $[Mg^{2+}]_i$ have been followed by ¹⁹F NMR at 16 phases in perfused ferret hearts paced at 1.25 Hz, at 30°.   After a 50 msec delay, $[Ca^{2+}]_i$ rose rapidly to a >1.5 μM maximum after 150 msec, then fell to the start value; the indicator was 5FBAPTA.   In contrast, pH_i and $[Mg^{2+}]_i$ (7.03; 1.2 mM) hardly changed. A decrease in developed pressure when the $[Ca^{2+}]_i$ indicator (but not the pH_i or $[Mg^{2+}]_i$ indic- ator) was loaded into hearts was substantially reversed by the addition of Zn^{2+} (to 50 μM) to the perfusion fluid: the Zn-5FBAPTA resonance appeared, connoting displacement of Ca^{2+} from the indicator, whose action on pressure may therefore be due to its Ca^{2+}-buffering effect on the myoplasm. Co-loading hearts with the $[Ca^{2+}]_i$ and pH_i indicators allowed several free cations in cytosol to be measured, offering a useful addition to the methods available for monitoring cardiac function and pharma- cology. *[Work done in Biochemistry Dept., Univ. of Cambridge.]*

==============

Section #C

SECRETORY AND OTHER PROCESSES INVOLVING Ca^{2+}
(not muscle or nerve)

#C-1

MEASUREMENT OF CYTOPLASMIC FREE CALCIUM IN SINGLE MOTILE CELLS

Frederick R. Maxfield[*], [†]Andrew Bush, [†]Peter Marks and Michael L. Shelanski

Department of Pathology, 630 W. 168th Street, Columbia University College of Physicians and Surgeons, New York, NY 10032, U.S.A.

and [†]Sackler Institute of Graduate Biomedical Science, New York University School of Medicine, New York, NY 10016, U.S.A.

Changes in cytosolic free calcium, $[Ca^{2+}]_i$, associated with cell motility have been measured using the fluorescent indicator dye fura-2. A computer-controlled microscope spectrofluorimeter was employed for measuring $[Ca^{2+}]_i$ changes throughout the cell, and digital image analysis for detecting localized gradients of $[Ca^{2+}]_i$ within the cell. Changes in $[Ca^{2+}]_i$ associated with phagocytosis have been observed in neutrophils and macrophages. In macrophages periodic transient increases in $[Ca^{2+}]_i$ were observed during phagocytosis; their duration was ~5 sec, and they were repeated at 20-40 sec intervals. During mitosis we saw long-lasting intracellular gradients of $[Ca^{2+}]_i$ as well as transient increases. A region of high $[Ca^{2+}]_i$ is centered near the mitotic spindle during anaphase. Transient increases were obtained throughout metaphase in close association with the start of cytokinesis.

Methods for making measurements on single cells with high time-resolution are discussed. Their strengths and limitations have been demonstrated by practical experience using several cell types. Without adequate precautions many types of serious artefacts can be encountered. Important considerations include the localization and chemical form of the indicator dye, the temporal and spatial resolution of the instrument, and the effects of illumination on the dye and the cells.

---

* to whom any correspondence should be addressed

The calcium ion has been recognized for many years as a major intracellular signal for the control of cell motility and force generation. Early studies in muscle demonstrated the ability of increased calcium concentrations to initiate contraction. Subsequently it was found that non-muscle cells also contain actin, myosin, and calcium-binding proteins which have a variety of effects on the actin cytoskeleton. In addition, shape and movement in non-muscle cells can be affected by the microtubule cytoskeleton, and the stability of this network can be altered by calcium acting through the calcium-binding protein calmodulin. In contrast to muscle, the biochemical mechanisms for transducing the calcium signal into mechanical force are relatively poorly understood [1, 2].

Movements in non-muscle cells often require very fine spatial and temporal control. Examples of these movements include the segregation of chromosomes during mitosis [3] and chemotaxis of phagocytic cells towards a target particle, followed by rapid engulfment of the particle [1]. For many years, changes in intracellular free calcium, $[Ca^{2+}]_i$, have been implicated in these motile events. However, only in the past few years have methods with sufficient speed and sensitivity been developed to measure $[Ca^{2+}]_i$ changes associated with motility in non-muscle cells.

In this article, the strategies employed in making such measurements of $[Ca^{2+}]_i$ are discussed. At the current stage of development these methods are still subject to serious limitations which can affect the validity of a measurement. Proper choice of equipment and design of the experimental protocol can minimize the likelihood of systematic errors. While the discussion in this article centres on motile cells, similar considerations apply generally to measuring $[Ca^{2+}]_i$ in single cells.

SOME GENERAL CONSIDERATIONS IN SINGLE-CELL MEASUREMENTS

There are several critical steps in carrying out an acceptable measurement of $[Ca^{2+}]_i$ in a single cell using fluorescent indicator dyes. First, the dye must be properly loaded into the cytoplasm, and its localization must be verified for the duration of the experiments. An instrument with adequate temporal (and, if necessary, spatial) resolution must be selected. Experimental procedures must be designed so that the information is obtained with minimal perturbation of the physiological system. Low concentrations of the dye must be used, and exposure to light minimized. Intense illumination of cells containing high dye concentrations will provide the best signal-to-noise ratios for $[Ca^{2+}]_i$ measurements. However, many biological processes are altered as a result of these harsh measurement conditions.

The first question to be addressed is whether single-cell measurements are required rather than measurement of $[Ca^{2+}]_i$ in a population of cells. As described below, single-cell measurements can provide information not available by other means. However, these methods are expensive and time-consuming by comparison with measurements in populations of cells, so the biological question must require single-cell information in order to justify the additional effort. Methods for the measurement of $[Ca^{2+}]_i$ in populations of cells include aequorin luminescence [4], NMR [5] and fura-2 fluorescence [6, 7]. $[Ca^{2+}]_i$ measurements can be made on large numbers of individual cells using flow cytometry with indo-1 [8]. [For flow cytometry see art. by M.G. Ormerod in Vol. 17.- *Ed.*]

In several contexts single-cell measurements have to be performed by fluorescence microscopy. The most obvious (and technically demanding) is the detection of intracellular free calcium gradients. In addition, single-cell measurements can detect cell-to-cell variation in the magnitude or timing of a response. This type of heterogeneity has now been seen in many systems even when a stimulus has been applied uniformly to an apparently homogeneous population of cells. In many cells the Ca^{2+} signal is very brief. Brief increases in Ca^{2+} which occur asynchronously within a cell population can easily be missed when average signals from a large number of cells are measured.

DYE SELECTION, LOADING AND VERIFICATION OF DISTRIBUTION

At present, fura-2 is the indicator most widely used for single-cell measurement of $[Ca^{2+}]_i$, having many advantages as described by R.Y. Tsien and co-workers [6, 7]. The dye has a good quantum yield, and is relatively resistant to photobleaching. This allows repeated measurements on the same cells using low levels of fluorescent illumination. The binding constant for $[Ca^{2+}]_i$ is ~225 nm, which is appropriate for measuring resting $[Ca^{2+}]_i$ and elevated levels up to the low μM range. There is a pronounced shift in the fluorescence excitation spectrum upon binding calcium (Fig. 1). This property is especially valuable for measurements on single cells, since $[Ca^{2+}]_i$ can be determined from the ratio of fluorescence intensity at two different wavelengths (usually 340 and 380 nm). This ratio is independent of dye concentration. $[Ca^{2+}]_i$ can be determined from the ratio, $R = I_{340}/I_{380}$, as follows [6]:

$$[Ca^{2+}]_i = K_d \cdot \frac{R - R_{free}}{R_{Ca} - R} \cdot \frac{I_{free}^{380}}{I_{Ca}^{380}}$$

where: K_d is the effective dissociation constant for Ca^{2+},
R_{free} and R_{Ca} are the ratio values at zero $[Ca^{2+}]$ and saturating $[Ca^{2+}]$ respectively, and
I_{free}^{380} and I_{Ca}^{380} are the intensities with 380 nm excitation.

Fig. 1. Calcium dependence of
fura-2 fluorescence.
Fura-2 (500 nM) was dissolved
in buffers at pH 7.05 containing
no calcium (2 mM EGTA) or 100 µM
Ca^{2+}.   Fluorescence excitation
spectra were obtained with
emission at 500 nm.

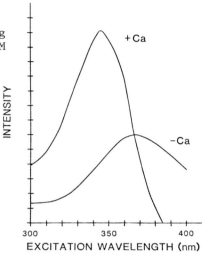

Fura-2 has good specificity for Ca^{2+} *vs.* most other
ions which are normally present in the cytoplasm, and the
dye properties are not greatly altered by small changes
in cytoplasmic pH [6].

Despite these favourable characteristics there are many
difficulties associated with the use of fura-2.   One of
the most serious is getting the indicator form of the dye
into the cytoplasm and keeping it there. Usually the acetoxymethyl ester of fura-2 (fura-2/AM) is added to the incubation
medium.   The fura-2/AM is membrane-permeant, and it diffuses
into the cytosol. The ester groups are removed by esterases,
and the indicator form of the dye is formed. With 5 carboxyl
groups, fura-2 will diffuse poorly across lipid bilayers,
and should ideally be trapped in the cytosol.

This pleasant scenario can break down at several stages
[9, 10].   First, the fura-2/AM is very poorly soluble in
water. Without adequate precautions, particles of fura-2/AM
will adhere to the cells.   These particles resist ester
hydrolysis, and a large percentage of the cell-associated
fluorescence can come from fura-2/AM rather than fura-2 [11].
Moreover, these particles can be phagocytosed, in which case
they will be trapped in intracellular organelles rather than
in the cytoplasm.   Even if fura-2/AM does enter the cytosol,
complete hydrolysis of the esters calls for an adequate
time (~15 min, depending on the cell type).

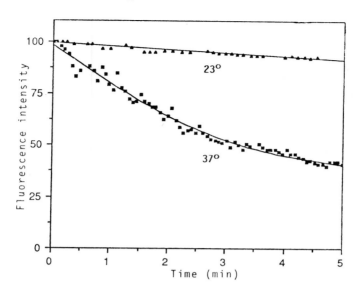

Fig. 2. Fura-2 loss from cells. PC12 cells were loaded with fura-2 acetoxymethyl ester (fura-2/AM; 20 μM) in Hepes-buffered saline, 5% (v/v) foetal calf serum, and 0.25% (w/v) Pluronic-F127 for 15 min at room temp. They were washed and incubated in Hepes-buffered saline for an additional 15 min at room temp. to ensure dye cleavage. The fluorescence intensity at 380 nm, normalized to 100% at time zero, was then measured with the microscope spectrophotometer.

One of the most vexing problems with fura-2 is that even the free acid form can be transported at high rates across biological membranes. This can lead to the accumulation of the dye within organelles and to its extrusion across the plasma membrane. This dye movement has characteristics similar to organic anion transport systems, and in macrophages it has been shown that dye loss can be blocked by probenacid, a drug which inhibits organic anion transport [12]. The movement of fura-2 across membranes is strongly temperature-dependent. The rate of dye loss in PC12 cells increases dramatically between 23° and 37° (Fig. 2). There is considerable variation in dye sequestration amongst various cell types.

These problems with dye loading and retention are serious, but usually methods can be developed to minimize their effect. For every experimental protocol a variety of tests should be used to monitor the localization of the dye and its chemical form. Hydrolysis of the esters can be monitored by spectroscopic changes in the dye [11]. If necessary, corrections can be made for the presence of unhydrolyzed dye using a third set of fluorescence filters [11]. However, it is certainly preferable to modify the dye-loading protocol so that such corrections are not required. Dye localization should be checked by observation of the cells with the fluorescence microscope. We have found that subcellular fractionation provides a very sensitive method for quantifying the amount of fura-2 trapped within organelles [13]. After dye loading, cells are chilled, homogenized and centrifuged. Cytoplasmic dye remains in the supernatant (120,000 **g**, 15 min).

By subcellular fractionation we have found as much as 50% of the total cell-associated dye in membrane-limited organelles using standard loading conditions [13]. Some workers have used digitonin permeabilization to test for trapping of dye within organelles [e.g. 14]. The plasma membrane is permeabilized by digitonin at low concentrations (say 0.005%); some organelles, e.g. the endoplasmic reticulum, remain intact, but many intracellular organelles are sensitive to digitonin even at very low concentrations. For example, dye trapped in endosomes would not be detected by the digitonin assay since endosomal compartments, which contain organic anion transport activity [12], when briefly treated with 0.01% digitonin release even large proteins (M_r >100,000) [15].

Dye-loading options.- The procedures used to load fura-2 into the cytoplasm must be tested for every cell type. A variety of loading protocols have been developed in various laboratories. A non-ionic detergent, Pluronic-F127, has been used to help solubilize fura-2/AM, and in many cases cells can be loaded easily and effectively using this procedure [16]. In Pt K2 kidney epithelial cells, we found that ~90% of the fura-2/AM entered the cytosol using Pluronic-F127 to disperse the dye [17], but only half without the detergent [13]. The low detergent concentration does not affect processes such as mitosis [17] or phagocytosis [18], but this should be checked carefully for each cell type. Alternatively, fura-2/AM can be dissolved in the presence of carrier proteins such as fatty acid-free albumin; undissolved fura-2/AM must then be removed by centrifugation.

Sometimes fura-2/AM loading fails even after trying various loading procedures. In these cases, the free acid can be loaded directly into the cytoplasm. Methods for introducing membrane-impermeant molecules into the cytosol [19] include pressure microinjection, dialysis through a patch-clamp pipette, electroporation [see Vol. 17, this series – *Ed.*], and scrape loading. For many cells, including macrophages and several types of transformed cell lines, the plasma membrane can be reversibly permeabilized by ATP^{4-} [12, 20], enabling gentle direct loading of fura-2 free acid into the cytosol [13].

Intracellular dye calibration.- A further difficulty is the intracellular calibration of the I_{340}/I_{380} intensity ratio vs. $[Ca^{2+}]_i$. Detailed descriptions of calibration methods using aqueous calcium buffers are available [6, 21, 22]. It is likely that conditions in the cytoplasm could alter the dye's spectroscopic properties or its response to calcium. For example, solvent conditions which increase the viscosity will increase the fluorescent signal at 380 nm excitation independently of $[Ca^{2+}]_i$. It is estimated that this can lead to an underestimation of resting $[Ca^{2+}]_i$ by 30-40% [16]. Other effects, such as binding of the dye to proteins or nucleic acids, or the unusual ionic environment in the nucleus, could also create different dye properties in the cell.

Ideally, one should calibrate the I_{340}/I_{380} ratio for fura-2 loaded into the cell and then equilibrated to known $[Ca^{2+}]_i$ values (e.g. using ionophores to collapse concentration gradients across membranes). This type of procedure has succeeded with some cell types [23]. However, the approach does not work well with all cell types [10, 17]. Moreover, in the absence of valid within-cell calibrations, it is possible that systematic errors in $[Ca^{2+}]_i$ values occur, as should be borne in mind. Generally, the values for $[Ca^{2+}]_i$ obtained using fura-2 agree well with values obtained by other methods.

In summary, the properties of fura-2 are such that it has limitations. Some of the problems have encouraged R.Y. Tsien and others to attempt to synthesize new molecules with more favourable properties. Despite these limitations, fura-2 provides a notably powerful tool for measuring calcium in single cells.

INSTRUMENTATION FOR SINGLE-CELL MEASUREMENTS

Most major research-microscope manufacturers now provide microscope optical systems which will pass 340 nm illumination on the fluorescence excitation light path. The system illustrated in Fig. 3 is based on an inverted microscope with capabilities for making photometric measurements using the photomultiplier or by image analysis using an image-intensified video system to obtain images. The inverted microscope allows easy access to the cells (e.g. to add stimuli to the incubation medium).

An important question is whether to use imaging techniques or photometric measurements. We have used both methods extensively, and each has its strengths and limitations. The major advantage of photometry is that it is simple to set up and use. One data point (intensity) is obtained per 340 nm or 380 nm exposure. Even at the fastest acquisition

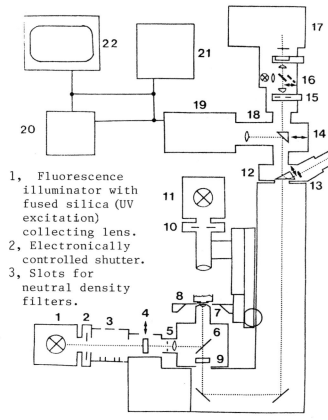

1, Fluorescence illuminator with fused silica (UV excitation) collecting lens.
2, Electronically controlled shutter.
3, Slots for neutral density filters.

Fig. 3. A fluorescence microscope (inverted; allows easy access to sample chamber) equipped for quantitative fluorescence microscopy. The diagram is based on the Leitz Diavert photometry system used in the authors' laboratories. Similar systems are available from other manufacturers.

Shutters, the filter wheel and the photometer are controlled by a PC, which records and analyzes intensity values. Photometry hardware and software were provided by Kinetek Corp. (Yonkers, NY).

4, Exchangeable excitation bypass filters in automated filter wheel.
5, Centrable field diaphragm for limiting area of illumination.
6, Dichroic beam splitter (reflects excitation wavelength, transmits the emission).
7, Objective (passes 340 nm excitation).
8, Cells growing on coverslip which forms lower surface of culture dish.
9, Emission filter.
10, Electronically controlled shutter.
11, Bright-field illuminator.
12, Exchangeable prism for reflecting light to eyepiece.
13, Electronically controlled shutter.
14, Movable partial reflecting mirror.
15, Diaphragm for limiting region of photometry measurement.
16, Lamp and removable mirror for adjusting measurement diaphragm.
17, Photomultiplier tube & housing.
18, Eyepiece for projecting image onto video camera.
19, Image intensifier video system.
20, Video tape recorder.
21, Image analysis system.
22, Video monitor.

rates a small personal computer (PC) can easily store the data and convert readings to [Ca^{2+}]$_i$ while the experiment is in progress.

Photometric measurements

The optics should be set up so that the illumination and measurement fields are masked by adjustable diaphragms set at image planes in the light path which are carefully aligned. This confocal arrangement of illumination and measurement fields provides strong rejection of stray light [24]. The measurement field may cover an entire cell, a group of cells, or a small region within a single cell.

A major advantage of photometry measurements is that the very high sensitivity of the detector allows data acquisition with minimal illuminative intensity. This limits photobleaching of the dye and photodamage to the cells. Even after 1000 [Ca^{2+}]$_i$ measurements on mitotic mammalian cells, the timing and morphology of mitosis seemed unaffected [17]. In making numerous measurements we have reduced the light intensity with neutral density filters and used electronic shutters to limit the light exposure to 0.1–0.5 sec per measurement. These precautions were necessary to obtain repeated measurements without damage to the cells.

The type of data obtained by these measurements is exemplified in Figs. 4 & 5. Fig. 4 shows changes in [Ca^{2+}]$_i$ as cells progress through mitosis: there are very brief increases in [Ca^{2+}]$_i$ which last for ~20 sec as well as more gradual changes of longer duration. Resting [Ca^{2+}]$_i$ in Pt K2 cells shows an S.D. of ±5 nM [13]. Fig. 5 shows the changes associated with phagocytosis of a complement–opsonized erythrocyte: [Ca^{2+}]$_i$ shows increases of periodic character, each lasting for only 5–30 sec. Both of the biological processes illustrated demonstrate the need for measuring [Ca^{2+}]$_i$ in single cells with high time resolution. In a population of cells, the asynchronous transient increases seen during phagocytosis would average out to a relatively small increase in [Ca^{2+}]$_i$. In mitosis only a few transient increases are observed during many minutes of continuous observation. These would almost certainly be missed if only a few measurements were made on each mitotic cell. In fact, we did not see the transient increases in cells when a few observations were made on each cell to obtain information on [Ca^{2+}]$_i$ localization [13].

The simplicity of photometric measurements makes them very attractive for most studies of [Ca^{2+}]$_i$ changes in single cells. However, when spatial information is also needed,

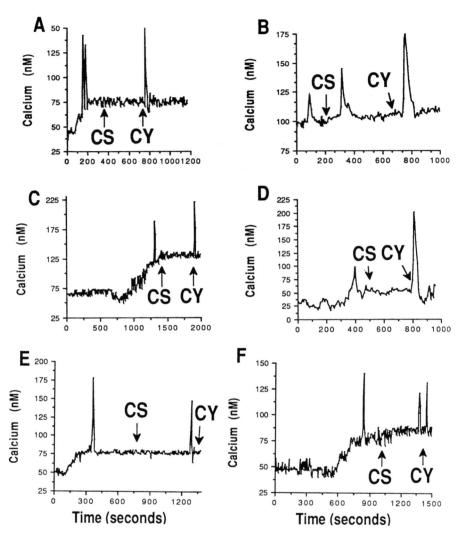

Fig. 4. Time courses of $[Ca^{2+}]_i$ in several typical dividing
Pt K2 cells (**A** to **F**) as they progress from metaphase to telophase.
Cells were loaded with fura-2/AM and placed on the microscope
stage, pre-equilibrated at 30–32° using an air–curtain incubator.
Cells at a random point in prometaphase or metaphase were selected
by bright-field observation. $[Ca^{2+}]_i$ measurements were made every
6–7 sec until telophase was complete, each accompanied by 3 sec of
bright-field observation to determine chromosome locations. **CS** =
time when the chromosomes can first be seen to be clearly separated; **CY** = point when the cell can first be seen to be constricting
at its waist (the onset of cytokinesis). *Adapted from ref. [17].*

Fig. 5. Calcium oscillations of neutrophils undergoing chemotaxis toward and phagocytosis of erythrocytes.

Sheep erythrocytes were attached to coverslip–bottomed plastic dishes designed for microscopy and were coated with anti–sheep IgM. Fura-2-loaded neutrophils were

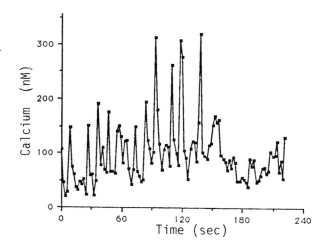

then allowed to attach to the dishes, which were then placed on the microscope stage at 30°, and chemotaxis was initiated by adding 10% fresh human serum to provide complement. Ratio measurements were made every 2 sec. The increase in amplitude of the oscillations occurs when phagocytosis of the erythrocyte begins.

one must use digital image analysis, as performed in several laboratories to examine $[Ca^{2+}]_i$ gradients within cells and to demonstrate heterogeneity of $[Ca^{2+}]_i$ response among a population of cells [13, 25, 28].

Digital image analysis

Video microscopy has been excellently described [29] by S. Inoue. Both the power and the difficulties associated with image processing arise from the enormous rate at which data can be generated. A standard 512 × 512 digital video image, which assigns each pixel an intensity from 0 to 0.255 (8 bits per pixel), contains 256 kilobytes of data. Operating at full video rate (30 frames/sec), a digital video system could generate ~7.5 megabytes of data per sec. Obviously, this overwhelms most small and mid-size computers. Some specialized digital storage systems are able to record digital images at video rates for a few minutes, but these are quite expensive at present.

Digital image analysis systems are produced by several manufacturers, and an analysis of specific systems is beyond the scope of this article. Generally, there is a trade-off between price and performance. A minimum requirement for convenient $[Ca^{2+}]_i$ measurements is a system which (1) has at least 512 × 512 resolution with 8 bits per pixel, (2) will automatically average the sequential frames in real time, and (3) will perform arithmetic operations on images (add, subtract, multiply and divide) with 16 bit results. The source code for the image processing software should be available for modification if needed for special applications.

Even with a powerful image processor, the amount of data generated requires that important decisions be made in the experimental design in order to store data as they are generated. One option is to record the images in analog on video tape or optical video disk for analysis at a later time. The major advantage of analog recording is that large numbers of images can be recorded economically. A disadvantage is that analog recording results in some degradation of the image [29; see below]. If images are to be stored directly in digital form, then a method must be devised to select a limited number of images so that the data storage capacity of the image-processing computer is not exceeded. A drawback to this approach is that important data can be irretrievably lost if the correct images are not digitized and stored.

In practice we have found that high-quality analog video recording devices provide the most convenient and economical method for image acquisition at the time of the experiment. At present we use a $\frac{3}{4}$" U-matic video cassette recorder (JVC CR 6650-U) and a Panasonic TQ 2028F monochrome video disk recorder. Both have manual gain control, which is required for recording quantitative data. Images stored on these devices can then be digitized and analyzed at a later time. Thousands of images from a single experiment are available for analysis without requiring massive digital storage. As shown in Fig. 6, recording does not significantly affect the linearity of the intensity response. The noise introduced into the intensity of a single pixel by recording is ~0.5% of the average intensity. The loss of resolution is 1-2 pixels at a sharp edge (corresponding to <0.5 μm in the cell). For most experiments, this level of image degradation is completely acceptable, considering the enormous increase in the number of images that are stored at the time of the experiment.

There are several image-acquisition devices with suffici-ent sensitivity to view fura-2 signals. Silicon-intensified target (SIT) video cameras have been used in several labora-tories. Image intensifiers can also be coupled with video cameras, and some comparisons with SIT cameras have been made [30]. A blue-green-sensitive image intensifier (KS 1380 from Video Scope, Washington DC) coupled with a high-resolution video camera is an excellent system for observing fura-2 fluoresence.

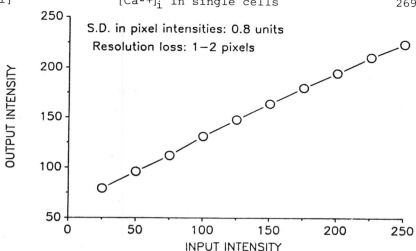

Fig. 6. Linearity and noise of video cassette recorder. For the test, an image containing a series of blocks (approx. 40 pixels × 40 pixels) of varying intensity was created on the image processor. The test image was recorded and digitized. The graphs show the intensity in the digitized images *vs.* that in the original test pattern. Evidently the response is very nearly linear over a wide range of input intensities.

The noise contribution from recording was estimated by measuring the S.D. of intensity of individual pixels within the blocks of uniform intensity. The intensity S.D. was nearly independent of input intensity. These data are for 8 frames averaged during digitization. Near the midpoint of the input range, the noise introduced by recording is ~1% of the signal intensity. *Effects on horizontal resolution* due to recording and digitizing were also evaluated from the test patterns. At the edge of a block, intensity in the original image would fall abruptly to zero. After recording and digitizing, the edges lose some sharpness. The intensity falls to ~30% of the block intensity in 2 pixels (out of 512 for one horizontal line). These 2 pixels correspond to ~260 nm in an image obtained with a 100 × objective.

Charged coupled devices (CCD's) have also been used in several laboratories for imaging fura-2 in cells [26]. CCD's have a much larger dynamic range than video cameras, i.e. they can record meaningful data over a wider intensity range. They have superb signal-to-noise properties, and they can integrate weak signals directly. Improvements in the sensitivity and read-out rates of CCD's can be expected over the next few years. At present it is not clear that they are preferable to image intensifiers when rapid measurements on living cells are required [30].

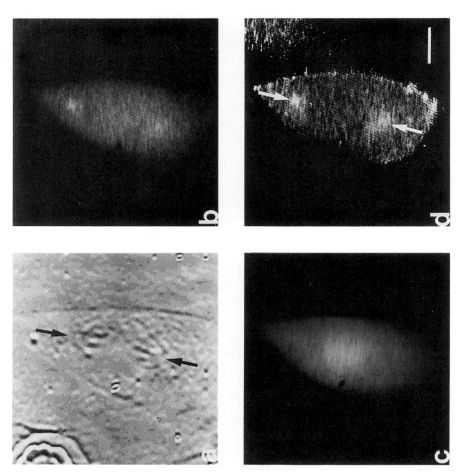

Fig. 7. Fura-2 fluorescence images of a Pt K2 cell in anaphase, depicting locally high calcium at the spindle poles. (**a**) Digitally processed, bright-field image of a cell in anaphase. *Arrows* point to the mitotic spindle poles. (**b**) & (**c**) Background-corrected fluorescence images of the same cell with excitation at 340 and 380 nm after loading with fura-2. Regions corresponding to the cell were identified using a threshold intensity in the 380-nm image, and $[Ca^{2+}]$ was calculated from I_{340}/I_{380} values. This image of calcium distribution was then redisplayed, (**d**). For this cell, $[Ca^{2+}]_i$ at the spindle poles (*arrows*) is 220 nM. The average $[Ca^{2+}]_i$ within the spindle mid-zone is 88 nM. *Bar = 5 μm. Adapted from ref. [13].*

Digital image acquisition and analysis has progressed rapidly over the past decade, and should continue to improve dramatically in power while the equipment cost declines. An example of the use of digital image processing is shown in Fig. 7. Pt K2 rat kangaroo epithelial cells were observed as they progressed through mitosis. During anaphase regions of high calcium were seen near the mitotic spindle poles. This area corresponds to one containing high concentrations of calmodulin, and the calcium-calmodulin may be important for regulating microtubule disassembly during anaphase [13].

Acknowledgements

We are grateful to Drs. B. Kruskal and R. Ratan for helpful discussions. Support by NHS grants GM 34770 (F.R.M.) and NS 15076 (M.L.S.) is acknowledged.

References

1. Southwick, F.S. & Stossel, T.P. (1983) *Semin. Hematol. 20*, 305-321.
2. Pollard, T.D. & Cooper, J.A. (1986) *Ann. Rev. Biochem. 55*, 987-1035.
3. Inoue, S. (1981) *J. Cell Biol. 91*, 131s-147s.
4. Blinks, J.R. (1985) in *Bioluminescence and Chemiluminescence: Instruments and Applications*, Vol. II (Van Dyke, K., ed.), CRC Press, Boca Raton, pp. 185-226.
5. Metcalfe, J.C., Hesketh, T.R. & Smith, G.A. (1985) *Cell Calcium 6*, 183-195.
6. Grynkiewicz, G., Poenie, M. & Tsien, R.Y. (1985) *J. Biol. Chem. 260*, 3440-3450.
7. Tsien, R.Y., Rink, T.J. & Poenie, M. (1985) *Cell Calcium 6*, 145-157.
8. Ransom, J.T., Di Giusto, D.L. & Cambier, J. (1987) *Meths. Enzymol. 141*, 53-63.
9. Almers, W. & Neher, E. (1985) *FEBS Lett. 192*, 13-18.
10. Malgaroli, A., Milani, D., Meldolesi, J. & Pozzan, T. (1987) *J. Cell Biol. 105*, 2145-2156.
11. Scanlon, M., Williams, D.A. & Fay, F.S. (1987) *J. Biol. Chem. 262*, 6308-6312.
12. Steinberg, T.H., Newman, A.S., Swanson, J.A. & Silverstein, S.C. (1987) *J. Cell Biol. 105*, 2695-2702.
13. Ratan, R.R., Shelanski, M.L. & Maxfield, F.R. (1986) *Proc. Nat. Acad. Sci. 83*, 5136-5140.
14. Morgen-Boyd, R., Stewart, J.M., Vavrek, R.J. & Hassid, A. (1987) *Am. J. Physiol. 253 (4 pt. 1)*, C588-598.
15. Yamashiro, D.J., Fluss, S.R. & Maxfield, F.R. (1983) *J. Cell Biol. 97*, 929-934.
16. Poenie, M., Alderton, J., Steinhardt, R. & Tsien, R.Y. (1986) *Science 233*, 886-889.

17. Ratan, R.R., Maxfield, F.R. & Shelanski, M.L. (1988) *J. Cell Biol.* *107*, 993-999.
18. Kruskal, B.A. & Maxfield, F.R. (1987) *J. Cell Biol.* *105*, 2685-2693.
19. McNeil, P.L., Swanson, J.A., Wright, S.D., Silverstein, S.C. & Taylor, D.L. (1986) *J. Cell Biol.* *102*, 1586-1592.
20. Heppel, L.A., Weisman, G.A. & Friedberg, I. (1985) *J. Membr.Biol.* *6*, 189-196.
21. Tsien, R.Y., Pozzan, T. & Rink, T.J. (1982) *J. Cell Biol.* *94*, 325-334.
22. Kruskal, B.A., Keith, C.H. & Maxfield, F.R. (1984) *J. Cell Biol.* *99*, 1162-1167.
23. Williams, D.A., Fogarty, K.E., Tsien, R.Y. & Fay, F.S. (1985) *Nature 318*, 558-561.
24. Piller, H. (1977) *Microscope Photometry*, Springer-Verlag, Berlin, 253 pp.
25. Poenie, M., Tsien, R.Y. & Schmitt-Verhalst, A.M. (1987) *EMBO J.* *6*, 2223-2232.
26. Connor, J.A., Cornwall, M.C. & Williams, G.H. (1987) *J. Biol. Chem.* *262*, 2919-2927.
27. Millard, P.J., Gross, D., Webb, W.W. & Fewtrell, C. (1988) *Proc. Nat. Acad. Sci.* *85*, 1854-1858.
28. Williams, D.A., Becker, P.L. & Fay, F.S. (1987) *Science 235*, 1644-1648.
29. Inoue, S. (1986) *Video Microscopy*, Plenum Press, New York, 584 pp.
30. Spring, K.R. & Lowy, R.J. (1989) *Meths. Cell Biol. 29A*, 269-290.

#C-2

PERMEABILIZED CELLS AS A MODEL FOR STUDYING
CALCIUM-DEPENDENT STIMULUS-SECRETION COUPLING

Ronald P. Rubin, Peter G. Bradford
and Jerry S. McKinney

Department of Pharmacology,
Medical College of Virginia,
Richmond, VA 23290-0524, U.S.A.

The permeabilized cell offers a unique and valuable tool to help elucidate steps in Ca^{2+}-dependent stimulus-secretion coupling. Isolated secretory cells rendered permeable to small molecules by incubation with certain detergents or by exposure to intense electric fields acquire pores in the p.m.[*] but intracellular organelles remain intact. Such preparations permit manipulation of the cytosolic environment, while maintaining cellular structure and functionality of stimulus-secretion coupling mechanisms. Thus, the roles of putative intermediates in the stimulus-secretion coupling pathway may be analyzed by their direct intracellular introduction, thereby bypassing the p.m. barrier. Examples of the utility of the permeabilized cell as a model for elucidating secretory mechanisms in isolated rat parotid acinar cells and rabbit neutrophils are presented.

It has been determined that: (a) physiologically relevant concentrations of cytosolic Ca^{2+} adequately stimulate secretion; (b) the diacylglycerol analogue 1,2-dioctanoyl-glycerol[⊗], but not cAMP, enhances the sensitivity of the secretory machinery to Ca^{2+}; and (c) binding of physiologically relevant concentrations of $Ins(1,4,5)P_3$[†] to a putative cellular receptor correlates with its ability to mobilize cellular Ca^{2+}. This article illustrates in some small measure the unique and valuable contributions rendered by the permeabilized cell in elucidating the various steps in Ca^{2+}-dependent exocytotic secretion.

Various *in vitro* model systems, from intact tissues to isolated cells and membrane fractions, have been employed over the years to study Ca^{2+}-dependent secretory mechanisms.

---

[*]*Abbreviations.-* p.m., plasma membrane; $InsP_3$, *my*oinositol 1,-4,5-triphosphate *(alternative abbreviations above* [†] *& in Figs.)*; diC8, *as for* ⊗ *above;* db-cAMP, dibutyryl cAMP.

Intact tissues, while preserving structural integrity, present problems in terms of diffusion of agonist and oxygen unless solutions are perfused directly via the circulation. Also, the preparations of slices and homogenates results in damage or disruption of tissue, so there is loss of cellular integrity and compartmentalization. Cell suspensions prepared by proteo-lytic dispersion* offer a distinct advantage over other in vitro preparations, in that the mechanism of Ca^{2+}-dependent exocytotic secretion can be studied independently of the influences of other cell types. Furthermore, they respond with a high degree of sensitivity, specificity and reproducibi-lity and offer a particularly useful tool for studying receptor-mediated cellular responses of a given secretory cell. In particular, the use of homogeneous isolated cell preparations has shed much light on the mechanisms of transmembrane signal-ling that utilize Ca^{2+}- and cAMP-dependent pathways.

Methodological approaches for rendering isolated cells permeable have lately become available, thus allowing the cellular barrier which protects the intracellular milieu from the external environment to be bypassed. Thus, a significant advantage of using permeabilized cells, rather than intact cells or isolated membranes, to study exocytosis is the ability to add charged molecules to the cell interior, while still maintaining a physiological setting. Moreover, the structural integrity of the permeabiliized cell and the functionality of receptor-effector coupling mechanisms may be maintained, while the cytosolic environment can be manipu-lated.

Cells can be permeabiliized without lysis or alteration in intracellular architecture with either detergents (e.g. saponin or digitonin) [1], intense electric fields [2], pore-forming toxins [3], or viral infection [4]. The patch-clamp technique at the single-cell level has also been used to study exocytosis since it also allows one to control the intracellular environment, e.g. cytosolic $[Ca^{2+}]$ [5]. However, this specialized technique is not readily available or mastered. Our procedures for permeabilizing isolated exocrine acinar cells and rabbit neutrophils, using intense electric fields or saponin, have been described previously in detail [6]. This article provides examples of the utility of these cell preparations in elucidating steps in Ca^{2+}-dependent secretion.

CALCIUM-ACTIVATED SECRETION

Considerable evidence exists affirming a pivotal role of Ca^{2+} in exocytosis, although the detailed mechanism(s)

---

*Note by Ed.- Vols. 8 & 11 (list at start of book) are pertinent. Permeabilization features in Vol. 17 (e.g. C.A. Pasternak).

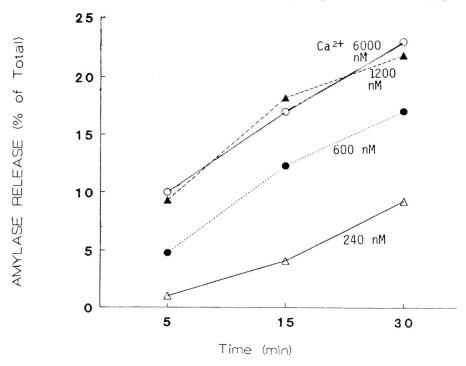

Fig. 1. Time-dependency of Ca^{2+}-evoked secretion from permeable rat-parotid cells. Acinar cells subjected to an intense electric discharge (4 kV/cm; time constant 55 µsec) were incubated at 37° with cytosolic $[Ca^{2+}]$ varied as shown. Aliquots were centrifuged through Nyosil silicone oil to separate cells from medium. Amylase in the supernatant was determined, and expressed as % of that present in cells + supernatant, with basal release subtracted.

responsible for Ca^{2+} activation of this fundamental process remain elusive. In permeabilized parotid acinar cells, the secretory apparatus is made responsive to the direct actions of Ca^{2+} without involvement of receptor-linked processes (Fig. 1). In such situations the release of secretory product from granules is elevated as cytosolic Ca^{2+} is raised within the range 10^{-7}-10^{-6} M; $[Ca^{2+}]$ is adjusted by adding varied amounts of Ca^{2+} to a constant EGTA concentration. By contrast, the release of cytoplasmic enzymes such as lactate dehydrogenase is not comparably enhanced (data not shown).

These findings indicate that, in a system energized by ATP, physiologically relevant concentrations of Ca^{2+} can directly activate the release of granular secretory product, but does not cause a concomitant leakage of cytosolic proteins.

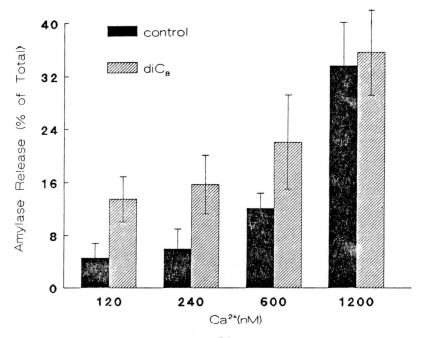

Fig. 2. The effect of diC8 on Ca^{2+}evoked secretion from permeable cells. Parotid acinar cells exposed to saponin (60 µg/ml, 5-10 min) were incubated for 5 min with varied cytosolic $[Ca^{2+}]$, with and without 100 µM diC8. Amylase release was assessed as in Fig. 1. Bars denote S.E.M.

It thus appears that Ca^{2+} ions plus ATP are adequate for expressing the releasing activity of permeabilized parotid cells. The effectiveness of Ca^{2+} as a secretagogue, in respect of the maximal extent of secretion, varies with the exposure time, as exemplified in Fig. 1 where 240 nM Ca^{2+} for 30 min is equivalent in effect to 1200 nM for 5 min.

In intact parotid acinar cells, Ca^{2+}-evoked amylase release is positively modulated by cAMP- and protein kinase-C-dependent pathways [7]. The Ca^{2+}-evoked amylase secretion in saponin-permeabilized parotid cells is also enhanced by diC8 even at basal Ca^{2+} levels (120 nM) (Fig. 2). This finding supports the general concept originally formulated by Nishizuka [8] that diacylglycerol-induced activation of protein kinase-C decreases the Ca^{2+} requirement for expressing its action on the secretory machinery. The action of diC8 contrasts with that of the stable GTP analogue GTP-γS, which is effective as a secretagogue only at elevated cytosolic Ca^{2+} concentrations (data not shown).

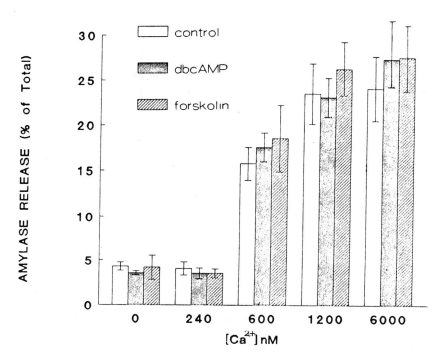

Fig. 3. The lack of effect of db-cAMP and forskolin on Ca^{2+}-evoked secretion from permeable cells as used in Fig. 1. Incubation was for 7 min with cytosolic $[Ca^{2+}]$ varied as indicated, with and without 1 mM db-cAMP or 60 μM forskolin. Bars denote S.E.M.

In saponin-permeabilized rat parotid cells, neither db-cAMP nor the adenylate cyclase activator forskolin can significantly augment Ca^{2+}-evoked secretion (Fig. 3). These findings fortify the notion that the role of cAMP in salivary amylase release is to modulate Ca^{2+} homeostasis, rather than altering Ca^{2+} sensitivity. Discrepant experimental findings regarding the interactions of Ca^{2+} and putative cellular messengers, <u>e.g.</u> guanine and cyclic nucleotides in permeabilized cells [9-11, & herein], stem not only from tissue variability, but also from the variable loss of these soluble cellular constituents, which in turn is dependent upon the selectivity of the permeabilization procedure.

IDENTIFICATION OF $InsP_3$ BINDING SITE

Since $InsP_3$ is highly charged and cannot permeate the p.m., the validity of the hypothesis that this water-soluble molecule is the cellular messenger responsible for mobilizing cellular Ca^{2+} during agonist-stimulated phosphoinositide

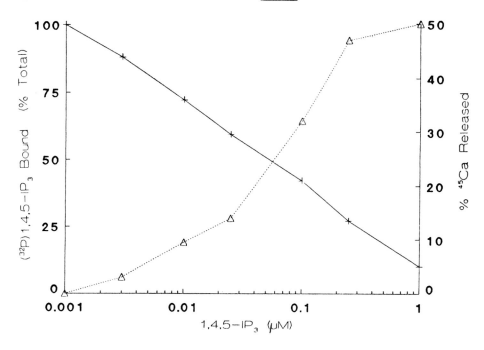

Fig. 4. Correlation of InsP$_3$-induced inhibition of [^{32}P]InsP$_3$
binding and activation of ^{45}Ca^{2+} release in permeable rabbit
peritoneal neutrophils. The binding in the presence of non-
radioactive InsP$_3$ at varied concentrations was determined as
described previously [12]; values (+) are % inhibition of
specific binding. ^{45}Ca^{2+} release was determined by measuring
the ^{45}Ca content of cells after filtration; data expressed as
% maximal release (Δ). *Modified from [12].*

breakdown must be tested on permeabilized cells and/or subcellular fractions. So we examined both InsP$_3$-stimulated
^{45}Ca^{2+} release and [^{32}P]InsP$_3$ receptor binding in the same
preparation of saponin-permeabilized rabbit neutrophils [12].
Adding InsP$_3$ to permeabilized neutrophils promotes a dose-dependent release of ^{45}Ca^{2+} from non-mitochondrial vesicular stores.
^{45}Ca^{2+} uptake into this pool at a cytosolic [Ca^{2+}] of 180 nM
is ATP-dependent and unaffected by mitochondrial inhibitors,
and the Ca^{2+} sequestered in this pool is released by InsP$_3$.

By studying the relation between the structure of inositol
polyphosphates and binding, we demonstrated a specific InsP$_3$
binding site within the endoplasmic reticulum that correlates
with biological activity. Indeed, InsP$_3$ binds to a specific
intracellular receptor in the permeabilized neutrophil, which
displays saturability and reversibility. Moreover (Fig. 4),

binding occurs over a concentration range comparable to that required to evoke $^{45}Ca^{2+}$ release from non-mitochondrial stores.

The InsP$_3$ content in the extracts derived from rabbit neutrophils stimulated by the chemotactic peptide fMet-Leu-Phe has been assayed in saponin-permeabilized neutrophils by its abilities to both compete with [^{32}P]InsP$_3$ binding to its intracellular receptor and to mobilize Ca^{2+} from cellular stores [13]. During stimulation, intracellular InsP$_3$ rises to a peak value >1 μM, which exceeds that needed to release intracellular $^{45}Ca^{2+}$ in rabbit neutrophils. This binding provides firm support for the concept that rises in cellular levels of InsP$_3$ elicited by receptor activation lead to the mobilization of cellular Ca^{2+} [14].

Analysis of a liver microsomal fraction revealed that the specific, high-affinity, intracellular binding site(s) on which InsP$_3$ exerts its Ca^{2+}-releasing effects is localized to the endoplasmic reticulum [15]. Impaired sensitivity of membrane fractions to the Ca^{2+}-releasing actions of InsP$_3$, as compared to permeabilized cells, may be associated with contamination with non-responsive fractions (p.m., mitochondria), accelerated IP$_3$ metabolism, or accelerated Ca^{2+} reuptake by the microsomal fraction.

SUMMARY AND CONCLUSIONS

Studies employing intact cells, homogenates and subcellular fractions have all contributed to our understanding of the biochemical mechanisms which underlie Ca^{2+}-dependent stimulus-secretion coupling. As documented in this brief article, permeabilized cells provide yet another very valuable means for studying this fundamental biological process. The physiological significance of the intracellular messengers Ca^{2+}, InsP$_3$ and diacylglycerol in the secretion process has been substantiated by the use of permeabilized cells, in conjunction with the employment of intact cells and subcellular fractions. Also, the coincidence of findings obtained in electrically permeabilized exocrine acinar cells and saponin-permeabilized rabbit peritoneal neutrophils validates the utility of the two permeabilization procedures.

With regard to future use of this model system, cell permeabilization has the potential for providing further revelations concerning the arsenal of putative cellular messengers that interact with Ca^{2+}, which includes cAMP, protein kinase-C, and arachidonic acid and its metabolites. Additionally, introduction of antibodies into permeabilized cells offers a potentially powerful tool to test for the functional involvement of molecules that may directly mediate exocytosis, e.g. actin, myosin, phospholipases and protein kinase-C.

However, despite the advantages of the permeabilized cell as an experimental tool, one must be aware of its limitations. For example, cell morphology may be altered after permeabilization and membrane function may be impaired because both sides of the p.m. are exposed to the same intracellular medium. Nevertheless, when the potential effects of residual intact cells are neutralized, e.g. by mitochondrial inhibitors, the continued use of the permeabilized cell should provide valuable clues regarding the still arcane role of Ca^{2+} in exocytotic secretion.

Acknowledgements

This work was supported by grants (AM 28029 and DE 05764) from the National Institutes of Health.

References

1. Burgess, G.M., McKinney, J.S., Fabiato, A., Leslie, B.A. & Putney, J.W., Jr. (1983) *J. Biol. Chem. 258*, 15336–15345.
2. Knight, D.E. & Scrutton, M.C. (1986) *Biochem. J. 234*, 497–506.
3. Bader, M., Thierse, D., Aunis, D., Ahnert-Hilger, G. & Gratzl, M. (1986) *J. Biol. Chem. 261*, 5777–5783.
4. Barrowman, M.M., Cockcroft, S. & Gomperts, M.D. (1987) *J. Physiol. 383*, 115–124.
5. Neher, E. (1988) *J. Physiol. 395*, 193–214.
6. Merritt, J.E., Bradford, P.G. & Rubin, R.P. (1987) in *In Vitro Methods for Studying Secretion* (Poisner, A.M. & Trifaro, J.M., eds.), Elsevier, Amsterdam, pp. 207–222.
7. Putney, J.W., Jr. (1986) *Ann. Rev. Physiol. 48*, 75–88.
8. Nishizuka, Y. (1984) *Nature 308*, 693–698.
9. Baldys-Waligorska, A., Pour, A., Moriarty, C.M. & Dowd, F. (1987) *Biochim. Biophys. Acta 929*, 190–196.
10. Haslam, R.J. & Davidson, M.M.L. (1984) *FEBS Lett. 174*, 90–95.
11. Kimura, T., Imamura, K., Eckhardt, L. & Schulz, I. (1986) *Am. J. Physiol. 250*, G698–G708.
12. Spät, A., Bradford, P.G., McKinney, J.S., Rubin, R.P. & Putney, J.W., Jr. (1986) *Nature 319*, 514–517.
13. Bradford, P.G. & Rubin, R.P. (1986) *J. Biol. Chem. 261*, 15644–15647.
14. Berridge, M.J. (1986) in *Phosphoinositides and Receptor Mechanisms* (Putney, J.W., Jr., ed.), Alan R. Liss, New York, pp. 25–45.
15. Spät, A., Fabiato, A. & Rubin, R.P. (1986) *Biochem. J. 233*, 929–932.

#C-3

SECRETORY GRANULE CALCIUM AND THE PROTEOLYTIC CONVERSION OF PROHORMONES

C.J. Rhodes, A.P. Dawson and J.C. Hutton[⊗]
Department of Clinical Biochemistry,
University of Cambridge, Addenbrookes Hospital,
Hills Road, Cambridge CB2 2QR, U.K.

and School of Biology, University of East Anglia,
Norwich, Norfolk NR4 7TJ, U.K.

Secretory granules in pancreatic islet β-cells store insulin after forming it from proinsulin, and release it through granule-membrane fusion with the p.m.[*] This article deals with granule isolation, and with intragranular ions (notably Ca^{2+}), calcium handling, and Ca^{2+}-dependent endopeptidases that process proinsulin. One aspect needing further study is the regulation of intragranular Ca^{2+} levels.

Many endocrine and exocrine cells store their major secretory products within membrane-limited intracellular organelles, the secretory granules or vesicles. Release of these products is achieved by fusion of the granule membrane with the cell's p.m. A typical example is the insulin-secreting β-cell of the pancreatic islets of Langerhans. The β-cell secretory granules (β-granules) [1] not only fulfil an intracellular storage function but also are involved in the generation of insulin from its inactive precursor proinsulin through limited proteolysis. This is achieved by β-granule Ca^{2+}-dependent endopeptidases [2, 3] and carboxypeptidase-H [4, 5]. Consistent with this Ca^{2+}-dependency of proteolysis is the presence of Ca^{2+} at mM levels in β-granules.

We describe here methods for the isolation of β-granules from a transplantable insulinoma source and the results of investigations aimed to address the question of how the

[⊗]to whom any correspondence should be sent (Cambridge address)
[*]*Abbreviations*.- p.m., plasma membrane; e.r., endoplasmic reticulum; MES - see overleaf; InsP$_3$, D-*myo*inositol 1,4,5-trisphosphate; d, density (g/ml); E-64, *trans*-epoxysuccinyl-L-leucyl-amido-(4-guanidino)butane.

intracellular Ca^{2+} is regulated.   The implication of such regulation on the prohormone conversion process is considered.

ISOLATION OF β-GRANULES

Transplantable rat insulinoma was propagated in rats of the New England Deaconess Hospital (NEDH) strain by s.c. injection of 0.1 ml of a suspension of tumour cells (~30 mg wet wt./ml) in Hank's physiological saline [6, 7].   Tumours were harvested 3-5 weeks later, at which point the animals were profoundly hypoglycaemic.   Insulinoma tissue was reckoned a better source of β-granules than isolated pancreatic islets, in view of the lack of contaminating endocrine pancreatic cells and of the 1000-fold greater yield obtainable from a single rat.

Fig. 1 outlines the β-granule isolation procedure. Tissue (~5-10 g) derived from 5-10 animals was homogenized (8-10 strokes, 4°) at pH 6.5 in 0.27 M sucrose/10 mM MES [2-(N-morpholino)ethanesulphonic acid]/1 mM EGTA, and centrifuged (MSE Coolspin).   For the lower gradient steps (Fig. 1) 19.2% (w/v) Nycodenz and 0.27 M sucrose were mixed in the proportions 1:1 and 1:3.   After centrifugation (Beckman L5-65 ultracentrifuge), the recovered interface fraction was washed twice in the homogenization buffer, mixed (1:8 by vol.) with 27% (v/v) Percoll in 0.27 M sucrose/10 mM MES pH 6.5, and centrifuged as shown (Sorvall RC-5B centrifuge).   A purified granule fraction was recovered (d 1.09-1.10), washed as shown, and suspended at 10 mg/ml protein concentration.   Marker enzyme analysis and electron microscopy [8, 9] showed this preparation to be essentially free from other subcellular organelles. It was either used immediately or stored at -70°.

From a similar homogenate, insulinoma microsomes were prepared with use of the Sorvall centrifuge (25,000 **g**, 20 min, 4°) to remove β-granules, mitochondria, lysosomes, nuclei and unhomogenized material, and then a Beckman Ti 70 rotor (15,000 **g**, 60 min, 4°) to give a pellet which was suspended (10 mg/ml) as for the granule fraction.

IONIC COMPOSITION OF ISOLATED β-GRANULES

Total Zn^{2+}, Ca^{2+} and Mg^{2+} in such β-granule fractions were determined by atomic absorption spectroscopy (Pye-Unicam SP90) after lysis in 0.1% (v/v) Triton X-100 and precipitation of high-mol. wt. material by 10% (w/v) trichloroacetic acid [8].   The $[P_i]$ in a similar supernatant was determined spectrophotometrically [10].   Fig. 2 shows the total ionic concentrations and the estimated free concentrations based upon differential centrifugation of granule lysates in the

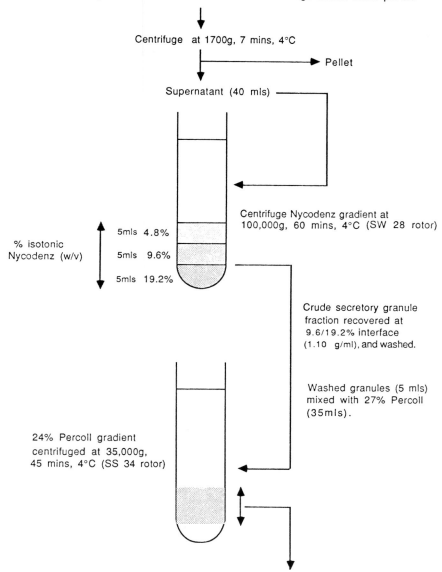

Potter-homogenised insulinoma tissue in isotonic homogenisation buffer pH 6.5

Centrifuge at 1700g, 7 mins, 4°C

→ Pellet

Supernatant (40 mls)

% isotonic
Nycodenz (w/v)

5mls 4.8%

5mls 9.6%

5mls 19.2%

Centrifuge Nycodenz gradient at
100,000g, 60 mins, 4°C (SW 28 rotor)

Crude secretory granule
fraction recovered at
9.6/19.2% interface
(1.10 g/ml), and washed.

Washed granules (5 mls)
mixed with 27% Percoll
(35mls).

24% Percoll gradient
centrifuged at 35,000g,
45 mins, 4°C (SS 34 rotor)

Secretory granules recovered and washed away from the
Percoll 5-6 times in 0.25M sucrose / 10mM MES pH 6.5

presence or absence of chelating agents and detergents [8]. There appeared to be very little free Zn^{2+}, the majority of granule Zn being associated with the insulin hexamer [8]. The β-granule Ca^{2+} concentration was very high (120 mM total), and has been estimated to account for 26% of the

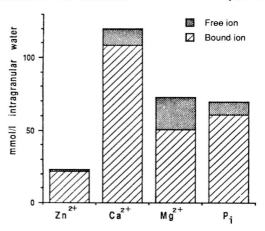

Fig. 2. The ionic
composition of
isolated β-granules.

total tissue content [8]; ~90-95% appeared to be in a bound
form, most likely associated with the intragranular P_i. A
ratio for β-granule Ca/P_i of 1.73 might indicate that the
Ca is complexed as insoluble Ca hydroxyapatite or amorphous
$Ca_3(PO_4)_2$; however, ~50% is released in an ionized form
upon granule lysis. It is therefore likely that $Ca(H_2PO_4)_2$
and $CaHPO_4$, which are more soluble, exist in the granule;
this would be more compatible with the acidic inner environment
of the organelle [9].   Ca may also be associated with some
intragranular adenine nucleotides or proteins.

     It should also be considered that the high intragranular
Mg^{2+} will compete with Ca^{2+} for binding to these various
sites.   The free Ca^{2+} concentration in β-granules may be
as high as 10 mM [8], suggesting the existence of a trans-β-
granule-membrane free $[Ca^{2+}]$ gradient of at least 10^4 in
the β-cell.   Such a gradient might result from the presence
of an energy-dependent Ca^{2+} transport mechanism either in
the mature granule or operating at some earlier stage of
β-granule biogenesis.

CALCIUM HANDLING IN ISOLATED β-GRANULES

     Ca movements from a suspension of freshly isolated osmoti-
cally intact β-granules and pancreatic β-cell microsomes were
measured at 30° with a Ca-sensitive electrode, viz. the neutral
ionophore ETH 1001/PVC/tetrahydrofluran solution dropped onto
1 mm plastic tubing, the internal solution therein being 10 mM
$CaCl_2$ in which a silver wire was suspended. This electrode
could measure pCa between 6 and 7 with a good response
time [11]. Isolated β-granules or microsomes (~200 μg protein)
were placed in 1.5 ml of the incubation buffer, consisting
of 150 mM sucrose, 50 mM KCl, 3% (w/v) polyethylene glycol
6000, 10 mM HEPES (K) pH 7.0, 1 mM dithiothreitol, 2.5 mM

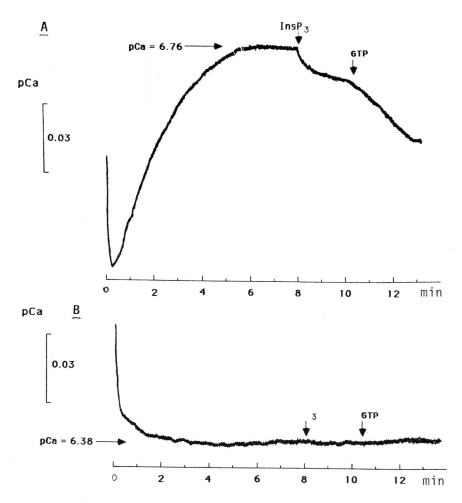

Fig. 3. Ca^{2+} mobilization from (**A**) pancreatic β-cell microsomes, and (**B**) β-granules, induced by 1 μM $InsP_3$ or 20 μM
GTP, as measured with a Ca^{2+}-sensitive electrode.

ATP-Na_2, 1 mM $MgCl_2$, 20 μg/ml creatine phosphokinase and
5 mM creatine phosphate (final concentrations). The final
pCa was 6.5. On addition of insulinoma microsomes there
was an immediate decrease in pCa, due to endogenous Ca^{2+}
in the preparation, which was followed by a progressive
increase in pCa over 5 min to reach a new steady pCa higher
than that of the original media.

Subsequent addition of either $InsP_3$ (to 1 μM) or GTP
(20 μM) caused a rapid release of microsomal Ca^{2+} (Fig. 3A),
consistent with previous studies [1, 12]. With β-granules

an initial fall in pCa was observed on addition of material
to reach a new lower steady pCa, but there was no evidence
of a Ca^{2+}-sequestering activity.   Neither $InsP_3$ nor GTP
had any effect on sequestration or mobilization of Ca from
these preparations (Fig. 3B), nor was there any effect from
inhibition of the β-granule proton pump by 10 μM tributyltin
or by the addition of the protonophore FCCP.

CALCIUM-DEPENDENT   PROINSULIN-PROCESSING ENDOPEPTIDASES

Pulse/chase labelling experiments on isolated pancreatic
islets show that there is a lag period of 25-30 min from
the initial synthesis of proinsulin until it is converted
to insulin [13, 14].   This time period coincides with the
time taken for the newly synthesized proinsulin to pass
from its site of synthesis on the rough e.r., through the
Golgi complex and into the newly forming β-granule compartment
[14].   The process of proinsulin conversion is initiated
by endoproteolytic cleavage C-terminally to two basic amino
acid sites on the proinsulin molecule's sequence: at Arg-31,
Arg-32 and at Lys-64, Arg-65 [3, 13]. Subsequent carboxypeptidase-H [4, 5] trimming of these basic residues yields the
final products - insulin and C-peptide. Two distinct β-granule
Ca^{2+}-dependent endopeptidases are involved: type-I cleaves
at the B-chain/C-peptide junction (Arg-31, Arg-32) of proinsulin, and type-II preferentially cleaves at the C-peptide/A-
chain junction [3].

These two enzyme activities can be separated by chromatography of a lysed β-granule preparation (15-30 mg protein)
on an anion exchange column [3].   Proinsulin processing
activity was determined by incubating 20 μl of the column
fractions for 2 h at 30° in 100 μl (final vol.) incubation
media containing 50 mM Na acetate (pH 5.5), 5 mM $CaCl_2$, 10 μM
E-64, 0.1 mM tosylphenylethyl-chloromethyl ketone (TPCK), 10 μM
pepstatin-A and ~30,000 dpm [^{125}I]proinsulin (15-20 ng). By
altering the $CaCl_2$ concentration in the presence of 2 mM
EDTA in this medium, these Ca^{2+}-dependent type-I and type-II
activities could be assayed at various free-Ca^{2+} concentrations
[3, 15]. At 5 mM Ca^{2+} both enzymes are fully active (Fig. 4);
however, they differed in sensitivity at lower [Ca^{2+}]: type-II is
half-maximally active at 50 μM Ca^{2+} where there is negligible
type-I activity, but type-I activity is half-maximally activated at 2.5 mM Ca^{2+}.   It was found that these activities
also differ in their pH requirements: both have a pH optimum
of 5.5, but the type-II enzyme is ~30% active at more neutral
pH's compared to only 1-2% for type-II activity [3].

Given that these type-I and type-II proinsulin processing
endopeptidases traverse the same intracellular pathway as

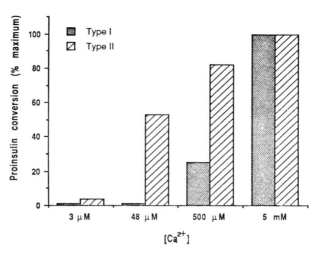

Fig. 4. Differential activation of pro-insulin-processing type-I and type-II endopeptidases by Ca^{2+}.

other β-cell proteins, it is postulated that type-II activity may be partially active in the *trans* Golgi-network environment of near-neutral pH [8] and lower Ca levels [16], as well as in the β-granule where it would be fully active (pH 5.5, mM Ca^{2+}). This expression of type-II activity at this point in the secretory pathway may have a function in the processing of other β-cell proproteins targetted to other cellular destinations such as the p.m., lysosomes, or proteins that are released constitutively [3]. The type-I enzyme's stringent requirements for mM $[Ca^{2+}]$ and acidic pH ensure that final conversion to insulin occurs exclusively in the β-granule environment, thus confining insulin production to the compartment where it is stored. This is important since the Zn-insulin complex is considerably less soluble than the Zn-proinsulin precursor [17].

Accordingly, a critical mechanism for initiating insulin production exclusively in the β-granule membrane would be the activation (or insertion) of Ca-transport proteins and the proton-translocating ATPase in the β-granule membrane at the level of β-granule biogenesis. The process by which high levels of Ca are introduced into the β-granule are not understood, although newly formed granules appear to be a major subcellular site of Ca accumulation [18].

DISCUSSION

Although it has frequently been postulated that the β-granule plays a regulatory role through sequestering or mobilizing Ca^{2+} [19, 20], we have no direct biochemical evidence for this function using isolated secretory granules. Ca^{2+} handling by such granules is distinct from that either in e.r. or mitochondria. The β-granule exhibits Ca-stimulated

ATP hydrolysis, but this is abolished by Mg^{2+} addition [8] and thus is not characteristic of a Ca^{2+} pump as in the p.m. or specialized Ca^{2+}-sequestering membranes, e.g. sarcoplasmic reticulum. A regulatory role of granule Ca^{2+} is unclear for cytosolic events but certainly applies to intragranular enzymic activities, e.g. the activation of the Ca^{2+}-endopeptidases involved in prohormone conversion. It remains to be settled how these organelles can attain very high $[Ca^{2+}]$, and what is the physical and chemical nature of Ca^{2+} bound within the granule. Also, since these enzymes may occur in other endocrine tissue and Ca^{2+} regulation of prohormone proteolytic conversion could permit differential processing to different bioactive peptides, it must be established whether intragranular Ca^{2+} can be modulated by physiological stimuli and to how far this in turn determines the rates and specificity of intragranular proteolytic processes.

Acknowledgements.- Dr B.H. Frank (Eli Lilly, Indianapolis) kindly gifted $[^{125}I]$proinsulin. The work was supported by the Wellcome Trust, Juvenile Diabetes Foundation International, British Diabetic Association and Nordisk Insulin Laboratories.

References (BJ signifies *Biochemical Journal)*

1. Hutton, J.C. (1984) *Experientia 40*, 1091-1098.
2. Davidson, H.W., Peshavaria, M. & Hutton, J.C. (1987) *BJ 246*, 277-286.
3. Davidson, H.W., Rhodes, C.J. & Hutton, J.C. (1988) *Nature 333*, 93-96.
4. Docherty, K.D. & Hutton, J.C. (1983) *FEBS Lett. 162*, 137-142.
5. Davidson, H.W. & Hutton, J.C. (1987) *BJ 245*, 575-582.
6. Chick, W.L., Warren, S., Chute, R.N., Like, A.A., Lauris, V. & Kitchen, K.C. (1977) *Proc. Nat. Acad. Sci. 74*, 628-632.
7. Hutton, J.C., Penn, E.J., Jackson, P. & Hales, C.N. (1981) *BJ 193*, 875-885.
8. Hutton, J.C., Penn, E.J. & & Peshavaria, M. (1983) *BJ 210*, 297-305.
9. Hutton, J.C. & Peshavaria, M. (1982) *BJ 204*, 161-170.
10. Itaya, K. & Ui, M. (1966) *Clin. Chim. Acta 14*, 361-366.
11. Dawson, A.P. (1987) *FEBS Lett. 185*, 147-150.
12. Berridge, M.J. (1987) *Ann. Rev. Biochem. 56*, 159-193.
13. Steiner, D.F, Kemmler, W., Clark,J.L., Oyer, P.E. & Rubenstein, A.H. (1972) in *Handbook of Physiology, Sect. 7, Endocrinology,* Vol. 1 (Greep, R.O. & Astwood, E.B., eds.), Am. Physiol.Soc. Wash. DC, pp.
14. Orci, L. (1985) *Diabetologia 28*, 528-546. [175-198.
15. Storer, A.C. & Cornish-Bowden, A. (1976) *BJ 126*, 1-5.
16. Herman, L., Sato, T. & Hales, C.N. (1973) *J. Ultrastruct. Res. 42*, 298-311.
17. Grant, P.T., Coombs, T.L. & Frank, B.H. (1972) *BJ 126*, 433-440.
18. Howell, S.L. & Tyhurst, M. (1976) *J. Cell Sci. 21*, 415-422.
19. Prentki, M. & Matschinsky, F.M. (1987) *Physiol. Rev. 67*, 1185-1248.
20. Hedeskov, C.J. (1980) *Physiol. Rev. 60*, 442-508.

#C-4

THE ROLE OF CALCIUM AND CALPAIN IN THE EXPRESSION OF PLATELET PROCOAGULANT ACTIVITY

E.M. Bevers, P.F.J. Verhallen, P. Comfurius and R.F.A. Zwaal

Department of Biochemistry, University of Limburg, P.O. Box 616, 6200 MD Maastricht, The Netherlands

Stimulation of blood platelets results in release of granule constituents and aggregation of the cells. An important though less known response is the formation of a procoagulant surface, involving a rearrangement of membrane phospholipids in such a way that their normal asymmetric distribution over both membrane leaflets becomes randomized. As a result, anionic phospholipids which were originally located in the membrane's cytoplasmic leaflet become exposed in the outer leaflet. Thereby there ensues the Ca^{2+}-mediated binding of two consecutive enzyme complexes of the coagulation cascade, greatly enhancing the rates of factor Xa and thrombin generation respectively.

The intracellular Ca^{2+} concentration plays an essential role in the procoagulant response. Substantial evidence shows that randomization of membrane phospholipids is inextricably correlated with activation of an endogenous Ca^{2+}-dependent thiol protease, platelet calpain. At least three different cytoskeletal proteins are the major substrates for this enzyme. We propose that membrane-cytoskeletal interactions play an active role in maintenance of phospholipid asymmetry in resting platelets. Activation of calpain by a rise in cytoplasmic Ca^{2+} causes a disturbance of these interactions, contributing to loss of membrane lipid asymmetry.

ROLE OF ANIONIC PHOSPHOLIPIDS IN COAGULATION

In the haemostatic process following vessel-wall injury, platelets and coagulation factors act in concert to ensure a rapid and efficient arrest of the bleeding. Adhesion, degranulation and aggregation of platelets leads to formation of a primary haemostatic plug, consisting of millions of platelets clumping together [1]. The platelet plug is consolidated by a network of insoluble fibrin strands formed as

the end-product of the coagulation cascade [2]. Through
another platelet response, the formation of a procoagulant
surface, coagulation is accelerated more than a million-fold.
In the execution of the platelet response, calcium ions
play a pivotal role, as will be described below.

ROLE OF ANIONIC PHOSPHOLIPIDS IN COAGULATION

Two sequential reactions of the coagulation cascade are
critically dependent on the presence of an anionic phospholipid
surface, viz. the formation of factor Xa by a complex of
factors IXa and VIIIa and the subsequent conversion of prothrombin to thrombin by a complex of factors Xa and Va. Formation
of these complexes occurs through a Ca^{2+}-mediated binding
of the γ-carboxy glutamic acid residues of the vitamin K-depen-
dent coagulation factors to the polar head-groups of the
anionic phospholipid molecules [3]. Anionic phospholipids
other than phosphatidylserine are less effective in binding
of these complexes, suggesting that the interaction is not
merely electrostatic. A chelation model for the association
of these proteins with phosphatidylserine-containing membranes
has been proposed [4, 5].

The effect of phospholipids on the kinetics of factor
Xa and thrombin formation have been studied extensively [review:
3]. In essence, phospholipids drastically decrease the K_m
for factor X and prothrombin, respectively, causing the reaction
rate to increase by several orders of magnitude. Since
the two reactions are sequential, the presence of a phosphatidylserine-containing phospholipid surface produces more than a
million-fold increase in the rate of thrombin formation and
hence in the rate of fibrin formation.

PLATELET PROCOAGULANT ACTIVITY

Although the presence of phosphatidylserine is specifically
required to provide a suitable catalytic surface for both
coagulation reactions, this phospholipid seems to be virtually
absent from the outer surface of the p.m.* of blood cells.
Indeed, the p.m. of resting platelets is characterized by
an asymmetric distribution of the different phospholipids
over both membrane leaflets [6, 7]. The outer leaflet
is composed of choline phospholipids (phosphatidylcholine and
sphingomyelin) and a minor amount of phosphatidylethanolamine.
Most extreme is the distribution of phosphatidylserine and
sphingomyelin, which are located almost exclusively at the
inner and outer leaflet of the p.m. respectively. Consequently,
resting platelets have a low procoagulant activity when tested

---

*Abbreviations: p.m., plasma mebrane; FI, fluorescence intens-
ity; TMA-DPH, trimethylammonium diphenylhexatriene.

in a system with highly purified coagulation factors. Activation of platelets, however, can cause a redistribution of
the phospholipids over both membrane leaflets. In our laboratory phospholipases were used to probe the outer surface
of activated platelets. These studies revealed that the
extent of phospholipid reorientation strongly depends on the
activation procedure.

ADP and epinephrine hardly affect the phospholipid
distribution, while thrombin causes a shift in the exposure
of phosphatidylethanolamine and a minor increase in that
of phosphatidylserine, confirming data of Schick and co-workers
[7]. An increased exposure of aminophospholipids including
phosphatidylserine is observed upon activation by collagen,
and this effect is even more pronounced when platelets are
stimulated by the combined action of collagen and thrombin.
This increased exposure is accompanied by a marked decrease
in choline phospholipids. The most dramatic changes in
phospholipid distribution were observed upon activation with
the non-physiological stimulator A23187 which acts as a Ca^{2+}-
ionophore. Since phosphatidylserine is virtually absent from
the outer surface of resting platelets, its increased exposure
after activation is a reflection of the extent of transbilayer
movement.

From the foregoing it will be evident that the exposure
of phosphatidylserine at the outer surface directly affects
the ability of platelets to enhance the rate of factor
Xa and thrombin formation. It was shown previously that
a quantitative relationship exists between the rate of thrombin
formation (determined by a prothrombinase assay) and the
amount of phosphatidylserine exposed (measured by phospholipases) as indicated in Fig. 1A.

INVOLVEMENT OF CALPAIN IN THE PLATELET PROCOAGULANT RESPONSE

Increasing evidence obtained in the last decade suggests
that the cytoskeleton plays an important role in the structural
organization of the surface membrane [10]. This organization
comprehends not only membrane proteins but also the lipids,
in particular the aminophospholipids. Accordingly, an
increased exposure of aminophospholipids upon activation of
platelets is presumably accompanied or preceded by alterations
in cytoskeletal structure.

Modification of the cytoskeleton can be achieved by
the action of an endogenous calcium-dependent thiol protease,
termed calpain [11, 12]. This enzyme, which has been demonstrated in many cells, can be activated by a rise in cytoplasmic
Ca^{2+} concentration. Platelet calpain consists mainly of

Fig. 1. Potency of
platelet stimulators
in inducing exposure
of phosphatidylserine
(**A**) and degradation
of cytoskeletal
proteins (**B**).

A. Relationship between
the exposure and the
rate of thrombin
formation for various
activated platelets.
See [8, 9] for experi-
mental details and
method descriptions.

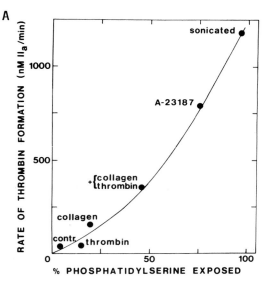

B. Protein patterns
of the various
activated platelets.
See [9] for methodo-
logy and conditions.

*Applicable to other
Figs. also:* being
on a nM/min basis,
rates are indepen-
dent of assay vol.
(5×10^6 platelets
per ml).

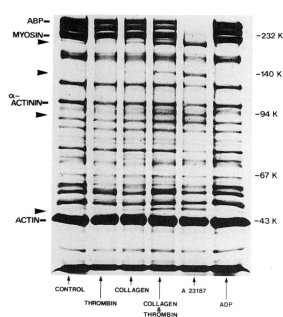

calpain I, requiring Ca^{2+} concentrations in the μM range
for optimal activity. In platelets, the primary targets
for calpain are the high mol. wt. cytoskeletal regulatory
proteins filamin (250 kDa), talin (235 kDa) and myosin
(200 kDa) [9, 11, 13]. In respect of inducing degradation
of cytoskeletal proteins, different agonists showed the same
order of potency (Fig. 1B) as for inducing exposure of phosphati-

Fig. 2.
Calcium
titration of
prothrombin-
ase activity
and *(lower
portion)*
calpain
activity in
intact
platelets,
incubated
in the
presence of
carefully
controlled
Ca^{2+}/HEDTA
buffers and
activated
by 0.3 μM
A23187.
See [14]
for assay
of enzyme
and of
protein
degradation.

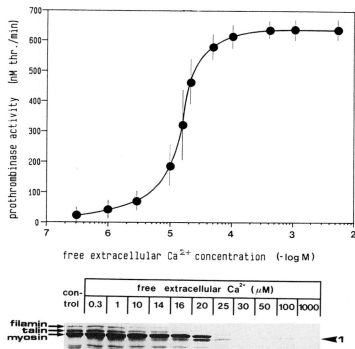

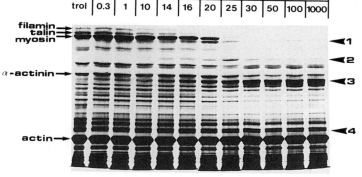

dylserine at the outer surface [9]. Increasing proteolysis
is observed in the order:
   ADP = epinephrine < thrombin < collagen < collagen + thrombin < A23187.

CALCIUM-DEPENDENCE OF THE PROCOAGULANT RESPONSE

In the absence of extracellular Ca^{2+} (EGTA present)
neither protein degradation nor exposure of phosphatidylserine
are still manifested using the above-mentioned agonists. This
indicates that release of Ca^{2+} from internal storage sites
is insufficient to raise the cytosolic Ca^{2+} concentration
to levels required for activation of calpain. The Ca^{2+}
dependence of both calpain activity and phosphatidylserine
exposure was examined (Fig. 2) by stimulating platelets with
Ca^{2+}-ionophore A23187 in the presence of carefully controlled
Ca^{2+}/HEDTA (hydroxyethylenediaminetriacetic acid) buffers at

Fig. 3. Time-course of Ca^{2+} influx (**A**) and prothrombinase activity (**B**) during incubation of platelets with 10 mM F^- at 37°, then (samples being diluted to avoid CaF_2 precipitation) +3 mM Ca^{2+}. See [15] for details. Curve **A** (mean ±S.E.M., 5 expts.) = fura-2 fluorescence increase just after Ca^{2+} addition. In **B** (mean ±S.E.M., 9 expts. in duplicate) the *inset* is activity as % of maximum *vs.* time after Ca^{2+} addition at 30 min after F^- addition.

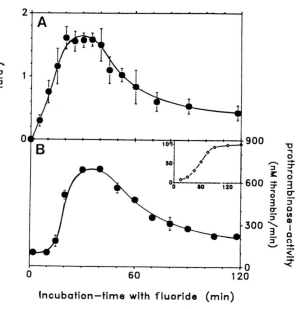

Incubation—time with fluoride (min)

various Ca^{2+} levels [14]. It was found (Fig. 2) that phosphatidylserine exposure developed with a steep increase between 5 and 50 µM free Ca^{2+} with a midpoint at 15 µM. It has to be emphasized that this approach to studying the Ca^{2+}-dependence assumes a rapid and efficient equilibration of extracellular and cytosolic Ca^{2+} levels. Moreover, the cytoskeletal degradation by calpain and phosphatidylserine exposure coincide in time course, irrespective of the agonist that was used to evoke the effect [14]. With Ca^{2+}-ionophore, the effects are achieved within 30 sec, whereas upon activation with collagen plus thrombin both effects developed on a time scale of 5 min.

More compelling evidence for a causal relationship between calpain activity and phosphatidylserine exposure was obtained by incubating platelets with fluoride: during 2 h the entry of extracellular Ca^{2+} is transiently facilitated (Fig. 3). Correspondingly, development of prothrombinase activity can be induced transiently upon addition of Ca^{2+}. In its absence no procoagulant surface is generated. In fact, procoagulant activity is demonstrated in the prothrombinase assay, which hinges on the presence of Ca^{2+} (Fig. 3, inset).

Activation of calpain in samples from the fluoride-treated platelets is also strictly dependent on the addition of extracellular Ca^{2+} (Fig. 4). No protein degradation is observed

Fig. 4. Relationship, as affected by leupeptin, between Ca^{2+}-induced prothrombinase (**A**) and calpain (**B, C**) activities of fluoride-treated platelets, incubated with 10 mM F$^-$ in the absence (●) or presence (○) of 0.5 mM leupeptin. **B**: degradation of cytoskeletal proteins by calpain after adding F$^-$ and then Ca^{2+}, judged by fragment formation (sum of protein bands indicated on right of **C**). Data were quantitated from the gels shown in **C**, after densitometric scanning [15]. Changes in the amount of fragments were counterbalanced by changes in the amount of substrates (not shown). **C**: leupeptin absent *(upper panel)* or present *(lower panel)*; fragments *(right)* **vs.** substrates *(left)* diminution indicate calpain activity. The same incubations served for **A** and **C/B**. See [15] for experimental details. Fig. 4 is a representative experiment.

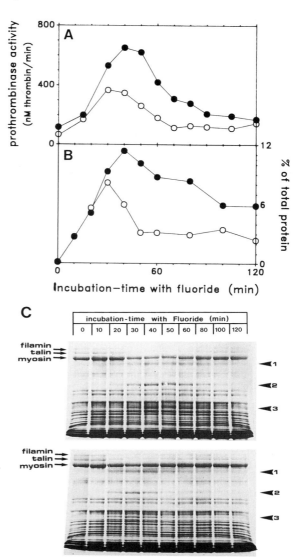

upon prolonged treatment with fluoride in the presence of EGTA. The apparent reversible exposure of procoagulant activity as well as the apparent reversible breakdown of cytoskeletal proteins merely reflects the transient increase in Ca^{2+}-permeability of the p.m. (Note that in platelets protein synthesis is negligible.) The identity in time course between Ca^{2+}-inducible prothrombinase activity and calpain activity strongly suggests a causal relationship between the two parameters. Moreover, using phospholipase to probe the lipid composition of the outer monolayer of the platelet membrane,

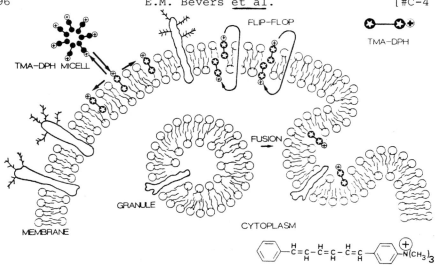

Fig. 5. Principle of the use of TRIMETHYLAMMONIUM DIPHENYLHEXATRIENE
trimethylammonium diphenylhexa-
triene (TMA-DPH). It equilibrates
over the aqueous phase and outer
TMA-DPH
leaflet of the p.m.. Its polar head-group prevents spontaneous migration to the inner leaflet. FI is linearly related
to the amount of p.m. accessible to the probe. **FI reflects**
too any transbilayer movement of the probe, increasing in parallel.

it was found that Ca^{2+}-inducible phosphatidylserine exposure
was similar in pattern to Ca^{2+}-inducible calpain and prothrombinase activities [15]. Supplementary evidence for the involvement
of calpain in the exposure of phosphatidylserine at the
outer surface of the platelet membrane came from use of
an inhibitor of calpain [14]. The presence of leupeptin
during fluoride treatment identically inhibited calpain activity and the development of a procoagulant surface (Fig. 4).

Ca^{2+}-DEPENDENCE OF TRANSBILAYER MOVEMENT

Transbilayer movement or flip-flop of phospholipids in
resting platelets is an extremely slow process with half-times
of several days rather than hours. The observation that
during platelet activation the aminophospholipids become exposed
at the membrane exterior within minutes needs postulation
of flip-sites along which a rapid migration of phospholipids
over both membrane leaflets becomes feasible. Local disturbances of the bilayer structure have been suggested to allow
a rapid flip-flop process [16, 17]. As Fig. 5 shows, to
obtain information on the existence of possible flip-flop
sites in the membrane during the activation process, we

made use of a fluorescent membrane probe, TMA-DPH [18].
Due to its specific properties, TMA-DPH will, after addition
to platelets, equilibrate between aqueous phase and membrane
within 1 min. The probe is essentially non-fluorescent in
water, but highly fluorescent in the apolar environment of
the lipid bilayer. The low quantum yield of the probe
in solution allows this fraction to be ignored, simplifying
the interpretation of the results. As long as membrane
integrity is ensured, the probe will essentially incorporate
into the outer leaflet of the p.m. Its charged polar head-group
will prevent a rapid exchange with the cytoplasmic leaflet
of the membrane or with the intracellular membranes; >95%
of the probe remains in the micellar form and does not
contribute significantly to fluorescence. Under these
conditions FI is linearly proportional to the amount of
membrane accessible to the probe.

Thrombin activation of platelets results in an FI increase
correlating with a p.m. increase (Fig. 6). This is caused
by fusion of the granule membrane with the p.m. as an integral
part of the secretory process, confirming data of Kubina
and co-workers [19]. Activation by the Ca^{2+}-ionophore ionomycin
[cf. art. #B-7 (Pozzan's laboratory), this vol.- Ed.] - A23187
is ruled out because of its intrinsic fluorescence - produces
a much larger increase in FI than thrombin, not due to
granule-release differences, which are insignificant; the most
likely explanation is that TMA-DPH has gained access to
the internal half of the p.m. as well as to the residual
intracellular membranes.

Sonication of ionomycin-activated platelets hardly
increases the relative FI. This strongly suggests that
during activation with ionophore all membrane sources within
the platelet become accessible to TMA-DPH. In contrast,
sonication of thrombin-stimulated and unstimulated platelets
markedly increases FI to a maximum corresponding to that
of ionomycin-stimulated platelets (Fig. 6).

Further evidence for the transbilayer movement (flip-flop)
of TMA-DPH during ionomycin stimulation came from dilution
experiments. Dilution of unstimulated and thrombin-stimulated
TMA-DPH-labelled platelets causes an efflux of probe from
the membranes in order to reach a new equilibrium distribution
with the aqueous phase. The efflux is evident in Fig. 6
as a rapid decline in FI (within 10 sec) which is similar
for both unstimulated and thrombin-stimulated platelets.
However, the level remains ~5 times higher than this when
ionomycin-treated platelets are similarly diluted. This
finding indicates the existence of a TMA-DPH pool which

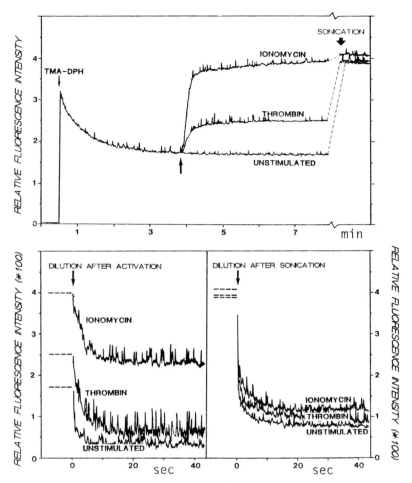

Fig. 6. FI of TMA-DPH in unstimulated and thrombin- and ionophore-stimulated platelets *(upper panel)* and the change after dilution, due to TMA-DPH efflux *(lower panel, left)*, as also performed after sonication *(right)*. Details given in [18].

is non-exchangable with the external environment. When the sonicated suspensions of all three preparations were diluted, FI decreased to the same level in each. This excludes the possibility that in ionomycin-treated platelets part of the probe has been irreversibly incorporated, *e.g.* by irreversible binding to proteins.

Ca²⁺-DEPENDENCE OF TMA-DPH FLIP-FLOP

If Ca^{2+}-mediated activation of platelet calpain is indeed responsible for an increased transbilayer movement of phosphatidylserine along flip-sites in the membrane, the appearance of these sites is expected to display a similar Ca^{2+}-dependence as demonstrated for protein degradation and prothrombinase activity. Using the same approach by which the Ca^{2+}-dependence of calpain and prothrombinase activities was determined, it was found that flip-flop of TMA-DPH was half-maximal at ~15-20 μM free Ca^{2+}, a value very similar to the Ca^{2+} concentration required for half-maximal activity of calpain and prothrombinase.

CONCLUSION

Formation of a procoagulant platelet surface, i.e. the exposure of phosphatidylserine, is the result of a process in which the asymmetric distribution of anionic phospholipids, as present in the p.m. of resting platelets, becomes partially randomized. Considering the rate at which a procoagulant surface is formed upon activation, randomization must be the result of a rapid transbilayer movement of the lipids. Regulation of transbilayer asymmetry of (anionic) phospholipids in resting as well as activated platelets is poorly understood. In the current model to explain maintenance of phospholipid asymmetry in erythrocyte membranes, two principles have been put forward: (1) a direct interaction of aminophospholipids with proteins of the cytoskeleton [20]; (2) the presence of an ATP-dependent aminophospholipid-specific translocase, capable of a rapid inward transport [21, 22].

Loss of phospholipid asymmetry in erythrocytes is thought to involve cytoskeletal reorganization, disabling interaction with p.m. phospholipids [23-25] and increased flip-flop rates via disturbances of the lipid bilayer structure [16].

Preliminary data (unpublished) from our laboratory indicate that the platelet cytoskeleton can bind phosphatidylserine in a specific and reversible manner. Moreover, findings by Devaux's group [21] and also ours (unpublished) strongly suggest the presence of an aminophospholipid specific translocase in platelets also.

Ample evidence, partly presented here, demonstrates that formation of a procoagulant surface needs platelet cytoskeletal modification, mediated by Ca^{2+}-dependent calpain activation. As a tentative model, we suggest that cytoskeletal degradation by calpain may play a dual role in the enhanced flip-flop of phosphatidylserine during platelet activation (Fig. 7). Binding of phosphatidylserine to proteins of the cytoskeleton is disturbed after degradation by calpain, allowing phosphatidylserine to participate in flip-flop. During, or as a result of, cytoskeletal degradation by calpain, phospholipid flip-flop is greatly accelerated (as measured by TMA-DPH)

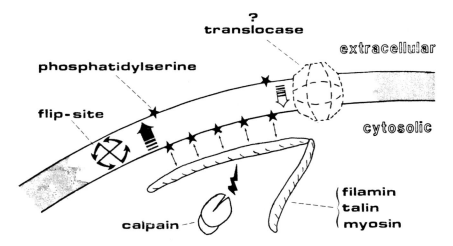

Fig. 7. Hypothetical model for the regulation of transbilayer lipid asymmetry in the platelet p.m.

by the formation of 'flip-sites'. We propose that the
loss of interaction between p.m. and cytoskeleton destabilizes
the membrane structure, producing local disturbances in the
lipid bilayer (flip-sites) whereby transbilayer movement of
phospholipids (including phosphatidylserine) can take place.
To what extent a specific translocase contributes in this
process remains to be established.

References

1. Zucker, M.B. & Nachmias, V.T. (1985) *Arterioscleroris 5*,
   2-18.
2. Jackson, C.M. & Nemerson, Y. (1980) *Ann. Rev. Biochem. 49*,
   765-811.
3. Tans, G. & Rosing, J. (1986) in *New Comprehensive
   Biochemistry*, Vol. 13 (Zwaal, R.F.A. & Hemker, H.C., eds.),
   Elsevier, Amsterdam, pp. 59-86.   [3033.
4. Resnick, R.M. & Nelsestuen, G.L. (1980) *Biochemistry 19*, 3028-
5. Rosing, J. Speijer, H. & Zwaal, R.F.A. (1988) *Biochemistry
   27*, 8-11.
6. Chap, H., Zwaal, R.F.A. & van Deenen, L.L.M. (1977)
   Biochim. Biophys. Acta 467, 146-164.
7. Schick, P.K., Kurica, K.B. & Chacko, G.K. (1976) *J. Clin.
   Invest. 57*, 1221-1226.
8. Bevers, E.M., Comfurius, P. & Zwaal, R.F.A. (1983)
   Biochim. Biophys. Acta 736, 57-76.
9. Comfurius, P., Bevers, E.M. & Zwaal, R.F.A. (1985)
   Biochim. Biophys. Acta 815, 143-148.

10. Niggli, V. & Burger, M.M. (1987) *J. Membr. Biol. 100*, 97-102.
11. Phillips, D.R. & Jakabova, M. (1977) *J. Biol. Chem. 252*, 5602-5605.
12. Pontremoli, S. & Melloni, E. (1986) *Ann. Rev. Biochem. 55*, 455-481.
13. Fox, J.E.B., Goll, D.E., Reynolds, C.C. & Phillips, D.R. (1985) *J. Biol. Chem. 260*, 1060-1066.
14. Verhallen, P.F.J., Bevers, E.M., Comfurius, P. & Zwaal, R.F.A. (1987) *Biochim. Biophys. Acta 903*, 206-217.
15. Verhallen, P.F.J., Bevers, E.M., Comfurius, P. & Zwaal, R.F.A. (1988) *Biochim. Biophys. Acta 942*, 150-158.
16. Bergmann, W.L., Dressler, V., Haest, C.W.M. & Deuticke, B. (1984) *Biochim. Biophys. Acta 769*, 390-398.
17. Comfurius, P., Bevers, E.M. & Zwaal, R.F.A. (1983) *Biochem. Biophys. Res. Comm. 117*, 803-808.
18. Verhallen, P.F.J. (1988) PhD. Thesis, University of Limburg, Maastricht, The Netherlands.
19. Kubina, M., Lanza, E., Cazenave, J.P., Laustriat, G. & Kuhry, J.G. (1987) *Biochim. Biophys. Acta 901*, 138-146.
20. Haest, C.W.M. (1982) *Biochim. Biophys. Acta 694*, 331-352.
21. Seigneuret, M. & Devaux, P.F. (1984) *Proc. Nat. Acad. Sci. 81*, 3571-3575.
22. Tilley, L., Cribier, S., Roelofsen, B., Op den Kamp, J.A.F. & Van Deenen, L.L.M. (1986) *FEBS Lett. 194*, 21-27.
23. Choe, H.R., Williamson, P., Rubin, E. & Schlegel, R.A. (1985) *Cell Biol. Int. Rep. 9*, 597-606.
24. Franck, P.F.H., Bevers, E.M., Lubin, B.H., Comfurius, P., Chiu, D.I., Op den Kamp, J.A.F., Zwaal, R.F.A., van Deenen, L.L.M. & Roelofsen, B. (1985) *J. Clin. Invest. 75*, 183-190.
25. Mohandas, N., Rossi, M., Bernstein, S., Ballas, S., Ravindrath, Y., Wyatt, J. & Mentzer, W. (1985) *J. Biol. Chem. 260*, 14264-14268.

#C-5

ROLE OF CALCIUM IONS IN ACTH-INDUCED LIPOLYTIC AND STEROIDOGENIC ACTIVITIES

H. Okuda, M. Sumida, T. Tsujita, H. Ninomiya and C. Morimoto

Department of Medical Biochemistry II,
School of Medicine, Ehime University,
791-02 Shigenobu, Ehime, Japan

The suggested key role of $[Ca^{2+}]_i$* in ACTH-induced lipolysis in adipocytes and in hormone-mediated steroidogenesis in adrenocortical cells has been investigated [1]. ACTH did not affect the s.a.'s of adipocyte lipolytic enzyme, viz. HSL, or of adrenal steroidogenic enzyme, viz. $P450_{11\beta}$; but the hormone greatly increased the availability of substrate (TG and DOC respectively) to each enzyme. Moreover, Ca^{2+} is significantly involved in the actions of ACTH.

Our findings include the following. (1) Lipolysis in epididymal adipocytes was markedly stimulated by ACTH in the presence of Ca^{2+}, with no change in the activity of extracted HSL - which appears to be present in a repressed state in the absence of ACTH and Ca^{2+}. (2) Stimulation of CS synthesis in adrenal cells in the presence of ACTH and Ca^{2+} has been reported. Ca^{2+} stimulates hydroxylation of DOC to CS in isolated mitochondria, whose phospholipids seem from phospholipase studies to be involved in $P450_{11\beta}$ activities and Ca^{2+} sensitivity.

We suggest that Ca^{2+} is an ACTH messenger both in adipocyte lipolysis and in adrenal-cell steroidogenesis. A Ca^{2+}-pump ATPase, which is located on microsomal membranes and functions by removing cytoplasmic Ca^{2+}, may regulate $[Ca^{2+}]_i$ and hence lipolysis and steroidogenesis. This ATPase, isolated from adrenocortical microsomes and resembling sarcoplasmic reticulum ATPase, reacts through a phosphorylated intermediate.

**Abbreviations.* - $[Ca^{2+}]_i$, intracellular free Ca^{2+}; BES, 2-[bis-(2-hydroxyethyl)amino]ethanesulphonic acid; BSA, bovine serum albumin; CS, corticosterone; DOC, deoxycorticosterone; FFA, free fatty acids; HSL, hormone-sensitive lipase; PC, phosphatidylcholine; TG, triglyceride; TES, 2-[tris(hydroxymethyl)methyl-amino]-1-ethanesulphonic acid. P450 connotes a cytochrome. A wet-weight basis is used for cells.

Mechanisms of hormonal action have been widely discussed, and metabolic step-ups by hormones are thought to be attributable to enhanced activities of enzymes. However, only a few authors have expressed interest in the role of the substrate in the activated metabolism of the stimulated cells. We have studied the mechanisms of lipolytic response in isolated adipocytes during stimulation by lipolytic hormones. In this system, the enzyme HSL has to interact efficiently with the water-insoluble substrate, fat, to effect the lipolysis. Herein we describe the experimental methods employed to demonstrate that the lipolytic hormones such as ACTH operate not on the enzyme, HSL, but on the substrate, fat, or on the interaction of the enzyme-substrate complex. We also discuss (a) common roles of Ca^{2+} as an intracellular mediator of the hormone, which is also involved in steroidogenesis in adrenal cells, and (b) a Ca^{2+} pump maintaining and regulating a physiological Ca^{2+} concentration in those cells.

FAT-CELL ACTH-INDUCED LIPOLYSIS, AND ISOLATED HSL ACTIVITY

To clarify the mechanisms of ACTH-induced lipolysis including the role of HSL and its TG substrates, we measured the activity of extracted HSL after incubating the fat cells, with or without ACTH, with a triolein emulsion as an artificial substrate.

Study of lipolysis.- Epididymal fat pads (1-2 g to furnish 20 samples) were isolated from several male Wistar-King rats (150-180 g) and dispersed for 40 min at 37° with collagenase (to 10 mg/ml) in 'solution **A**', viz. 135 mM NaCl, 5 mM KCl, 2.8 mM glucose, 10 mM TES; the buffer pH was set at 7.4. The cells thus obtained [2] (10^6/ml), filtered through nylon mesh (~0.5 μm pores), were tested for lipolysis by incubating in medium **A** with varying [Ca^{2+}], with or without 1 μg/ml ACTH[1-24]. In some experiments the cells were pre-incubated with 100 U./ml of phospholipase-C or -D for 10 min at 37° to examine the role of phospholipids in the cells, and before the lipolysis incubation the cells were washed with **A**. Assays were performed for FFA [3] and glycerol [4] released into the medium. In some experiments, Ca^{2+} was replaced by other divalent cations, e.g. Mg^{2+}, Ba^{2+}, Cu^{2+}, Zn^{2+} or Mn^{2+}.

Determining the activity of extracted HSL.-From cells (0.7 g) following the lipolysis assay, HSL was extracted either mildly (teflon homogenizer, several strokes at room temp.) or vigorously (Polytron set at 7.5; 20 sec, 10°), in 1 ml pH 7.4 buffer, containing **either** TES, NaCl and KCl as above, plus 1 mM $MgCl_2$ and 2.5% BSA, **or** 20 mM Tris-HCl, 255 mM sucrose and 1 mM EDTA. Centrifugation (5,000 rpm, 3,000 **g**; 15 min, 10°) gave a fat layer, supernatant and pellet. Assay of each

(0.1 ml) was by incubating for 1 h at 37° after additions totalling 0.1 ml – 0.12 mg [H^3]trioleate (0.5 µCi), 1.8% gum arabic and 1.4% BSA – and ultrasonic emulsification (Tomy-seiko, model UR-200P, set at 1). The released ^{3}H-FFA were scintillation-counted by the modified method of Belfrage & Vaughan [5]. Substances to be evaluated for effects on lipolysis were added prior to emulsification.

Lipolysis in intact fat cells *vs.* extracted HSL

In the above experiments, without ACTH, HSL activity was highest, viz. 1143 ±23 nmol/h per g wet wt. cell equivalent (mean ±S.E.M.) in the supernatant fraction prepared with the teflon homogenizer and Tris-HCl buffer. All other fractions had HSL activity only half that of this supernatant, indicating that the extraction methods markedly influence HSL activity.

The supernatant fraction with the highest HSL activity was used, in parallel with intact cells, to evaluate the effect on lipolytic activity of ACTH present in the intact-cell incubation (Table 1). In intact cells lipolysis was maximal with both ACTH and Ca^{2+} present, whereas extracted HSL had much higher activities in all instances, whether or not the cells had been exposed to ACTH and/or Ca^{2+}.

Effects of various factors on the HSL activities

If indeed the lipase itself intrinsically has high activity inside the intact cells notwithstanding observed differing results in the incubations, possible inhibitory machinery that exists in the cells and depresses the activity warrants investigation. One of the authors (Brockman & Tsujita [6]) has suggested that the possible inhibitory machinery may exist on the surface of the lipid droplets that comprise the HSL-substrate, since phospholipids when in high amount interfered with various lipase activities. Accordingly we separated and analyzed endogenous lipid droplets from fat cells: ~99% consisted of TG, and the remaining 1% of phospholipid, cholesterol, carbohydrates and proteins. These components when mixed into the substrate emulsion prior to sonication did not affect the lipase activities. However, when exogenous phospholipids, egg phosphatidylcholines, were added to the substrate emulsion, lipase activity was inhibited by >70% (Fig. 1). Thus, phospholipids in the lipid droplets are suggested to comprise an inhibitory machinery restricting the lipolytic action of HSL. This possibility is also supported by the fact that incubation of hypertonically treated fat cells (fat globules) with 100 µU./ml of phospholipase-C (10 min, 25°) raised lipolytic activities ~4-fold (unpublished observations).

Table 1. Effect of ACTH (1 μg/ml) and CaCl$_2$ (1 mM) in fat-cell incubates on lipolysis (endogenous lipid) and on HSL activity (emulsified [^{3}H]trioleate). Details in text. Rates are nmol glycerol or (HSL) oleate per h per g cell wet wt., as means ± S.E.M.; n = 3.

ACTH	Ca^{2+}	Lipolysis	HSL activity
−	−	78 ±36	630 ±33
−	+	38 ±26	728 ±33
+	−	110 ±40	667 ±17
+	+	378 ±46	801 ±25*

*P <0.05 _vs_. ACTH but no Ca^{2+}

Fig. 1. Effect of egg PC on HSL from rat fat cells. The PC was added in varying amount as shown, before the sonication. ^{3}H-FFA released was the measure of lipolysis (see text). The error bars represent S.E.M.'s.

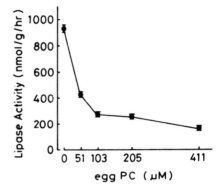

Ca^{2+} effect on the HSL–fat interaction.- This interaction has to take place to initiate the lipolytic reaction. In our model system the interaction, including binding and lipolysis, was significantly affected by Ca^{2+}. When the HSL and the dispersed fat were mixed and homogenized with varying [Ca^{2+}]$_e$, and then the mixture was separated centrifugally into the fat-layer and the supernatant, the HSL lipolytic activity with high [Ca^{2+}]$_e$ (1-3 mM) increased in the former and decreased in the supernatant [7], implying a shift of HSL from the supernatant to the fat layer. Thus it was suggested that phospholipids might interfere with the HSL-fat interaction, but the Ca^{2+} might stimulate the ability of the HSL to bind to the fat.

From these results we suggest that the enzyme HSL exists initially in the cell as its active form. However, to exert its ability to stimulate lipolysis it must interact suitably with the substrate. Consequently, the target of the hormonal action is not merely the HSL itself but probably the optimal interaction of the HSL and substrate.

Ca^{2+} ROLE IN STEROIDOGENESIS BY MITOCHONDRIA ISOLATED FROM BOVINE ADRENOCORTICAL TISSUE

To evaluate the role of Ca^{2+} as the intracellular mediator common to various tissues including adipocytes and adrenal cortex, mitochondria from the latter were examined for the Ca^{2+}-dependency of steroidogenesis, since ACTH stimulation was found to raise $[Ca^{2+}]_i$ [8].

Measurement of steroidogenesis in mitochondria, isolated from bovine adrenocortical tissue [9] and thawed after 1 month at $-40°$, entailed incubation (1 mg protein/ml) for 1 h at $37°$ in a pH 6.8 solution containing 0.25 M sucrose, 20 mM KCl, 10 mM BES, 100 µg DOC, 0.8 mg NADPH, and Ca^{2+} in varied amount (or another divalent cation). The product, CS, was assayed fluorimetrically [10].

The same solution but with DOC omitted (1 ml, $25°$) was used to measure the reduction rate of P450 by 0.3 mg mitochondria pre-treated with CO for 3 min. Detection of the increase in differential absorption (490 $vs.$ 450 nm) was by a dual-wavelength spectrophotometer (Hitachi 557).

Ca^{2+} requirement in the electron-transfer step

In the incubation system, there was strong Ca^{2+}-dependence of CS synthesis from DOC. The following rates (mean ±S.E.M.; 3 independent samples) were found for **steroidogenesis,** and (see below) for **$P450_{11\beta}$ reduction** (expressed respectively as nmol CS/h and *pmol P450/min* per mg protein).-

Ca^{2+} <1 µM (none added):   17.4 ±0,   *81 ±8;*
Ca^{2+} added, to 1 mM: 215 ±2,   *240 ±40.*

Since the system for DOC-hydroxylation involving P450 is composed of the electron-transfer enzymes adrenodoxin reductase, adrenodoxin and $P450_{11\beta}$, we investigated which was the Ca^{2+}-requiring step. As shown above *(italics),* Ca^{2+} markedly stimulates the P450-reduction step. Moreover, neither for steroidogenesis nor for $P450_{11\beta}$ *reduction* could Ca^{2+} be replaced by other divalent cations (1 mM).-

Mg^{2+}: 32 ±3, *49 ±24;* Sr^{2+}: 38 ±8, *61 ±12;* Mn^{2+}: 48 ±1, *98 ±12;*
Zn^{2+}: 3.3 ±0, *0 ±0;* Co^{2+}: 8.0 ±0, *61 ±12;* Ni^{2+}: 5.3 ±0, *49 ±12;*
Cd^{2+}: 4.4 ±0, *44 ±4;* Hg^{2+}: 4.3 ±0, *0 ±0.*

Ca^{2+}-dependent reduction of the mitochondrial P450 was strongly inhibited (by fully 60%) if the mitochondria were treated with phospholipase-C (10 mU./ml; 10 min, $24°$), but with phospholipase-D ~95% of the activity remained (Fig. 2). Thus, the phosphate groups in the phospholipid molecules integrating the mitochondrial membrane were likely to be involved in the Ca^{2+}-stimulated reaction.

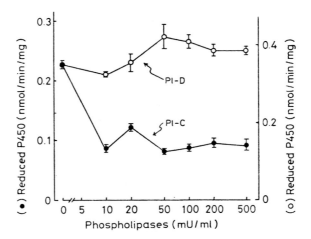

Fig. 2. Rate of $P450_{11\beta}$ reduction in CO-treated bovine
adrenocortical mitochondria as affected by phospholipase-C
(Pl-C) or -D (Pl-D), each at mU./ml levels shown.   0.92 mM
NADPH + 0.1 mM $CaCl_2$ present, + mitochondria, 1 mg protein.
During incubation at 25° the reduction was followed as stated
in the text. Bars represent S.E.M. of 3 samples.

These results emphasize that regulation of both lipolysis
and steroidogenesis is crucially influenced not merely by
the activity levels of HSL and $P450_{11\beta}$ but also by other
factors such as the phospholipid. In addition, Ca^{2+} possibly
functions as the common intracellular mediator of ACTH,
making the interaction of substrate and the enzymes more
efficient.

MICROSOMAL Ca^{2+}-ATPase AS ADRENOCORTICAL Ca^{2+} REGULATOR

Ca^{2+}-dependent ATPase of skeletal-muscle sarcoplasmic
reticulum has been well characterized and shown to function
as a potent intracellular Ca^{2+} pump [12]. In microsomal
fractions isolated from bovine adrenocortical tissue, we have
demonstrated activities of ATP-dependent Ca^{2+} uptake and phosphorylation of Ca^{2+}-ATPase with properties similar to those
of skeletal and other muscle [12]. Although the activities
are much lower than for skeletal muscle, presumably the
ATPase functions as the intracellular Ca^{2+} regulator and
strongly influences steroidogenesis (cf. above).

Measuring microsomal Ca^{2+} uptake and phosphorylation.-
Post-mitochondrial fractions obtained by differential centrifugation from fresh bovine adrenal cortex were applied to
a sucrose density gradient [13]; from 50 g wet wt. of tissue
the yield of microsomes was 20 mg protein.

For Ca^{2+} uptake the microsomes (1 mg) were incubated in 1 ml containing 0.1 M KCl, 20 mM BES, 1 mM $MgCl_2$, 1 mM ATP and various Ca^{2+} concentrations (with 10^6 cpm of ^{45}Ca) at pH 6.8 for 2 min at 30°. The microsomes were then filtered on a Millipore membrane (diam. 2.5 cm, pore size 0.8 μm) and washed with 5 ml ice-cold 1 mM EGTA 3 times within 30 sec. The amount of ^{45}Ca remaining on the filter was determined with a liquid scintillation counter.

Phosphorylation of the microsomal proteins was assayed by incubating 0.5 mg microsomes for 10 sec at 0° in the same reaction solution (1 ml) as for Ca^{2+} uptake without or with 10 μM Ca^{2+}, except for omission of ^{45}Ca and inclusion of 10 μM ATP with 10^6 cpm of $[\gamma-^{32}P]ATP$. The phosphorylation was terminated by adding 1 ml 10% trichloroacetic acid, and the denatured microsomes were centrifugally washed with 3 x 5 ml 3% (v/v) perchloric acid containing 1% polyphosphate. Micro-somal aliquots (50 μg) were solubilized in 1% SDS and subjected to polyacrylamide gel electrophoresis (PAGE) to study which proteins were phosphorylated. The ^{32}P in the denatured microsomes or in the gels was determined by liquid scintillation counting.

Microsomal Ca^{2+} uptake and phosphorylation

The Ca^{2+} uptake rate by the microsomes was found to be $[Ca^{2+}]$-dependent, and was paralleled by $[Ca^{2+}]$-governed protein phosphorylation (Fig. 3). PAGE analysis of the micro-somal ^{32}P-labelled proteins, which were acid-stable, showed that a protein of M_r ~105,000 was phosphorylated only in the presence of Ca^{2+}, and the phosphorylation was highly susceptible to hydroxylamine. These and other results [13] also coincided well with the characteristics of the phosphoryl-ated intermediate of skeletal muscle sarcoplasmic reticulum.

CONCLUSION

To summarize, we propose that insoluble-enzyme reactions, e.g. lipolysis and steroidogenesis, are governed not by enzyme level but by the circumstances of the enzymes and their substrates including their interactions. Moreover, Ca^{2+} has a key role in the reactions as the common intracellular mediator for ACTH.

Acknowledgments

This work was supported by a research grant from the Ministry of Education of Japan. We thank Drs. M. Katagiri and K. Suhara (Kanazawa University) for their valuable comments.

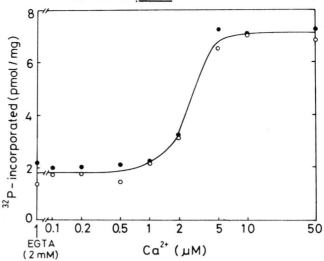

Fig. 3. Ca^{2+}-dependency of phosphorylation in bovine adrenocortical microsomes, using [γ-^{32}P]ATP as described in the
text but with a range of Ca^{2+} concentrations (0.1-50 μM) or
with 2 mM EGTA. Incubation for 5 (o) or 10 (●) sec.

References

1. Okuda, H., Tsujita, T. & Kinutani, M. (1986) *Pharmacol.
    Res. Comm 284*, 877-893.
2. Rodbell, M. (1964) *J. Biol. Chem. 239*, 375-380.
3. Dole, V.P. (1956) *J. Clin. Invest. 35*, 150-154.
4. Warnick, G.R. (1986) *Meth. Enzymol. 129*, 101-123.
5. Belfrage, P. & Vaughan, M. (1969) *J. Lipid Res. 10*,
    341-344.
6. Brockman, H.L. & Tsujita, T. (1988) *FASEB J. 2*, A1368
    (Abstract).
7. Masaka, M. Okuda, H. & Fujii, S. (1972) *J. Biochem. 72*,
    1565-1566.
8. Kojima, I. & Ogata, E. (1986) *J. Biol. Chem. 261*, 9832-
    9838.
9. Omura, T., Sanders, E., Estabrook, R.W., Cooper, D.Y. &
    Rosenthal, O. (1966) *Arch. Biochem. Biophys. 117*, 660-673.
10. Silber, R.H., Busch, R.D. & Oslapas, R. (1958) *Clin.
    Chem. 4*, 278-285.
11. Kawamura, M. (1974) *Jikeikai Med. J. 21*, 151-162.
12. Takenaka, H., Sumida, M. & Hamada, M. (1988)   *Studies of
    Enzymes 2* (Kuby, S.A., ed.), CRC Press, Boca Raton, FL, in press.
13. Sumida, M., Hamada, M., Shimowake, A., Morimoto, C. &
    Okuda, H. (1988) *J. Biochem. 104*, 687-692.

#ncC

NOTES and COMMENTS relating to

SECRETORY AND OTHER PROCESSES INVOLVING Ca^{2+}
(not muscle or nerve)

'COMMENTS' are on pp. 329-332, starting with Forum discussions
on the preceding main articles, then on Note #ncC.1

#ncC.1

A Note on

THE CALCIUM MESSENGER IN THE ACTIVATION OF ALDOSTERONE SYNTHESIS: FUNCTIONAL CHARACTERIZATION OF THE SUBCELLULAR ORGANELLES INVOLVED

A.M. Capponi, M.F. Rossier, E. Johnson, E. Davies and M.B. Vallotton

Division of Endocrinology, University Hospital, CH-1211 Geneva, Switzerland

The octapeptide hormone, angiotensin II, is a physiological stimulator of aldosterone production in the adrenal zona glomerulosa. It is now established that angiotensin II triggers steroidogenesis by activating the Ca^{2+} messenger system [1]. Indeed, as a result of hormone activation, marked and complex changes in $[Ca^{2+}]_i$* occur [2, 3]. Very specific and powerful fluorescent probes, <u>e.g.</u> fura-2 [4], have rendered possible the measurement of $[Ca^{2+}]_i$ in single cells. Angiotensin II, at a maximally stimulating concentration (10^{-8} M), induces a transient rise in $[Ca^{2+}]_i$ in adrenal zona glomerulosa cells. After an initial and very rapid peak, $[Ca^{2+}]_i$ oscillates regularly for several minutes, with a mean period of 17 sec. The oscillations then slowly fade away, but $[Ca^{2+}]_i$ remains higher than the initial prestimulation levels. Removal of extracellular Ca^{2+} during continuous angiotensin II stimulation leads to almost immediate suppression of hormone-induced $[Ca^{2+}]_i$ oscillations, a finding which suggests an involvement of Ca^{2+} influx in this oscillatory process. In contrast, the initial $[Ca^{2+}]_i$ rise results exclusively from Ca^{2+} mobilization from an intracellular pool sensitive to $InsP_3$.

We have described, as now outlined, the functional characteristics of the organelle which is the source of the initial $[Ca^{2+}]_i$ transient, as well as the organelle which constitutes one of its sites of action.

The source of $[Ca^{2+}]_i$.- Electropermeabilized bovine adrenal glomerulosa cells, when placed in an 'intracellular' medium containing ATP, sequester Ca^{2+}, reducing its ambient concentra-

---

*$Abbreviations.-$ $[Ca^{2+}]_i$, intracellular (cytosolic) free Ca^{2+} concentration; CaM, calmodulin; e.r, endoplasmic reticulum; $InsP_3$, inositol 1,4,5-trisphosphate; Ptd, phosphatidyl.

tion to ~200 nM — as measured with a Ca^{2+}-sensitive minielect-
rode [5]. The organelle responsible for this Ca^{2+} sequestration
releases Ca^{2+} in response to $InsP_3$ (EC_{50} = 0.65 μM). This
organelle is distinct from the mitochondria, since both pumping
and $InsP_3$-induced release of Ca^{2+} still occur in the presence
of ruthenium red, a specific inhibitor of mitochondrial Ca^{2+}
influx.

Subcellular fractionation of bovine adrenal cortex was
achieved by differential centrifugation and discontinuous
sucrose density gradients. The specific [^{3}H]$InsP_3$ binding
capacity was measured in the various fractions and correlated
with $InsP_3$-induced Ca^{2+} release. The $InsP_3$ binding site
was separate from both the plasma-membrane and the e.r.,
determined by its sulphatase C activity and by mRNA content.
Hence $InsP_3$ binds to, and releases Ca^{2+} from, a vesicular,
non-mitochondrial organelle distinct from the e.r.

The site of action of $[Ca^{2+}]_i$.- In order to demonstrate
directly the role of Ca^{2+} in steroidogenesis, we examined
the effects of Ca^{2+} on aldosterone and corticosterone production
in electropermeabilized (2500 V/cm, 0.1 sec, × 10) bovine
adrenal glomerulosa and fasciculata cells placed in a superfusion
chamber [6]. When the superfusion medium was supplemented
with buffered Ca^{2+} (0.1-10 μM), a concentration-dependent and
sustained release of aldosterone output was observed in glomerulosa cells. The maximum response (3 times the basal production
rate) was achieved with 1-2 μM Ca^{2+}, and the EC_{50} for Ca^{2+}
was 0.5 μM. Qualitatively and quantitatively similar results
were obtained for corticosterone production in permeabilized
fasciculata cells. The presence of $NADP^+$ was obligatory
for a Ca^{2+}-induced stimulation of aldosterone and corticosterone
production, and these responses were completely blocked by
ruthenium red (1 μM) and by W7, a CaM inhibitor.

Thus, permeabilized adrenal cortical cells retain the
ability to produce steroids. In addition, Ca^{2+} influx into
the mitochondria and Ca^{2+}/CaM-dependent reactions are critical
steps in the activation of steroidogenesis in the adrenal
cortex.

These studies provide a further direct link between
$[Ca^{2+}]_i$ and the biological responses induced in the adrenal
cortex by steroidogenic, Ca^{2+}-mobilizing activators such as
angiotensin II (Fig. 1) and K^+ ions, and yield information
as to the nature of the intracellular organelles involved
in this response.

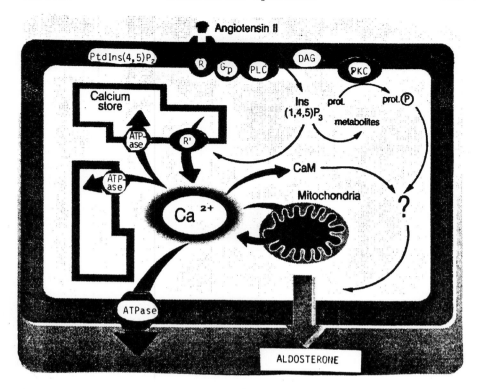

Fig. 1. The Ca^{2+} messenger system in angiotensin II-induced aldosterone production: involvement of various subcellular organelles. $InsP_3$ releases Ca^{2+} from a non-mitochondrial pool. The released Ca^{2+} activates mitochondrial and CaM-dependent mechanisms, before being extruded into the extra-cellular medium or sequestered into an $InsP_3$-sensitive pool. R/R', angiotensin II/$InsP_3$ receptor; DAG, diacylglycerol; PLC, phospholipase C; PKC, Ca^{2+}-sensitive, phospholipid-dependent protein kinase (protein kinase C).

References

1. Capponi, A.M., Lew, P.D., Jornot, L. & Vallotton, M.B. (1984) *J. Biol. Chem. 259*, 8863-8869.
2. Quinn, S.J., Williams, G.H. & Tillotson, D.L. (1988) *Proc. Nat. Acad. Sci. 85*, 5754-5758.
3. Johnson, E.I.M., Capponi, A.M. & Vallotton, M.B. (1989) *J. Endocrinol.*, in press.
4. Grynkiewicz, G., Poenie, M. & Tsien, R.Y. (1985) *J. Biol. Chem. 260*, 3440-3450.
5. Rossier, M.F., Krause, K.H., Lew, P.D., Capponi, A.M. & Vallotton, M.B. (1987) *J. Biol. Chem. 262*, 4053-4058.
6. Capponi, A.M., Rossier, M.F., Davies, E. & Vallotton, M.B. (1988) *J. Biol. Chem. 263*, 16113-16117.

#ncC.2

A Note on

RESTING [Ca²⁺]ᵢ IN RAT CHROMAFFIN CELLS: EVIDENCE FOR AN OSCILLATORY BEHAVIOUR

[1]A. Malgaroli, [2]T. Pozzan and [1]J. Meldolesi

[1]Department of Pharmacology, School of Medicine,
H.S. Raffaele, University of Milan,
Via Olgettina 60, 20132 Milano, Italy

and [2]Institute of General Pathology, School of Medicine,
University of Padua, Via Loredan 16, Padua, Italy

Up to very recently measurements of cytosolic free calcium concentration, $[Ca^{2+}]_i$, were carried out only in cell suspensions, and the results suggested remarkable stability of level with time. However, the development of single-cell $[Ca^{2+}]_i$ measurements by the use of a Ca^{2+}-sensitive fluorescent probe, fura-2, demonstrated that in many cellular systems 'resting Ca^{2+}' is constantly undergoing transient changes [1] (cf. arts. earlier in this Sect.). An interesting model is furnished by endocrine tissues known to spontaneously secrete hormones in large amount, as $[Ca^{2+}]_i$ plays an important role in hormone secretion.

Primary cultures of chromaffin cells purified from rat adrenal medulla, which spontaneously secrete large amounts of catecholamine [2], exhibited $[Ca^{2+}]_i$ fluctuations in a high proportion of cells (70%). The frequency of these oscillations varied from cell to cell and also in the same cell (range: 0-0.1 Hz; average: 0.015 Hz). The amplitude ranged between 0.2 and 1 μM over basal values ~50-120 nM.

As electrophysiological data [3] demonstrated that rat chromaffin cells fire action potentials spontaneously (with a major sodium component), we evaluated the involvement of Ca^{2+} channels in $[Ca^{2+}]_i$ oscillations. In our hands the chelation of extracellular Ca^{2+} with EGTA in excess did not cause immediate suppression of the spontaneous Ca^{2+} fluctuations, which continued for as long as 30-120 sec although attenuated. On the other hand, the phenomenon seems to be modulated by membrane potential, since a small increase in $[K^+]_o$ (from 5.5 to 15 mM) augmented the frequency of the oscillations, while greater increases (>15 mM) were inhibitory.

The most ready explanation of this behaviour is that voltage-operated Ca^{2+} channels and intracellular Ca^{2+} stores coparticipate in the generation of $[Ca^{2+}]_i$ oscillations. In order to elucidate which is the initiator, $[Ca^{2+}]_i$ and electrophysiological combined experiments are now being undertaken.

References

1. Malgaroli, A., Vallar, I., Elahi, F.R., Pozzan, T., Spada, A. & Meldolesi, J. (1987) *J. Biol. Chem.* *262*, 13920-13927.
2. Kidokoro, Y. & Ritchie, A.K. (1980) *J.Physiol.* *307*, 199-210.
3. Brandt, B.L., Hagiwara, S., Kidokoro, Y. & Miyazaki, S. (1976) *J. Physiol.* *263*, 417-439.

#ncC.3

A Note on

SPATIAL AND TEMPORAL ASPECTS OF AGONIST-INDUCED CHANGES IN CYTOSOLIC CALCIUM IN BOVINE ADRENAL CHROMAFFIN CELLS REVEALED BY FLUORESCENT IMAGING

Timothy R. Cheek, [†]Anthony J. O'Sullivan, Roger B. Moreton, Michael J. Berridge and [†]Robert D. Burgoyne

AFRC Unit of Insect Neurophysiology and Pharmacology, Department of Zoology, University of Cambridge, Cambridge CB2 3EJ, U.K.

and [†]MRC Secretory Control Research Group, The Physiological Laboratory, University of Liverpool, Liverpool L69 3BX, U.K.

The trigger for secretion of catecholamines from adrenal chromaffin cells is a rise in $[Ca^{2+}]_i$* [1]. Stimuli such as nicotine and high $[K^+]$, which open voltage-dependent Ca^{2+} channels, are potent secretagogues whereas agonists such as muscarinic agents, bradykinin and angiotensin, all of which raise $[Ca^{2+}]_i$ by mobilizing stored calcium *via* the breakdown of inositol lipids, stimulate little or no secretion [2, 3]. In order to gain more insight into the relationship between Ca^{2+} and secretion, we have examined the nature of the Ca^{2+} response in single fura-2-loaded cells using fluorescence imaging techniques.

In resting cells, $[Ca^{2+}]_i$ was 78.9 ±1.86 nM (± S.D.; n = 148) and appeared homogeneously distributed throughout the cytoplasm. Both nicotine (10 µM) and high K^+ (55 mM) caused a rapid and marked transient elevation in $[Ca^{2+}]_i$: maximum responses above basal were respectively 192 nM (n = 19) and 267 nM (n = 17). Onset and decay times varied between cells. The presence of excess extracellular EGTA totally abolished the Ca^{2+} response to both agonists. The majority of the $[Ca^{2+}]_i$ rises in response to either agonist did not occur uniformly throughout the cytosol: in larger cells (>9 µm) the initial rise in $[Ca^{2+}]_i$, ~40 mM above basal, was specifically located directly beneath the plasma membrane. The Ca^{2+} then 'in-filled' until, after several seconds, Ca^{2+} at an elevated concentration was uniformly distributed in the cell. This response presumably resulted

*$[Ca^{2+}]_i$ is the concentration of intracellular Ca^{2+}.

from influx of external Ca^{2+}. This influx phase was followed by a further marked rise in $[Ca^{2+}]_i$, the maximum response in 50% of the cells being recorded at one pole of the cell. This second phase of the response may have been due to release of calcium stores either by Ca^{2+}-induced Ca^{2+} release or by $InsP_3$ [4]. Consistent with this notion is the observation from electron micrographs that the endoplasmic reticulum is often restricted to one side of these cells.

The response of chromaffin cells to less potent secreta-gogues is strikingly different from that outlined above. Only 41% of cells examined responded to muscarinic agents with any detectable $[Ca^{2+}]_i$ rise, whose magnitude and pattern showed considerable variation: some cells responded with a delayed and modest (40 nM above basal) rise, whereas others showed a more rapid, pronounced increase (200 nM above basal). Many of the responsive cells showed various degrees of oscil-latory behaviour. Many of the $[Ca^{2+}]_i$ rises were spatially restricted such that they appeared to originate from a particu-lar pole of the cell before spreading over the rest of the cell to give a uniform distribution. This is consistent with calcium being released from an internal store located at a specific point within the cell. The fact that some cells are capable of responding to muscarinic agents with a maximum $[Ca^{2+}]_i$ rise comparable to that of nicotine without apparently stimulating the cell to secrete suggests that magnitude of response is only one important aspect of a Ca^{2+} signalling mechanism which comprises several components. Temporal and spatial aspects and oscillatory behaviour all vary considerably between agonists, and all may be equally as important as magnitude in determining the efficacy of a given agonist in eliciting a secretory response from these cells [5].

References

1. Baker, P.F. & Knight, D.E. (1981) *Phil. Trans. Roy. Soc.* *296*, 83-103.
2. Cheek, T.R. & Burgoyne, R.D. (1985) *Biochim. Biophys. Acta* *846*, 167-173.
3. O'Sullivan, A.J. & Burgoyne, R.D. (1989) *Biosci. Rep.*, in press.
4. Eberhard, D.A. & Holz, R.W. (1987) *J. Neurochem. 49*, 1634-1643.
5. O'Sullivan, A.J., Cheek, T.R., Moreton, R.B., Berridge, M.J. & Burgoyne, R.D (1989) *EMBO J. 8*, 401-411.

#ncC.4

A Note on

APPROACHES TO INVESTIGATING THE ROLE OF INTRACELLULAR Ca²⁺ IN THE REGULATION OF PANCREATIC ENZYME SECRETION

R.L. Dormer and ⊗G.R. Brown

Department of Medical Biochemistry,
University of Wales College of Medicine,
Heath Park, Cardiff CF4 4XN, U.K.

Secretion of digestive enzymes and electrolytes from pancreatic acinar cells is stimulated by cholinergic agonists or cholecystokinin and related peptides. Direct evidence that Ca^{2+} is a primary intracellular regulator of this process has come from studies, utilizing intact cells, which fulfil the following general criteria (reviewed in [1]).-

1. Physiological stimuli increase $[Ca^{2+}]_i$* rapidly enough to precede the onset of secretion.

2. Introduction of Ca^{2+} into the cell, by use of ionophores or by otherwise permeabilizing the cell, stimulates secretion directly, with a positive correlation between the amount of secretion and $[Ca^{2+}]_i$.

3. Introduction of Ca^{2+} chelators into intact cells inhibits the stimulation of secretion.

It is accepted that the balance of Ca^{2+} in the cell is maintained by a number of active Ca^{2+} pumps which sequester Ca^{2+} at intracellular sites or extrude it from the cell (E. Carafoli, #ncB.1 in this vol.). In addition, current evidence has suggested that the Ca^{2+} which triggers pancreatic secretion, following physiological stimulation, is initially released from an internal store, extracellular Ca^{2+} being required to maintain a stimulated rate of secretion [see 1]. Studies entailing cell breakage, either completely in order to isolate defined subcellular fractions, or partially by selectively permeabilizing the outer membrane, have led to the hypothesis that a Ca^{2+}-activated, Mg^{2+}-dependent ATPase

---

⊗now at Univ. of British Columbia, Pharmacology Divn., Faculty of Pharmaceutical Sciences, E. Mall, Vancouver, Canada V6T 1W5.

Abbreviations.- $[Ca^{2+}]_i$, cytoplasmic free Ca^{2+} concentration; Ab, antibody; e.r., endoplasmic reticulum; p.m., plasma membrane(s); CHAPS, 3-[(3-cholamidopropyl)–dimethylammonio]-1-propane-sulphonate.

Purified membranes
↓
Purified protein constituents
↓
┌──────────── Antibodies (Ab's) ────────────┐
↓   ↓
Immunolocalization   Test for effects on activity
(fixed cell sections)   ↓
                            Introduce into living cells

Scheme 1. Strategy for studying regulation of intracellular Ca^{2+} in intact cells.

(Ca^{2+}-ATPase), located in the rough e.r., is important in two respects: in maintaining $[Ca^{2+}]_i$ in unstimulated cells and in accumulating the Ca^{2+} store, released by the action of inositol-1,4,5-trisphosphate, following activation of acinar cells by physiological secretagogues [see 1].

The techniques used do, however, have limitations. Subcellular fractionation disrupts organelle structure and inter-organelle contacts and may thereby prevent recognition of previously unrecognized subcompartments. In the absence of specific inhibitors, cell permeabilization studies do not establish which organelles are involved in the processes studied. We are currently developing techniques to use the information from broken-cell techniques to study which mechanisms are most important in regulating the cell Ca^{2+} balance and the precise localization of the internal Ca^{2+} store in intact cells.

Scheme 1 outlines our strategy. Isolated membranes, e.g. rough e.r. and p.m., purified from acinar cells (for p.m. see art. #D-4 in Vol. 17, this series) are the starting material for purifying membrane proteins such as the Ca^{2+}-ATPase, which our data suggest are important in intracellular Ca^{2+} regulation. Ab's raised to these proteins may then be used in two ways.- (1) Using fixed cells, subcellular sites may be immunolocalized. (2) Ab's are selected for their ability to modify the activity of the protein and then introduced into living cells to test their effects on Ca^{2+} homeostasis and secretory activity. We now outline progress made towards these goals with reference to the Ca^{2+}-ATPase of rough e.r. membranes purified from rat pancreatic acinar cells [2].

Purification of rough e.r. Ca^{2+}-ATPase

This ATPase is a minor protein of the e.r. compared to its counterpart in sarcoplasmic reticulum, and is fairly labile although partially stabilized by dithiothreitol (50-60%

loss of the activity on overnight storage at -70°). **Step 1.-**
Rough e.r. membranes isolated as described [2, 3] were treated
with 0.5 M KCl, 1 mM puromycin, 0.1 mM $CaCl_2$, 0.2 M $NaHCO_3$
(pH 8.0) for 15 min at 37°, followed by centrifugation at
165,000 **g** for 30 min and one wash in the same medium. The
final pellet contained <2% of the RNA and 4.1 ±3.3% (mean
±S.E.M., n = 5) of the amylase activity of the original
rough e.r., demonstrating removal of the majority of the
ribosomes and luminal content of the membranes. **Step 2.-**
Active Ca^{2+}-ATPase was most successfully solubilized by non-
ionic or zwitterionic detergents. All experiments were carried
out at a rough e.r. protein concentration of ~3 mg/ml in
the presence of 20% (v/v) glycerol since this resulted in
a 7-fold greater yield in the solubiliized supernatant
(100,000 **g**, 45 min) than in the absence of glycerol. With
Triton X-100, 0.5 mg/mg rough e.r. protein gave maximal activity
in the supernatant, 40.0 ±12.3% (n = 4) of the original
rough e.r. With *N*-octylglucopyranoside, 62.8 ±15.6% (n =
6) was solubiliized at 2 mg detergent/mg protein. CHAPS
at 3 mg/mg protein solubiliized 50.4 ±14% (n = 3).

Steps **1.** and **2.** combined resulted in varying yields
of Ca^{2+}-ATPase (11.3 - 47.3%) with increases in specific activity
of up to 5.0-fold compared with the original rough e.r.
No gain in purity of the active enzyme has so far been
achieved by chromatographic steps applied to solubilized
protein.

Raising monoclonal antibodies

In view of the difficulties encountered in purifying
active Ca^{2+}-ATPase a strategy was employed in which an impure
preparation was used as the antigen for injection into Balb/c
mice. Spleen cells were removed from animals showing a
positive reaction in an ELISA for rough e.r. proteins (unpublished work) and fused with plasmacytoma cells (SP2 line).
Having established hybridomas secreting Ab's to rough e.r.
proteins, culture medium from the cells was tested for effects
on Ca^{2+}-ATPase activity. The results in Fig. 2 show media
from 3 wells which were typical of the majority in having
no consistent effect on this activity; the fourth (4F6)
showed concentration-dependent inhibition. The latter cells
were cloned by limiting dilution and two clones were found
which secreted Ab's that inhibited Ca^{2+}-ATPase activity by
36% and 64% at a 1:10 dilution of the culture medium. On
recloning and expanding these cells, the effect was lost;
but these experiments suggest that the selection of Ab's
having specific effects on ATPase or other activities involved
in Ca^{2+} regulation, using the monoclonal technique, is a
promising way forward (see also below).

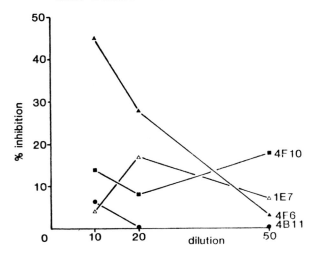

Fig. 1. Ca²⁺-ATPase activity as affected by culture medium
from hybridomas secreting Ab's against rough e.r. proteins.
The activity in rough e.r. membranes was measured (method: [2])
in the presence of dilutions of culture media supernatants
and expressed as % inhibition *vs.* activity in membranes
incubated in equivalent dilution of medium not exposed to
cells. Each point is the mean of triplicate assays. Notations
represent different wells of a cell culture plate.

Introduction of antibodies into living cells

To introduce proteins and other impermeant molecules
into acinar cells, the method that has so far proved the
most useful is incubation in hypotonic media. This was
first used to develop a method for measuring $[Ca^{2+}]_i$ using
Ca^{2+}-activated photoproteins [4] and to demonstrate that incorporation of Ca^{2+}-chelators inhibited cholinergic stimulation
of enzyme secretion [5].

Aequorin uptake in hypotonic medium was maximal in 5-10 min
and half-maximal at ~30 sec. Aequorin or [³H]inulin was
associated with acinar cells in 2- to 3-fold greater amount
when incubated for 10 min in hypotonic compared to isotonic
medium [4]. However, soluble cell contents are lost: 8-fold
more lactate dehydrogenase was released following a 10-min
incubation in hypotonic compared to isotonic medium, but
there was no increase in secretion of the granule-bound
enzyme amylase. Nevertheless the viability of cells incubated
in hypotonic medium was not compromised, as assessed by
ATP content or secretory reponsiveness [5]. Preliminary
data (not shown) have demonstrated that with hypotonic medium
for incubating acinar cells, fluorescein-labelled immunoglobu-
lins become associated with the cells to a greater extent.

Summary

From data obtained on subcellular fractions we are working towards applying the knowledge gained to observe the behaviour of membrane transport systems in intact cells. The aim is to define the precise location of these potential sites of Ca^{2+} uptake and release, and their interaction in maintaining the cell Ca^{2+} balance and stimulating secretion. Crawford et al. ([6], & #B-3, this vol.) have recently used an immunological approach similar to ours: they used purified membranes as antigen and, from effects on function, selected Ab's to produce a MAb which inhibits Ca^{2+} transport by platelet internal membranes.

Acknowledgements

The work was supported by the Medical Research Council.

References

1. Dormer, R.L., Brown, G.R., Doughney, C. & McPherson, M.A. (1987) *Biosci. Rep. 7*, 333-344.
2. Brown, G.R., Richardson, A.E. & Dormer, R.L. (1987) *Biochim. Biophys. Acta 902*, 87-92.
3. Richardson, A.E. & Dormer, R.L. (1984) *Biochem. J. 219*, 679-685.
4. Dormer. R.L. (1983) *Biosci. Rep. 3*, 233-240.
5. Dormer, R.L. (1984) *Biochem. Biophys. Res. Comm. 119*, 876-883.
6. Hack, N., Wilkinson, J.M. & Crawford, N. (1988) *Biochem. J. 250*, 355-361.

#ncC.5

A Note on

CALCIUM IONS, ION CHANNELS AND CONTROL OF SECRETION

O.H. Petersen

MRC Secretory Control Research Unit,
Department of Physiology, University of Liverpool,
P.O. Box 147, Liverpool L69 3BX, U.K.

The $[Ca^{2+}]_i$* level is a key regulator of the secretion of both macromolecules (by exocytosis) and fluid. Ion channels in the plasma membrane play a crucial role in the control of $[Ca^{2+}]_i$ and are also directly involved in electrolyte and fluid secretion [1].

Direct investigations of ion channels are carried out by the patch-clamp technique for single-channel or whole-cell current recording [2]. It is now also possible to change the intracellular composition of diffusible molecules in individual experiments while measuring the membrane ionic currents [3].

Many gland cells (e.g. most endocrine cells) have the ability to fire action (spike) potentials, and in these cases Ca^{2+} enters the cell in a controlled manner through voltage-gated Ca^{2+}-selective pores [1]. Gating of the Ca^{2+} channels can also be modulated by secretagogues. In insulin-secreting cells carbohydrate stimulation evokes membrane depolarization (via closure of ATP/ADP-sensitive K^+ channels) and this increases the probability of Ca^{2+}-channel opening, but glycolysis also markedly increases the mean open time of the Ca^{2+} channels at a given membrane potential as well as lowering the voltage threshold [4]. Ca^{2+}-activated and voltage-sensitive K^+ channels, with a high conductance, provide an important negative feedback since they will be switched on by the combination of depolarization and increase in $[Ca^{2+}]_i$. Activation of these high-conductance K^+ channels repolarizes the cell membrane, thereby closing the voltage-sensitive Ca^{2+} channels. Thereafter $[Ca^{2+}]_i$ falls and this removes the stimulus for the opening of the high-conductance K^+ channels, again allowing the membrane to depolarize and in this way the cycle continues [1, 4].

---

*Terms used.- $[Ca^{2+}]_i$, concentration of intracellular free Ca^{2+}; InsP, an inositol phosphate: thus, $InsP_3$ = the 1,4,5-trisphosphate.

In secretory cells that are unable to support action potentials (e.g. most exocrine cells), voltage-gated Ca^{2+} channels have not been found. Secretagogues (hormones and neurotransmitters) evoke release of Ca^{2+} from intracellular stores mediated by the generation of $InsP_3$, and also promote Ca^{2+} uptake into the cell from the outside through a pathway not yet characterized in detail but known to be regulated by both $InsP_3$ and $InsP_4$ [5, 6]. Continuous stimulation with various secretagogues can in this way give rise to a sustained elevation of $[Ca^{2+}]_i$ [5]. The increased $[Ca^{2+}]_i$ switches on the high-conductance voltage-sensitive K^+ channels in the basolateral membrane as well as Cl^--selective channels in the luminal membrane [5]. The opening of these Ca^{2+}-sensitive ion channels allows Cl^- uptake through the basolateral membrane via a $Na^+/K^+/Cl^-$ co-transport mechanism, and the Na^+/K^+ pump is thereby activated [5]. The operation of these four membrane transport proteins, finely controlled by the actions of internal Ca^{2+} on the K^+ and Cl^- channels, results in the formation of a NaCl-rich fluid representing the primary secretion in, e.g., salivary glands and the exocrine pancreas [5].

References

1. Petersen, O.H. & Findlay, I. (1987) *Physiol. Rev. 67*, 1054-1116.
2. Petersen, O.H. & Petersen, C.C.H. (1986) *News Physiol. Soc. 1*, 5-8.
3. Janch, P., Petersen, O.H. & Läuger, P. (1986) *J. Membr. Biol. 94*, 99-115.
4. Petersen, O.H. (1988)   *ISI Atlas of Science: Biochemistry, 1*, 144-149.
5. Petersen, O.H. & Gallacher, D.V. (1988) *Ann. Rev. Physiol. 50*, 65-80.
6. Morris, A.P., Gallacher, D.V., Irvine, R.F. & Petersen, O.H. (1987) *Nature 330*, 653-655.

#ncC

COMMENTS related to

SECRETORY AND OTHER PROCESSES INVOLVING Ca^{2+} (not muscle or nerve)

Comments on #C-1: F.R. Maxfield -[Ca^{2+}]$_i$ IN SINGLE MOTILE CELLS
#C-2: R.P. Rubin - STIMULUS-SECRETION COUPLING

F. R. Maxfield, replying to E. M. Bevers: the technique is applicable down to ~200 nm diam. (endosomes; **T.J. Rink** said it works for platelets as well). Moreover (**reply to R. Huch**), as the constraining time limit lies at ~30 msec, Ca^{2+} gradients within muscle cells at the moment of activation might in principle be feasible. **Reply to A. Capponi**, who asked how a moving cell is kept in focus: the cells, in our case, are moving only horizontally and there are no changes in thickness along the z-axis; hence focussing presents no problem. **H.J. Hilderson asked Maxfield** whether, with InsP's as an analogy, it is the amplitude or the frequency of the oscillation that the cells are 'looking at' during phagocytosis or chemotaxis. **Answer.-** We don't know; but as the cells are approaching their target the amplitude is rising. **E. Johnson remarked** that, since attachment of neutrophils to an opsonized surface is a phagocytic stimulus, it would be of interest to establish the distinction (and any differences in the profile of [Ca^{2+}]$_i$ changes) between cell attachment to the surface and specific phagocytosis of an opsonized particle. **Comments by P H. Cobbold.-** An alternative to your imaging technology is the cryogenic CCD, which can have similar noise (dark current) and quantum effectiveness to photomultipliers. Read-out noise (~7 electrons per pixel) is low, and integration on-chip removes video read-out noise; computer-controlled pixellation can allow spatial/temporal optimization. The much higher 'grey levels' attainable should also improve precision compared with video detectors.

B.F. Trump asked R.P. Rubin, concerning the mechanism of protein kinase C stimulation, whether Ca^{2+} and the kinase go through a common final path (e.g. the cytoskeleton). **Reply.-** We don't know. However, high enough levels of Ca^{2+} may activate the kinase. **P.H. Cobbold remarked** that it would be interesting to know whether the secretion pattern is alterable in permeabilized cells by inducing Ca^{2+} oscillations, e.g. by use of 'caged calcium'. **C.A. Pasternak remarked** that viruses when used to permeabilize cells are acting in a 'toxin-like' manner; virus infection as such does not permeabilize cells.

Comments on #**C-5**: M. Sumida & co-authors - ACTH EFFECTS
          #**ncC.1**: A.M. Capponi *et al*. - ALDOSTERONE STUDIES

Sumida, **answering K.E. Suckling**, in respect of the general belief that the hormone-specific lipase from adipose tissue and cholesteryl ester hydrolase from adrenal cortex are thought to be identical.- We have not tried on adipose tissue similar experiments to those on adrenal cortical preparations. **Further response**.- We are investigating how the known ACTH-induced effects on phosphorylation of the lipases by cAMP-dependent mechanisms relate to the Ca^{2+}-dependent processes we describe.

Comment by Suckling to A.M. Capponi.- It was shown some years ago that another step in steroidogenesis promoted by Ca^{2+} is the transport of cholesterol to the mitochondria. **Response**.- Indeed the case; the effects in permeabilized cells are less than in intact cells, and there are many sites where Ca^{2+} may exert its effect. **Response to Rubin**: indeed our primary $InsP_3$-binding site is not the e.r., but it has not yet been definitely characterized. **J.P. Huggins asked Capponi** whether he was sure that mitochondria were indeed being affected, since ruthenium red affects many Ca^{2+}-dependent processes. **Reply**.- We know that ruthenium red does not affect Ca^{2+} release from an InsP-specific pool. **O.H. Petersen remarked** on the possible applicability of a report by Hescheler *et al*. in 1988 *(EMBO J.)* concerning angiotensin II activation of a voltage-specific channel mediated by a G-protein. **Capponi, replying to T. Pozzan**.- We feel it unlikely that the Ca^{2+} spikes due to angiotensin II arise from release of Ca^{2+} from intracellular stores rather than influx.

For comment on #**ncC.5** (O.H. Petersen, *re* InsP's), *see* p. 251.

SOME LITERATURE PERTINENT to #C THEMES, noted by Senior Editor
- with **bold type** for test material or other 'keyword'
- *also pertinent: some citations in* #**ncB**

In canine tracheal **epithelium** (Cl^--secreting), intracellular calcium regulates basolateral K^+ channels.- Welsh, M.J. & McCann, J.D. (1985) *Proc. Nat. Acad. Sci. 82*, 8823-8826.

Inhibition of EGF binding in rat **pancreatic acini** by palmitoyl-carnitine: regulation independent of Ca^{2+} and protein kinase C.- Brockenbrough, J.S. & Korc, M. (1987) *Cancer Res. 47*, 1805-1810.

'Cyclic AMP raises cytosolic Ca^{2+} and promotes Ca^{2+} influx in a clonal **pancreatic β-cell line** (HIT T-15)'.- Prentki, M., Glennon, M.S., Geschwind, J-F., Matschinsky, F.M. & Corkey, B.E. (1987) *FEBS Lett. 220*, 103-107.

'Modulation by cytosolic pH of calcium and rubidium fluxes in rat **pancreatic islets**'.- Best, L., Yates, A.P., Gordon, C. & Tomlinson, S. (1988) *Biochem. Pharmacol. 37*, 4611-4615.

In isolated **pancreatic islets**, unsaturated fatty acid stimulation of insulin release is effected partly via Ca^{2+} fluxes and partly, as inhibited selectively by dantrolene, *not* via fluxes - possibly by inhibition of the effects of protein kinase C on exocytosis.- Metz, S.A. (1988) *Biochem. Pharmacol. 37*, 2237-2245.

Insulin release induced by a P_2-purinergic receptor activator (ADP/ATP analogues) in isolated **pancreatic islets** involves increased Ca^{2+} uptake, probably via voltage-sensitive channels.- Petit, P., Manteghetti, M., Puech, R. & Loubatieres-Mariani, M.M. (1987) *Biochem. Pharmacol. 36*, 377-380.

Ca^{2+} mobilization in **pancreatic islets** by arachidonic acid, in comparison with $InsP_3$.- Wolf, B.A., Turk, J., Sherman, W.R. & McDaniel, M.L. (1986) *J. Biol. Chem. 261*, 3501-3511.

Ca^{2+} role in **Leydig cell** steroidogenesis: $[Ca^{2+}]_i$ stimulation by LH, LHRH agonist, and cAMP.- Sullivan, M.H.F. & Cooke, B.A. (1986) *Biochem. J. 236*, 45-51.

A $[Ca^{2+}]_i$ transient rise (studied by a modified aequorin procedure) precedes the sustained rise in aldosterone production in bovine **adrenal capsule** strips (zona glomerulosa) exposed to angiotensin II or an aminopeptidase-resistant analogue.- Apfeldorf, W.J. & Rasmussen, H. (1988) *Cell Calcium 9*, 71-80 [cf. Rasmussen citation in #ncB].

Effect of pertussin-toxin islet-activating protein on catecholamine release, Ca^{2+} mobilization and $InsP_3$ formation in cultured **chromaffin cells**.- Sasakawa, N., Yamamoto, S., Nakaki, T. & Kato, R. (1988) *Biochem. Pharmacol. 37*, 2485-2487.

'Evidence that prostaglandins activate calcium channels to enhance basal and stimulation-evoked catecholamine release from bovine adrenal **chromaffin cells** in culture'.- Koyama, Y., Kitayama, S., Dohi, T. & Tsujimoto, A. (1988) *Biochem. Pharmacol. 37*, 1725-1730.

PGE_2 synthesis in superfused **renal cortical tubular cells** in response to (e.g.) antiotensin II or A 23187 is Ca^{2+}-dependent (via phospholipases) but not involving voltage-dependent Ca^{2+} channels.- Wuthrich, R.P. & Vallotton, M.B. (1986) *Biochem. Pharmacol. 35*, 2297-2300.

'**Renal** and **cardiovascular** effects of atrial natriuretic factor' (a Commentary): probably via alterations in $[Ca^{2+}]_i$ homeostasis and cGMP.- Maack, T. & Kleinert, H.D. (1986) *Biochem. Pharmacol. 35*, 2057-2064.

Hormone release from rat **pituitary tumour** (GH_3) cells: cAMP and $[Ca^{2+}]_e$ involvement in regulation.- Sletholt, K., Haug, E. & Gautvik, K.M. (1987) *Biosci. Rep. 7*, 93-105.

In mouse **fibroblasts**, the intracellular pH rise induced by growth factors (PDGF, PMA, BK) is mediated not by increased $[Ca^{2+}]_i$ but by the Na^+/H-exchanger.- Ives, H.E. & Daniel, T.O. (1987) *Proc. Nat. Acad. Sci. 84*, 1950-1954.

'Human **neutrophil** activation with interleukin-1. A role for intracellular calcium and arachidonic acid lipoxygenation'. - Metabolite(s) from latter, and $[Ca^{2+}]_i$, mediate IL-1-induced granule exocytosis. - Smith, R.J., Epps, D.E., Justen, J.M., Sam, L.M., Wynalda, M.A., Fitzpatrick, F.A. & Yein F.S. (1987) *Biochem. Pharmacol. 36*, 3851-3858.

Section #D

MITOCHONDRIAL CALCIUM

#D-1

STUDY BY X̲-RAY PROBE AND ELECTRON MICROSCOPY OF MITOCHONDRIA EXPOSED TO CALCIUM IONS

Werner Krause

Department of Ultrastructural Pathology and Electron
  Microscopy, Institute of Pathological Anatomy,
Charité Clinic, Humboldt-University,
Schumannstrasse 20/21, 1040 Berlin, G.D.R.

In our model studies, mitochondria isolated from rat heart and liver were incubated in a medium containing various concentrations of Ca^{2+} [1]. Calcium may accumulate in mitochondria during 10 min at 37° under appropriate conditions, the variations tried being 1.5 or 3.0 mM Ca^{2+}, and succinate-ATP-phosphate, succinate-ATP or ATP only. Mitochondria contained several precipitates between the outer and inner membrane and also in the intracristal spaces. The choice of conditions influenced the size and localization of electron-dense calcium-containing particles. Ca^{2+}-ATPase is cytochemically demonstrable in only some of the mitochondria. X-ray microanalysis verified the existence of calcium in the electron-dense particles seen in the isolated heart and liver mitochondria, and largely supported the conclusions based on ultrastructural patterns.

The lower Ca^{2+} concentration predominantly induced a mitochondrial swelling process which occurs stepwise. Ultimately the inner membrane and cristae are damaged. There is a close relationship between the swelling and the breakdown of mitochondria. Possibly a strong Ca^{2+} influx activates the mitochondrial phospholipase A2 which is one of the potential 'triggers' for ATP-dependent proteolysis [2].

Mitochondria accumulate calcium by an energy-dependent process [3]. Calcium plays a pivotal role in the regulation of metabolic processes in the cell including the activation of enzymes for degradation of cell components [4, 5]. It is well established that isolated mitochondria when exposed to Ca^{2+} can actively transport and sequester it internally in large amount [3, 6]. For applying e.m.* alone or in

---

*Abbreviations.- e.m., electron micro|scopy, |scopic, |graph;
RCI, Respiratory Control Index.

conjunction with X-ray microanalysis, mitochondria are ideal. Most models which seek to explain the mechanism by which cellular Ca^{2+} is regulated assign an important role fo mitochondria. They offer a notable opportunity for an experimental approach to structure-function relationships, in particular those involved in active transport, and metabolic control mechanisms at a subcellular level.

Mitochondrial isolation and O₂-consumption measurement

It has now been attempted to choose isolation methods that furnish mitochondria which are representative of mitochondria *in vivo*, with attention to oxidative phosphorylation as manifested by the RCI. Rat-heart mitochondria were isolated by Ultra-Turrax (Jahnke & Kunkel AG) and Nagarse (Serva) treatments according to the method of Palmer et al. [7]. The medium was 210 mM mannitol/70 mM sucrose/10 mM EDTA/2 mM Tris, pH 7.4, For liver we modified the method of Schneider & Hogeboom [8]. We used the Potter S-homogenizer (Braun, Melsungen), with 0.3 M sucrose/10 mM EDTA/5 mM HEPES, pH 7.4. After isolation, both heart and liver mitochondria were washed twice with the isolation medium but omitting EDTA.

Polarographic measurements (YSI 53, Yellow Springs) were made to estimate the mitochondrial RCI with both glutamate and malate as substrates [9]. The assay medium, O_2-saturated, consisted of 75 mM KCl/25 mM Tris/5 mM $MgCl_2$/12.5 mM KH_2PO_4/ 6 mM glutamate/6 mM malate, pH 7.4. The RCI's obtained were 5.6 ±0.3 (S.D.) for liver and 5.0 ±0.4 for heart. The mitochondria were of aggregated appearance (Figs. 1 & 2)⊗

CALCIUM ACCUMULATION BY MITOCHONDRIA

The aim was to map sites of calcium accumulation in isolated mitochondria. This raised the following questions.- (1) What was the spatial distribution of the calcium transferred into the mitochondria? (2) Was calcium diffusely distributed, or was it localized to a membrane or to only a portion thereof?

Our isolated mitochondria were incubated for 10 min at 37° in 113 mM sucrose/5 mM Tris/10 mM KH_2PO_4/5 mM succinate/ 3 mM ATP (pH 7.4) containing various Ca^{2+} concentrations: (a) 1-2 mM, (b) 2-4 mM. We employed 'massive loading' in the presence of relatively high concentrations [10]. Between 2 and 4 mM Ca^{2+} we observed accumulation predominantly in the mitochondria. Fig. 3 shows large dense precipitates of calcium formed mostly along the membranes in the presence of substrate and ATP. With phosphate present also (Fig. 4), electron-dense precipitates are more localized in the matrix.

⊗Pellets washed, Os-fixed, then (not for X-ray probe) U acet./Pb citrate

Some mitochondrial preparations were capable of oxidative phosphorylation but not of Ca^{2+} accumulation. Possibly the Ca^{2+}-transporting system in the inner membrane was destroyed during the isolation procedure. These findings are similar to those of Ota [11].

Ca^{2+} uptake by isolated mitochondria was observed by e.m. as a function of time. The calcium accumulation, already evident after 5 min, must be crucial in the regulation of cellular activity [12] since it occurred before oxidative phosphorylation decreased.

Ca^{2+}-ATPase

Prior to the configurational transition, Ca^{2+} was found to inhibit respiration with all NAD^{2+}-requiring substrates, and ATPase activity became evident [13]. ATP-dependent accumulation of calcium reflected the e.m. localization of the ATPase as shown by a cytochemical method [14] (Fig. 5; traces of Na^+ and Mg^{2+} present in the electron-dense Ca phosphate precipitates). Evidently not all mitochondria show reaction-product deposits. This points to functional differences amongst mitochondria in respect of Ca^{2+} uptake [15]. The presence of osmiophilic densities as distinct from precipitates might signify two different types of mitochondrial electron-dense particles, one containing Ca and the other probably lipid.

SWELLING AND BREAKDOWN OF MITOCHONDRIA

A major consequence of Ca^{2+}-uptake in the presence of 1-2 mM Ca^{2+} is the induction of a configurational transition observable by e.m. The Ca^{2+}-induced swelling of isolated mitochondria is a process which proceeds in a series of discrete steps. Vesicles are formed in the matrix. They shift to the outer membrane and initiate the swelling process. Ultimately the inner membrane and cristae are damaged (Figs. 6 & 7). Clearly the cristae are the source of the vesicles (Fig. 8). In a control experiment, mitochondria incubated for 10 min in the absence of Ca^{2+} remained in the aggregated configuration.

We suggest that Ca^{2+} transport occurs via the vesicles to the outer membrane for activation of phospholipase A2. Possibly Ca^{2+} influx activates this enzyme, which is located mainly in the outer membrane [16] and is one of the potential 'triggers' for ATP-dependent proteolysis.

In accord with other authors we suggest that the role of phospholipase A2 in swelling is to catalyze the hydrolysis of phospholipid at a key locus (see Figs. 6 & 7). Such a preferential site of hydrolysis might explain the observation that not all mitochondrial phospholipid is hydrolyzed at

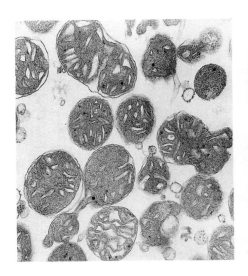

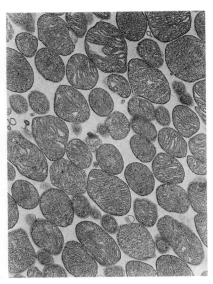

Fig. 1. Isolated liver mitochondria: mostly in the aggregated (condensed) configuration.

Fig. 2. As for Fig. 1, but heart.

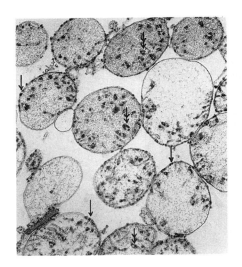

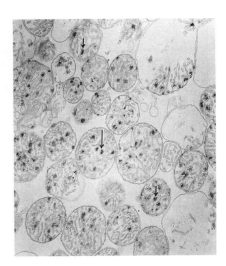

Fig. 3. Liver mitochondria. Ca-
particles* between inner and
outer membrane (↓) and in intracristal space (⤓).

Fig. 4. Heart mitochondria.
Ca-particles* mostly in matrix
(↓); sparse in the membranes.

*─────────
Ca-rich electron-dense particle

Fig. 1: × *19,000; others* × *25,000.*

Right: **Fig. 5.** Heart mitochondria: Ca^{2+}-
ATPase cytochemical product in matrix (↓) but
not in all mitochondria. ×22,000. *See below.*

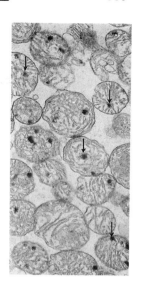

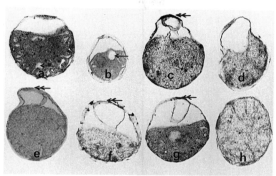

Above: **Fig. 6.** Liver mitochondria. Ca^{2+}-
induced swelling starts (**a-b**; ←). Vesicles
formed in matrix, then (**c-g**) shift
through the light part to outer membrane, with which the vesicle surrounding membrane connects closely (←←),
initiating steps in swelling with (**h**) an
intermediate orthodox configuration.
×22,000.

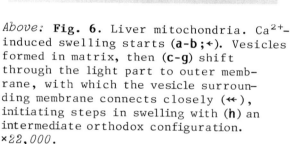

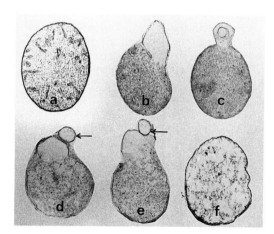

Above: **Fig. 7.** Liver mitochondria.
a, normal conformation. Swelling
proceeds (**b-e**), and the vesicles
are possibly extruded (**d-e**; ←).
f: swelling complete; inner membrane and cristae damaged. ×24,000.

Above: **Fig. 8.** Heart mitochondria, shortened incubation (5 min). Vesicles are
evidently built up from the
cristae. ×30,000.

For Ca^{2+}-ATPase on pellets: 10 min
at 37° in pH 7.2 Tris/MgCl₂/NaCl/
ATP/KH₂PO₄/CaCl₂ (3 mM).

the same rate [17]. We suggest the following.- (1) The critical sites of hydrolysis are in the outer membrane, because the vesicles move to it from the matrix and initiate the swelling process. (2) Swelling of the outer membrane is connected with breakdown of the inner membrane and cristae. (3) Both membranes may have to be hydrolyzed for swelling and degradation under these conditions: after swelling of the outer one the highly folded inner one becomes unfolded and could be accessible to ATP-dependent proteolysis [2]. Our results confirm the finding of Rapoport et al. [18] that the mitochondrial ATP-dependent proteolytic system is responsible for the regular turnover of the inner membrane, whereas the outer membrane may be degraded by the corresponding system present in the cytosol.

Mitochondrial heterogeneity.- After 10 min incubation with Ca^{2+} some mitochondria still resembled the stock suspension, and others were partly or totally swollen — which could be due to damage to some of the mitochondria during isolation or to physiological differences such as age, discussed by Baudhuin & Berthet [19]. No method is available to separate mitochondria according to age, since no relevant criteria are known.

INHIBITION OF OBSERVED EFFECTS

No Ca-rich electron-dense particles were found when Ca^{2+}-transport was inhibited by 3 mM La^{2+}, 3 mM Mg^{2+}, 5 mM EDTA or 4 mM tetracaine. La^{2+} and likewise ruthenium red could prevent Ca^{2+} movement into mitochondria. It is well established, through testing the effect of several inhibitors that compete with Ca^{2+} for binding to the outer membrane, that extent of binding and rate of Ca^{2+} accumulation are closely related [20].

Mg^{2+} inhibited competitively the Ca^{2+}-induced transition, most mitochondria remaining in the aggregated conformation as in Figs. 1 & 2. EDTA inhibited the phospholipase A2 activity - as do ATP and ADP, which might prevent swelling [22]. Various local anaesthetics are competitive inhibitors of the binding of various mono- and di-valent cations to sub-mitochondrial particles [22]; tetracaine inhibited the phospholipase by nearly 87% [23].

CONJOINT USE OF X-RAY PROBE MICROANALYSIS

As a tool to complement transmission e.m. we employed the electron probe X-ray microanalyzer, an instrument which can reveal the distribution of chemical elements (Ca being the most-studied) in cells or organelles. We analyzed electron-dense particles or reaction-products (cf. Figs. 3-5), taking

Fig. 9. X-ray probe micro-
analytical trace for the
electron-dense particles shown
in Figs. 3-5. The calcium
kα emission is clearly
manifest. The conditions ([24];
0.1 μm probe, ~0.1 μm section
thickness) gave sensitivity
adequate to detect Ca^{2+} in the
electron-dense precipitates;
in some samples, element estimation is inaccurate due to
changes in precipitate composition during fixation,
dehydration and embedding.
The electron-dense precipitates were destroyed by the
electron beam if the X-ray
collection time exceeded
250 sec.

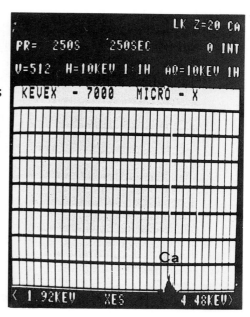

samples of mitochondria before and after Ca^{2+} addition.
Fixation for e.m. was done with 1% OsO_4, pH 7.4, at ~4°
(ice-bath) or, with no difference in the ultrastructure,
at room temperature. Ultrathin and semi-thick unstained
sections were examined with a Siemens Elmiskop 102 e.m.
connected to a KEVEX 7000 system. The accelerating voltage
was 40 or 80 kV, and the live counting time 250 sec.

Fig. 9 shows the X-ray spectrum of the electron-dense
particles with the calcium kα peak at 3.69 keV. (In an
earlier study [1] we had demonstrated a high level of calcium
in these particles by a semi-quantitative chemical analysis.)
Ratio calculations show K^+ also, some Na^+, and sometimes Mg^{2+} too.

CONCLUDING COMMENTS

We have, then, verified by X-ray probe microanalysis
that calcium is present in the electron-dense particles seen
in the isolated mitochondria. Respiratory-rate stimulation by
ADP declined to 50% with Ca accumulation, in the early phase,
and RCI's decreased to zero. The changes in RCI parameters
reflect basic alterations in functional elements of the mitochondrial inner membrane.

Acknowledgement

The author thanks Mrs. D. Behrisch for giving technical
assistance.

References

1. Krause, W., Daimon, T., Uchida, K. Mimura, T. & Takahashi, I. (1984) *Teikyo J. 7*, 375-381.
2. Rapoport, S.M. & Schewe, T. (1986) *Biochim. Biophys. Acta 864*, 471-495.
3. Lehninger, A.L., Carafoli, E. & Rossi, C.S. (1967) *Adv. Enzymol. 29*, 259-320.
4. Trump, B.F. & Berezesky, I.K. (1983) *Surv. Synth. Path. Res. 2*, 165-169. *[Cf. #E-8, this vol.- Ed.]*
5. Farber, J.L. (1981) *Life Sci. 29*, 1289-1295.
6. Chance, B. (1965) *J. Biol. Chem. 240*, 2729-2748.
7. Palmer, J.W., Tandler, B. & Hoppel, C.L. (1977) *J. Biol. Chem. 252*, 8731-8739.
8. Schneider, W.C. & Hogeboom, G.H. (1950) *J. Biol. Chem. 183*, 123-128.
9. Horrum, M.A., Jennett, R.B., Ecklund, R.E. & Tobin, R.B. (1986) *Mol. Cell Biochem. 71*, 79-86.
10. Sordahl, L.A., Besch, H.R., Allen, J.C., Crown, C., Lindenmayer, G.E. & Schwartz, A. (1971) *Meth. Achiev. Exp. Path. 5*, 287-346.
11. Ota, A. (1979) *Int. J. Biochem. 10*, 1033-1037.
12. Rossi, C.S. & Lehninger, A.L. (1964) *J. Biol. Chem. 239*, 3971-3980.
13. Hunter, D.R., Haworth, R.A. & Southard, J.H. (1976) *J. Biol. Chem. 251*, 5069-5077.
14. Hajos, F., Sotonyi, P., Kerpel-Fronius, S., Somogyi, E. & Bujdoso, G. (1974) *Histochemistry 40*, 89-95.
15. Shariro, I.M., Burke, A. & Lee, N.H. (1976) *Biochim. Biophys. Acta 451*, 583-591.
16. Nachbaur, J. & Vignais, P.M. (1968) *Biochem. Biophys. Res. Comm. 33*, 215-321.
17. Waite, M., van Deenen, L.L.M., Ruigrok, T.J.C. & Elbers, P.F. (1969) *J. Lipid Res. 10*, 599-608.
18. Rapoport, S.M., Dubiel, W. & Müller, M. (1982) *FEBS Lett. 147*, 93-96.
19. Baudhuin, P. & Berthet, J. (1967) *J. Cell Biol. 35*, 631-642.
20. Azzone, G.F., Massari, S., Rossi, E. & Scarpa, A. (1969) *FEBS Symp. 17*, 301-314.
21. Krause, W. & Halangk, W. (1977) *Acta Biol. Med. Germ. 36*, 381-387.
22. Scarpa, A. & Azzi, M. (1968) *Biochim. Biophys. Acta 150*, 473-479.
23. Scherphof, G., Scarpa, A. & van Toorenenbergen, A. (1972) *Biochim. Biophys. Acta 270*, 226-240.
24. Caswell, A.H. (1979) *Int. Rev. Cytol. 56*, 145-156.

#D-2

THE ROLE OF INTRAMITOCHONDRIAL Ca²⁺ AND
MITOCHONDRIAL Ca²⁺ TRANSPORT IN MAMMALIAN TISSUES

James G. McCormack* and **†Richard M. Denton**

Department of Biochemistry, University of Leeds,
Leeds LS2 9JT, U.K.

and †Department of Biochemistry, University of Bristol
Medical School, University Walk, Bristol BS8 1TD, U.K.

Mammalian mitochondria contain three key regulatory dehydrogenases (DH's ®) which can be activated several-fold by increases in Ca²⁺ in the range ~0.1-10 µM, with $K_{0.5}$ values for Ca²⁺ of ~1 µM, viz. PDH, NAD⁺-ICDH and OGDH. Techniques have been developed to assay the Ca²⁺-sensitive properties of these DH's within intact mitochondria isolated from a variety of different mammalian sources. Under appropriate conditions the enzymes can be activated as the extramitochondrial concentration of Ca²⁺ is raised within the expected physiological range. Recent studies with fura-2-loaded rat heart mitochondria have shown that the enzymes can serve as probes for intramitochondrial Ca²⁺. Thereby evidence has been obtained to support the concept that hormones and other agents which bring about increases in cytosolic Ca²⁺ to activate energy-requiring processes in this compartment also, in consequence, increase intramitochondrial Ca²⁺ and activate these matrix enzymes and hence promote energy production. It is therefore concluded that the principal physiological function of the mitochondrial Ca²⁺-transport system is to regulate matrix Ca²⁺ by relaying changes in cytosolic Ca²⁺ into this compartment.

Many of the hormones and other agents which, in the cytoplasm of mammalian cells, activate energy-requiring processes such as contraction or secretion by causing increases in cytosolic Ca²⁺ are also known to bring about the compensatory stimulation of mitochondrial oxidative metabolism [1, 2]. There is now much evidence to suggest that the mechanism for the latter response largely lies in the Ca²⁺-dependent activation of three intramitochondrial DH's, each of which

---
*to whom any correspondence should be sent (at the Leeds address)
®Abbreviations besides those defined in text (e.g. the media KTP, PECST, STE).- In dehydrogenase (DH) context: IC, isocitrate (threo-D$_s$-); OG, 2-oxoglutarate; P, pyruvate (hence PDH, etc.).

occupies a key regulatory site in oxidative metabolism, as the result of consequential increases in the concentration of Ca^{2+} in the mitochondrial matrix [see 1-4]. It is therefore envisaged that mammalian cells have evolved a mechanism to extend the second-messenger role of Ca^{2+} into the mitochondrial matrix to thus allow an accompanying mechanism for stimulated oxidative flux and ATP production under conditions where ATP demand is enhanced [5], but without any need to decrease cellular or mitochondrial ATP or NADH concentrations.

Significantly, in many instances this scheme is entirely compatible with observations made in whole-tissue preparations exposed to such hormones, where no decreases or even increases in ATP/ADP can be observed, together with increases in NAD(P)H [see 1, 2]. Thus it is also most likely that the principal function of the mitochondrial Ca^{2+}-transport system in mammalian cells under normal physiological conditions is to relay changes in cytosolic Ca^{2+} to the matrix and thus control matrix Ca^{2+} and not, as was thought earlier, to buffer or set extramitochondrial Ca^{2+} concentrations [see 6].

PROPERTIES OF THE Ca^{2+}-SENSITIVE INTRAMITOCHONDRIAL ENZYMES

The Ca^{2+}-activated intramitochondrial enzymes are PDH, NAD-ICDH and OGDH, all found only in the matrix in mammalian cells. [Art. #ncD.1 is pertinent.- *Ed.*] Ca^{2+} activates PDH by causing increases in the amount of active, non-phosphorylated PDH (PDHa), principally through activation of PDH phosphate phosphatase [7] by lowering its K_m for Mg^{2+} and diminishing the inhibitory effects of the phosphorylation sites 2 and 3 on PDH on the dephosphorylation of the activity-controlling site 1 [8], but perhaps also by some inhibition of PDHa kinase [9]. Ca^{2+} activates NAD-ICDH and OGDH by causing marked decreases in their respective K_m values for IC and OG [10, 11]. Ca^{2+} causes several-fold activation of each enzyme. The effective Ca^{2+} concentration range was originally thought to be very similar in each case, ~0.1-10 μM with $K_{0.5}$ values for Ca^{2+} (half-maximally effective concentrations) of ~1 μM. However, it is now clear that the $K_{0.5}$ values for NAD-ICDH and OGDH can be decreased or increased as the ATP/ADP ratio is decreased or increased [12], and in the case of NAD-ICDH at high ATP/ADP, $K_{0.5}$ values of up to about an order of magnitude greater than those for PDH and OGDH can be observed [12]; this may therefore be a device to allow regulation of NADH formation over a wider range of matrix [Ca^{2+}]. Sr^{2+} can mimic Ca^{2+} action but at ~10-fold higher concentrations in each case [11]. These Ca^{2+}-sensitive properties have been found in extracts of all vertebrate tissues so far studied, but not in invertebrates [13].

All these DH's are irreversible and are key sites in the production of matrix NADH, the principal substrate for mitochondrial (and hence cellular) ATP production. All are also activated by decreases in ATP/ADP and NADH/NAD⁺, and OGDH and NAD-ICDH can also be activated by pH decreases in the physiological range [5]; the substrate K_m's also manifest all these effects [see 5]. However, regulation by Ca²⁺ can override these 'intrinsic' effects in keeping with its proposed hormone-messenger role described above. Incidentally, control by nucleotide ratios can also be observed in some invertebrates [13].

The various assay methods used to examine the properties (summarized above) of the enzymes in tissue or mitochondrial extracts, or after purification, are detailed in the above-cited references. All are relatively straightforward. Of more importance has been the development of techniques, detailed below, to assay for the Ca²⁺-sensitive properties of the enzymes within intact mitochondria, as this has led to the observations in support of the main hypothesis at the start of this article. However, the possibility of changes in the concentrations of other effectors of the enzymes (see above) must be kept in mind when changes in activation are being ascribed to matrix Ca²⁺. An essential pre-requisite for most of the studies described below has been the development of methods to partially purify mitochondrial fractions from rapidly disrupted mammalian tissues under conditions where artefactual Ca²⁺ redistribution can be minimized.

PREPARATION OF MITOCHONDRIA

The same overall strategy has now been used in our laboratories with rat heart [3], liver [14], kidney [15], white [16] and brown [17] adipose tissue, and also skeletal muscle [see 15], and should also be applicable to most other mammalian tissues. Rapid disruption is achieved using a Polytron tissue homogenizer (Kinematica Gmbh, Kriens-Luzern, Switzerland; 0.5-2 cm diam. probes) by employing 2 or 3 bursts of 2-3 sec duration at about Mark 4. Tissue is homogenized into at least 4 vol. of ice-cold STE medium (250 mM sucrose, 20 mM pH 7.4 Tris, 2 mM EGTA) with the additional presence of 1% (w/v) defatted albumin for most tissues, or 3% + 7.5 mM glutathione for the adipose tissues. The EGTA present is sufficient to prevent any artefactual Ca²⁺-uptake during mitochondrial preparation [18], and experiments using mitochondria pre-loaded with ⁴⁵Ca have established that very little Ca²⁺ is lost during this process [3, 4], as even in the presence of Na⁺ (which promotes mitochondrial Ca²⁺-egress [see 1, 2]) this is very slow at 0-4° (see Table 3, later in the text).

The homogenates are diluted 2- to 3-fold with STE medium and centrifuged at 1000 **g** for 90 sec. The mitochondria are then sedimented at 10,000 **g** for 5-10 min. Each pellet is then suspended in 6 ml STE, and 1.4 ml Percoll is added and the mitochondria sedimented as a loose pellet at 10,000 **g** for 10 min before being washed in STE, recentrifuged and finally suspended in STE at ~40-60 mg protein/ml; the yield should be ~20-40% [18]. The major advantage of the Percoll step (which allows 1.5- to 6-fold purification) is that it removes much of the substantial contaminating material (endoplasmic reticulum, Golgi apparatus, plasma membrane) [19]. This is of particular importance in measuring total mitochondrial Ca content after hormone treatment following the realization that the InsP$_3$-sensitive Ca-pool is associated with endoplasmic reticulum [see 20].

ASSAY OF THE Ca^{2+}-SENSITIVE ENZYMES IN INTACT MITOCHONDRIA

(1) The pyruvate dehydrogenase (PDH) system

The effects of matrix Ca^{2+} on PDH phosphate phosphatase within intact mitochondria are best assayed indirectly by following changes in the steady-state PDH$_a$ content as extramitochondrial Ca^{2+} is altered, under conditions where PDH$_a$ kinase activity is likely to be low [see 17, 21]. Typical incubations are carried out in 0.5-1 ml (~0.5-1 mg mitochondrial protein/ml) at 25-34° for 3-6 min (but long enough to establish new steady-states), in 'KTP' medium (typically 125 mM KCl, 20 mM pH 7.3 Tris, 5 mM KH$_2$PO$_4$) together with appropriate respiratory substrates. The latter (e.g. 5 mM OG with 0.2 mM L-malate or 5 mM succinate) serve to keep intramitochondrial ATP high for PDH$_a$ kinase. As buffers (usually 2-5 mM) to provide the required extramitochondrial Ca^{2+} concentrations, use is made (details in [18]) of EGTA.Ca, or of HEDTA.Ca (Ca salt of **N**-hydroxyethylethylenediaminetriacetic acid).

With intact rat-liver mitochondria, or heart mitochondria from starved rats, it is also helpful to add pyruvate (0.5-2 mM) or its analogue dichloroacetate, or ADP (0.2-1 mM) with oligomycin (5 μg/ml), to inhibit PDH$_a$ kinase to some extent and so make Ca^{2+} effects on the phosphatase more evident [see 14, 22]. With uncoupled mitochondria [17], instead of respiratory substrates the KTP media would contain ATP.Mg (2-5 mM) and oligomycin (5 μg/ml) along with 1 μM FCCP (carbonyl cyanide p-trifluoromethoxyphenylhydrazone).

After incubation, the mitochondria are rapidly sedimented (microfuge; 10,000 **g**, 20 sec) and quickly frozen in liquid N$_2$; samples can be stored for several weeks at -70° before extraction and assay. Alternatively, for direct extraction and assay, incubations can be stopped by using an excess (say 5 vol.) of an ice-cold medium [50 mM pH 7.1 HEPES, 3 mM EGTA, 25 mM NaF, 0.1% Triton X-100, 1 mM dichloroacetate, 1 mM dithiothreitol (DTT)] which will break the mitochondria directly under conditions where PDH interconversion by the phosphatase and kinase is prevented (see below and [23]).

Another recent system for studying Ca^{2+} effects on PDH (also on NAD-ICDH and OGDH) is toluene-permeabilized mitochondria where PDH_a can be continuously monitored and the activities of the kinase and phosphatase can be looked at individually (see [8, 12] and #ncD.1, this vol.).

Frozen pellets are disrupted and extracted by passage up-and-down in a 500 µl microsyringe with 150-300 µl ice-cold 'PECST' medium [100 mM pH 7.3 K_2HPO_4, 2 mM EDTA, 0.1% (v/v) Triton X-100, and - added on the day of use - 1 mM Cleland's reagent (DTT) and rat serum, 50 µl/ml] and then dropping into liquid N_2 (2-3 days' pre-assay storage allowable). The EDTA prevents PDH interconversion, as both the phosphatase and the kinase require Mg^{2+}; the Triton disrupts the mitochondrial inner membrane and solubilizes the enzyme, and the serum helps prevent proteolysis of PDH [see 18]. Before assay, the sample is thawed (to 0°) and spun (10,000 g, 1 min) to remove broken mitochondrial membranes.

Assays for extracted PDH are best performed by use of a 'following enzyme' which will measure and also remove the acetylCoA produced (which is a potent end-product inhibitor of the enzyme), suitably -

(1) pyruvate + NAD⁺ + CoA $\xrightarrow{PDH}$ CO_2 + NADH + acetylCoA

(2) acetylCoA + AABS $\xrightarrow{AAT}$ acetylated AABS + CoA,

where AABS is p-(p-aminophenylazo)benzene sulphonic acid (Pfaltz & Bauer, Stamford, CT 06902) which has a mM extinction coefficient of 6.5 at 460 nm where it loses colour on acetylation, and AAT is arylamine acetyltransferase prepared from pigeon liver acetone powder [details: 18]. The dye is prepared as a Na or Tris salt in water at 1 mg/ml (solution stable at room temp.). The assay buffer is 100 mM pH 7.8 Tris/0.5 mM EDTA/1 mM $MgSO_4$ to which is freshly added (per ml) AABS solution (20 µl) and mercaptoethanol (0.3 µl). To 1.5 ml of buffer in each cuvette is added 20 µl of a substrate mix [36 mg thiamine pyrophosphate (TPP), 23 mg NAD⁺, 9 mg pyruvate and 7.5 mg CoA; storable at -20° for a fortnight], and 20 µl AAT (60-100 m-Units, where 1 U. catalyzes the conversion of 1 µmol/min) and 10-200 µl of sample (1-10 m-Units). The decrease in absorbance is then followed, usually at 30°. There may be a short time-lag (1-3 min) before linear rates develop if there is any malate present (<u>e.g.</u> from the incubations: see above), as the extracts will contain fairly high activities of malate DH and citrate synthase.

It is usual to express the amount of PDH_a as % of the total PDH activity, which is determined after converting all the PDH phosphate into PDH_a by incubating 10-50 µl of the extract with 20-40 µl semi-purified PDH phosphate phosphatase from pig heart (preparative details in [18]) in the presence of 25 mM $MgCl_2$ and 1 mM $CaCl_2$ at 30° until full activation is achieved (5-15 min). Typically total PDH per mg protein should range from 30 to 50 m-U. (liver) or 100-120 m-U. (heart). If this method is inconvenient, an alternative strategy to estimate total PDH is to incubate some of the mitochondrial sample in KTP media as above but with only FCCP present and for 5-10 min, after which time and with no generation of matrix ATP, all of the PDH in the sample should be converted to PDH_a.

The above type of assay can also be modified for use on low-activity samples by using the AAT to acetylate cresyl violet acetate, as this results in a shift of its fluorescence spectrum (Ex/Em, maxima) from 575/620 to 475/575 nm. By fluorimetric assay of acetylated dye the sensitivity can be improved ~10-fold [24]. Alternatively, if the above quench-stop method [23] of terminating incubations is used, the PDH can be assayed by using 0.2 ml of the quenched mitochondria with 0.8 ml PDH assay mixture (pH 7.1; 50 mM HEPES, 1.1 mM $MgSO_4$, 4 mM DTT, 2.5 µM rotenone, 2 mM NAD^+, 0.18 mM TPP, 0.08 mM CoA, 0.08 mM EGTA, 16.7 mM L-lactate, and 2.5 U. of lactate-DH) and measuring at 340 nm the absorbance increase due to NADH production.

For tissue extracts the above PDH assay methods can also be used, but the AAT-based methods are perhaps preferable in that both acetylCoA and NADH (in the absence of rotenone) will be removed and both are potent end-product inhibitors of PDH. The assay of $^{14}CO_2$ production from [1-^{14}C]pyruvate also suffers from acetylCoA build-up and the further problem that many samples have to be taken to check linearity; moreover there can be considerable non-enzymatic decarboxylation [see 18].

(2) 2-Oxoglutarate dehydrogenase (OGDH)

In many instances [see 15] the Ca^{2+}-sensitive properties of OGDH within various types of intact mitochondria can be simply demonstrated by measuring O_2 uptake (Clark-type electrode) by the mitochondria (0.5-1 mg protein/ml) when the restrictions imposed on this by the respiratory chain are overcome either by uncoupling the mitochondria or by adding an excess of ADP. Under such conditions, circumstances can be found where OGDH is the rate-limiting step for the oxidation of OG and thus marked effects of Ca^{2+} (controlled by Ca-buffers) can be observed at non-saturating concentrations of the substrate. L-malate or malonate is usually included as a transport partner for OG, and media such as the basic KTP, at 25-37°, are most often used [15-17].

Ca^{2+} effects on OGDH in intact mitochondria have also been demonstrated by monitoring the production of $^{14}CO_2$ from α-[1-^{14}C]OG in liver or kidney mitochondria incubated in the absence of ADP [14, 15] or, in the latter case, its presence; the technique is also more widely applicable to other mitochondria and other conditions [unpublished observations by J.G. McCormack]. Typically mitochondria are preincubated for 2-5 min in a KTP medium at 0.5-1 mg protein/ml at 30°, and then duplicate 100 µl samples are added to small soda-glass test-tubes (30 × 6 mm) resting in 0.5 ml of 2-phenylethylamine in the base of a standard glass scintillation vial. The vial is then sealed with a rubber septum through which the additions of radioactive substrate are made to the small test-tube by micro-syringe, followed 1-5 min later by 50 µl of 20% (v/v) $HClO_4$ to stop the reaction; the reaction time should be chosen so that rates of decarboxylation are still linear. The samples are left gently shaking for at least 1 h before ^{14}C in the phenylethylamine is counted.

The third and perhaps most flexible and useful method for studying the Ca^{2+}-sensitive properties of OGDH in intact mitochondria is to use either a fluorimeter (ex, 340-360 nm; em, 460 nm) or a dual-wavelength spectrophotometer (e.g. 345-370 nm wavelength pair) to monitor mitochondrial NAD(P)H production induced by adding OG [see 14, 15]. The samples are continuously stirred in a thermostatted cuvette unit (30°) and incubated in KTP and Ca-buffer solutions as above, together with other appropriate additions to achieve the desired responses [see 14, 15, 18]. For example, respiratory chain inhibitors can be used at varying concentrations to enhance NAD(P)H build-up or, conversely, ADP or NH_4Cl can be used to dissipate the accumulation to levels where effects of Ca^{2+} are more readily observed. An experiment with heart mitochondria is exemplified in Fig. 1 below.

(3) NAD⁺-isocitrate dehydrogenase (ICDH)

The effects of Ca^{2+} on NAD-ICDH within intact mitochondria from white [16] and brown [17] adipose tissue and kidney [15] can be readily demonstrated using an O_2-electrode as above, except that IC (or citrate) is presented as the substrate and hydroxymalonate is used instead of malonate. Unfortunately NAD-ICDH cannot be readily assessed in rat-heart mitochondria owing to their having a very low activity of the tricarboxylate carrier [21]. Also, the $^{14}CO_2$-trapping technique is not satisfactory for NAD-ICDH due to problems with isotopic exchange by the NADP-linked enzyme [14]. However, effects of Ca^{2+} on the reduction of NAD(P) induced by IC (or citrate) can be observed in liver [14] and kidney [15] mitochondria as described above.

THE LOADING OF RAT HEART MITOCHONDRIA WITH FURA-2

Evidently the capability for measuring Ca^{2+} effects on the activity status of these DH enzymes in intact mitochondria allows their use as probes for matrix $[Ca^{2+}]$ (see below). However, recently it has become possible to directly and continuously monitor this key parameter in intact rat-heart mitochondria by using the fluorescent Ca^{2+}-indicator fura-2 [see 25-27]. Brain mitochondria also appear able to become loaded [28], and the technique may well be more widely applicable. Rat-heart mitochondria can be loaded [27] by incubation in the final STE suspension medium (above) at ~40 mg protein/ml for 5 min at 30° after adding (to 10 μM) the acetoxymethyl (AM) ester of fura-2, followed by a brief (1-2 sec) vortexing. Under such conditions suitable amounts of the free acid are entrapped in the matrix, yet the functional bioenergetic parameters of the mitochondria do not appear to be at all compromised [27]. The loaded mitochondria can then be used immediately or else returned to ice where they appear to retain >90% of the entrapped fura-2 for up to 5 h [26]. Incubations in the fluorimeter are similar to those for OGDH except for the nm settings to monitor changes in matrix Ca^{2+}: 340/500 for fluorescence increases, 380/500 for decreases [29]. Thereafter responses are measured and signals calibrated on the basis of now well-established methodology [25-27 & 29]; Triton X-100 (0.1%), or ionomycin (2 μM) + FCCP (1 μM), serves to completely or selectively permeabilize the inner membrane. The isobestic point for the fura-2 complex (365/500 or 365/460) serves to follow OGDH activity similarly to Ca^{2+} measurement (see Fig. 1); PDH is also measurable [27].

STUDY OF MITOCHONDRIAL Ca^{2+} TRANSPORT USING THE Ca^{2+}-SENSITIVE MATRIX ENZYMES OR FURA-2 TO MONITOR MATRIX Ca^{2+}

The Ca^{2+}-transport system of mitochondria comprises an electrophoretic Ca^{2+}-uniporter, driven by the membrane potential set up through proton extrusion by the respiratory chain, and probably two egress mechanisms, the main one being an electroneutral Na^+/Ca^{2+} exchanger which is likewise in effect driven by the proton-motive gradient due to subsequent Na^+/H^+ exchange, and the lesser probably involving direct $Ca^{2+}-H^+$ exchange [see 1, 2, 6]. Ca^{2+}-uptake can be inhibited physiologically by Mg^{2+} and artificially by ruthenium red, and activated by spermine [see 1, 2], and the Na^+-dependent egress can be inhibited physiologically by extramitochondrial Ca^{2+} and artificially by diltiazem and other drugs more normally used as blockers of the plasma-membrane Ca^{2+} channel [see 1, 2]; the overall activity of the uptake pathway is ~10-fold in excess of the egress pathways.

The above techniques are readily applicable to study of these effectors in respect of Ca^{2+} distribution across the inner membrane (Table 1). This is also exemplified in Fig. 1, where OGDH is assayed within intact rat-heart mitochondria with fluorimetric monitoring of OG-induced NAD(P)H production under various conditions. It should be noted that the various effectors do not affect the ranges of matrix Ca^{2+} to which the enzymes respond [27] and that, in particular, with physiological concentrations of Na^+ and Mg^{2+} present the enzymes are activated as the extramitochondrial concentration of Ca^{2+} is raised within the expected physiological range [1, 2, 21]. Such approaches have the major advantage that the Ca^{2+}-transport system and matrix Ca^{2+} can be studied in the presence of physiological (and buffered) extramitochondrial $[Ca^{2+}]$ and at normal physiological loads of mitochondrial Ca^{2+} [see 5]; mitochondria do not buffer the extramitochondrial concentration of Ca^{2+} in such circumstances [see 5].

Of great interest is the ability to demonstrate in the fura-2-loaded mitochondria, incubated with physiological concentrations of Na^+ and Mg^{2+}, that with low extramitochondrial $[Ca^{2+}]$ (<400 nM) there in fact appears to be a <1.0 gradient of $[Ca^{2+}]$ (in:out), whereas with higher $[Ca^{2+}]$ the gradient increases, indicating a highly cooperative nature of transmission of the Ca^{2+} signal into the matrix under such likely physiological conditions (Fig. 2). Also notable is that the total Ca content of the mitochondria over the activatory range for the enzymes is ~0.5-4 nmol/mg protein, this being within the range found for tissues *in situ* by X-ray probe microanalysis [see 1, 2].

THE EFFECTS OF HORMONES ON INTRAMITOCHONDRIAL Ca²⁺

The ability to assay for the Ca^{2+}-sensitive properties of PDH and OGDH within intact rat-heart [3] and liver [4] mitochondria has allowed convincing evidence to be obtained in support of the postulated role (above) of intramitochondrial Ca^{2+}. The approach adopted involved the rapid preparation (see above) of mitochondria from control and hormone-treated tissues (*i.e.* hormones which are known to raise cytosolic Ca^{2+}) followed by incubation under various conditions where

Fig. 1. Effects on Ca^{2+} distribution across rat-heart mitochondrial inner membrane of effectors of the mitochondrial Ca^{2+}-transport system, the matrix $[Ca^{2+}]$ being monitored by OGDH (**A**) or entrapped fura-2 (**B**) loaded as described in text (none in the **A** mitochondria). Mitochondria (~2 mg protein) were incubated at 30° in the fluorimeter cuvette (see text; **A**, 340/460 nm; **B**, 340/500 nm) in 2 ml KTP pH 7.3 buffer containing initially 0.5 mM malonate + 1 mM EGTA and OG - **A**, 100 μM (<u>i.e.</u> non-saturating); **B**, 10 mM (saturating). After a 2 min stabilization (not shown), the following were added sequentially *(arrows):* **a:** (giving main curve, ———) 2 mM EGTA + 2 mM $CaCl_2$ (added as a buffer solution pre-equilibrated for pH [see 18]; resultant free extramitochondrial Ca^{2+} ~105 nM), **or** (curve - - -) 1 μM ruthenium red + 2 mM EGTA & 2 mM $CaCl_2$; **b:** (———) 10 mM NaCl, **or** (- - -) 10 mM NaCl + 250 μM diltiazem; **c:** 2 mM $MgCl_2$; **d:** 0.5 mM spermine.

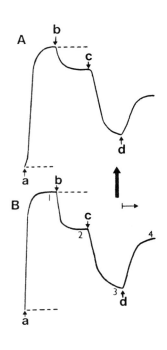

The **heavy arrow** *in the y-axis direction* indicates increases in fluorescence (arbitrary units), and the *barred arrow in the x-axis direction* indicates 1 min and the direction of additions. See [27] for further details. In **B** the calculated values for matrix $[Ca^{2+}]$ (see text) at the points indicated were (nM): **1**, 1780; **2**, 650; **3**, 65; **4**, 410.

Table 1. $K_{0.5}$ (nM) values for extramitochondrial Ca^{2+} in its activation of rat-mitochondrial PDH and OGDH. Data from [21] and [14] (details therein). N.D., not determined.

Incubation conditions	*Heart:* PDH	OGDH	*Liver:* PDH	OGDH
Control (coupled)	40	20	100	120
+ Na^+ (10-15 mM)	190	80	180	100
+ Mg^{2+} (0.5-1 mM)	170	100	360	420
+ Na^+ and Mg^{2+}	460	330	480	560
Uncoupled	980	940	1060	N.D.

Fig. 2. Ca^{2+} gradient across the inner
membrane of fura-2-loaded [see text] rat-
heart mitochondria at different extramitochondrial [Ca^{2+}] levels. Incubation
was as for **B** in Fig. 1 except that initially
only 0.5 mM EGTA was present together with
10 mM NaCl + 2 mM MgCl$_2$. Post-stabilization
additions *(arrows)*: **a**, 1 mM EGTA + 1 mM
CaCl$_2$ [resultant free extramitochondrial
Ca^{2+} ~125 nM]; **b**, 2.5 mM EGTA + 2.5 mM CaCl$_2$
[440 nM]; **c**, 1.5 mM HEDTA + 0.5 mM CaCl$_2$
[900 nM]. *Dashed lines* - - - - show effects
of Triton addition at the points indicated.
Annotations as in Fig. 1; fuller details
in [27]. Calculated values for matrix
[Ca^{2+}] (nM): **1**, 30; **2**, 410; **3**, 1400.

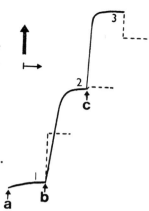

Table 2. PDH activity (% of total PDH) and OGDH activity
(% of V_{max}) in incubated rat-heart mitochondria as affected
by Ca^{2+} pre-loading of mitochondria or adrenaline pre-
treatment of tissue.   Data from [3] (details therein).

Pre-treatment	Incubation additions	PDH, untreated	PDH, pre-treated	OGDH, untreated	OGDH, pre-treated
Ca^{2+} pre-loading of mito-chondria	None	11	26*	20	35*
	Na$^+$ (10 mM)	11	12†	19	18†
	Na$^+$ + diltiazem$^⊗$	13	28*	18	34*
	Excess Ca^{2+} (150 nM)	49	50	N.D.	N.D.
Pre-exposure of tissue to adren-aline	None	8	20*	23	35*
	Na$^+$	8	7†	25	24†
	Na$^+$ + diltiazem	9	20*	23	32*
	Excess Ca^{2+}	45	47	N.D.	N.D.

$^⊗$300 µM. Significant effects ($P<0.05$ or better): *pre-treatment; †Na$^+$.

the Ca^{2+}-sensitive enzymes can be assayed as indicators of
hormone-induced alterations in matrix [Ca^{2+}]. Hopefully the
further development of the use of fura-2 may eventually
allow the direct measurement of this key parameter in intact
cells [see 27].

The feasibility of the above approach can readily be
tested by loading isolated mitochondria with sufficient Ca^{2+}
to cause PDH$_a$ increases similar to those obtained with the
hormones acting on the tissues [see 3, 4] and then 're-preparing'
them by the above-described methods. It can then be demonstrated
(Table 2) that the behaviour of PDH and OGDH in such artificially

Table 3. Effects of Na^+ and diltiazem on ^{45}Ca egress from pre-loaded rat-heart mitochondria incubated at different temperatures. Values are for ^{45}Ca content of 're-isolated' mitochondria after incubation, as % of initial (0 min) content. Data from [3] (details therein).

Incubn. time, min	Incubn. temp.	No Na^+: % ^{45}Ca	+ Na^+ (10 mM): % ^{45}Ca
4	0°	104	100
120	0°	100	88
4	20°	96	50*
4	30°	87	15*
4	37°	81	51*
1	30°	93	54*
8	30°	83	2*
+ 300 µM diltiazem:			
4	30°	88	90

*Significant effect of Na^+

Ca-loaded and re-prepared mitochondria in subsequent incubations, under various conditions designed to assess the role of matrix Ca^{2+}, closely matched that of the enzymes in mitochondria from hormone-treated tissues. In particular, the enzyme activations persist in mitochondria incubated at 30° in the absence of Na^+, but are rapidly lost on addition of Na^+ (to cause Ca^{2+} egress) or enough extramitochondrial Ca^{2+} to saturate the Ca^{2+}-dependent activations of the enzymes; moreover, these effects of Na^+ can be blocked by diltiazem. This can also be demonstrated directly with mitochondria pre-loaded with ^{45}Ca [3, 4, 18] (Table 3) or prepared from tissues perfused with ^{45}Ca [J.G. McCormack, unpublished]. Table 3 shows that very little Ca^{2+} is lost from the matrix under conditions similar to mitochondrial preparation (i.e. at 0° with Na^+ present), or when the mitochondria are incubated at 30° with Na^+ absent, or at 30° in the presence of Na^+ and diltiazem. In contrast, Ca^{2+} is rapidly lost at 30° in the presence of Na^+ (Table 3). Total mitochondrial Ca content is also increased by the hormones (in the 0.5-4 nmol/mg protein range) and changed by the mitochondrial incubations in the expected manner [see 19]. The use of an approach such as that outlined above has also allowed the demonstration that the activation of PDH by insulin in white adipose tissue is independent of changes in matrix $[Ca^{2+}]$ [see 16].

CONCLUSIONS AND IMPLICATIONS

There is now convincing evidence in heart and liver (Table 2) for the concept that when cytosolic Ca^{2+} is increased by external stimuli, the intramitochondrial Ca^{2+} also increases

- perhaps in an amplified manner (Fig. 2) - and results in the activation of the Ca^{2+}-sensitive matrix DH's. This is likely to be a widespread mechanism of matching increased energy-demand (raised cytosolic Ca^{2+}) with increased energy supply (increased intramitochondrial Ca^{2+}) in mammalian cells. The experimental approaches outlined above should prove useful in establishing this, and also another intriguing possibility - that hormones may also bring about effects on components of the Ca^{2+}-transport system [see 1, 2]. The observations which have been made using such approaches argue strongly in favour of the contention that the primary physiological role of the mitochondrial Ca^{2+}-transport system of mammalian cells is to relay changes in cytoplasmic Ca^{2+} to the matrix and *not*, contrary to the supposed key role hitherto widely believed, to buffer cytosolic Ca^{2+} with the mitochondria behaving as Ca^{2+}-stores or Ca^{2+}-sinks, or even in furnishing Ca^{2+} in activated cells [see 6]. Such buffering behaviour, which is inherent in the ~10-fold greater capacity of the uptake pathway (normally working much below its capacity) compared with the egress pathways, may be reserved for conditions of abnormal cellular Ca^{2+}-influx and resulting supraphysiological concentrations of cytosolic Ca^{2+} such as may occur in pathophysiological circumstances, e.g. the re-perfusion period following ischaemia [see 1] (& see other arts. in this vol.- *Ed.*). However, it should be noted that the derangement of mitochondrial bioenergetic functions may result if they take up excessive Ca^{2+} (amplified by M. Crompton in art. #D-3).

Acknowledgements

J.G.McC. is a Lister Institute Research Fellow. Work in the authors' laboratories has been supported by grants from the MRC, the British Heart Foundation and the British Diabetic Association.

References

1. Denton, R.M. & McCormack, J.G. (1985) *Am. J. Physiol. 249*, E543-E554.   [1-72.
2. Hansford, R.G. (1985) *Rev. Physiol. Biochem. Pharmacol. 102*,
3. McCormack, J.G. & Denton, R.M. (1984) *Biochem. J. 218*, 234-
4. McCormack, J.G. (1985) *Biochem. J. 231*, 597-608.   [247.
5. McCormack, J.G. & Denton, R.M. (1986) *Trends Biochem. Sci. 11*, 258-262.
6. Nicholls, D.G. & Akerman, K.E.O. (1982) *Biochim. Biophys. Acta 683*, 57-88.
7. Denton, R.M., Randle, P.J. & Martin, B.R. (1972) *Biochem. J. 128*, 161-163.

8.   Midgely, P.J.W., Rutter, G.A., Thomas, A.P. & Denton, R.M. (1987) *Biochem. J. 241*, 371-377.

9.   Cooper, R.H., Randle, P.J. & Denton, R.M. (1974) *Biochem. J. 143*, 625-641.

10. Denton, R.M., Richards, D.A. & Chin, J.G. (1978) *Biochem. J. 176*, 899-906.

11. McCormack, J.G. & Denton, R.M. (1979) *Biochem. J. 180*, 533-544.

12. Rutter, G.A. & Denton, R.M. (1988) *Biochem. J. 252*, 181-189.

13. McCormack, J.G. & Denton, R.M. (1981) *Biochem. J. 196*, 619-624.

14. McCormack, J.G. (1985) *Biochem. J. 231*, 581-595.

15. McCormack, J.G., Bromidge, E.S. & Dawes, N.J. (1988) *Biochim. Biophys. Acta 934*, 282-292.

16. Marshall, S.E., McCormack, J.G. & Denton, R.M. (1984) *Biochem. J. 218*, 249-260.

17. McCormack, J.G. & Denton, R.M. (1980) *Biochem. J. 190*, 95-105.

18. McCormack, J.G. & Denton, R.M. (1989) *Meths. Enzymol. 174*, in press.

19. Assimacopouos-Jeannet, F.D., McCormack, J.G. & Jeanrenaud, B. (1986) *J. Biol. Chem. 261*, 8799-8804.

20. Berridge, M.J. & Irvine, R.F. (1984) *Nature 312*, 315-321.

21. Denton, R.M., McCormack, J.G. & Edgell, N.J. (1980) *Biochem. J. 190*, 107-117.

22. McCormack, J.G., Edgell, N.J. & Denton, R.M. (1982) *Biochem. J. 202*, 419-427.

23. Hansford, R.G. & Castro, F. (1985) *Biochem. J. 227*, 129-136.

24. Solomon, M. & Stansbie, D. (1984) *Anal. Biochem. 141*, 337-343.

25. Lukács, G.L., Kapus, A. & Fonyó, A. (1988) *FEBS Lett. 229*, 219-223.

26. Davis, M.R., Altschuld, R.A., Jung, D.W. & Brierley, G.P. (1987) *Biochem. Biophys. Res. Comm. 149*, 40-45.

27. McCormack, J.G., Browne, H.M. & Dawes, N.J. (1989) *Biochim. Biophys. Acta*, in press.

28. Komulainen, H. & Bondy, C.H. (1987) *Neurochem. Int. 10*, 50-64.

29. Cobbold, P.H. & Rink, T.J. (1987) *Biochem. J. 248*, 313-328.

#D-3

PULSED-FLOW MEASUREMENTS OF THE PERMEATION STATE OF A Ca²⁺-ACTIVATED PORE IN MITOCHONDRIA

Martin Crompton

Department of Biochemistry, University College London, Gower Street, London WC1E 6BT, U.K.

Liver and heart mitochondria contain a Ca²⁺-activated pore that admits sucrose and other small molecules to the matrix. The pore closes on Ca²⁺ removal; under optimal conditions pore closure is complete in 5 sec. Pore reversibility provides the basis for a solute entrapment technique to measure the kinetics of pore opening and closure. Pore closure is too fast to be measured by manual procedures, and a rapid pulsed-flow entrapment technique has been developed for this purpose. The same device may be used to measure the rates of solute permeation via the pore.

Recent studies in this laboratory have provided evidence that heart and liver mitochondria contain a reversible Ca²⁺-activated pore, which when open uncouples oxidative phosphorylation and renders the inner membrane freely permeable to solutes of low mol. wt. [1-3]. Although its physiological role is as yet obscure, the pore may be an important element in tissue re-perfusion injury, since pore opening depends on three factors – Ca²⁺ overload, oxidative stress and increased [P$_i$]* – which are all recognized features of this type of injury [4-6]. Pore opening would have very severe consequences for cell recovery, since the cellular capacity to maintain a high phosphorylation state of ATP, which is necessary, directly and indirectly, for active Ca²⁺ extrusion from the cell and accumulation by the endo(sarco)plasmic reticulum, would be compromised. It should be borne in mind that pore opening not merely prevents oxidative phosphorylation but effectively exposes uncoupled F$_1$-ATPase to the cytoplasm.

Release of low-mol. wt. matrix constituents (<u>e.g.</u> Mg²⁺, adenine nucleotides) and dissipation of imMP* offer possible means of detecting pore opening. Release of matrix solutes clearly cannot be used for the measurement of pore closure

---

*Abbreviations.- [P$_i$], inorganic phosphate concentration; imMP (*in place of author's* Δψ), inner mitochondrial membrane potential; TPP, thiamine pyrophosphate. Cf. other arts. in this vol.

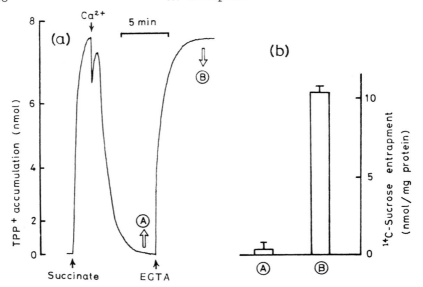

Fig. 1. Reversible pore opening. (a) Liver mitochondria
(1 mg protein/ml) were suspended at 25° in 120 mM KCl/10 mM
HEPES pH 7.0/5 mM P_i/9 µM TPP$^+$/1 µM rotenone/5 mM sucrose
containing 3 µCi ^{14}C/ml. Succinate (to 5 mM), CaCl$_2$ (30 nmol/
mg protein) and EGTA (1 mM) were added as indicated. Samples
were withdrawn at times A and B, centrifuged, and the pellet
was washed by re-suspension and re-sedimentation. (**b**) The
^{14}C-sucrose contents of the washed mitochondria are shown.

or the distribution between open and closed pore forms and
the rates of interconversion. In this laboratory we have
developed a versatile assay based on the entry and entrapment
of ^{14}C-labelled solutes in the matrix space, that can be
applied to all these questions. In the following, various
important aspects of the assay and its uses are considered.

THE REVERSIBILITY OF PORE OPENING

The basis of the entrapment assay is that Ca^{2+}-induced
pore opening is fully reversed on Ca^{2+} chelation, and the
^{14}C-solute that has entered becomes entrapped [1-3]. This
is shown in Fig. 1, where pore-state reversibility is conven-
iently 'visualized' by means of TPP$^+$ uptake, an index of
the magnitude of imMP. On commencement of respiration (with
succinate) TPP$^+$ was accumulated in response to imMP (Fig. 1a).
Addition of Ca^{2+} induced a biphasic depolarization; the initial
rapid phase is attributable to electrophoretic Ca^{2+} entry,
and the slower subsequent phase to pore opening. Ca^{2+}
chelation with EGTA-induced pore closure and the restoration
of imMP.

Fig. 2. Time course of heart
mitochondrial pore opening
measured by ^{14}C-sucrose
entrapment. See text for
experimental design. Data are
means ±S.E.M. (4 expts.).
The o values are for ^{14}C-sucrose
added 5 sec before pore closure,
rather than initially (●).

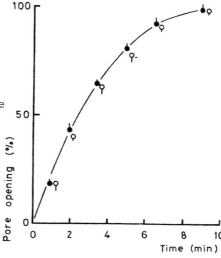

Measurement of ^{14}C-sucrose entrapment confirms that reversible depolarization is due to pore opening/closure. Mitochondria separated from the incubation medium after EGTA addition
retain ^{14}C-sucrose, whereas negligible amounts were retained
when the mitochondria were separated before EGTA addition
(Fig. 1b). The amount of ^{14}C-sucrose retained (10 nmol/mg
protein) corresponds to ^{14}C-sucrose equilibration and entrapment within a matrix space of ~2 µl/mg protein, which is
about the matrix space of swollen mitochondria.

PORE OPENING PERMITS ACCESS TO THE MATRIX SPACE

Although mitochondrial preparations are contaminated with
endo(sarco)plasmic reticulum and other organelles, there is
strong evidence that pore opening/closure leads to solute
entrapment in the mitochondrial matrix. In particular, the
procedure may be used to entrap markers of the matrix compartment. The Ca^{2+}-indicator arsenazo III is entrapped with similar
facility to ^{14}C-sucrose, and may be used to monitor Ca^{2+}
fluxes into and out of the matrix, _e.g._ succinate-supported
Ca^{2+} accumulation and Na^+-dependent Ca^{2+} release [7, 8].
The capacity of mitochondria to oxidize NAD-linked substrates,
_e.g._ glutamate, is lost on pore opening, but is restored
when pore closure is induced in the presence of NAD [1].
On the other hand, pore opening does not allow loss of
matrix proteins [1].

THE TIME COURSE OF PORE OPENING

In Fig. 2 the ^{14}C-sucrose entrapment procedure has been
applied to the measurement of the time course of pore opening
induced by Ca^{2+} plus P_i. The values shown —●— were obtained

by adding ^{14}C-sucrose together with Ca^{2+} and P_i at zero time. At intervals thereafter samples were transferred to 1 mM EGTA/1 mM ADP which causes complete pore closure within 5 sec (see next section), so that ^{14}C-sucrose that had entered became entrapped. The samples were then diluted 10-fold in 120 mM KCl/5 mM HEPES (pH 7.0) containing ^{3}H-sucrose, centrifuged, and the pellet content of ^{14}C *minus* ^{3}H was determined, thereby correcting for extramitochondrial sucrose and furnishing entrapped ^{14}C-sucrose. Essentially the same time course was obtained when ^{14}C-sucrose was added 5 sec before pore closure (—o—). This is an important feature, since it indicates that once each mitochondrion converts to the open-pore state it rapidly develops a high permeability to sucrose and, consequently, that the measured entrapments provide a true time-course of mitochondrial conversion to the open-pore state.

Maximal rates of pore opening in heart mitochondria (half-time: 1-2 min at 25°) require >10 mM P_i and ~30 nmol Ca^{2+}/mg protein; 10 nmol Ca^{2+}/mg protein yields about half-maximal rates. P_i may be replaced by a redox couple oxidant, *e.g.* t-butylhydroperoxide [2, 3].

MEASUREMENT OF PORE CLOSURE

Pore closure associated with Ca^{2+} chelation is much faster than pore opening and necessitates a rapid mixing device for measurement under optimal conditions. Fig. 3 shows the pulsed-flow device, adapted from that of Fersht & Jakes [9], which is adequate for the kinetics of pore closure. The device consists of 5 syringes (S), each attached to a 3-way valve (V, 90° plug) and connected to two 4-way mixing chambers (M). S_1 and S_2 are evacuated simultaneously by a piston acting on bar B_1 and driven by compressed air controlled by a solenoid valve (not shown). Likewise, S_3 and S_4 are evacuated simultaneously, and S_5 separately. A flow-rate of 7 m/sec in M_1 and M_2 provides efficient mixing for time delays down to ~50 msec between pulses.

In operation, the contents of S_1 (mitochondria in the open-pore state) and S_2 (^{14}C-solute and $^{3}H_2O$) are mixed in M_1 by 180° opposed flows and collected in T_1, where the ^{14}C-solute enters the matrix. After a time delay, evacuation of S_3 and S_4 mixes the contents of T_1 with those of S_4 (EGTA), and the mixture collects in T_2, where pore closure begins. After a further time delay, evacuation of S_5 drives the contents of T_2 into a flat-bottomed receiver R containing a 10-fold volume of dilution medium. The diluted samples are then centrifuged, and the pellet content of ^{14}C *minus* ^{3}H determined.

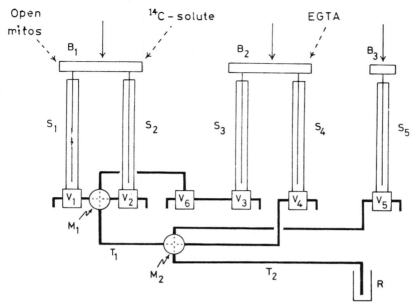

Fig. 3. The pulsed-flow device. Abbreviations (see text for amplification): B, sliding metal bar; V, valve; S, syringe; M, mixing chamber; T, tubing; R, receiver.

The essence of the technique is that dilution would affect matrix ^3H_2O but not ^{14}C-solute (<u>e.g.</u> sucrose) in the matrix of the re-sealed (closed pore) fraction of mitochondria, leading to excess ^{14}C over ^3H in the pellet with respect to the supernatant. The magnitude of this excess is a measure of the degree of re-sealing at the instant of dilution.

A typical time-course is shown in Fig. 4 (lower curve) for EGTA-induced pore closure determined from the entrapment of ^{14}C-mannitol. The data points are best fitted according to the expression:

$$\text{(Internal solute)}_t / \text{(Internal solute)}_\infty = 1 - e^{-krt} \quad \text{[eqn. A}$$

in which k_t, the apparent rate constant for re-sealing (pore closure), has the value 0.1 sec^{-1}. The illustrative data therefore conform to the simple model of exponential re-sealing of mitochondria in which the pores of any single mitochondrion close simultaneously.

The basic assumption of the entrapment technique is that on dilution all matrix ^{14}C-solute in the fraction of mitochondria that has not re-sealed is diluted. This will be valid as long as the rate of solute permeation <u>via</u> the

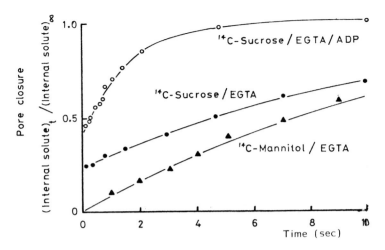

Fig. 4. Kinetics of pore closure in heart mitochondria measured by pulsed flow. Subscripts signify times t and ∞. The experimental points are best fitted with theoretical curves (see text).

open pore greatly exceeds the rate of re-sealing; this evidently holds in the case of mannitol. When solute permeation is rate-limiting, a more general expression may be applied:

$$\text{(Internal solute)}_t / \text{(Internal solute)}_\infty = 1 - e^{-k_r t} + e^{-k_p/k_r} \cdot e^{-k_r t}$$

[eqn. **B**

This takes into account the rate constant for solute permeation, k_p. When $k_p \gg k_r$, as with mannitol, eqn. **B** reduces to eqn. **A**. As shown in Fig. 4 (middle curve), when ^{14}C-sucrose entrapment is measured, the apparent re-sealing curve does not intersect ordinate zero; this reflects, for the fraction that did not re-seal, limiting sucrose permeation on dilution. The sucrose data points are best fitted with a curve according to eqn. **B** with $k_r = 0.1 \text{ sec}^{-1}$ and $k_p = 0.13 \text{ sec}^{-1}$. Thus both sucrose and mannitol entrapments yield the same rate constant for re-sealing.

Fig. 4 also shows that the rate of EGTA-induced re-sealing is greatly increased by ADP; these data points are best fitted according to eqn. **B** with $k_r = 0.7 \text{ sec}^{-1}$. Thus EGTA/ADP induces complete re-sealing within 5 sec, and this combination is used in the measurement of the time course of pore opening (Fig. 2).

MEASUREMENT OF RATE OF SOLUTE PERMEATION <u>VIA</u> THE OPEN PORE

The basic pulsed-flow entrapment technique may be used to measure the rates of solute permeation. In this case

Fig. 5. Entry of sucrose
and mannitol into open-pore
heart mitochondria measured
by pulsed-flow. Subscripts
refer to times t and ∞.
The data points (means ±S.E.M.,
4 determinations) are best
fitted with theoretical
curves (see text).

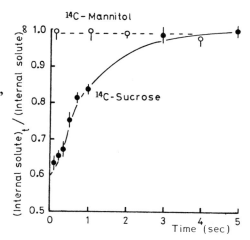

the mitochondria are re-sealed with EGTA/ADP, the time delay
in T_2 is made to exceed 5 sec (to allow complete pore closure),
and the time delay in T_1 is varied. It is readily shown
that with exponential mitochondrial re-sealing (as demonstrated
in Fig. 4) solute entrapment with time would be given by:

$$\text{(Internal solute)}_t/\text{(Internal solute)}_\infty = 1 - e^{-k_p^* t} \cdot e^{-k_p/k_r} \quad \text{[eqn. } \mathbf{C}$$

where k_p and k_p^* are the rate constants for solute permeation
in the presence and absence of EGTA.

Fig. 5 shows sucrose entrapment data best fitted according
to eqn. **C** with $k_p^* = 0.9 \text{ sec}^{-1}$ and $k_p/k_r = 0.9$. Thus in the
presence of Ca^{2+}, sucrose permeates the open pore with a
half-time of ~800 msec at 25°, which is ~9-fold faster than
in the absence of Ca^{2+} (before pore closure, Fig. 4). Mannitol,
however, permeates too quickly for the time course to be
resolved; the mannitol entrapment data are best fitted according
to eqn. **C** with k_p^* and k_p/k_r both >4.0. Thus mannitol permeates
the open pore at least 4-fold faster than sucrose.

RELATIVE MERITS OF SUCROSE AND MANNITOL IN ENTRAPMENT STUDIES

The use of the more rapidly permeating mannitol as
entrapped marker offers a simple interpretation of entrapment
data. However, the use of sucrose yields the same rate
of re-sealing when permeation is taken into account, and
is advantageous for two reasons.- (1) In the entrapment
procedure, inevitably a certain time elapses between sample
dilution and mitochondrial sedimentation. Any losses of
entrapped solute during this interval must be accounted for.
With sucrose such losses are negligible and may be ignored
(<5%/min at 25°). Appreciable losses of entrapped mannitol
do occur, however (half-time ~2 min for mannitol loss at

25°). Mannitol losses may be corrected for by varying the
the interval between dilution and centrifugation and back-
extrapolating to zero time; the mannitol entrapment data
in Figs. 4 and 5 have been thus corrected. (2) Sucrose
provides additional information about k_p.

EFFECT OF MATRIX COMPOSITION ON PORE BEHAVIOUR

Reversible pore opening allows the low mol. wt. solute
composition of the matrix to be manipulated according to
that of the resulting medium, and should find wide application.
To date we have utilized the technique for the entrapment
of the Ca^{2+} indicator arsenazo III [7, 8] and of Ca^{2+} buffers
[1]. When mitochondria in the open pore form are re-sealed
with Ca^{2+} buffers, the buffer becomes entrapped in the matrix
and maintains internal free $[Ca^{2+}]$ constant at a pre-determined
level. Provided that the free $[Ca^{2+}]$ is sufficiently low
to induce significant re-sealing, the re-sealed fraction then
re-opens at a rate determined by the internal free $[Ca^{2+}]$.
This technique has shown that the first-order rate constant
for pore opening is directly proportional to free $[Ca^{2+}]$
up to 10 μM, and is far from maximal at 24 μM [1]. In
principle, the technique may be extended to investigate the
role of other entrapped solutes in pore opening.

INHIBITION OF PORE ACTIVITY

The solute entrapment technique has been used in the
search for pore inhibitors. Of the many compounds screened
that interact with Ca^{2+}-binding proteins, by far the most
potent is cyclosporin [3]. Complete inhibition of sucrose
permeation is attained with ~60 pmol cyclosporin A/mg protein
(= 60 nM cyclosporin). Its use is instructive since ~10-fold
more cyclosporin is required for inhibition of imMP dissipation
than sucrose permeation. It seems that minimal pore opening
may allow sufficient rapid H^+ penetration to depress imMP.
This underlines the fact that although imMP dissipation and
uncoupling are easy assays to apply, they may over-estimate
pore opening, in particular at acid pH values [1].

References (BJ signifies Biochemical Journal)

1. Al Nasser, I. & Crompton, M. (1986) *BJ 239*, 19-29.
2. Crompton, M., Costi, A. & Hayat, L. (1987) *BJ 245*, 915-918.
3. Crompton, M., Ellinger, H. & Costi, A. (1988) *BJ 255*, 255-260.
4. McCord, J. (1985) *New Engl. J. Med. 312*, 159-165.
5. Poole-Wilson, P.A., Harding, D.P., Bourdillon, P.D.V. &
    Tones, M.A. (1984) *J. Mol. Cell. Cardiol. 16*, 175-185.
6. Kammermeier, H., Schmidt, P. & Jungling, E. (1982) *J. Mol.
    Cell. Cardiol. 14*, 267-277.
7. Al Nasser, I. & Crompton, M. (1986) *BJ 239*, 31-40.
8. Hayat, L. & Crompton, M. (1987) *BJ 244*, 533-538.
9. Fersht, A.R. & Jakes, R. (1975) *Biochemistry 14*, 3350-3356.

#ncD

NOTES and COMMENTS relating to

MITOCHONDRIAL CALCIUM

'COMMENTS' are on pp. 375-376, starting with Forum discussion on the preceding main articles

#ncD.1

A Note on

THE STUDY OF Ca^{2+}-SENSITIVE DEHYDROGENASES WITHIN MITOCHONDRIA PERMEABILIZED BY CONTROLLED TOLUENE TREATMENT

P.J.W. Midgley, G.A. Rutter and R.M. Denton

Department of Biochemistry,
School of Medical Sciences, University of Bristol,
Bristol BS8 1TD, U.K.

The three Ca^{2+}-sensitive DH's[*] which occur within mammalian mitochondria have been studied extensively using purified enzymes and mitochondrial extracts. NAD$^+$-ICDH [1, 2] and OGDH [3, 4] are activated by a lowering of the K_m value for their substrates, *threo*-D$_s$-isocitrate and 2-oxoglutarate respectively. PDH (a complex), on the other hand, is activated through an increase in its active, dephosphorylated form, largely through stimulation of PDHP-phosphatase [5]. There is now much evidence to suggest that Ca^{2+} ions may stimulate not only ATP-requiring cytoplasmic processes but also intra-mitochondrial oxidative metabolism, through activation of these three DH's, in order to supply the increased demand for ATP [6].

The study of isolated enzymes, however, may not always reveal precisely their properties within the mitochondria, and the study of enzymes within intact mitochondria is difficult to carry out with any precision. To overcome this problem, we have permeabilized mitochondria from both heart and adipose tissue (rat), using a method developed from that introduced by Matlib <u>et al</u>. [7]. We have shown that such mitochondria retain their matrix enzymes [8, 9] while allowing small molecules to diffuse freely across the inner membrane, hence allowing matrix enzymes to be studied *in situ*.

Preparation of permeabilized mitochondria.- As amplified elsewhere [8, 9], intact mitochondria are prepared and suspended at 20 mg/ml protein concentration in medium containing 250 mM sucrose, 20 mM Tris/HCl (pH 7.4), 2 mM EGTA and 8.5% (w/v) polyethylene glycol 6000. Toluene is added to the suspension (to 0.6% v/v), which is agitated gently for 2 min; the

[*]*Abbreviations.-* DH, dehydrogenase; IC, isocitrate; OG, 2-oxo-glutarate; P, pyruvate. PDHP = pyruvate dehydrogenase phosphate.

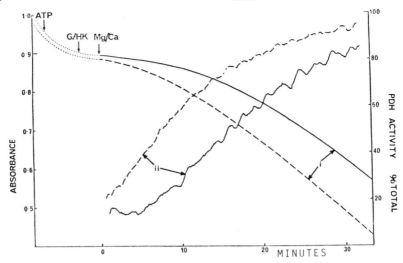

Fig. 1. Effect of Ca^{2+} on re-activation of phosphorylated PDH by endogenous phosphatase in toluene-permeabilized mitochondria from rat epididymal fat pads. Inactive PDHP was generated by incubating permeabilized mitochondria with Mg-ATP, and excess ATP was removed by adding hexokinase and glucose (**G/HK**). Re-activation was initiated by adding $MgCl_2$ with (----) or without (——) $CaCl_2$ to give 0.18 mM Mg^{2+} ± 100 μM Ca^{2+}. The curves are time courses for A_{460} decrease corresponding to acetylCoA formation (**i**) and for the increase in PDH activity (**ii**) obtained by differentiating the (**i**) curves.

mitochondria are resuspended in fresh medium after separation from the toluene by centrifugation (10,000 **g**, 4 min).

Dehydrogenase studies.- Development of permeabilized mitochondria has allowed the continuous assay of PDH to be carried out while all the PDH system's components remain within the mitochondrial matrix. Hence direct measurement of the effects of Ca^{2+} on the time course of PDHP re-activation by endogenous phosphatase has become possible. Fig. 1 shows an experiment (details in legend) in which intramitochondrial PDH, inactivated by Mg-ATP, was re-activated by Mg^{2+} with either <1 nM Ca^{2+} or with 0.1 mM Ca^{2+}. The noteworthy finding is the marked re-activation time-lag if Ca^{2+} is absent but not if it is present. The most likely explanation is that multiple phosphorylation of PDH is inhibiting re-activation, while the presence of Ca^{2+} in some way overcomes this inhibition [10]. Similar results have been obtained with rat-heart mitochondria [9].

Further studies of PDHP-phosphatase have indicated that its $K_{0.5}$ is ~0.4 µM [10]. The effect of Ca^{2+} is to increase the sensitivity of the enzyme to Mg^{2+} at subsaturating Mg^{2+} levels, in direct contrast to results with purified enzyme in which the effect is at saturating Mg^{2+}, with little or no effect on the $K_{0.5}$ for Mg^{2+}. The sensitivities of OGDH and NAD-ICDH have been studied under various conditions [9], the $K_{0.5}$ for OGDH being ~1 µM whereas that for NAD^+-ICDH was in the range 5-43 µM. The sensitivities of both these enzymes to Ca^{2+} have been shown to decrease with increasing ATP/ADP ratios [9].

Acknowledgements

These studies were supported by grants from the MRC and the British Diabetic Association. G.A.R. holds an MRC Postgraduate Scholarship.

References

1. Denton, R.M., Richards, D.A. & Chin, J.G. (1978) *Biochem. J. 176*, 899-906.
2. Aogaichi, T., Evans, J., Gabriel, J. & Plant, G.W.E. (1980) *Arch. Biochem. Biophys. 195*, 30-34.
3. McCormack, J.G. & Denton, R.M. (1979) *Biochem. J. 180*, 533-544.
4. Lawlis, T.E. & Roche, V.B. (1980) *Mol. Cell. Biochem. 32*, 147-152.
5. Denton, R.M., Randle, P.J. & Martin, B.R. (1972) *Biochem. J. 128*, 161-163.
6. Denton, R.M., McCormack, J.G., Midgley, P.J.W. & Rutter, G.A. (1987) in *Krebs' Citric Acid Cycle - Half a Century and Still Turning* (Kay, J. & Weitzman, P.D.J., eds.), Biochemical Society, London, pp. 127-143.
7. Matlib, M.A., Shannon, W.A. & Srere, P.A. (1977) *Arch. Biochem. Biophys. 179*, 396-407.
8. Thomas, A.P. & Denton, R.M. (1986) *Biochem. J. 238*, 93-101.
9. Rutter, G.A.& Denton, R.M. (1988) *Biochem. J. 252*, 181-189.
10. Midgley, P.J.W., Rutter, G.A., Thomas, A.P. & Denton, R.M. (1987) *Biochem. J. 241*, 371-377.

#ncD.2

A Note on

THE CALCIUM-SPECIFIC ELECTRODE: ITS ROLE IN ASSESSING THE CLINICAL AND TOXICOLOGICAL PROPERTIES OF THERAPEUTIC AGENTS

A. Markham, R.M. Morgan, A.R. Baydoun and A.J. Sweetman[*]

Department of Pharmacology, Sunderland Polytechnic, Sunderland SR1 3SD, U.K.

and [*]John Dalton Faculty of Technology, Manchester Polytechnic, Manchester M1 5GD, U.K.

The importance of intracellular calcium (Ca^{2+}_i) distribution in the regulation of cellular activity is now well established. In addition, there is now increasing evidence regarding the critical role of Ca^{2+} in cellular pathophysiology, particularly ischaemic tissue damage [1] and hyperactivity associated with airways smooth muscle [2].

In order to relate biochemical or pathophysiological events to changes in cellular Ca^{2+} fluxes a number of techniques have been employed, e.g. $^{45}Ca^{2+}$ isotope exchange and atomic absorption [3]. We now report the use of a Ca^{2+}-specific electrode in determining the action of Ca^{2+} antagonists or agonists on the transport of this ion in rat heart mitochondria.

Methods.- Tightly coupled heart mitochondria were isolated from female Wistar rats using a modification of the method of Varcesi et al. [4]. Ca^{2+} movements were followed by a modification of the method of Crompton et al. [5], using a Corning Ca^{2+}-specific electrode coupled to a Petracourt PM10 pH-meter and a BBC SE 120 pen recorder. Protein was measured spectrophotometrically by the method of Gornall et al. [6]. All results are the means (±S.E.M.) of at least 4 different experiments.

RESULTS

Mitochondrial Ca^{2+} influx is known to occur through an electrophoretic antiporter pathway which, in turn, is dependent on mitochondrial MP (membrane potential, $\Delta\psi$; *Editor's abbreviation*). Studies on ruthenium red (RR; 2-12 nM) confirmed the dye to be a non-competitive inhibitor of the uniporter pathway, reducing the influx rate from 259.1 ±15.4 to 32.7 ±2.6 nmol Ca^{2+}/min per mg protein (IC_{50} = 4.5 ±9.37 nM).

Replacement of RR with either verapamil (12.5-100 µM) or diltiazem (50-400 µM) resulted in similar but less potent inhibitory effects on the uniporter pathway (IC_{50}: verapamil, 19.5 ±2.0 µM; diltiazem, 93.8 ±3.8 µM). However, unlike RR, these inhibitory effects were not accompanied by Ca^{2+} efflux, thus indicating an indirect action resulting from incorporation of the antagonists into the lipid phase of the mitochondrial membrane and a decrease in MP.

Except with the Ca^{2+} agonist BAY K8644, inhibition of the influx rate was always accompanied by a corresponding decrease in total Ca^{2+} uptake. Decreases produced by verapamil were unrelated to changes in the equilibrium between Ca^{2+} influx and efflux, thus indicating a decrease in uniporter activity rather than blockade. Bay K8644 produced a concentration-dependent increase in the total amount of Ca^{2+} taken up by the mitochondrion (Fig. 1): at 60 µM this increase was from 248 ±8.4 to 406.9 ±17.6 nmol Ca^{2+} per mg protein (EC_{25} = 18.9 ±1.4 µM). Use of RR (25 nM) to inhibit the uniporter pathway (thus unmasking the efflux pathway) confirmed that the slow build-up of matrix Ca^{2+} induced by Bay K8644 was related to inhibition of the Na^{+}-Ca^{2+} antiporter of the efflux pathway.

Additional studies on efflux confirmed both specific and non-specific effects. Specific stimulation of Ca^{2+} efflux, from mitochondria pre-loaded with the cation, without disruption of energy metabolism was found to occur in the presence of nitrendipine (40-120 µM; Fig. 2) and low concentrations of DL-palmitoylcarnitine (PC; <10 µM; EC_{50} = 0.11 ±0.007 µM). Nitrendipine (100 µM) increased the efflux rate from 2.9 ±0.05 to 114.2 ±6.2 µM/min per mg protein (EC_{50} = 57.3 ±1.3 µM). Previous combination studies using PC and Ba^{2+} (10-40 µM) confirmed these specific effects to be a direct result of stimulation of the Na^{+}-Ca^{2+} antiporter system [7]. Non-specific efflux was also inducible using high concentrations (>100 µM) of either antagonists or agonist, this effect resulting from the collapse of MP following uncoupling of oxidative phosphorylation.

DISCUSSION

These data confirm that for measuring rapid ion fluxes across the mitochondrial membrane the Ca^{2+}-electrode is a sensitive technique which enables discrimination between different mitochondrial carrier systems. However, the technique described did not allow for pre-incubation with drug, thus necessitating the use of higher concentrations than those encountered *in vivo*.

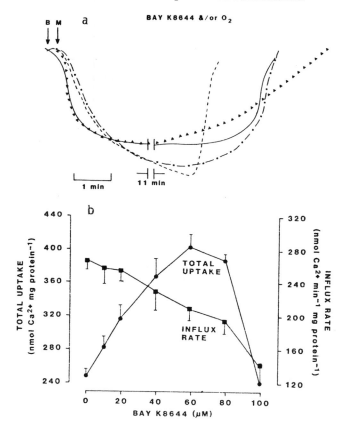

Fig. 1. Effects of Bay K 8644 (**B**) on mitochondrial Ca^{2+} influx rate and total Ca^{2+} uptake, ascertained using a Ca^{2+}-specific electrode coupled to a Petracourt PM10 pH meter and a BBC SE 120 pen recorder. The medium contained ($\mu mol/10$ ml) sucrose, 2500; succinate (Tris-salt), 50; KH_2PO_4, 20; Tris-HCl pH 7.4, 50. After **B** addition (100–1000 nmol) if applicable, Ca^{2+} uptake was initiated by adding 2.5 mg mitochondrial protein (**M**) at 37°. **a**): Ca^{2+} fluxes - control (——), + 600 nmol **B** (- - - -), control + O_2 (▲▲▲), and **B** + O_2(—■—■—). **b**): effects of **B** on Ca^{2+} influx (■) and total Ca^{2+} uptake (●). The results are the mean of 5 separate experiments ±S.E.M.

The ability of the therapeutic agents used in these studies to discriminate between different Ca^{2+} carrier systems confirms the existence of specific mitochondrial sites of action, and indicates a role for the mitochondrion in the cardioprotective action of these agents. Furthermore, these studies confirm that marked differences in Ca^{2+} handling exist between the plasma membrane [8] and the mitochondrial membrane.

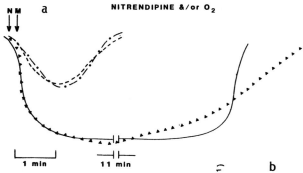

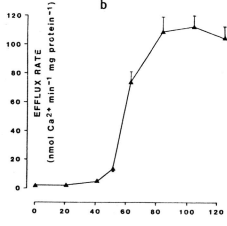

Fig. 2. Effect of nitrendipine
(**N**) on Ca^{2+} efflux (conditions
as for Fig. 1). **a**): after any **N**
(300 nmol) and/or O_2 addition,
mitochondria (**M**) were added;
——, control; ---, + **N**; ▲▲▲,
control + O_2; —■——■--, + **N** and
O_2. **b**): induction of efflux
by adding **N** (200-1200 nmol)
at the 'set point' (influx in
equilibrium with efflux) after
the mitochondria had accumulated a Ca^{2+} load of 242.1 ±10.9
nmol/mg protein.

References

1. Regitz, V., Paulson, D.J., Hodach, R.J., Little, S.E.,
   Schaper, W. & Shug, A.L. (1984) *Basic Rev. Cardiol. 79*,
   207-217.
2. Saeed, S.A. & Burka, J.F. (1983) *Prostaglandins 27*, 74S.
3. Bygrave, F.L. (1977) in *Mitochondrial Calcium Transport*.
   In *Current Topics in Bioenergetics* (Sanadi, D.R., ed.),
   Academic Press, New York, pp. 260-318.
4. Vercessi, A., Rejnaforje, B. & Lehningher, A.L. (1978)
   J. Biol. Chem. 253, 6379-6385.
5. Crompton, M., Campano, M. & Carafoli, E. (1976) *Eur. J.
   Biochem. 69*, 453-462.
6. Gornall, A.G., Bardawill, C.J. & David, M.M. (1949) *J.
   Biol. Chem. 177*, 751-766.
7. Baydoun, A.R., Markham, A., Morgan, R.M. &
   Sweetman, A.J. (1988) *Biochem. Pharmacol. 37*, 3103-3107.
8. Spedding, M. & Mir, A.K. (1987) *Br. J. Pharmacol. 92*,
   457-468.

#ncD

COMMENTS related to
MITOCHONDRIAL CALCIUM

Comments on #D-1: W. Krause - Ca^{2+} EFFECTS ON MITOCHONDRIA
#D-2: J.G. McCormack - MITOCHONDRIAL Ca^{2+} **ROLE**
#D-3: M. Crompton - A Ca^{2+}-ACTIVATED PORE

W. Krause, answering B.F. Trump.- The fact that agents such as K$^+$ produce mitochondrial swelling may not imply that it depends on Ca^{2+}. **Trump.**- Massive Ca^{2+} accumulation occurs only in certain types of cell injury, <u>e.g.</u> by HgCl$_2$, where injury does not occur initially. **T. Pozzan asked J.G. McCormack** whether fura-2/AM hydrolysis in mitochondria appears to occur specifically with heart and not with liver. **Reply.**- The hydrolysis is much more marked with heart. **Replies to A.K. Campbell.**- Our mitochondria (a) are free from lysosomes, and (b) leak very little fura-2 if fresh.

C.A. Pasternak asked M. Crompton (reply: conditions were different) how his results were reconcilable with McCormack's suggestion that 1 μM cytosolic Ca^{2+} causes activation of mitochondrial function — which would be difficult to sustain if 1 μM Ca^{2+} permeabilizes mitochondria. **Crompton, answering Trump.**- We assume that Ca^{2+} has to enter to cause the pore, but need to prove this. **Trump:** This is very pertinent to cell injury because swelling occurs under conditions where Ca^{2+} cannot be taken up at all. **Reply to A.P. Allshire.**- We have not investigated whether pore opening and resealing occurs in mitoplasts.

SOME LITERATURE PERTINENT TO #D THEMES, noted by Senior Editor -- with **bold type** for test material or other 'keyword'

'Regulation of the mitochondrial **matrix volume** *in vivo* and *in vitro*. The role of calcium!- Halestrap, A.P., Quinlan, P.T., Whipps, D.E. & Armston, A.E. (1986) *Biochem. J. 236*, 779-787.

In **hepatocytes**, senecione (a toxic alkaloid) appears to inhibit Ca^{2+} sequestration in both extra-mitochondrial and mitochondrial compartments, possibly by inactivating free -SH groups and by oxidizing pyridine nucleotides respectively.- Griffin, D.S. & Segall, H.J. (1989) *Biochem. Pharmacol. 38*, 391-397.

Reappraisal of earlier data disfavours the view that there is a Ca^{2+} (Sr^{2+})3-hydroxybutyrate symport in rat-**liver** mitochondria.- Moody, A.J., West, I.C., Mitchell, R. & Mitchell, P. (1986) *Eur. J. Biochem. 157*, 243-249.

Administration of phenformin decreased [Ca^{2+}] in **liver** mitochondria even with novel isolation conditions that minimized a Ca^{2+} redistribution artefact. The altered mitochondrial [Ca^{2+}] may impair Ca^{2+}-sensitive NAD^+-dependent mitochondrial dehydrogenases.- Gettings, S.D., Reeve, J.E. & King, L.J. (1988) *Biochem. Pharmacol. 37*, 281-289.

'Structural dependency of the inhibitory action of benzodiazepines and related compounds on the mitochondrial Na^+-Ca^{2+} exchanger', as studied with **heart**.- Chiesi, M., Schwaller, R. & Eichenberger, K. (1988) *Biochem. Pharmacol. 37*, 4399-4403.

'The Ba^{2+} sensitivity of the Na^+-induced Ca^{2+} efflux in **heart** mitochondria: the site of inhibitory action'.- Lukács, G.L. & Fonyo, A. (1986) *Biochim. Biophys. Acta 858*, 125-134.

'Na^+-independent, pyridine nucleotide-linked efflux of Ca^{2+} from preloaded rat **heart** mitochondria: induction by chlortetracycline'.- Sokolove, P.M. (1987) *Biochem. Pharmacol. 36*, 4020-4024; later [(1988) *37*, 803-812] the aglycone of adriamycin (cardiotoxic) was shown to cause, in mitochondria, Ca^{2+} release and Ca^{2+}-dependent adverse effects.

Mitochondrial aspects of **cell injury** feature on p. 426 of 'Intracellular calcium homeostasis'.- Carafoli, E. (1987) *Annu. Rev. Biochem. 56*, 395-433. For cytosolic Ca^{2+} rises induced by agents that increase p.m. Ca^{2+} permeability, the low-affinity Ca^{2+}-uptake by mitochondria represents a vital safety device: through co-accumulation of P_i, large amounts of calcium can be stored long-term as hydroxyapatite in the matrix without serious functional impairment. This may enable the injured cell to survive. *Cf. this vol., art.* #ncB.1.

'Regulation of the mitochondrial matrix volume *in vivo* and *in vitro*. The role of calcium'.- Halestrap, A.P., Quinlan, P.T., Whipps, D.E. & Armston, A.E. (1986) *Biochem. J. 236*, 779-787. The matrix volume increase due to α-agonists and vasopressin needs external Ca^{2+}: probably [Ca^{2+}]$_i$ rises within mitochondria and, associated with K^+ entry, they swell. **Hepatocytes** were the test material.

Section #E

CELL PROTECTION; PATHOLOGY, INCLUDING CARDIOVASCULAR

#E-1

THE PROTECTIVE ROLE OF EXTRACELLULAR CALCIUM IONS

C.A. Pasternak

St. George's Hospital Medical School,
Cranmer Terrace, London SW17 ORE, U.K.

Many cytotoxic agents act by forming non-specific pores across the p.m.* of susceptible cells. The leakage of ions and low mol. wt. compounds out of such damaged cells is inhibited by raising Ca^{2+} in the incubation medium. In intact animals, simultaneous injection of cytotoxic agents and Ca^{2+} has the same result. Protection by Ca^{2+} has been shown to be exerted at the extracellular side of the p.m.

This book mostly concerns the actions of Ca^{2+} as a trigger and modulator of intracellular events. But Ca^{2+} has an extracellular role also. Besides its function as a coenzyme for processes such as fat digestion in the intestine or blood coagulation in the vascular system, it also acts to protect cellular membranes against externally inflicted injury. This role, first suggested in 1914 for plant cells [1] and since assumed to be more or less true for animal cells, may be summarized as follows:-

Cellular roles of calcium
1. Intracellular Ca^{2+}: $\sim 10^{-7}$ M
   Increase to 10^{-6}-10^{-5} M triggers movement (muscle contraction; cell motility), secretion (exocytosis in neural and endocrine cells), activation (lymphocyte stimulation, egg fertilization), <u>etc</u>.
2. Extracellular Ca^{2+}: $\sim 10^{-3}$ M
   Protects cell membranes against injury.

Our studies over the past 15 years have focused on the extracellular role of Ca^{2+} ([2-11], & art. in Vol. 17, this series). We have shown that Ca^{2+} protects against a wide variety of membrane-damaging cytotoxic agents of viral, bacterial or animal origin, against endogenous immune mechanisms and against certain synthetic compounds. Its protective role on cell membranes in the absence of extraneous agents has been confirmed.

---

* p.m., plasma membrane

In this article, we summarize two results recently obtained. First, cells in intact animals can be protected against cytotoxic agents by raising extracellular Ca^{2+}. Second, the protective action of extracellular Ca^{2+} - in contrast to its role in stimulating recovery from cytotoxic attack [12-17] - is exerted not by increasing intracellular Ca^{2+} but by a direct action at the extracellular surface of the p.m.

METHODS AND MATERIALS

For experiments with intact animals, lactating mice at 4-5 days after parturition were used. Cytotoxic agents (*Staphylococcus aureus* α-toxin [18], kindly donated by Dr Joyce de Azavedo, or UV-irradiated haemolytic Sendai virus [19]) was injected through the nipple of one of the abdominal mammary glands, and cytotoxic agent plus Ca^{2+}, or a control solution, was injected into the other gland. Some hours later the animals were killed, and sections of the mammary glands cut, fixed and stained with haematoxylin-eosin.

For experiments with isolated cells, use was made of human or rabbit erythrocytes - with measurement of K^+ leakage and haemolysis - or of Lettré cells which had been passaged i.p. in mice and which were analyzed for leakage of K^+ and, the cells having been pre-labelled with [^{3}H]choline, of phosphoryl[^{3}H]choline [19, 20].

RESULTS WITH INTACT ANIMALS

Injection of purified *S. aureus* α-toxin into the mammary glands of lactating mice produces morphological changes typical of those seen when live bacteria are injected [21]. These include swelling and destruction of the epithelial cells lining the alveoli, leading to collapse of the alveolar ducts themselves (Fig. 1C). Injection of 100 μmol of Ca^{2+} at the same time as the toxin largely prevents these changes (Fig. 1D), i.e. the gland resembles that after injection of saline alone (Fig. 1A) or saline plus 100 μmol of Ca^{2+} (Fig. 1B). Injection of Ca^{2+} several hours prior to the toxin likewise protects, as does injection of a lower dose of Ca^{2+} (10 μmol) although protection is more patchy, being restricted to certain areas of the gland (not illustrated).

Essentially similar results are obtained when a haemolytic virus, UV-treated so as to render it non-infective, is injected, as shown in Fig. 1 (E, virus alone; F, virus + 100 μmol Ca^{2+}). When a non-haemolytic preparation of Sendai virus is used, little cytological damage is seen. This confirms that morphological changes induced by haemolytic Sendai virus result from its pore-forming, toxin-like action [22].

RESULTS WITH ISOLATED CELLS

If intracellular Ca^{2+} of erythrocytes is somehow raised, K^+ leakage is induced. This effect, known as the Gardos effect after its discoverer [see art. by D. Allan in Vol. 17 - Ed.], is analogous to the opening by intracellular Ca^{2+} of K^+ channels in excitable cells [23]. K^+ leakage from erythrocytes can therefore be used as an indirect measure of changes in intracellular Ca^{2+}.

When erythrocytes are treated with S. aureus α-toxin, K^+ leakage and haemolysis are induced. Both are abolished by the presence of 30 mM Ca^{2+} (Fig. 2A), which itself causes no K^+ leakage whereas as little as 10 μM Ca^{2+} induces K^+ leakage when added to erythrocytes in the presence of the ionophore A23187 (Fig. 2B). This shows that the protective effect of 30 mM Ca^{2+} cannot have been exerted through an increase in intracellular Ca^{2+} since such an increase is unlikely to induce and to prevent K^+ leakage at the same time. Other pore-forming agents such as haemolytic Sendai virus or melittin give the same result.

With Lettré cells an increase in intracellular Ca^{2+} induced by A23187 (as assessed by an increased uptake of Ca^{2+}) does not lead to K^+ leakage. In this instance, therefore, the protective effect of Ca^{2+} on metabolite leakage was assessed in the absence or presence of A23187. Its addition to Lettré cells exposed to S. aureus α-toxin or other haemolytic agent did not affect the protective action of Ca^{2+} (Fig. 3); if anything, A23187 gave more, not less, leakage.

DISCUSSION

Two conclusions can be drawn from the results presented here. First, the protective action of Ca^{2+} that has been demonstrated for many different cell types [5, 6] exposed to a range of different cytotoxic, haemolytically active agents [7-11, 24-26] can be shown to be exerted in an intact animal model also. Increasing plasma Ca^{2+} to levels high enough to combat the damage caused in man by such agents is not a feasible approach, (a) since elevated Ca^{2+} is toxic to many organs including the heart, and (b) since plasma Ca^{2+} is under tight endocrine control in any case. However, the fact that Zn^{2+} protects against such damage in vitro [25, 26] and in vivo [27] - at concentrations <10% of those required for protection by Ca^{2+} - opens up the possibility of zinc therapy in some instances [28].

The second conclusion to be drawn from these experiments is that the protective effect of Ca^{2+} is exerted at

[continued on p.384

A

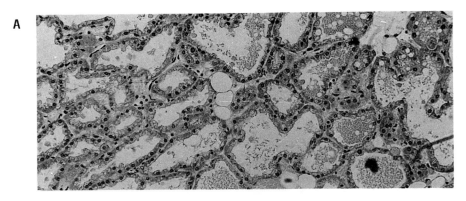

B

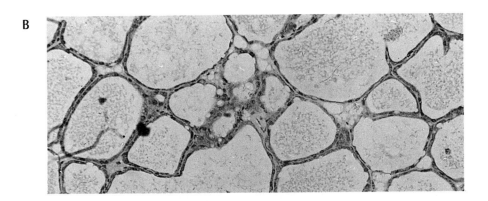

C

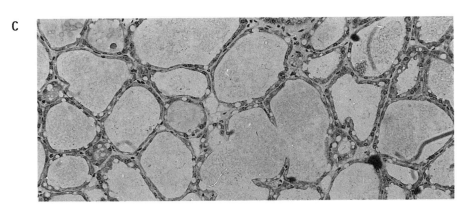

Fig. 1. Micrographs (× 90) showing protection by Ca^{2+} against cytopathic changes in mouse alveolar cells. Lactating mice injected with saline alone (**A**) or + 100 μmol Ca^{2+} (**B**), or with *S. aureus* α-toxin alone (**C**) or + 100 μmol Ca^{2+} (**D**), or with

[continued opposite

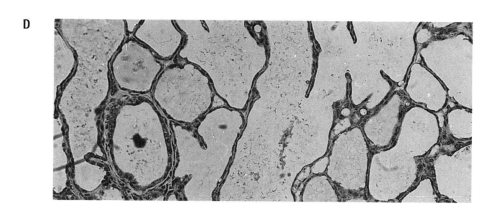

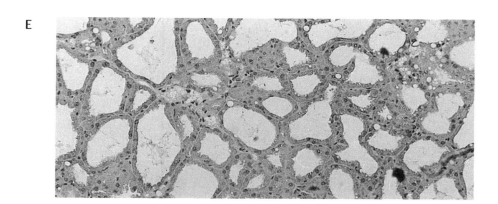

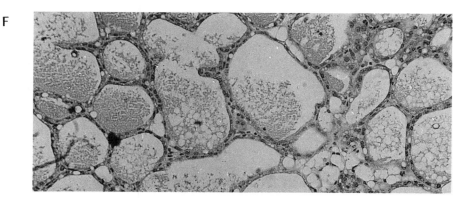

haemolytic Sendai virus alone (E) or + 100 µmol Ca^{2+} (F).
See text; the mammary glands were removed and sections cut
and stained 6 h after injection. (Mice killed by cervical dislocation.)
Expt. performed by D. Mahadevan, M. Frost and J.A. Vincent.

Figs. 2 & *(below)* **3.**
Protection against
S. aureus α-toxin-induced
damage by Ca²⁺ at 37°.
Open *symbols:* Ca²⁺ absent.
Closed *symbols:* Ca²⁺ **present.**

Fig. 2.   Erythrocytes.
Upper panel: cells exposed
to the toxin (□, ■); buffer,
o, ● (no toxin). 'Leakage' is
an alternative way of expressing the cation ratio.
*Lower panel (**note** the reduced
concentration of Ca²⁺):* cells
exposed to A23187 [*not* to
toxin] (△; + Ca²⁺, ▲).
●: Ca²⁺, without A23187.

*Note that exten-
sive haemolysis
occurred if, and
only if, α-toxin
was present.*

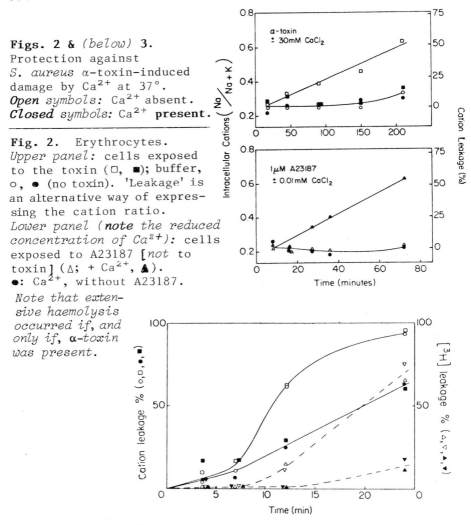

Fig. 3.   Lettré cells, pre-incubated with [³H]choline.
Where the exposure to toxin was in the presence of Ca²⁺
(closed symbols), Ca²⁺ was 5 μM. In some experiments A23187,
at 1 μM, was also present (cation leakage; □ & ■) (phosphoryl-
[³H]choline: *inapplicable*). *From [9], courtesy of Biochem. J.*

the extracellular side of the p.m.   It is therefore distinct
from the role of Ca²⁺ in inducing recovery of cells exposed
to a pore-forming agent such as activated complement [13-17].
There are, then, two independent mechanisms that limit damage
to cells by cytotoxic pore-forming agents: damage ensues
only when the   protective   capacity of extracellular Ca²⁺
is overcome, and in that situation an increase in intracellular

Ca^{2+} (resulting from the non-specific increase in cellular permeability) helps cells to recover from the toxic insult. Only when cells are exposed to high and sustained levels of cytotoxic agent does irreversible injury and lysis ensue.

It should be noted that Zn^{2+} or H^+, which also protect against cell damage caused by pore-forming agents [25, 29], do so from the extracellular side also. In this instance it is unlikely that an increase in intracellular Zn^{2+} or H^+ is able to promote recovery. But such experiments have not as yet been carried out.

Acknowledgements

The help of many colleagues, especially G.M. Alder, C.L. Bashford, D. Mahadevan and H. Salazinsky is gratefully acknowledged. Financial support from The Cell Surface Research Fund enabled this work to be carried out.

References

1. True, R.H. (1914) *Am. J. Bot. 1*, 255-273.
2. Pasternak, C.A. & Micklem, K.J. (1974) *Biochem. J. 140*, 405-411.
3. Impraim, C.C., Micklem, K.J. & Pasternak, C.A. (1979) *Biochem. Pharmacol. 28*, 1963-1969.
4. Pasternak, C.A. & Micklem, K.J. (1981) *Biosci. Rep. 1*, 431-448.
5. Pasternak, C.A. (1984) in *Membrane Processes: Molecular Biological Aspects and Medical Applications* (Benga, G., Baum, H. & Kummerow, F., eds.), Springer-Verlag, New York, pp. 140-166.
6. Pasternak, C.A. (1984) *J. Biosci. 6*, 569-583.
7. Bashford, C.L., Alder, G.M., Patel, K. & Pasternak, C.A. (1984) *Biosci. Rep. 4*, 797-805.
8. Pasternak, C.A., Bashford, C.L. & Micklem, K.J. (1985) *Proc. Int. Symp. Biomol. Struct. Interactions: Suppl., J. Biosci. 8*, 273-291.
9. Pasternak, C.A., Alder, G.M., Bashford, C.L., Buckley, C.L., Micklem, K.J. & Patel, K. (1985) in *The Molecular Basis of Movement through Membranes* (Pasternak, C.A. & Quinn, P.J., eds.), *Biochem. Soc. Symp. 50*, 247-264.
10. Pasternak, C.A. (1986) *Cell Calcium 7*, 387-397.
11. Pasternak, C.A. (1987) *Archiv. Virol. 93*, 169-184.
12. Micklem, K.J., Nyaruwe, A., Alder, G.M. & Pasternak, C.A. (1984) *Cell Calcium 5*, 537-550.
13. Campbell, A.K. & Morgan, B.P. (1985) *Nature 317*, 164-166.
14. Ramm, L.E., Whitlow, M.B., Koski, C.L., Shin, M.L. & Mayer, M.M. (1983) *J. Immunol. 131*, 1411-1415.
15. Carney, D.F., Koski, C.L. & Shin, M.L. (1985) *J. Immunol. 134*, 1804-1809.

16. Morgan, B.P., Dankert, J.R. & Esser, A.F. (1987) *J. Immunol. 138*, 246-253.

17. Morgan, B.P., Luzio, J.P. & Campbell, A.K. (1986) *Cell Calcium 7*, 399-411, **and** art. #E-2, *this Vol.*

18. McNiven, A.C., Owen, P. & Arbuthnott, J.P. (1972) *J. Med. Microbiol. 5*, 113-122.

19. Impraim, C.C., Foster, K.A., Micklem, K.J. & Pasternak, C.A. (1980) *Biochem. J. 186*, 847-860.

20. Pasternak, C.A. (1987) in *Cells, Membranes and Disease including Renal [this Series, Vol. 17]* (Reid, E., Cook, G.M.W. & Luzio, J.P., eds.), Plenum, N. York, pp. 189-198.

21. Anderson, J.C. & Mason, A.J. (1974) *Res. Vet. Sci. 16*, 23-26.

22. Pasternak, C.A. (1987) in *The Role of Calcium in Biological Systems*, Vol. 4 (Anghileri, L.J., ed.), CRC Press, Boca Raton, FL, pp. 99-113.

23. Schwartz, W. & Passow, H. (1983) *Ann. Rev. Physiol. 45*, 359-374.

24. Harshman, S. & Sugg, N. (1985) *Infect. Immunol. 47*, 37-40.

25. Bashford, C.L., Alder, G.M., Menestrina, G., Micklem, K.J., Murphy, J.J. & Pasternak, C.A. (1986) *J. Biol. Chem. 261*, 9300-9308.

26. Pasternak, C.A. (1987) *BioEssays 6*, 14-18.

27. Pasternak, C.A. & Mahadevan, D. (1988) *Ind. J. Biochem. Biophys. 25*, 1-7.

28. Pasternak, C.A. (1987) *Biosci. Rep. 7*, 81-91.

29. Bashford, C.L., Alder, G.M., Graham, J.M., Menestrina, G. & Pasternak, C.A. (1988) *J. Membr. Biol. 103*, 79-94.

#E-2

THE PARADOXICAL ROLE OF CALCIUM IN CELL INJURY INDUCED BY MEMBRANE PORE-FORMERS

B.P. Morgan , [†]J.P. Luzio and A.K. Campbell

Department of Medical Biochemistry,
University of Wales College of Medicine,
Cardiff CF4 4XN, U.K.

and [†]Department of Clinical Biochemistry,
University of Cambridge,
Addenbrooke's Hospital, Cambridge CB2 2QR, U.K.

Disturbances in $[Ca^{2+}]_i$[*] are important in the mediation of cell injury caused by diverse toxic factors, notably the group of toxins which act by forming physical or functional pores in the membranes of target cells, thereby allowing transmembrane passage of ions and small molecules. An important pore-forming toxin is the cytolytic MAC of complement. Formation of the MAC on a cell causes a rapid increase in $[Ca^{2+}]_i$, which in nucleated cells stimulates recovery processes and causes cell activation. Cell death occurs only when recovery mechanisms are overwhelmed. Whether changes in $[Ca^{2+}]_i$ caused by other pore-forming toxins, e.g. the T-cell perforins and bacterial toxins, also stimulate recovery processes and cell activation is unsettled. Extracellular calcium may also play a part in protecting cells from lysis by pore-forming agents.

At the start of this book, Campbell surveys the century-old concept that $[Ca^{2+}]_i$ affects cell function, with relevance to disease [1, 2]. $[Ca^{2+}]_i$ increases accompany many forms of cell injury, and may ultimately cause cell death [e.g. 3]; but it is now clear that moderate increases are not always detrimental to the cell, and may protect it against killing by a toxic agent or stimulate activation of various cellular processes [4, 5]. This apparently paradoxical situation – a single effector, $[Ca^{2+}]_i$, having both cell-damaging and protective effects – is particularly evident when one examines the mode of action of the group of toxic agents which act by forming physical or functional pores in cell membranes. Because of the very large transmembrane gradient of Ca^{2+},

---

[*]*Abbreviations.* - $[Ca^{2+}]_i$, intracellular free calcium concentration; MAC, membrane attack complex.

relatively minor disturbances by these pore-forming toxins
will result in a significant Ca^{2+} flux and $[Ca^{2+}]_i$ increase.
The present aim is to present the evidence for a role of
calcium in cell damage and activation initiated by pore-forming
toxins. The article, following up our article in a previous
vol. [6], concentrates on the most comprehensively studied
of these, the MAC of complement, but considers other pore-
formers also [as surveyed by Pasternak, #C-1, in Vol. 17,
this series - Ed.].

THE MEMBRANE ATTACK COMPLEX (MAC) AND OTHER PORE-FORMING TOXINS

The MAC is the primary example of the above-mentioned
group of pore-formers, now listed (not comprehensively).-

Endogenous agents	MAC of complement
	Perforins (cytolysins) from cytotoxic T-cells, NK (natural killer) cells, etc.
Animal toxins	Melittin (from honey bee venom)
	Barbatolysin (from red ant venom)
	Paradaxin (from Moses sole secretions)
Bacterial toxins	Streptolysin O
	Staphylococcal α-toxin and δ-toxin
Other pore-forming agents	Haemolytic viruses (e.g. Sendai, Newcastle disease)
	Polylysine
	Detergents

These all act, at least in part, by allowing ions and small
molecules to cross the target-cell membrane, either via true
protein-lined pores or by causing a 'leaky patch' in the
lipid bilayer [7, 8].

The MAC is formed on the target cell by non-proteolytic
self-assembly of the terminal 5 components of complement,
C5-C9. The final lesion visualized on the membrane has
the appearance of a cylinder, formed mainly from between
12 and 18 C9 molecules arranged in a ring polymer [9, 10].
It has been assumed that the functional pore caused by
complement MAC is identical with this rigid ring-like structure;
but its relevance to MAC function has recently been challenged
by data from several experimental approaches [11-13]. Whatever
the exact nature of the lesion produced, the MAC is highly
efficient at lysing aged erythrocytes. Nucleated cells are,
however, much more resistant to killing by the MAC [14,
15], and the relevance of Ca^{2+} to this resistance is considered
in the following section.

Of the other pore-forming toxins listed, the perforins
are the most similar to the MAC. They are similar in
molecular size to, and have some sequence homologies with,

MAC component proteins, suggesting a common evolutionary origin [16, 17]. Monomeric perforin molecules secreted by the cytotoxic cell insert into the target-cell membrane and polymerize to form ring-like lesions similar in appearance to the MAC [17, 18]. The calcium-dependence of perforin-mediated cell killing is discussed later.

The other pore-forming toxins listed are much less similar in physicochemical characteristics to the MAC. The staphylococcal toxins are proteins which self-associate to form pores in target-cell membranes which, at least for the α-toxin, resemble the MAC and poly-perforin [19]. Melittin, the most-studied animal toxin, is a 26-amino acid amphipathic polypeptide which binds to and disrupts target-cell membranes, causing lysis [20] – recently shown to be mechanistically similar if not identical to lysis by the MAC with which melittin has sequence homologies that may be pertinent to function [21]. The pore-forming toxins, despite their diverse origins and structures, therefore have much in common in their mechanisms of cell damage.

CALCIUM AND THE MAC

We showed earlier [22] that formation of MAC's on erythrocyte ghosts entrapping the Ca^{2+}-activated photoprotein obelin causes a rapid increase in $[Ca^{2+}]_i$. When one considers that the MAC causes increased membrane permeability and that a large transmembrane gradient of Ca^{2+} exists, it is perhaps not surprising that the MAC causes a $[Ca^{2+}]_i$ increase. Three features of this increase are, however, noteworthy: (1) $[Ca^{2+}]_i$ increases rapidly to a plateau in the μM range and only increases further after several minutes (Fig. 1); (2) the increase is detectable before release of entrapped marker molecules of similar molecular radius to the hydrated Ca^{2+} ion from the ghost (Fig. 1); (3) $[Ca^{2+}]_i$ increases even at levels of MAC attack that cause no detectable cell death [23]. Seemingly MAC exhibits some degree of selectivity for Ca^{2+}, and erythrocyte ghosts are able, at least in the short-term, to maintain $[Ca^{2+}]_i$ at sub-lethal levels in the face of influx via the MAC.

That nucleated cells are relatively resistant to MAC-induced lysis, as mentioned above [14, 15], is due to their possession of specific recovery mechanisms, involving MAC removal by vesiculation and endocytosis [6, 24, 25]. We showed with neutrophils that recovery from attack by homologous MAC's is, at least in part, Ca^{2+}-dependent [26], being inhibited by removal of extracellular Ca^{2+} and additionally by chelation of intracellular Ca^{2+} (Fig. 2). In this case, then, a rise in $[Ca^{2+}]_i$, by stimulating recovery processes, protects

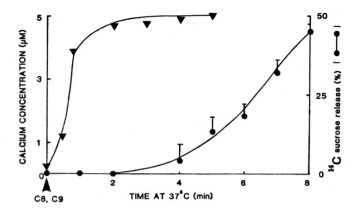

Fig. 1. $[Ca^{2+}]_i$-increasing action of the MAC. Antibody-sensitized erythrocyte ghosts (GA) entrapping obelin and $[^{14}C]$-sucrose were incubated with C8-depleted serum to form GAC1-7 intermediates. To a portion of the cells in a luminometer at 37°, C8 and C9 were added. Obelin luminescence was measured and used to calculate $[Ca^{2+}]_i$ [27]. The remainder of the cells were incubated at 37° with C8 and C9, and portions removed at intervals for measurement of $[^{14}C]$sucrose release. The values for each parameter are the mean of 3 separate measurements.

the cell against lysis, and preventing this rise results in increased killing. Similar protective effects of increased $[Ca^{2+}]_i$ have since been described in other cell types [3, 28].

As complement membrane attack on nucleated cells is not universally lethal, the possibility arises that the MAC may induce more subtle non-lethal effects on these cells. Since our early description of the stimulation of reactive oxygen metabolite production from neutrophils by non-lethal amounts of the MAC [27], a number of non-lethal, stimulatory effects of the MAC on a wide variety of cell types have been described [e.g. 29] – as listed in Table 1, which further indicates that many of these effects are dependent on a rise in $[Ca^{2+}]_i$. Thus Ca^{2+} is not only involved in recovery from complement membrane attack, but also mediates many of the non-lethal effects of the MAC on cells.

CALCIUM AND OTHER PORE-FORMERS

Any agent causing increased membrane leakiness will cause an increased flow of Ca^{2+} into the cell down its concentration gradient. Thus other pore-forming agents would be predicted to cause an increase in $[Ca^{2+}]_i$, and this has been demonstrated

Fig. 2. Role of
$[Ca^{2+}]_i$ in neutrophil
recovery, shown
with erythrocyte
ghosts entrapping
obelin and fused
with neutrophils.
Upper diagram. -
Formation of non-
lethal amounts of
the MAC on these
neutrophil hybrids
resulted in a rapid,
transient increase
in $[Ca^{2+}]_i$ (calc.; ●).
Chelation of extra-
cellular Ca^{2+} com-
pletely abolished
this rise (○).
Lower diagram. -
Removal of MAC's from
the cell surface,
monitored using
radiolabelled anti-MAC
antibody, was part-
blocked by chelation
of extracellular Ca^{2+}
(○) but almost
completely blocked
by chelation of
intracellular Ca^{2+} (□).
Modified from ref. [26].

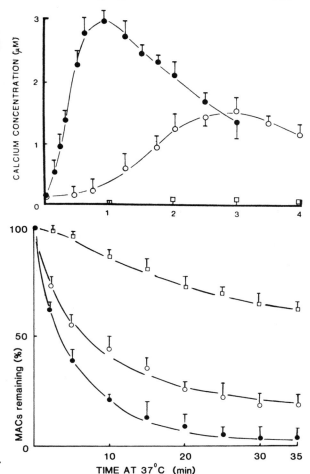

for haemolytic viruses, *Staph. aureus* **α**-toxin and detergents
(refs. in [30]). Target-cell killing by perforins needs
Ca^{2+} at several distinct stages: the initial exocytotic
secretion of perforin monomers, the assembly of the polyperforin
lesion on the target cell, and in all probability the lethal
event in the target cell are all calcium-dependent [18,
31]. Whether changes in $[Ca^{2+}]_i$ induced by any or all
of these agents stimulate recovery processes analogous to
those described for the MAC remains to be established, though
Ca^{2+}-dependent cell activation has been demonstrated for several
of them [19, 32].

EXTRACELLULAR CALCIUM AND PORE-FORMERS

The foregoing sections have described the dual role
of increases in $[Ca^{2+}]_i$, acting both as a cell-damaging factor
and as a stimulus for cell recovery and activation following

Table 1. Non-lethal effects of the MAC. In Ca^{2+}-dependence column, × means not investigated. AAM, arachidonic acid metabolite (PG, LT, TBX); PL, phospholipase; ROM, reactive oxygen metabolite; activ[n], activation; prod[n], production.

Cell	Effect	Ca^{2+}	Ref.
Rat neutrophil	ROM prod[n]	Yes	27
	AAM prod[n]	Yes	29
Human neutrophil	ROM prod[n]	Yes	33
	AAM prod[n]	Yes	34
	Vesiculation	Yes	25
Platelet	Prothrombinase activ[n]	Yes	35
	Vesiculation	Yes	5
	AAM prod[n]	No	36
Monocyte	Proliferation	Yes	36
	AAM & ROM prod[n]	Yes	36
Glomerular epithelial cell	Vesiculation	×	37
	PL activ[n] and $InsP_3$ prod[n]	Yes	38
	AAM prod[n]	×	39
Glomerular mesangial cell	AAM & interleukin prod[n]	×	40
	ROM prod[n]	×	41
KAT cell	Vesiculation	Yes	42
EAT cell	AAM prod[n]	×	43
Oligodendrocyte	AAM prod[n]	×	44
Synoviocyte	ROM prod[n]	Yes	45

attack by pore-forming toxins. A further factor influencing the outcome of cell attack by pore-formers is the extracellular Ca^{2+} concentration, $[Ca^{2+}]_e$. Pasternak and co-workers have provided abundant evidence that increases in $[Ca^{2+}]_e$ inhibit pore formation and hence cell lysis by a diverse range of toxins, including haemolytic viruses, melittin, *Staph. aureus* α-toxin and the MAC [30, 32] (& arts. in this vol. and in Vol. 17 – *Ed.*). They suggest that increased $[Ca^{2+}]_e$ prevents Ca^{2+} entry into cells by closing pores in the membrane and inhibiting new pore formation, and have demonstrated that other divalent cations, *e.g.* Zn^{2+}, have similar effects [31].

CONCLUDING REMARKS

Calcium therefore plays a paradoxical role in cell injury caused by toxins which act by forming pores in membranes. Severe attack by these toxins will result in large increases in $[Ca^{2+}]_i$, overwhelming recovery mechanisms and resulting in cell death. During less severe attack, however, calcium will act to protect cells in two ways: (1) extracellular Ca^{2+} will inhibit pore formation and close pores already present on the membrane, and (2) increased $[Ca^{2+}]_i$ resulting from leakage through the lesion (and perhaps release from intracellular stores) will stimulate recovery processes. This latter protective mechanism is so far only known to exist for the MAC, but in view of the functional and structural similarities within the group of toxins described here it is likely that similar mechanisms will be identified for other pore-formers. Ca^{2+}-mediated cell activation has been shown to result from non-lethal attack by the MAC and several other pore-formers. For several of these toxins these non-lethal effects may be of more relevance to their actions *in vivo* than their cytolytic activities.

Acknowledgements. - We thank the Wellcome Trust, the MRC and the Arthritis & Rheumatism Council for support. B.P.M. is a Wellcome Trust Senior Clinical Research Fellow.

References

1. Eagle, H. (1956) *Arch. Biochem. Biophys. 61*, 356-366.
2. Campbell, A.K. (1983) *Intracellular Calcium: its Universal Role as Regulator*, Wiley, Chichester, 656 pp.
3. Schanne, F.A.X., Kane, A.B., Young, E.E. & Farber, J.L. (1979) *Science 206*, 700-702.
4. Morgan, B.P., Luzio, J.P. & Campbell, A.K. (1986) *Cell Calcium 7*, 399-411.
5. Sims, P.J. & Wiedmer, T. (1986) *Blood 68*, 556-561.
6. Luzio, J.P., Abraha, A., Richardson, P.J., Daw, R.A., Sewry, C.A., Morgan, B.P. & Campbell, A.K. (1987) in *Cells, Membranes and Disease, including Renal* [Vol. 17, this series] (Reid, E., Cook, G.M.W. & Luzio, J.P.), Plenum, N. York, 199-209.
7. Bhadki, S. & Tranum-Jensen, J. (1984) *Phil. Trans. Roy. Soc. B 306*, 311-324.
8. Bernheimer, A.W. & Rudy, B. (1986) *Biochim. Biophys. Acta 864*, 123-141.
9. Podack, E.R. & Tschopp, J. (1982) *Proc. Nat. Acad. Sci. 79*, 574-580.
10. Tschopp, J., Müller-Eberhard, H.J. & Podack, E.R. (1982) *Nature 298*, 534-537.
11. Morgan, B.P., Patel, A.K. & Campbell, A.K. (1987) *Biochem. Soc. Trans. 15*, 659-660.
12. Bhadki, S. & Tranum-Jensen, J. (1985) *Complement 2*, 10 (abs.)
13. Esser, A.F. (1983) in *Membrane-mediated Cytotoxicity*, U.C.L.A. Symp. Series V. 56, A.R. Liss, New York, pp. 411-422.
14. Green, H. & Goldberg, B. (1960) *Ann. N.Y. Acad. Sci. 87*, 352-362.
15. Koski, C.L., Ramm, L.E., Hammer, C.H., Mayer, M.M. & Shin, M.L. (1983) *Proc. Nat. Acad. Sci. 80*, 3816-3820.
16. Young, J.D-E., Cohn, Z.E. & Podack, E.R. (1986) *Science 233*, 184-190.
17. Tschopp, J., Masson, D. & Stanley, K.K. (1986) *Nature 322*, 831-834.

18. Tschopp, J. & Jongeneel, C.V. (1988) *Biochemistry 27,* 2641-2646.
19. Bhadki, S., Seeger, W., Suttorp, N. & Tranum-Jensen, J. (1985) in *The Pathogenesis of Bacterial Infections,* Bayer Symp. VIII, Springer-Verlag, Berlin, pp. 268-279.
20. Hider, R.C., Khader, F. & Tatham, A.S. (1983) *Biochim. Biophys. Acta 728,* 206-214.
21. Laine, R.O., Morgan, B.P. & Esser, A.F. (1988) *Biochemistry 27,* 5308-5314.
22. Campbell, A.K., Daw, R.A., Hallett, M.B. & Luzio, J.P. (1981) *Biochem. J. 194,* 551-560.
23. Morgan, B.P. (1984) *Ph.D. Thesis, Univ. of Wales,* pp. 253-274.
24. Campbell, A.K. & Morgan, B.P. (1985) *Nature 317,* 164-166.
25. Morgan, B.P., Dankert, J.R. & Esser, A.F. (1987) *J. Immunol. 138,* 246-253.
26. Morgan, B.P. & Campbell, A.K. (1985) *Biochem. J. 231,* 205-208.
27. Hallett, M.B., Luzio, J.P. & Campbell, A.K. (1981) *Immunology 44,* 569-576.
28. Carney, D.F., Hammer, C.H. & Shin, M.L. (1988) *J. Immunol. 137,* 263-268.
29. Imagawa, D.K., & Barbour, S.E., Morgan, B.P., Wright, T.M., Shin, H.S. & Ramm, L.E. (1987) *Mol. Immunol. 24,* 1263-1271.
30. Pasternak, C.A. (1986) *Cell Calcium 7,* 387-397.
31. Bashford, C.L., Alder, G.M., Menestrina, G., Micklem, K.S., Murphy, J.J. & Pasternak, C.A. (1986) *J.Biol. Chem. 261,* 9300-9308.
32. Pasternak, C.A. (1987) *BioEssays 6,* 14-19.
33. Morgan, B.P. (1988) *Immunology 63,* 71-77.
34. Seeger, W., Suttorp, N., Hellwig, A. & Bhadki, S. (1986) *J. Immunol. 137,* 1286-1293.
35. Wiedmer, T., Esmon, C.T. & Sims, P.J. (1986) *J. Biol. Chem. 261,* 14587-14592.
36. Betz, M., Seitz, M. & Hansch, G.M. (1987) *Int. Arch. Allergy Appl. Immunol. 82,* 313-316; also (Hansch et al.) 317-320.
37. Camussi, G., Salvidio, G., Biesecker, G., Brentjens, J. & Andres, G. (1987) *J. Immunol. 149,* 2906-2914.
38. Cybulsky, A.V., Salant, D.J., Quigg, R.J., Badalamenti, J. & Bonventre, J.V. (1988) *FASEB J. 2,* A627.
39. Hansch, G.M., Betz, M., Gunther, J., Rother, K.O. & Sterzel, B. (1988) *Int. Arch. Allergy Appl. Immunol. 85,* 87-93.
40. Lovett, D.H., Hansch, G.M., Goppelt, M., Remsch, K. & Gemsa, D. (1987) *J. Immunol. 138,* 2473-2480.
41. Adler, S., Johnson, R.J., Ochi, R.F., Pritzl, P. & Couzer, W.G. (1986) *J. Clin. Invest. 77,* 762-767.
42. Kim, S-H., Carney, D.F., Hammer, C.H. & Shin, M.L. (1987) *J. Immunol. 138,* 1530-1536.
43. Imagawa, D.K., Osifchin, N.E., Paznekas, W.A., Shin, M.L. & Mayer, M.M. (1983) *Proc. Nat. Acad. Sci. 80,* 6647-6651.
44. Shirazi, Y., Imagawa, D.K. & Shin, M.L. (1987) *J. Neurochem. 48,* 271-278.
45. Morgan, B.P., Daniels, R.H., Watts, M.J. & Williams, B.D. (1988) *Clin. Exp. Immunol. 73,* 467-472.

#E-3

DEGENERATION OF CULTURED RAT BRAIN NEURONS INDUCED BY EXCITOTOXIC AGENTS: ROLE OF CALCIUM IONS

Petrus J. Pauwels*, Harrie P. van Assouw and
Josée E. Leysen

Department of Biochemical Pharmacology,
Janssen Research Foundation,
B-2340 Beerse, Belgium

Neurons from embryonic rat brains were cultured in serum-free, chemically defined medium. Addition of Glu-like[†] agonists to these cultures caused neuronal cell death. The neuronal degeneration can be quantified by measuring LDH efflux into the medium. The toxic effects of Glu-like agonists were observed early during the development of the neuronal cultures. The excitotoxic action appeared to be neuron-specific, since astrocyte cultures were not affected. The neurons could be protected from the Glu-like toxicity by reducing the extracellular Ca^{2+} concentration. The role of Ca^{2+} in excitotoxicity is discussed.

Various types of Ca^{2+} channels exist in neurons including different voltage-specific or receptor-operated Ca^{2+} channels. In CNS neurons Ca^{2+} can also enter through ion channels activated by excitatory amino acids. Those operated by NMDA are particularly permeable to Ca^{2+}. Those activated by KA have a limited permeability to Ca^{2+} [1]. Excessive or prolonged activation of excitatory amino acid receptors can lead to neuronal cell death. It is thought that the neuronal degeneration associated with cerebral ischaemia, epilepsy, Alzheimer's disease and Huntington's chorea may result from the release of abnormal quantities of Glu-like agonists *in situ* [2]. To approach the Ca^{2+}-dependency underlying excitotoxic neuronal degeneration [3-6], cultured rat-brain neurons were exposed to various neurotoxic stimuli in the presence of Ca^{2+}-chelators and Ca^{2+}-blockers.

*addressee for any correspondence
[†]*Abbreviations.* - AM, antimycin; DMEM, Dulbecco's modified Eagle's Medium; EAA, excitatory amino acid; Glu (or GLU), glutamate; KA, kainic acid; LDH, lactate dehydrogenase; NMDA, *N*-methyl-D-aspartate; OUA, ouabain; TTX, tetrodotoxin.

SERUM-FREE CULTURE OF NEURONS

For preparing neuronal cultures [details: 7], the hippo-campal formation was dissected from brains of 17-day-old rat embryos, placed immediately in Ca^{2+}-and Mg^{2+}-free Hank's Balanced Salt Solution, and mechanically dissociated by repeti-tive suction through a Pasteur pipette. Enzymatic tissue dissociation was avoided since less viable cultures were obtained. The resulting cell suspension was centrifuged (400 **g**, 10 min), washed once with DMEM, and filtered through a layer of nylon (pore diam. 40 μm). Cells were plated (7×10^5/ml) in polylysine-coated (0.001%) 24-well tissue culture plates (Nunc; well size 1.76 cm^2) with 1 ml DMEM/Ham's F12 (3:1 v/v) plus heat-inactivated horse serum (to 20%). After 24 h, cultures were switched to serum-free chemically defined CDM R12 medium [8] based on the DMEM/Ham's F12 medium buffered with 25 mM HEPES. Partial replacement with fresh medium was done weekly. Trypan blue exclusion tests generally gave viability values >95%. Cultures were maintained at 37° in air/CO_2 (95:5).

The serum-free chemically defined CRM R12 medium minimized the growth of non-neuronal cells and promoted neuronal differ-entiation. Romyn et al. [8] have discussed the presence in serum of factors that are growth-stimulating or are deleteri-ous to cell growth and functioning. Another argument against the use of serum in neuronal cultures is that *in vivo* serum directly encounters neuronal cells only in the case of injury. Moreover, serum also favours glial cell survival and prolifer-ation [9]. In the total absence of serum, glial growth was found to be minimal and sensitivity to neurotoxic stimuli was observed early during the development of the neuronal culture, in contrast with the neurotoxic effects reported by Rothman [10].

NEUROTOXICITY IN SERUM-FREE NEURONAL CULTURES

Cell damage was monitored by measuring the efflux into the culture medium of the cytosolic enzyme LDH; the phenomenon seemed to be non-specific since astrocyte cultures were not affected by the various neurotoxic agents [7]. Fig. 1 shows the development of sensitivity of the neuronal cultures of increasing age *in vitro* to the Glu receptor agonists Glu, KA and NMDA, to the Na^+,K^+-ATPase pump inhibitor OUA, and to the anoxic agent KCN. Exposure to 1 mM Glu, NMDA or KCN for 20 min did not induce morphological degeneration in cultures <7 days old. However, prolonged exposure to these 'triggers' was followed by progressive neuronal degener-ation. Released LDH attained 11%, 20% and 20% after a 24 h exposure to NMDA, Glu and KCN respectively. The 4-day cultures were more sensitive to 1 mM KA and 20 μM OUA, which gave

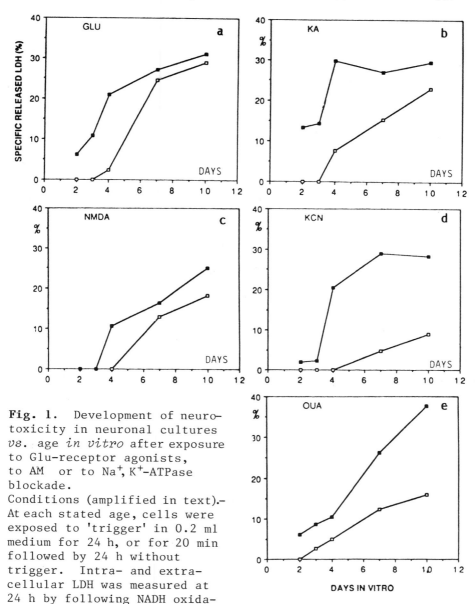

Fig. 1. Development of neurotoxicity in neuronal cultures
vs. age *in vitro* after exposure
to Glu-receptor agonists,
to AM   or to Na$^+$,K$^+$-ATPase
blockade.
Conditions (amplified in text).–
At each stated age, cells were
exposed to 'trigger' in 0.2 ml
medium for 24 h, or for 20 min
followed by 24 h without
trigger.   Intra- and extracellular LDH was measured at
24 h by following NADH oxida-
tion [7] using an automatic EPOS–5060 (Eppendorf, Hamburg) linked
to a Macintosh personal computer.   Values are expressed as
specifically released LDH after subtracting released–LDH
values for control cultures (not exposed to a trigger). All
values are means of 6 to 9 separate cultures from 2 or 3 independent culture set-ups. **a,** 1 mM Glu; **b,** 1 mM KA; **c,** 5 mM NMDA;
d, 5 mM KCN; **e,** 10 μM OUA. Trigger exposure: *open symbols,*
20 min; *closed symbols,* 24 h.

8% and 5% LDH release respectively after a 20-min exposure
followed by a 24 h incubation without trigger (Fig. 1, b
& e). Maximal neuronal damage was obtained with 10-day
neuronal cultures exposed for 20 min to Glu: LDH release
attained 31% 24 h later (Fig. 1, a).

Fig. 2 shows the time courses of neurotoxin-induced damage
in 7-day cultures. Half-maximal damage with 1 mM Glu,
1 mM KA or 5 mM NMDA was attained between 7 and 10 min
of exposure followed by 24 h without trigger. The onset
of 5 mM KCN- and 10 μM OUA-induced damage was slower: it
was half-maximal at 27 and 30 min respectively. Toxic blockade
of the respiratory chain by AM instead of KCN was prominent:
damage with 20 μM AM was maximal in <5 min.

ROLE OF CALCIUM IONS IN NEUROTOXICITY

Incubation of 7-day cultures with the Ca^{2+}-chelator EGTA
(2 mM) present for 24 h did not affect their viability:
LDH release into the medium was in the same range as for
control cultures (Table 1). Cultures exposed to neurotoxic
triggers with 2 mM EGTA present showed no sign of neuronal
degeneration, nor did neurotoxins induce LDH release. The
findings were similar with 2 mM EDTA. This confirms that
removal of free extracellular Ca^{2+} selectively reduces neuronal
cell loss [11].

Since the neurotoxic effects were Ca^{2+}-dependent, experiments were repeated in the presence of the partial Na^+-channel
agonist veratridine (10 μM), which modifies the Na^+-channel
so that Ca^{2+} ions can pass through it [12]. Released LDH
was increased after 16 h, in the same range as for 1 mM
Glu-exposed cultures. Addition of Glu or any other tested
neurotoxic trigger to veratridine-treated cultures led to
virtually the same damage as observed without veratridine.

Blockade of the Na^+-channels by 1 μM TTX did not affect
LDH release in non-triggered cultures. TTX gave a slight
protection against OUA-induced neurotoxicity: LDH release
decreased by 36%. The inability of TTX to afford total
protection suggests that the process responsible for delayed
neurotoxicity does not require action-potential generation.
According to Peacock & Walker [13], spontaneous electrical
activity and transmitter release can be generated by voltage-
sensitive Ca^{2+}-currents in cultured hippocampal neurons. These
currents would not be affected by TTX.

Since extracellular Ca^{2+} seems to be required for neurotoxicity, various compounds with Ca^{2+}-blocking properties were
tested for their ability to prevent the Ca^{2+}-neurotoxicity.

Fig. 2. Time-courses of
neurotoxicity in 7-day
neuronal cultures after
exposure to Glu-receptor
agonists, to AM and to
Na^+,K^+-ATPase pump blockade.
Conditions generally as in
text and Fig. 1 legend.
Incubation of the washed
cultures with trigger in
0.2 ml medium, for the
times shown, was followed
by washing and incubation
up to 24 h. Specifically
released LDH values (the
values for non-exposed
controls having been sub-
tracted) for trigger-
exposed (<24 h) cells are
expressed as % of those
for cells treated with
trigger for 24 h. All
values are means of 3
separate cultures.
a, 1 mM Glu, 1 mM KA and
5 mM NMDA;
b, 20 µM AM and 10 µM OUA.

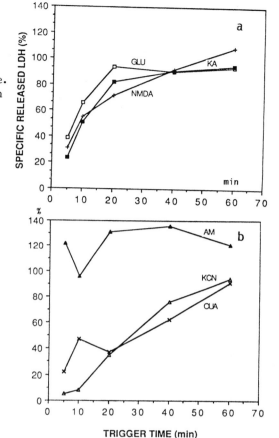

As shown in Table 2, only slight effects were observed with
nitrendipine, verapamil, cinnarizine, flunarizine, lidoflazine
and pimozide at the high level of 10 µM. This suggests
that in neuronal cultures the channels at which Ca^{2+}-blockers
act do not mediate the neurotoxicity of Ca^{2+}. However,
if this was an important mechanism of Ca^{2+} entry, then
depolarization of the cultures with a high-K^+ buffer extra-
cellularly should also lead to excessive Ca^{2+}-entry and delayed
deterioration. This was not observed by Rothman et al.
[14].

Other possible mechanisms for neurotoxic Ca^{2+} entry.-
There might be direct activation of inositol phospholipid
metabolism and mobilization of intracellular Ca^{2+} as shown
by Sugiyama et al. [15], or through receptor-operated channels.
Antagonism of the NMDA subclass of Glu receptors can substantially attenuate Glu-induced neuronal loss [11, 16-19]. This

Table 1. Released LDH, as % of total (index of neurotoxicity), in 7-day neuronal cultures treated for 16 h with Ca^{2+} chelators (EGTA, EDTA) or Na^+-channel toxins (veratridine, TTX) and triggered with 1 mM Glu, 1 mM KA, 5 mM NMDA, 5 mM KCN and 10 μM OUA. For LDH measurement see Fig. 1 legend. All values are means ±S.D. of 4 separate cultures.

Trigger:	None	Glu	KA	NMDA	KCN	OUA
No treatment	4.8 ±0.6	22.3 ±2.4	19.5 ±0.8	18.3 ±1.2	22.7 ±2.2	19.5 ±1.6
2 mM EGTA	3.0 ±0.3	4.6 ±0.8	5.5 ±0.5	9.7 ±1.2	3.7 ±1.1	2.8 ±0.3
2 mM EDTA	4.3 ±0.6	5.4 ±0.2	6.1 ±1.1	7.4 ±0.6	3.7 ±0.5	3.7 ±0.4
10 μM veratridine	22.8 ±1.9	24.6 ±0.7	20.5 ±4.1	18.9 ±0.9	26.7 ±0.4	20.8 ±2.4
1 μM TTX	6.3 ±0.5	21.6 ±1.3	20.5 ±3.2	20.8 ±0.1	20.6 ±0.6	14.1 ±0.2

Table 2. Protection (%) by various Ca^{2+}-blocking compounds (10 μM; 15-min pre-triggering exposure) of cultures (4) as in Table 1 against neurotoxicity, assessed by LDH release after induction by triggering agents during 16 h.

	D r u g	*Trigger:* Glu	KA	NMDA	KCN	OUA
Selective	I. Verapamil	25 ±2	36 ±1	50 ±4	40 ±4	-
for slow	II. Nifedipine	0	30 ±4	18 ±3	17 ±3	-
Ca^{2+}	II. Nitrendipine	27 ±3	23 ±4	45 ±7	21 ±3	-
channels	III. Diltiazem	0	3 ±0	0	0	
Non-selective	IV. Cinnarizine	41 ±3	29 ±1	43 ±3	31 ±3	-
for these	IV. Flunarizine	15 ±2	31 ±5	0	9 ±1	56
channels	IV. Lidoflazine	14 ±1	7 ±1	15 ±2	7 ±1	35 ±7
Others	Fluspiriline	0	0	5 ±1	0	11 ±2
	Pimozide	0	0	18	0	38 ±8

implies that activation of these receptors is at least partially responsible for delayed neuronal cell death. According to Rothman et al. [14] three different mechanisms can be considered for the Ca^{2+}-dependent neurotoxicity of Glu.— (1) Ca^{2+} entry may allow neurons to express more functional postsynaptic receptors for EAA [20, 21]. This would make more neurons much more sensitive to the effects of EAA. (2) Elevation of intracellular Ca^{2+} in presynaptic neurons may stimulate the secretion of more transmitter with each impulse [22]. (3) Ca^{2+} influx could interfere with Glu uptake. If any of these mechanisms operates after Ca^{2+} influx, the level of receptor activation will be amplified. The explanation of delayed Glu-neurotoxicity then becomes circular: transient exposure to Glu produces late neuronal degeneration because it makes the neurons more susceptible to their store of endogenous Glu or other compounds such as aspartate, quinolinate, homocysteate and *N*-acetylaspartylglutamate [11].

Glu receptor blockade may be one strategy to protect neurons from neurotoxicity. The therapeutic utility of NMDA-antagonists in hypoxia or ischaemia still has to be proven. Prevention of pathological accumulation of EAA is perhaps a better strategy. This may be attainable by decreasing Glu synthesis or release, or by increasing cellular uptake of Glu. Cell culture may, therefore, be a tool for characterizing the pharmacology of hypoxic neuronal injury. When combined with appropriate *in vivo* testing, the present approach may provide a convenient means for identifying compounds potentially able to reduce hypoxic neuronal injury in the clinical setting.

References

1. Murphy, S.N., Thayer, S.A. & Miller, R.J. (1987) *J. Neurosci.* **7**, 4145-4158.
2. Schwarcz, R., Foster, A.C., French, E.D., Whetsell, W.O. & Kohler, C. (1984) *Life Sci.* **35**, 19-32.
3. Garthwaite, G. & Garthwaite, J. (1986) *Neurosci. Lett.* **66**, 193-198.
4. Garthwaite, G., Hajos, F. & Garthwaite, J. (1986) *Neurosci.* **18**, 437-447.
5. Hajos, F., Garthwaite, G. & Garthwaite, J. (1986) *Neurosci.* **18**, 417-436.
6. Choi, D.W. (1987) *J. Neurosci.* **7**, 369-379.
7. Pauwels, P.J., van Assouw, H.P. & Leysen, J.E. (1989) *Cellular Signalling* **1**, 45-54.
8. Romyn, H.J., van Huizen, F. & Wolters, P.S. (1984) *Neurosci. Biobehav. Rev.* **8**, 301-334.
9. McCarthy, K.D. & de Vellis, J. (1980) *J. Cell Biol.* **85**, 890-902.
10. Rothman, S.M. (1983) *Science* **220**, 536-537.
11. Choi, D.W., Koh, J-Y. & Peters, S. (1988) *J. Neurosci.* **8**, 185-196.
12. Jacques, Y., Frelin, C., Vigne, P., Romey, G., Parjari, M. & Lazdunski, M. (1981) *Biochemistry* **20**, 6219-6225.
13. Peacock, J.H. & Walker, C.R. (1983) *Dev. Brain Res.* **8**, 39-52.
14. Rothman, S.M., Thurston, J.H. & Hauhart, R.E. (1987) *Neuroscience* **22**, 471-480.
15. Sugiyama, H.I., Ito, I. & Hirono, C. (1987) *Nature* **325**, 531-533.
16. Simon, R.P., Swan, J.H., Griffiths, T. & Meldrum, B.S. (1984) *Science* **226**, 681-683.

17. Wieloch, T. (1985) *Science 230*, 681-683.
18. Goldberg, M.P., Weiss, J. & Choi, D.W. (1986) *Soc. Neurosci. Abstr. 12*, 64.
19. Weiss, J., Goldberg, M.P. & Choi, D.W. (1986) *Brain Res. 380*, 186-190.
20. Baudry, M., Bundman, M.C., Smith, E.K. & Lynch, G.S. (1981) *Science 212*, 937-938.
21. Lynch, G. & Baudry, M. (1984) *Science 224*, 1057-1063.
22. Miledy, R. (1973) *Proc. Roy. Soc. B 183*, 421-425.

#E-4

CALCIUM TRANSPORT IN CONDITIONS OF FREE-RADICAL PATHOLOGY OF CELLULAR MEMBRANES

Yu.I. Gubskiy

Institute of Pharmacology and Toxicology,
Public Health Ministry of the Ukrainian S.S.R.,
Eugene Potye Street 14,
252057 Kiev, U.S.S.R.

The methodological aspects of investigation of peroxidative reactions and calcium-transport systems in hepatic and myocardial cells activated by free-radical LP* are reviewed. In conditions of AOD deficiency there is marked activation of NADPH- and ascorbate-dependent LP in hepatic subcellular structures, particularly the e.r. membranes. Non-enzymic LP is similarly activated in myocardial s.r. in AOD. As compared with the AOD model, the CCl₄-stimulated LP in hepatocytes is accompanied by a more profound increase in content of diene conjugates and in damage to the monooxygenase system.

Activation of LP in hepatic e.r. membranes leads to drastic diminution of ATP-dependent ⁴⁵Ca accumulation and Ca-ATPase activity. Ca-pump function and phosphorylation in myocardial s.r. membranes are also disturbed. The administration of antioxidants, α-tocopheryl acetate and ionol, to albino rats with AOD promoted a decrease in the rates of enzymic and non-enzymic LP as well as normalization of Ca-transport systems of hepatocytes and myocardiocytes and restoration of the compartmentalization, disturbed in free-radical pathology, of intracellular calcium.

Free-radical peroxidation of polyenic acyls of membrane phospholipids is believed to be the molecular basis underlying such fundamental biological and pathological processes, and their manifestations, as senescence, cancerogenesis, and the generalized response of viable cells to such damaging influences

---

*Abbreviations.- AOD, antioxidant deficiency; LP, lipid peroxidation; e.r./s.r., endoplasmic/sarcoplasmic reticulum; MDA, malonyl dialdehyde.

as ionizing radiation, intoxications, and cardiac ischaemia
and re-perfusion. The LP-promoted modification of physicochemical properties of the lipid bilayer may be accompanied
by drastic changes in membrane systems for transporting ions,
particularly Ca^{2+}. Consequently, the disorganization of intracellular calcium compartmentalization is a biochemical event
that leads to significant disturbances of Ca^{2+}-dependent
biochemical reactions and physiological functions. In this
context the adequacy of methods applied for estimating peroxidative reactions and Ca^{2+}-transport systems is cardinal.

PEROXIDATIVE REACTIONS IN LIVER AND MYOCARDIUM
MEMBRANE STRUCTURES IN FREE-RADICAL PATHOLOGY

Quantitative estimation of LP activity in biological
structures commonly entails measuring one of the following
types of product:- (1) final products of the degradation
of fatty acids, e.g. MDA, ethane, pentane or other volatile
hydrocarbons; (2) intermediates of free-radical reactions,
e.g. diene conjugates or hydroperoxides: ESR spectroscopy
may be used for measurement; (3) certain fluorescent compounds
(estimated in lipid extracts) formed by *in situ* reaction
of 2-alkenals and 2,4-alkadienals with amino groups of tissue
proteins and amino acids.

The investigations performed in our laboratory have shown
that the LP stimulation *in vivo* induced by keeping female
rats for 2 months on vitamin E-free laboratory chow [1]
produced varying effects on LP parameters estimated by different
methods. The ratios between the activities of NADPH-dependent
and ascorbate-dependent (non-enzymic) LP, determined by MDA
accumulation *in vitro* (Fig. 1), varied greatly in the individual
preparations of hepatic microsomes from control rats. In
the microsomes from AOD rats, the non-enzymic LP showed
the greater increase in activity (Fig. 1)[2]. Similarly to
hepatic microsomal fractions, in myocardial s.r. fractions
ascorbate-dependent LP (non-enzymic) predominated (Table 1).

The marked stimulation of LP, estimated by MDA accumulation
during incubation of hepatic subcellular fractions *in vitro*,
was also observed in mitochondria, plasma membranes and nuclear
chromatin of AOD rats. As compared with increased capacity
of hepatic membrane vesicles for NADPH- and ascorbate-dependent
LP, a moderate increase in the levels of diene conjugates
was found in e.r. membranes. We failed to detect a significant
build-up of fluorescent compounds (expressed as amount per
mg of microsomal lipids) in vitamin E deficiency, presumably
due to non-accumulation of aldehyde products *in vivo*.

Fig. 1. Activities of enzymic and non-enzymic LP (expressed as MDA accumulation *in vitro* with NADPH or ascorbate present) in the hepatic microsomes of (1) control, and (2) AOD rats. Δ, NADPH-dependent LP; o, ascorbate-dependent LP. Values are mean ±S.D. (in Fig. 2 also; the 'units' are mcmol). *From ref. [2].*

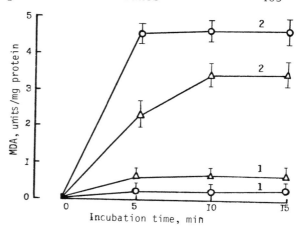

Table 1. Peroxidation of lipids (expressed as MDA accumulation *in vitro*: μmol/mg protein per 30 min) in rat myocardial s.r. in AOD. For each group n = 5; ± values are S.D.'s. *From ref. [7]. Abbreviations:* d't, dependent; asc., ascorbate.

Group	NAPH-d't LP	asc.-d't LP	*Ratio,* asc.-d't/NADPH-d't LP
Control	5.06 ±0.32	0.76 ±0.27	0.16 ±0.03
AOD	3.87 ±0.66	1.53 ±0.27*	0.40 ±0.02*

*$P < 0.05$

In contrast to AOD, the CCl_4-modelled free-radical pathology of hepatocytes as examined with microsomal fractions *in vitro* was not accompanied by an increased capacity for prooxidant-stimulated LP, whereas the diene conjugate level showed a marked increase, commencing one hour after intoxication. 'Resistance' manifested by hepatic microsomes, in chemical injury, to the prooxidant effects of NADPH and ascorbate is thought to be attributable to a sharply diminished content of polyunsaturated fatty acids, especially $C_{20:4}$ and $C_{18:3}$, the principal substrates of LP in biomembranes.

CALCIUM TRANSPORT IN HEPATIC AND MYOCARDIAL SUBCELLULAR STRUCTURES IN FREE-RADICAL PATHOLOGY

Stimulation of LP *in vivo* and *in vitro* caused drastic changes in Ca^{2+}-transport systems in vesiculated membrane structures of liver and myocardium [3-7].

Measuring ^{45}Ca incorporation into isolated microsomal fractions in the presence of ATP [8], we have established that there is significant damage to active calcium transport

Table 2. Properties of rat hepatic e.r. Ca^{2+}-pump in AOD [6].
The values for pump-efficiency parameters are mean ±S.D.
(10-11 animals), expressed for the $^{45}Ca^{2+}$ and ATPase measurements as nmol/mg protein per min. Ca^{2+}/ATP signifies the Ca^{2+}
transport system's 'efficiency': no. of Ca^{2+} ions moved (ATP-dependent process) *vs.* ATP molecules hydrolyzed by Ca^{2+}-ATPase. *From [6]*.

Group	ATP-dependent $^{45}Ca^{2+}$ accumulation	Ca^{2+}-ATPase activity	Ca^{2+}/ATP
Control	400 ±50	881 ±54	0.57 ±0.10
AOD	81 ±62*	641 ±85*	0.21 ±0.05*
AOD + α-tocopheryl acetate	325 ±68**	814 ±45**	0.42 ±0.07**
AOD + ionol	129 ±72	894 ±24**	0.19 ±0.07

*P <0.05 *vs.* control group   **P <0.05 *vs.* AOD group

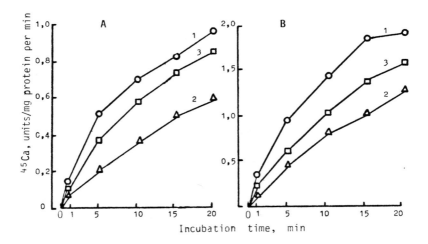

Fig. 2. ATP-dependent Ca^{2+} transport in rat myocardial s.r.
(expressed as $^{45}Ca^{2+}$ accumulation by s.r. vesicles *in vitro*):
A: without additional phosphorylation of membrane proteins;
B, in conditions of phosphorylation of membrane proteins by
cAMP-dependent protein kinase. **1**, control; **2**, AOD;
3, AOD + ionol. *From ref.* [7]. The soluble kinase was from rabbit
myocardium: 0.2 mg/mg s.r. protein, added just before ATP, + cAMP to 10 µM.

across the membranes of the hepatic e.r. (Table 2) and myocardial s.r. (Fig. 2A) in AOD-induced LP. Ca^{2+}ATPase and Ca^{2+}/ATP,
which Table 2 shows were decreased in hepatic e.r., are further
considered below, and likewise the effects of antioxidant
administration.

Another pattern of change was discerned for the absorption of Ca^{2+} by membrane structures investigated without ATP stimulation, i.e. there was some increase in Ca^{2+} uptake in hepatic microsomes derived from AOD rats. This may accord with the raised concentration in lipoperoxidated membranes of polar molecules such as hydroperoxides, this rise accounting for the somewhat increased calcium binding.

The key component of the Ca^{2+} pump of s.r. and hepatic e.r. is the membrane-bound Ca^{2+}-ATPase. The enzyme from skeletal muscle and myocardium is known to be severely damaged during LP activation in vivo and in vitro [3, 4]. According to Martonosi & Feretos [8], the Ca^{2+}-ATPase of s.r. as measured in the presence of 10^{-4}-10^{-5} M Ca^{2+} adequately reflects the function of s.r. transport Ca^{2+}-ATPase. Using similar experimental conditions, we have shown significant modification of hepatic microsomal Ca^{2+}-ATPase kinetic properties in AOD-induced LP stimulation.

The estimation of ATP-dependent calcium transport in myocardial s.r. in AOD also revealed a sharp decrease in $^{45}Ca^{2+}$ accumulation (Fig. 2), which accords with modifications previously found in Ca^{2+}-transport systems and ultrastructural organization of the s.r. membranes during lipid peroxidation [3, 4]. Disruption of Ca^{2+}-pump function in LP activation is assumed to be attributable largely to changes in the molecular organization of the membrane lipid bilayer, and especially to selective modification of phosphatidylethanolamine containing hexa-, penta- and tetra-enoic fatty acid residues in position 2 [9]. By use of high-resolution ^{1}H-NMR spectroscopy and ESR spectroscopy of spin probes, Klaan et al. [4] found a drastic increase in the polarity of the s.r. membrane lipid bilayer induced by accumulation of LP products. The resulting rise in highly organized lipids ('clusters') is accompanied by a decrease in the fluid lipids and in the quantity of phospholipids interacting with protein components of the membrane.

The considerable loss of Ca^{2+}-pump efficiency during LP stimulation (manifested by the Ca^{2+}/ATP decrease shown in Table 2) may result from the increasing outward leakage of calcium through the newly formed hydrophilic membrane pores. According to Vorobets et al. [10], estimation of $^{45}Ca^{2+}$ efflux from the cation-loaded vesicles into calcium-free medium is an adequate method for measuring passive permeability of membrane vesicles. The evaluation of $^{45}Ca^{2+}$-efflux from the hepatic microsomal vesicles derived from AOD rats revealed a marked increase in their passive permeability, which may be partly responsible for the disturbance of calcium compartmentalization in this pathological condition.

A disturbance of the phosphorylation of regulatory membrane proteins has been postulated as an additional molecular mechanism of Ca^{2+}-pump damage under LP stimulation. It is well known that the phosphorylation of regulatory proteins of s.r. membranes, such as phospholamban, is effected by treatment with exogenous protein kinases – a biochemical approach applied in our experiments [7]. Thereby it was shown that in myocardial s.r. from AOD-rats, both the initial rate of cAMP-dependent membrane proteins and ATP-dependent Ca^{2+}-transport were markedly decreased as compared with controls (Fig. 2).

There is increasing evidence that CCl_4-induced LP causes a key modification of hepatic e.r. membrane functions, including Ca^{2+} transport. To confirm that there are drastic modifications of calcium compartmentalization in liver cells following chemical injury, we have used atomic absorption spectroscopy to study the distribution of cation within subcellular fractions [5]. It was established that the chemical damage to hepatocyte membranes led to considerable increase of Ca^{2+} concentration in the fractions studied, especially in the cytosol fraction (105,000 **g** supernatant). Taking account of possible disadvantages of the method used, it is likely that this rise in cytosol Ca^{2+} may result from disturbances both in calcium elimination through plasma membranes and in its sequestration in the e.r. and mitochondria.

CCl_4-induced decreases of ATP-dependent $^{45}Ca^{2+}$ accumulation and Ca^{2+}-ATPase activity observed in our experiments are in agreement with the data reported above on the peroxidative modification of hepatic e.r. membranes and with the well-known inhibition of monooxygenases in this pathological condition. Using HPLC and TLC, James et al. [9] have shown a 6-fold increase of diene conjugates in an e.r. phosphatidylserine fraction – the phospholipid playing an important role in the organization of the Ca^{2+}-transport system [11].

EFFECT OF ANTIOXIDANTS ON PARAMETERS OF CALCIUM TRANSPORT IN FREE-RADICAL PATHOLOGY

For investigating possible biochemical mechanisms of the changed properties of hepatic and myocardial cell membranes in free-radical pathology, we used a methodological approach entailing comparative study of α-tocopheryl acetate and a synthetic LP inhibitor, ionol (4-methyl-2,6-di-tertbutyl-phenol). Ionol and especially α-tocopheryl acetate, given orally for 4 days to vitamin E-deficient rats, largely normalized LP, monooxygenase activity and some Ca^{2+}-transport parameters in myocardial s.r. and (Table 2) in hepatic e.r. although ionol had no restorative effect on Ca^{2+}/ATP [6, 7].

The ineffectiveness of ionol in restoring the Ca^{2+}/ATP in hepatic e.r. may be attributable to the outward leakage of Ca^{2+} ions through the peroxidized bilayer of microsomal vesicles. Ionol has nevertheless proved to be effective in restoring Ca^{2+}-ATPase activity concomitantly with the inhibitory action of antioxidant on oxidation of enzyme SH-groups that occurs in conditions of AOD [6, 7].

Using fluorescent probes we have shown considerable restoration by ionol of microviscosity of e.r. membrane lipids, which was drastically decreased in AOD-rat membranes. The results obtained confirm the close involvement of damage to membrane lipid constituents in the disturbances of calcium transport and compartmentalization in free-radical cellular pathology.

References

1. Edwin, E.E., Diplock, A.T., Bunyan, J. & Green, J. (1961) *Biochem. J. 79*, 91–105.
2. Gubskiy, Yu.I. (1987) *Dokl. AN UkrSSR, series B, N2*, 72–74 *(in Russian)*.
3. Kagan, V.E., Arkhipenko, Yu.V. & Kozlov, Yu.P. (1983) *Biokhimija 48*, 158–166 *(in Russian)*.
4. Klaan, N.K., Azizova, O.A., Sibeldina, L.A., Arkhipenko, Yu.V. & Kagan, V.E. (1983) *Biokhimija 48*, 626–633 *(in Russian)*.
5. Gubskiy, Yu.I., Zadorina, O.V., Tkachyuk, V.V. & Igrunov, L.P. (1986) *Vopr. Med. Khim. 32*, 80–85 *(in Russian)*.
6. Gubskiy, Yu.I., Kursky, M.D., Zarodina, O.V.,Fedorov, A.N. & Tkachyuk, V.V. (1987) *Dokl. AN UkrSSR, series B, N1*, 68–71 *(in Russian)*.
7. Gubskiy, Yu.I., Kalinsky, M.I., Rudnitskaya, N.D., Zadorina, O.V. & Kursky, M.D. (1988) *Ukr. Biokhim. Zhurn. 60*, 46–51 *(in Russian)*.
8. Martonosi, A. & Feretos, R. (1986) *J. Biol. Chem. 239*, 648–658.
9. James, J.L., Moody, D.E., Chan, C.H. & Smuckler,E.A. (1982) *Biochem. J. 206*, 203–210.
10. Vorobets, Z.D., Puzyrev, S.Yu. & Kursky, M.D. (1986) *Biokhimija 51*, 1295–1301 *(in Russian)*.
11. Brownig, J.L. & Seelig, J. (1980) *Biochemistry 19*, 1262–1270.

EDITOR'S NOTE: Chemical Abstracts searching will disclose English translations, particularly in Biochemistry (USSR).

#E-5

CALCIUM, CHOLESTEROL METABOLISM AND ATHEROSCLEROSIS

Keith E. Suckling

Smith, Kline & French Research Ltd.,
The Frythe, Welwyn, Herts. AL9 6AR, U.K.

An observation in the early 1980's, that nifedipine (a calcium antagonist) dosing of cholesterol-fed rabbits reduces the arterial cholesterol content, matched improved understanding of events that accompany atherosclerotic plaque development in the artery wall and of events in blood and liver related to the hypercholesterolaemia that is crucial for atherosclerosis. Subsequent in vivo studies, giving conflicting results, have focused on calcium-antagonist actions on relevant tissues rather than on the role of calcium itself. Possible ways in which antagonists might affect plaque development and cholesterol metabolic balance may include a linkage to the second messenger system. They may prevent cell proliferation (smooth muscle cells) and - in common with agents that increase intracellular cAMP - cholesteryl ester accumulation (macrophages). The effects may occur through modulation of certain enzyme activities and of cell-surface lipoprotein receptors.

In liver, calcium-sensitive processes have been related to cholesterol synthesis, bile-acid synthesis and secretion, and lipoprotein secretion. Studies on tissues relevant to cholesterol metabolism and atherosclerosis have tended to use methods devised for other tissues, and led to inappropriate model systems such as the cholesterol-fed animal. Studies of specific cell types in culture can define mechanisms but may not illuminate the longer-term factors that are probably cardinal for atherosclerosis development.

As one of the major causes of death in the western world, atherosclerosis has come more and more to the forefront of research in the pharmaceutical industry. The public, particularly in the U.S.A., has become more aware of the problems of heart disease, to which atherosclerosis is a major contributor. There has been a sudden burst of optimism because recent studies have shown that drugs can reduce

heart-disease mortality and even induce regression of atherosclerosis [1, 2]. The drugs in question all affect cholesterol
metabolism and tend to limit its deposition in arteries
and promote its loss from the body.

In simple terms, atherosclerosis is thought to develop
over several decades through the following stages [3].- The
arterial endothelium becomes permeable to particles such as
lipoproteins; this may occur by injury which may be very
minor, e.g. by turbulent flow at bifurcations. An environment
is generated, perhaps due to secretion of growth factors by
blood-derived cells such as platelets and macrophages, that
promotes the migration of smooth muscle cells from the arterial
media into the intima. During this stage macrophages and
smooth muscle cells take up large quantities of lipid from
lipoproteins, and the characteristic early lesion, the fatty
streak, appears. With time the fatty streak grows, and
smooth muscle cells become more prominent and secrete collagen
and glycosaminoglycans. The lesion now develops a fibrous
cap. Lipids continue to accumulate and cholesterol crystals
appear in the centre of the lesion. The function of the
artery may now be seriously impaired. Finally cells at
the centre of the lesion die as calcium mineralization takes
place.

CALCIUM ANTAGONISTS AS A POSSIBLE BASIS FOR THERAPY

For a number of years there has been increasing interest
in the possibility of treating atherosclerosis through
mechanisms affected by calcium. Probably the main driving
force for this has been pragmatic - the observation in
a number of laboratories that calcium antagonists of several
types can inhibit the deposition of cholesterol in the arteries
of certain animal models of atherosclerosis [4]. One of
the most widely used models is the cholesterol-fed rabbit.
Such rabbits rapidly develop a massive hypercholesterolaemia,
and deposition of cholesterol in the arteries follows. Such
animals respond to calcium antagonists with a reduction in
deposition of cholesterol in their arteries [4]. Doses
are typically ~0.3 mg/kg per day over up to 3 months.

Another rabbit model of atherosclerosis develops a different kind of hypercholesterolaemia spontaneously. This animal,
known as the Watanabe Heritable Hyperlipidemic (WHHL) rabbit,
has a genetic deficiency similar to one known in man in
which active low-density lipoprotein (LDL) receptors are absent
from the surface of cells. This results in an impaired
ability of the liver to clear LDL from the blood and in a
build-up of plasma LDL concentration. Cholesterol accumulates

in the arteries because other non-regulated lipoprotein receptors are available in the arterial cells. The pathology
of the lesion that develops in the arteries of the WHHL
animals is identical to that of the cholesterol-fed rabbits,
although the nature of the hypercholesterolaemia is different
[5]. However, the WHHL rabbits do not respond to calcium
antagonists (20 mg nifedipine twice a day) in the same way:
no reduction in the accumulation of cholesteryl ester is
found in the arteries of these animals [6]. It is clear
that if we use calcium antagonists as our major tool in
studying the interactions between calcium-dependent systems
and the major tissues involved in the development of atherosclerosis, we must be very aware of the limitations of the animal
models that are chosen.

Nevertheless, there is evidence that calcium-dependent
regulatory systems do have a role in the metabolism of
cells associated with the atherosclerotic plaque (and also
with the liver; see below). In this article we shall examine
some examples where links have been made between cholesterol
and lipoprotein and calcium-dependent systems.

Hypercholesterolaemia is a major contributor to the
development of atherosclerosis; but other factors are involved
in the growth of the plaque. Particularly important here
is the proliferation of smooth muscle cells and the migration
of macrophages into the arterial intima [3]. These events
may not be directly linked with cholesterol and lipoprotein
metabolism, but they are almost certainly calcium-dependent
events of importance, e.g. in association with growth factors.
We shall confine ourselves here to discussing the meeting
of cholesterol and lipoprotein metabolism and calcium-dependent systems.

IN VIVO STUDIES

In studying a complex disease such as atherosclerosis,
whole-animal studies in models of the disease are essential.
Many of the studies to be discussed here were performed
in simpler experimental systems such as cells in culture
and can give only an indication of what mechanisms may
operate. Care is needed in extrapolating them to the whole
animal. One interesting recent study in an animal model
of atherosclerosis has shown that in the normocholesterolaemic
rabbit, calcium antagonists (nifedipine, 10 mg/kg twice a
day) inhibit the proliferation of arterial smooth muscle
cells in the aorta injured with a balloon catheter. The
effect was not limited to organic calcium antagonists; lanthanum
produced a similar effect. It is interesting that antihyper-
tensives such as prazosin that are not calcium antagonists

had a similar effect, suggesting that other physiological
factors may be involved in the reduction of cell growth
[7]. Similar observations had been made before in the
cholesterol-fed rabbit. Thus some of the key events in
the pathology of atherosclerosis need not be linked to hypercholesterolaemia despite the weight of many *in vivo* studies,
including newer ones of increased complexity.

In a new method in which ^{45}Ca efflux was measured it
was shown that intracellular calcium increased >4-fold *vs.*
controls in the cholesterol-fed rabbit aorta, due to an
increase in permeability of cells to Ca^{2+} [8]. Nilvaldipine
at 0.01-10 mg/day produced dose-dependent inhibition of the
intimal thickening of rabbit carotid artery. The migration
of smooth muscle cells in a Boyden chamber was inhibited
by nilvaldipine with an IC_{50} of ~3.3 x 10^{11} M [9].

CELL STUDIES, AND THE CHOLESTEROL SCENE

To understand mechanisms we need to go from animal
models to simpler systems such as cell culture. The stimulated
growth of cultured aortic smooth muscle cells due to the
presence of hypercholesterolaemic serum was inhibited by
verapamil (50 µM) [10]. Earlier studies showed shorter-term
effects of calcium antagonists on these and other cell types
[11-13]. To discuss these and other studies in more detail
we need to outline the pattern of cholesterol metabolism in
the cells of interest to us [14].-

CELLULAR FUNCTIONS IN CHOLESTEROL METABOLISM

   A. *Sources of cholesterol metabolism*
Endogenous synthesis (HMG-CoA reductase).
Hydrolysis of intracellular stores of cholesteryl ester.
From blood: receptor-mediated uptake of lipoproteins
      - regulated (<u>e.g.</u> liver);
      - unregulated (<u>e.g.</u> <u>via</u> scavenger receptor).
Exchange of free cholesterol with plasma membrane.

   B. *Metabolic transformations and transport*
Esterification (ACAT).
Hydrolysis of esters (lysosomal & cytoplasmic
   cholesteryl ester hydrolase).
Secretion in lipoproteins (liver & intestine).
Other secretory mechanisms (reverse cholesterol transport).
Secretion into bile (liver).
Steroid hormone synthesis.

Cells need cholesterol for growth, since it is an important
component of membranes. Certain other cell types require
cholesterol for metabolic purposes, <u>e.g.</u> for conversion in

steroid hormone-producing tissues into their characteristic hormones and, in liver, for oxidation in the synthesis of bile acids. The latter process is of particular importance since it is the only way by which cholesterol can be lost from the body in quantity.

Cells can obtain cholesterol from several sources. One special case is the intestinal epithelial cell which obtains some cholesterol by absorption from the lumen of the intestine. Many cells take up cholesterol from the blood in lipoproteins through the mechanism of receptor-mediated endocytosis. This process is particularly well characterized for LDL [15]. The expression of the receptor on the cell surface is under metabolic control. Cells that are already rich in cholesterol down-regulate the number of receptors on the cell surface.

Alternatively, cholesterol can be synthesized in the cell, rate-limited by 3-hydroxy-3-methyl-glutaryl (HMG) CoA reductase. This enzyme has been thoroughly studied [16]. Cells can also take up free cholesterol from the blood by exchange. The quantitative importance of this process is not well understood, and it may be more significant than most workers acknowledge. Within the cell, cholesterol can follow a number of paths. These depend upon the function of the cell in cholesterol metabolism.

Recent studies in our laboratory have attempted to define the pathways open to cholesterol within liver and adrenal cortex [17-21]. In most cells excess cholesterol appears to be esterified with long-chain fatty acids such as oleic acid. This process is catalyzed by acyl CoA : cholesterol transferase (ACAT) [14]. Cholesteryl esters are a form of intracellular storage of cholesterol. In some cell types, particularly steroid hormone-producing cells, certain types of macrophage, and smooth muscle cells, the cholesteryl ester can be hydrolyzed, releasing the cholesterol for hormone synthesis or, in the arterial cells, for uptake by an extra-cellular acceptor. The enzyme catalyzing the hydrolysis, cholesteryl ester hydrolase, is cytoplasmic and is subject to regulation through a cAMP-dependent mechanism, being activated by phosphorylation [14]. Some cells can therefore exhibit a cycle of synthesis and hydrolysis of cholesteryl ester that can be modulated by factors which affect the intracellular cAMP concentration [14-21].

ALTERATIONS IN CHOLESTEROL METABOLISM

The deposition of excess cholesteryl ester that is charac-teristic of an early atherosclerotic lesion could result from excess uptake of LDL or other lipoproteins, e.g. a

modified LDL such as oxidized LDL [22]. Removal of the
accumulated cholesteryl ester may be promoted if ACAT activity
is inhibited and/or cholesteryl ester hydrolase activity is
enhanced. There is evidence that calcium antagonists can
influence the process of accumulation of cholesteryl ester.
This is inhibited by verapamil (50 μM) in cultured macrophages
[23, 24], not due to a direct effect on ACAT: the drug
was thought to prevent the transport of cholesterol to the
cellular site of esterification, the endoplasmic reticulum,
thus preventing the expected rise in measured ACAT activity.

Such a mechanism requires that the cells have a means
to relieve themselves of the additional unesterified cholesterol that results from the inhibition of the ACAT reaction.
This loss of excess cholesterol is thought to occur by
the uptake from cells by acceptors in the plasma such as
red blood cells or HDL. Nifedipine (2 μM) was shown to
increase cholesterol efflux from macrophages by a mechanism
independent of the presence of acceptor HDL [24]. The
drug increased the rate of uptake of modified LDL, and
stimulated the synthesis of apolipoprotein E, a component
of certain HDL fractions. The mechanisms of these effects
were not determined. In addition, an increase in intracellular
cAMP concentration was observed with increases in the activity
of cholesteryl ester hydrolase and of ACAT. However, in
bovine adrenal cortex ACAT activity falls in the presence
of factors that stimulate cholesteryl ester hydrolase activity
[14].

The results from the macrophage study [24] are not
entirely consistent with those from the adrenal cortex.
Possibly the influx of cholesterol from the modified LDL
into the macrophage is sufficient to overcome the inhibition
of ACAT activity that might otherwise be observed. Nifedipine
has also been reported to inhibit cholesterol synthesis directly
and also ACAT activity [12]. It has been suggested that
the inhibition of the deposition of cholesteryl ester in
cultured macrophages by calcium channel blockers is unrelated
to their calcium-entry-blocking effects [25].

In other experiments, compounds that stimulate the
production of intracellular cAMP, e.g. certain prostaglandins,
have been shown to stimulate cholesteryl ester hydrolase
activity but not ACAT [26]. In smooth muscle cells nitrendipine
has been shown to promote prostacyclin secretion [27] and
nifedipine causes an increase in cAMP with a corresponding
increase in cholesteryl ester hydrolase [28]. Unlike many
of the other drug studies described above, the drug was
used at low concentrations (i.e. 0.3 μM).

The uptake of LDL is also stimulated by calcium channel blockers. Verapamil (10-50 μM) increased the expression of LDL receptors on bovine aortic endothelial and smooth muscle cells [11]. Similar results were obtained with verapamil and diltiazem, but not with nifedipine or flunarizine, in human fibroblasts and HepG2 (a hepatoma cell line). Effects were also observed in fibroblasts from heterozygous familial hypercholesterolaemic patients but not in fibroblasts from homozygotes, these patients having an almost complete deficiency in LDL receptor activity [29]. These effects were shown to be due to the synthesis of LDL receptors since they could be blocked by cycloheximide. In fibroblasts calmodulin antagonists were shown to stimulate the synthesis of the LDL receptor [30]. Thus inhibition of calcium-dependent functions tended to stimulate the uptake of cholesterol into cells. This is paradoxical if the overall effect *in vivo* of blocking calcium channels is to promote efflux of cholesterol. Provided that the cell has a sufficient capacity for processing the cholesterol that enters it through the receptor-mediated pathway, i.e. for presenting it in a form that can be taken up by extracellular acceptors, such an increased uptake would present no problem. However, if the secretion of cholesterol were inhibited, then it is clear that calcium-directed drugs could lead to an undesirable net accumulation of cholesterol in the artery.

INVOLVEMENT OF CALCIUM

In many of these studies a direct link with cellular calcium concentrations is clearly missing. Most of the published effects and hypotheses deal with effects of drugs acting on calcium channels and to some extent connect these with other second-messenger systems. There is a need for establishing the molecular basis for the phenomena described in the literature in more detail. The concentrations and doses of calcium antagonists used in many of the studies we have discussed are higher than those required for their normal pharmacological action. Different classes of calcium antagonists, and different compounds within a class, can have varying effects in a system such as cholesteryl ester deposition in cultured cells [25, 29]. Conceivably some of the observed effects are due to phenomena that do not depend directly upon the calcium-entry blocking activity of these compounds [6, 25, 31, 32].

The cells of the artery have naturally attracted the greatest interest since it is here that the atherosclerotic lesion develops. However, it is in the liver that most of the endogenous synthesis of cholesterol takes place and that cholesterol is eliminated from the body. A number of studies have provided connections between cholesterol metabolism and calcium-dependent processes in the liver. Cholesterol synthesis is known to be regulated through synthesis of and degradation of the enzyme HMG-CoA reductase. This occurs through regulation of the expression of the gene for this enzyme, similarly to expression of the gene for the LDL receptor.

A short-term regulatory mechanism also exists by which HMG-CoA reductase is inhibited by phosphorylation by a specific kinase [16]. Interestingly, the kinase system that regulates cholesterol synthesis has recently been shown also to regulate fatty acid synthesis, another acetylCoA-consuming process [33]. This process is not controlled directly through cAMP: the effects of agents such as glucagon that increase the intra-cellular cAMP level are probably exerted on the protein phosphatase that hydrolyzes the phospho-HMG-CoA reductase [16]. The phospho-form of the enzyme is thought to be more susceptible to degradation; hence covalent modification of the enzyme leads to its inactivation in two ways. *In vitro* evidence shows that several protein kinases including protein kinase C and the calmodulin-dependent protein kinase can also phos-phorylate HMG-COA reductase, also leading to inactivation of the enzyme. The effects of these other kinases have not yet been demonstrated *in vivo* or in intact cells; hence their relevance to the normal physiological process must still be proved [16]. Calmodulin antagonists have also been shown to inhibit later stages of cholesterol synthesis at the Δ^{24}-reductase step [34].

Another regulatory system in the liver that appears to be sensitive to calcium is the rate-limiting enzyme of bile acid synthesis, cholesterol 7α-hydroxylase. Evidence in the literature, by no means uncontroversial, suggests that this enzyme is stimulated by phosphorylation. Recent studies suggest that the phosphatase which deactivates this enzyme may be sensitive to intracellular calcium [35].

Other hepatic systems in which calcium has been implicated require the participation of many components of the cell. The complex cellular response that prevents lipoprotein-derived cholesterol from reaching ACAT as discussed above is another example of these more broadly-based effects. VLDL, the main lipoprotein secreted by the liver, is packaged in several stages by the cell's secretory apparatus, from the endoplasmic reticulum to the Golgi stack, and this secretion is inhibited by calcium antagonists (verapamil, 0.2 mM; EGTA, 5 mM). Co^{2+} had a similar effect at concentrations up to 8 mM. The evidence suggested that calcium antagonists inhibited the secretory pathway [36]. This is clearly a more complex phenomenon than the more precisely targeted effects on enzyme that we have just considered, and the concentrations of drugs used in the studies are very high.

Another complex system is the secretion of bile, which is a mixture of unesterified cholesterol, bile acid and phosphatidylcholine. cAMP has been shown by several groups to stimulate the secretion of bile acids from hepatocytes in primary culture [37-39]. This effect is blocked by the presence of verapamil (40 μM) but not by the calcium ionophore A23187 (1 μM) [40]. As with many other of the studies we have just outlined, the precise molecular mechanism of this effect is unknown.

CONCLUSIONS

The studies reviewed here show that a single, widely used experimental tool, the calcium antagonists, has provided a wealth of circumstantial evidence for the interaction of cholesterol metabolism with calcium-dependent systems in the development of hypercholesterolaemia and atherosclerosis. Some mechanisms may be regarded as plausible, but not proved. Changes in the calcium concentration in macrophages and smooth muscle cells may promote the hydrolysis of cholesteryl ester and its release to the circulation. The primary effect that stimulates these responses is not known. These events could be related to the described effects on lipoprotein receptors which clearly occur through protein synthesis and degradation.

From a methodological point of view several broad conclusions can be reached. More penetrating experimental approaches than the pharmacological need to be attempted, but it is important that a relevant model system be chosen. Cell culture systems will be suitable for some purposes, but the complexities of the disease process necessitate also a more integrated system. This should allow studies to be focused on regulation in the unusual subendothelial environment that is characteristic of the developing atherosclerotic plaque.

Acknowledgement

I thank Trevor Hallam for his helpful comments from another perspective.

References

1. National Institutes of Health Consensus Development Conf.
    Statement (1985) *J. Am. Med. Ass. 253*, 2080-2086.
2. Blankenhorn, D.H., Nessim, S.A., Johnson, R.L.,
    Sanmarco, M.E., Azer, S.P. & Cashin-Hemphill, L. (1987)
    J. Am. Med. Ass. 257, 3233-3240.
3. Ross, R. (1986) *New Engl. J. Med. 314*, 488-500.
4. Weinstein, D.B. & Heider, J.G. (1987) *Am. J. Cardiol. 59*,
    163B-172B.
5. Rosenfeld, M.E., Tsukada, T., Gown, A.M. & Ross, R.
    (1987) *Arteriosclerosis 7*, 9-23.
6. Watanabe, N., Ishikawa, Y., Okamoto, R. & Fukuzaki, H.
    (1987) *Artery 14*, 283-294.
7. Jackson, C.L., Bush, R.C. & Bowyer, D.E. (1987)
    Atherosclerosis 69, 115-122.
8. Strickberger, S.A., Russek, L.N. & Phair, R.D. (1988)
    Circ. Res. 62, 255-261.
9. Nomoto, A., Hirosumi, J., Sekiguchi, C., Mutch, S.,
    Yamaguchi, I. & Aoki, H. (1987) *Atherosclerosis 64*, 255-261.
10. Stein, O., Halperin, G. & Stein, Y. (1987)
    Arteriosclerosis 7, 585-592.
11. Stein, O., Leitersdorf, E. & Stein, Y. (1985)
    Arteriosclerosis 5, 35-44.
12. Shirai, K., Takeshiro, H., Morisaki, N., Saito, Y. &
    Yoshida, H. (1986) *Ther. Res. (Japan) 5*, 1098-1102.
13. Nilsson, J., Sjölund, M., Palmberg, L., Von Euler, A.M.,
    Jonzon, B. & Thyberg, J. (1985) *Atherosclerosis 58*, 109-122.
14. Suckling, K.E. & Stange, E.F. (1985) *J. Lipid Res. 26*,
    647-671.
15. Brown, M.S. & Goldstein, J.L. (1986) *Science 232*, 34-46.
16. Beg, Z.H., Stonik, J.A. & Brewer, H.B. (1987) *Metabolism
    36*, 900-917.
17. Sampson, W.J., Suffolk, R.A., Bowers, P., Houghton, J.D.,
    Botham, K.M. & Suckling, K.E. (1987) *Biochim. Biophys.
    Acta 920*, 1-8.
18. Sampson, W.J., Botham, K.M., Jackson, B. & Suckling, K.E.
    (1988) *FEBS Lett. 227*, 179-182.
19. Sampson, W.J., Houghton, J.D., Bowers, P., Suffolk, R.A.,
    Botham, K.M., Suckling, C.J. & Suckling, K.E. (1988)
    Biochim. Biophys. Acta 960, 268-274.
20. Ochoa, B. & Suckling, K.E. (1987) *Biochim. Biophys. Acta
    918*, 159-167.
21. Jamal, Z., Suffolk, R.A., Boyd, G.S. & Suckling, K.E.
    (1985) *Biochim. Biophys. Acta 834*, 230-237.
22. Mitchison, M.J., Ball, R.Y., Carpenter, K.L.H. &
    Parums, D.V. (1988) in *Hyperlipidaemia and Atherosclerosis*
    (Suckling, K.E. & Groot, P.G., eds.), Academic Press,
    London,   pp. 117-134.

23. Bujo, H., Morisaki, N., Shirai, K., Saito, Y. & Yoshida, Y. (1987) *Domyaku Koka 15*, 179-184.
24. Schmitz, G., Robenek, H., Beuck, M., Krause, R., Shurek, A. & Niemann, R. (1988) *Arteriosclerosis 8*, 46-56.
25. Daugherty, A., Rateri, D.L., Schonfeld, G. & Sobel, B.E. (1987) *Br. J. Pharmacol. 91*, 113-118.
26. Hajjar, D.P. & Salisbury, B.G.J. (1986) *Path. Immunopath. Res. 5*, 437-454.
27. Grodzinska, L., Basista, M., Basista, E., Slawinski, M., Swies, J., Stachura, J. & Ohlrogge, R. (1988) *Arzneimittelforsch. 37*, 412-415.
28. Etingin, O.R. & Hajjar, D.P. (1985) *J. Clin. Invest. 75*, 1554-1558.
29. Paoletti, R., Bernini, F., Fumagalli, R., Allorio, M. & Corsini, A. (1988) *Ann. N.Y. Acad. Sci. 522*, 390-398.
30. Eckardt, E., Filipovic, I., Hasĩlik, A. & Buddecke, E. (1988) *Biochem. J. 252*, 889-892.
31. Heider, J.G., Weinstein, D.B., Pickens, C.E., Lan, S. & Su, C-M. (1987) *Transplantation Proc. 19 Suppl. 5*, 96-101.
32. Chobanian, A. (1987) *J. Hypertension 5 (Suppl. 4)*, S43-S48.
33. Carling, D., Zammit, V.A.& Hardie, D.G. (1987) *FEBS Lett. 223*, 217-222.
34. Filipovic, I. & Buddecke, E. (1987) *Lipids 22*, 261-265.
35. Holsztynska, E.J. & Waxman, D.J. (1987) *Arch. Biochem. Biophys. 256*, 543-549.
36. Nossen, J.O., Rustan, A.C.& Drevon, C.A. (1987) *Biochem. J. 247*, 433-439.
37. Botham, K.M. & Boyd, G.S. (1983) *Eur. J. Biochem. 136*, 313-319.
38. Sundaram, G.S., Rothman, V. & Margolis, S. (1983) *Lipids 18*, 443-447.
39. Botham, K.M. & Suckling, K.E. (1986) *Biochim. Biophys. Acta 889*, 382-385.
40. Botham, K.M. (1987) *Biochim. Biophys. Acta 922*, 46-53.

#E-6

CALCIUM FLUXES ON REOXYGENATION OF MYOCARDIAL TISSUE

P.A. Poole-Wilson

National Heart and Lung Institute,
Dovehouse Street, London SW3 6LY, U.K.

This article bears on two questions concerning the understanding of myocardial ischaemia: (1) whether the apparent myocardial damage is an inherent consequence of ischaemic events or results from biochemical events at the moment of reperfusion, and (2) how the latter events come about, and what role they play in inducing cell necrosis. For the ultimate therapeutic aim of such work, animal models may be questionable. Some features of reperfusion damage have recently been shown in man.

Conclusive answers to these questions are hard to obtain, for reasons which are considered. The most persuasive evidence for the existence and importance of reperfusion damage comes from experiments where interventions (particularly low $[Ca^{2+}]_e^*$) at the time of reperfusion alter the subsequent recovery of the myocardium. The major mechanisms currently proposed to account for reperfusion damage are ionic exchange mechanisms and the generation of radicals which then either damage cell-membrane lipids or possibly actual channel proteins. Pertinent approaches and difficulties in interpretation are discussed.

A large net gain of calcium is evident in heart muscle on reperfusion or reoxygenation of the muscle after a prolonged period of ischaemia or hypoxia [1-5]. The magnitude of the net gain is related to the recovery of normal systolic and diastolic muscle tension. Numerous authors have argued that calcium overload in heart muscle is closely linked to the development of myocardial necrosis in the context of ischaemia or hypoxia. There is some evidence that calcium overload is an important mechanism in other pathological states such as hypertrophy, heart failure or cardiomyopathy [1]. Calcium overload may be an accompaniment of necrosis of

---

*Usual nomenclature for free cations, thus: $[Ca^{2+}]_e$ = extracellular Ca^{2+} concentration; Ca^{2+}_i = intracellular (cytosolic).
Abbreviations.- AM, acetoxymethyl ester (of the probe indo-1); BAPTA, 1,2-*bis*(O-aminophenoxy)ethane-N,N,N',N'-tetraacetic acid.

myocardial tissue, but the accumulation of small amounts of calcium does not always lead to cell death [6], and other mechanisms may cause tissue necrosis in heart muscle in the absence of calcium overload.

Controversy surrounds the pathological mechanism of calcium accumulation [2, 7, 8]. If the mechanism were known then it might be possible to prevent necrosis. This would be particularly relevant to the protection of the heart following thrombolytic therapy for myocardial infarction. A current therapeutic objective is to delay tissue necrosis during a period of ischaemia or to inhibit those processes linked to the development of necrosis at the time of reperfusion.

CALCIUM EXCHANGE DURING HYPOXIA

Calcium fluxes have been studied during and after hypoxia or ischaemia in animals *in vivo* and in many isolated myocardial preparations *in vitro* [1-5]. The experimental conditions have varied greatly. Important variables are the duration of the hypoxic or ischaemic period, the amount of tissue subserved by the occluded artery, temperature, heart rate, perfusion with blood or physiological solution, rapid or slow reperfusion, the presence of collaterals, and the nature of the substrate present in the perfusate [9-11].

Despite the confounding effects of these variables, similar results have been reported. During hypoxia or low-flow ischaemia, the total tissue calcium in heart muscle does not increase [1, 12]. If the period of hypoxia or ischaemia is so prolonged that resting tension increases substantially and no recovery occurs on reoxygenation or reperfusion, then there is an increase in tissue calcium [13]. On reperfusion or reoxygenation tissue calcium increases immediately. The recovery of mechanical function is an indirect measure of tissue necrosis and relates to the size of the calcium gain. This gain is due to increased influx of Ca^{2+} and not to a reduced efflux [4, 9, 12]. The calcium gain occurs at a time when extracellular markers do not have access to the intracellular fluid volume. Furthermore, an efflux of K^+ does not occur at the moment of reperfusion when calcium gain is evident, again suggesting that the sarcolemma is intact [14].

The Ca^{2+} influx is not affected by alteration of heart rate, by depolarization, by the presence of a higher $[K^+]_e$, or by Ca^{2+}-antagonists [12]. Increased calcium accumulation is inhibited by the presence of nickel or cyanide [1]. It is suggested, in relation to this finding, that nickel inhibits the movement of Ca^{2+} through the sarcolemma, although it is possible that nickel enters the cell and blocks the uptake of Ca^{2+} by mitochondria. The removal of Ca^{2+} from the extracellular space at the time of reperfusion [15, 16] prevents Ca^{2+} uptake, and on restoring $[Ca^{2+}]_e$ to normal there is an improved recovery of both function and tissue ATP and creatine phosphate.

These observations have led to the hypothesis that during ischaemia, low-flow ischaemia or hypoxia, mechanisms prevent the uptake of Ca^{2+}. Conceivably the low intracellular pH inhibits the movement of Ca^{2+} into mitochondria where most of the accumulation occurs, or the carrier mechanism for the entry of Ca^{2+} into the mitochondria is not functional because of the low concentration of ATP. On reperfusion or reoxygenation a channel is opened so that Ca^{2+} moves freely into the cell and is taken up by the mitochondria. It is possible that such a channel is open during ischaemia and that the Ca^{2+} moves into the cytosol but is only able to be taken up by the mitochondria on reperfusion or reoxygenation. The key questions are the nature of the channel through which the Ca^{2+} passes, how it might be modified, and the $[Ca^{2+}]_i$ during ischaemia and on reperfusion.

MECHANISMS FOR Ca^{2+} INFLUX ON REPERFUSION AND REOXYGENATION

Three hypotheses have been put forward. (1) Radicals generated on reperfusion may alter the lipids in the cell membrane or even the channel proteins themselves. (2) Na^+ may accumulate in the myocardium during hypoxia or ischaemia and on reperfusion Na^+/Ca^{2+} exchange is stimulated [17]. (3) Proteases and lipases may modify the cell membrane or channel proteins during ischaemia or hypoxia and for some reason allow Ca^{2+} influx to occur on reperfusion or reoxygenation [18].

Radicals are generated on reperfusion of the myocardium. Radical-generating systems can mimic the Ca^{2+} influx which occurs on reperfusion. The systems within cardiac muscle which normally protect the cell from radical damage have a reduced functional capacity after a period of ischaemia [19]. Agents such as superoxide dismutase and catalase which might be expected to reduce radical damage have been reported to have both a beneficial or no effect [20]. The source of radicals may be the xanthine oxidase system in endothelial cells, or white cells although Ca^{2+} influx is demonstrable in experimental models in the absence of white cells and in species where xanthine oxidase is absent. Other sources of radicals are mitochondria and the cell membrane itself.

Lazdunski et al. [17] have proposed that the extracellular acidosis during ischaemia inhibits Na^+/H^+ exchange. On reperfusion when the extracellular pH rises, protons are ejected from the cell in exchange for Na^+. The accumulation of Na^+ within the cell stimulates Na^+/Ca^{2+} exchange, thus accounting for Ca^{2+} influx. A major difficulty with this hypothesis is that Ca^{2+} influx occurs on reoxygenation as on reperfusion and in the former extracellular acidosis is mild or absent. However, there may exist alternative control mechanisms which allow Na^+/Ca^{2+} exchange to take place at the moment of reperfusion and reoxygenation. Ca^{2+} influx on reoxygenation is influenced by the extracellular Na^+. Some authors have been unable to demonstrate Na^+ efflux at the time of Ca^{2+} influx [21], and recent unpublished work indicates that Ca^{2+} influx is not inhibited by lithium. $[Na^+]_i$ has been measured in a variety of tissues from several species under different conditions. In general, after a substantial period of ischaemia or hypoxia $[Na^+]_i$ has been shown to increase [22]. This means that during low-flow ischaemia or hypoxia some mechanism must exist for inhibiting Na^+/Ca^{2+} exchange.

CYTOSOLIC CALCIUM CONCENTRATION, $[Ca^{2+}]_i$

A key variable in any hypothesis accounting for Ca^{2+} influx on reoxygenation or reperfusion is $[Ca^{2+}]_i$ during ischaemia and on reperfusion. Current data are complex and conflicting [1]. Using aequorin to detect Ca^{2+}_i, Ca^{2+} in the ferret has been reported not to change early in hypoxia [23] but to increase later or on reoxygenation. Cell shortening in isolated single myocytes may precede the rise in $[Ca^{2+}]_i$ [24]. During ischaemia [25, 26] and measuring $[Ca^{2+}]_i$ with BAPTA in the ferret, $[Ca^{2+}]_i$ does not rise early in ischaemia but increases 4-fold after 20 min and declines on reperfusion. In the rabbit using indo-1/AM to detect Ca^{2+}, $[Ca^{2+}]_i$ increased by one-quarter within 90 sec of the onset of ischaemia and declined on reperfusion [27, 28]. At present there is no agreement between authors using similar or different methods on precisely how diastolic or systolic $[Ca^{2+}]_i$ alters during hypoxia or ischaemia. Numerous factors affect these two variables, in particular intracellular pH and phosphate concentration.

STUDIES ON ISOLATED MYOCYTES

Many studies have been reported in the last few years on the effect of hypoxia or metabolic poisons on isolated myocytes ([29-34] & this vol., especially #E-7 - Ver Donck). Two major problems exist with this preparation. (1) Because oxygen consumption is low the effects of hypoxia are difficult to demonstrate in myocytes, particularly if quiescent. (2) Some of the effects observed may be due to the damage caused to the cell membrane and the intercalated disc when the extracellular matrix is disrupted in the preparation of cells. For example, many of these cells contract with shortening, beginning at the intercalated disc. This may be related to damage at the disc allowing Ca^{2+} influx. Blebbing is a common phenomenon in isolated myocytes, and although blebbing has been reported in intact tissue it does not seem to be a common phenomenon.

CONCLUSIONS

On reoxygenation of the myocardium there is a net accumulation of calcium which is related to the development of cell necrosis. The net gain of calcium is due to an increased influx, whose molecular basis has still not been determined. The major current hypotheses are that it is the result of damage to the cell membrane or to channel proteins by the action of lipases, oxygen radicals or cell swelling, or alternatively that Ca^{2+} passes into the cell by physiological mechanisms stimulated by $[Na^+]_i$ accumulation.

References

1. Poole-Wilson, P.A., Harding, D.P., Bourdillon, P.D.V. & Tones, M.A. (1984) *J. Mol. Cell Cardiol. 16*, 175-187.
2. Poole-Wilson, P.A. (1987) *Clin. Physiol. 7*, 439-453.
3. Hearse, D.J. (1977) *J. Mol. Cell Cardiol. 8*, 605-616.
4. Shen, A.C. & Jennings, R.B. (1972) *Am.J. Path. 67*, 441-452.
5. Fleckstein, A., Janke, J., Doring, H.J. & Leder, O. (1984) in *Recent Advances in Studies on Cardiac Structure and Metabolism* (Dhalla, N.S., ed.), Univ. Park Press, Baltimore, 563-580.
6. Murphy, E., Aiton, J.F., Horres, C.R. & Liebermann, M. (1983) *Am. J. Physiol 245*, C316-C321.
7. Allen, D.G. & Orchard, C.H. (1987) *Circ. Res. 60*, 153-168.
8. Mak, I.T., Kramer, J.H. & Weglicki, W.B. (1986) *J. Biol. Chem. 261*, 1153-1157.
9. Harding, D.P. & Poole-Wilson, P.A. (1980) *Cardiovasc. Res. 14*, 435-445.
10. Yamazaki, S., Fujibayashi, Y., Rejogopalan, R.E. & Corday, E. (1986) *J. Am. Coll. Cardiol. 7*, 564-572.
11. Schaper, W. (1984) in *Therapeutic Approaches to Myocardial Infarct Size Limitation* (Hearse, D.H. & Yellon, D.M., eds.), Raven Press, New York, pp. 79-90.
12. Bourdillon, P.D.V. & Poole-Wilson, P.A. (1982) *Circ. Res. 50*, 360-368.
13. Nayler, W.G., Poole-Wilson, P.A. & Williams, A. (1979) *J. Mol. Cell Cardiol. 11*, 683-706.
14. Crake, T. & Poole-Wilson, P.A. (1986) *J. Mol. Cell Cardiol. 18 (Suppl. 4*, 31-35.
15. Shine, K.I. & Douglas, A.M. (1983) *J. Mol. Cell Cardiol. 15*, 252-260.

16. Kuroda, H., Ishiguro, S. & Mori, T. (1986) *J. Mol. Cell Cardiol. 18*, 625-633.

17. Lazdunskai, M., Frelin, C. & Vigne, P. (1985) *J.Mol. Cell Cardiol. 17*, 1029-1042.

18. Guarnieri, C., Flamigni, F. & Calderera, C.M. (1980) *J. Mol. Cell Cardiol. 12*, 797-808.

19. Poole-Wilson, P.A. & Tones, M.A. (1987) in *Myocardial Ischaemia* (Dhalla, N.S., Innes, I.R. & Beamish, R.E., eds.), Martinus-Nijhoff, Berlin, pp. 123-131.

20. Poole-Wilson, P.A. & Tones, M.A. (1988) *J. Mol. Cell Cardiol. 20 (Suppl. II)*, 15-22.

21. Kleber, A.G. & Wilde, A.A.M. (1986) *J. Mol. Cell Cardiol. 18 (Suppl. 4)*, 27-30.

22. Allen, D.G. & Orchard, C.H. (1983) *J. Physiol. 339,* 107-122.

23. Cobbold, P.H. & Bourne, P.K. (1984) *Nature 312*, 444-446.

24. Steenbergen, C., Murphy, E., Levy, L. & London, R.E. (1987) *Circ. Res. 60*, 700-707.

25. Marban, E., Kitakaze, M., Kusuoka, H., Porterfield, J.K. & Yue, D.T. (1987) *Proc. Nat. Acad. Sci. 84,* 6005-6009.

26. Lee, H-C., Smith, N., Mohabir, R. & Clusin, W.T. (1987) *Proc. Nat. Acad. Sci. 84,* 7793-7797.

27. Lee, H-C., Mohabir, R., Smith, N., Franz, M.R. & Clusin, W.T. (1988) *Circulation 78*, 1047-1059.

28. Piper, H.M., Schwartz, P., Spahr, R., Hutter, J.F. & Spieckermann, P.G. (1984) *Pflügers Archiv. 401*, 71-76.

29. Altona, J.C., Van der Laarse, A., Bloy, S. & Van Treslong, C.H.F. (1984) *Cardiovasc. Res. 18*, 99-106.

30. Altschuld, R.A., Hostetler, S.H. & Brierley, G.P. (1981) *Circ. Res. 49*, 307-316.

31. Borgers, M. & Piper, H.M. (1986) *J. Mol. Cell Cardiol. 18,* 439-448.

32. Cheung, J.Y., Leaf, A. & Bonventre, J.V. (1986) *Am. J. Physiol. 250*, C18-C25.

33. Schwartz, P., Piper, H.M., Spahr, R. & Spieckermann, P.G. (1984) *Am. J. Path. 115*, 349-361.

34. Silverman, H., Capogrossi, M.C., Nichols, C., Lakatta, E.G., Lederer, W.J. & Stern, M. (1988) *J. Mol. Cell Cardiol. 20 (Suppl. 5)*, S.36.

#E-7

ROLE OF CALCIUM IN PATHOPHYSIOLOGY OF ISOLATED CARDIOMYOCYTES: PROTECTION BY CALCIUM ANTAGONISTS

Luc Ver Donck, Godelieve Vandeplassche and Marcel Borgers

Department of Cardiovascular Pathophysiology,
Janssen Research Foundation,
Turnhoutseweg 30, B-2340 Beerse, Belgium

Single-cell preparations from several organs are now widely used in various fields of fundamental research, including cardiovascular investigations where, as considered in this article, isolated cardiomyocytes have already proved their value. Cardiomyocytes, isolated enzymatically from adult rat heart, were subjected to various pathological stimuli: veratrine, activated oxygen species, digitalis-intoxication or charged amphiphiles. Each stimulus induced a very characteristic shape change from elongated rod-shaped to hypercontracted round cells. Depending on the nature of the pathological stimulus applied, shape changes of nearly all cardiomyocytes occurred within 1 to 60 min, and were dependent upon an influx of extracellular Ca^{2+} suggesting the occurrence of intracellular Ca^{2+}-overload. Pre-treatment of the cardiomyocytes with Ca^{2+}-antagonists resulted in prevention of shape changes, as assessed by determining the % of rod-shaped cells remaining after exposure to the stimulus. A marked difference in efficacy of the Ca^{2+}-antagonists was observed: slow-channel blockers (e.g. verapamil) were less potent when compared to Ca^{2+}antagonists not having affinity for the slow Ca^{2+}-channel.

Numerous experimental *in vivo* and *in vitro* models are available for use in diverse studies. In cardiovascular research, models using *in situ* working hearts as well as isolated perfused hearts and excised muscle strips are commonly used. These experimental models have provided many valuable findings and new insights. Yet a number of questions have remained unanswered and fostered further exploration of the basic constituent of the myocardium and other organs, viz. the single cell.

In 1976 a pioneer paper by Powell & Twist [1] described the isolation of viable Ca^{2+}-tolerant single cardiomyocytes from adult rat hearts. Their technique has been copied and extended in many laboratories in order to investigate various problems in cardiology, which generally could not be examined in whole hearts. Indeed, single-cell preparations offer several advantages over intact organs: the homogeneity of the population allows evaluation of the characteristics of individual cells in the absence of external influences such as neuronal and hormonal modulation; in addition, myocyte-microvasculature interactions and mechanical influences of neighbouring tissues are eliminated.

In contrast to multicellular preparations, in which it is very difficult to obtain adequate potential control for electrophysiological studies, voltage-clamp studies can easily be performed on individual cardiomyocytes. Isolated cells may also provide material of excellent quality for biochemical and subcellular fractionation studies. Finally, since millions of cells are obtained from a single heart, the use of isolated cardiomyocytes substantially reduces the number of laboratory animals needed to obtain a given amount of experimental information.

This article describes some experimental set-ups with the isolated cardiomyocytes model in order to investigate the pathophysiology of Ca^{2+} in the myocardium and its application in pharmacology, _i.e._ evaluation of cardioprotection by Ca^{2+}-antagonists.

EXPERIMENTAL

Isolation of cardiomyocytes

Fig. 1 shows schematically the procedure used to isolate single myocardial cells from adult heart [2]. In brief, the heart is quickly excised from the anaesthetized animal and perfused for 5 min with nominally Ca^{2+}-free KRH buffer [(mM) NaCl, 125; KCl, 2.6; KH_2PO_4, 1.2; $MgSO_4.7H_2O$, 1.2; HEPES, 10; glucose, 5.5; pH 7.4] gassed with 100% O_2. Then the heart is perfused for 30 min with KRH buffer supplemented with 25 μM $CaCl_2$, 0.1% bovine serum albumin (BSA, fatty acid-free, fraction V, Calbiochem) and 0.06% crude collagenase (Wako Chemicals). Next, the heart is cut up into small pieces and the cells are separated from each other by gentle mechanical agitation, centrifuged (25 **g**, 90 sec), washed and resuspended in KRH containing 50 μM $CaCl_2$ and 2% BSA (Serva). Then the Ca^{2+} concentration is raised stepwise to 0.5 mM. The cells are layered twice over a 10 cm high column of KRH containing 1 mM $CaCl_2$ and 2% BSA, and allowed to sink (10 min), then washed several times with BSA-free KRH + 1 mM $CaCl_2$.

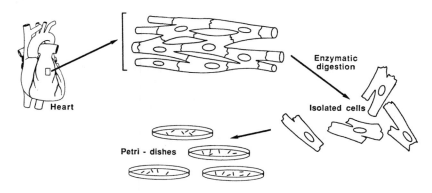

Fig. 1. Scheme for isolating Ca^{2+}-tolerant cardiomyocytes from adult hearts (hamster, rat, cat, rabbit) and attaching them to petri-dishes. See text for amplification.

Aliquots of the cell suspension are placed in petri-dishes (Falcon) and allowed to attach to their bottoms for 5-15 min. Finally, unattached cells are washed out and the dishes with firmly attached rod-shaped cells are stored in an O_2-chamber prior to the experiments. This procedure, with a suitable enzymatic perfusion flow-rate, is in routine use for several species: rat, hamster, cat and rabbit at, respectively, 9, 3.5, 38 and 38 ml/min. Investigations reported here all refer to rat cardiomyocytes.

Pathological stimuli

To investigate the role of Ca^{2+} in the pathophysiology of the isolated cardiomyocyte, a series of experimental conditions inducing intracellular Ca^{2+}-overload was developed. The criterion is the characteristic shape change from a rod-shaped to a hypercontracted cardiomyocyte when it becomes overloaded with Ca^{2+}. Standard incubation conditions in all the experiments were: KRH with 1 mM $CaCl_2$, pH 7.4 at room temperature.

(a) Veratrine.- This alkaloid (Sigma, 100 µg/ml), a toxin prolonging the open state of the Na^+-channel [4, 5], is added to the incubation medium for 5 min [2, 3, 5].

(b) Digitalis intoxication.- Ouabain (Sigma, 100 µM) is added to the KRH buffer and the cardiomyocytes are paced to synchronous rhythmic contractions by electrical pulses, generated by a home-built stimulator (50 V, 10 msec pulse duration, 3 Hz) and guided in the medium <u>via</u> Pt electrodes (Fig. 2A); stimulation is maintained for 25 min [5].

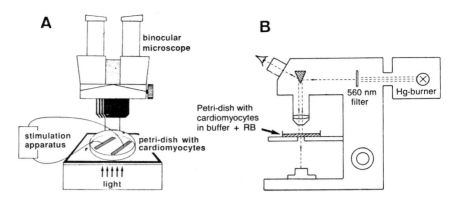

Fig. 2. Equipment for certain experiments (during which the cardiomyocytes can be observed throughout). See text.
A: electrical stimulation (in digitalis intoxication).
B: O_2^1 generation in the cell incubation medium by photo-oxygenation of rose bengal.

(c) **Activated O_2 species** are generated *in vitro* in two different ways:
- (1) Hypoxanthine (Sigma, 1 mM in the incubation medium) is enzymatically reduced by xanthine oxidase (Biozyme Labs., U.K.; 0.025 U/ml), yielding uric acid and superoxide anion (O_2^-). Superoxide immediately dismutates to H_2O_2 which generates hydoxyl radical ('OH) in the presence of transition metals [6]. Cardiomyocytes are exposed to the free radical-generating system for 60 min [5].
- (2) Photo-oxygenation of the light-sensitive dye rose bengal (Janssen Chimica) yielded singlet oxygen (O_2^1). Rose bengal (50 nM) is added to the incubation medium and the petri-dish with cardiomyocytes is placed on the stage of a fluorescent microscope (epifluorescence mode; Orthoplan, Leitz) and illuminated through the objective lens (4 ×), by 560 nm filtered light from a Hg-burner (50 W, Osram) (Fig. 2B), for 45–65 sec; the effect is observed 5 min after the onset of illumination [7].

(d) **Charged amphiphiles.**– Cardiomyocytes in petri-dishes are exposed for 60 min to Polymyxin B (100 µM), an amphiphilic compound with a positively charged head group [8].

Pharmacological investigations

In view of the central role of Ca^{2+} in these pathological stimuli (see RESULTS, below) the effects of various representatives of Ca^{2+}-antagonists were investigated (Table 1). The dihydropyridines nifedipine and nicardipine were dissolved

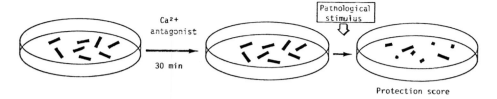

Fig. 3. Experimental protocol for investigating effects of Ca^{2+}-antagonists against Ca^{2+}-overload in cardiomyocytes induced by various pathological stimuli.

Table 1. Classification [9] of the Ca^{2+}-antagonists examined.

SELECTIVE	*NON-SELECTIVE*	*for slow Ca^{2+} channels*
Nifedipine	Flunarizine	
Nicardipine	Cinnarizine	
Verapamil	Lidoflazine	
Diltiazem		

in 100% ethanol (10 mM stock solution) and handled in the dark throughout. For other compounds the stock solution (1 mM) was in water containing 2.5% lactic acid and 2.5% ethanol. Appropriate dilutions were made in the incubation medium.

Fig. 3 indicates the protocol for evaluating the effects of Ca^{2+}-antagonists, each kept 30 min at room temperature with the cardiomyocytes in petri-dishes prior to exposure to the pathological stimulus. The drug effect is evaluated by scoring the number of rod-shaped cells in a well-marked area of the dish (~100–200 cells in view) before and after the stimulus. The % ratio for rod-shaped cells ([after/before] × 100) is the index for effect comparisons.

RESULTS

Cell isolation.- The isolation procedure generally yielded 60–90% viable Ca^{2+}-tolerant quiescent rod-shaped cardiomyocytes. Seeding the cell suspension in petri-dishes resulted in a selection of rod-shaped cells by washout of Ca^{2+}-intolerant rounded cells: a population of nearly 100% Ca^{2+}-tolerant cells was obtained.

Effect of pathological stimuli.- Cells exposed to one of these underwent a characteristic change from a rod-like to a hypercontracted rounded shape (Fig. 4A). This shape change coincided with vigorous uncontrolled contractions and was dependent upon the presence of extracellular Ca^{2+}: shape

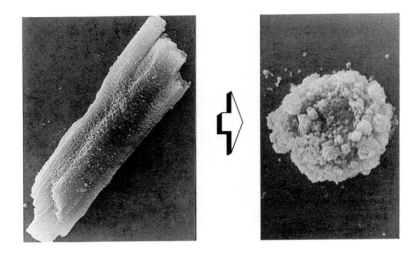

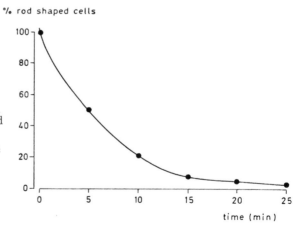

Fig. 4. Effect of a pathological stimulus on the morphology of an isolated rat cardio-myocyte. **A,** *above:* scanning e.m. (×920) of the shape change (rod→hypercontracted rounded; see text). **B,** *right:* time course of shape change in a population of isolated cardiomyocytes during digitalis intoxication.

changes were not observed in Ca^{2+}-free media ·(EGTA added to 0.1 mM), even after prolonged exposure to the pathological stimulus. This suggests that intracellular Ca^{2+}-overload takes place at the same time as the shape change. Cells did not all hypercontract at the very same moment, but there was a certain staggering of shape change among cells, especially evident when pathological stimuli with longer exposure times (>5 min) were applied. Fig. 4B exemplifies such a shape-change time course in a cell population challenged by digitalis intoxication.

Table 2. Cells exposed to pathological stimuli: % protection (mean ±S.E.M.) by Ca^{2+}-antagonists [concentration, M (& n, = no. of experiments) stated in heading for each stimulus].

Stimulus: [Antagonist]:	Veratrine $[2 \times 10^{-7}]$ (n = 6)	Ouabain $[2 \times 10^{-7}]$ (n = 4)	Superoxide anion $[5 \times 10^{-6}]$ (n = 6)	Singlet oxygen $[2 \times 10^{-7}]$ (n = 6)	Polymyxin B $[1 \times 10^{-6}]$ (n = 5)
Solvent	6.0 ±1.7	4.0 ±0.7	2.0 ±0.9	3.0 ±1.4	8.6 ±1.8
Flunarizine	24.2 ±6.0	33.0 ±5.7	22.8 ±0.9	55.8 ±3.8	33.2 ±2.1
Cinnarizine	34.0 ±10.5	30.8 ±6.6	19.5 ±2.8	51.3 ±1.4	48.8 ±6.9
Lidoflazine	34.8 ±2.0	18.5 ±2.7	28.8 ±1.7	50.0 ±1.3	26.0 ±2.2
Verapamil	7.2 ±2.4	35.8 ±3.8	22.7 ±2.0	12.3 ±1.5	10.4 ±2.4
Diltiazem	4.3 ±1.4	5.5 ±1.6	15.5 ±2.3	11.2 ±1.1	13.8 ±1.5
Nifedipine	3.7 ±0.4	49.5 ±2.5	9.7 ±0.9	10.5 ±1.6	7.8 ±1.0
Nicardipine	6.7 ±2.1	46.3 ±3.3	53.5 ±2.3	22.0 ±1.3	18.0 ±4.2

Protection by Ca^{2+} antagonists.- Table 2 summarizes the results obtained with various antagonists and pathological stimuli. It is evident that varying degrees of protection against stimulus-induced shape changes are obtained. Generally viewed, Ca^{2+}-antagonists with no affinity for the slow Ca^{2+}-channel are protective in all types of experiment. On the other hand, the genuine slow-channel blockers generally are less effective or provide no protection at all, except in the digitalis intoxication test.

DISCUSSION

The use of isolated cardiomyocytes is now widely accepted and of proved value, as a model for the intact myocardium, in various fields of cardiological study as highlighted in a review [10]. The situations now modelled are intracellular Ca^{2+}-overload in isolated cardiomyocytes and its pharmacological obviation by Ca^{2+}-antagonists. It is well known that loss of Ca^{2+}-homeostasis occurs in several clinical and experimental disease states of the myocardium, including the Ca^{2+} paradox [11], catecholamine cardiotoxicity [12, 13], digitalis intoxication [13, 14] and ischaemic-reperfusion injury [15, 16].

The dependence on extracellular Ca^{2+} suggests that the models described here all share induction of an excessive Ca^{2+}-influx as a common pathway giving rise to cellular hypercontracture. The way this influx is induced differs according to the model applied. In digitalis intoxication, a combination of intracellular Ca^{2+}-accumulation via Na^+/K^+-ATPase inhibition and Na^+/Ca^{2+} exchange [17], along with the Ca^{2+}-influx due to electrical pacing of the cells, finally causes Ca^{2+}-overloading. Exposure of quiescent cells to ouabain (no stimulation) does not result in shape changes! The slow-channel blockers are protective, but this is most likely due to inhibition of cellular excitability [5]. On the other hand, Ca^{2+}-antagonists with no affinity for the slow Ca^{2+}-channel are almost equally protective without affecting the cells' excitability [5]. These findings suggest a mechanism of action for the latter drugs which is quite different.

This is confirmed by the other experimental models where the blockers non-selective for slow channels afford a high degree of protection when compared to the genuine slow-channel blockers.- Veratrine-induced Ca^{2+}-overload probably occurs _via_ Na^+/Ca^{2+} exchange or _via_ influx of Ca^{2+}-ions through the modified Na^+-channel [2]; activated O_2 species induce lipid peroxidation resulting in a loss of the ion-selective barrier function of the sarcolemma, allowing Ca^{2+} to enter the cell [7]; the charged amphiphile polymyxin B reduces the number of negatively charged sites at the cytoplasmic face of the sarcolemma, thereby reducing sarcolemmal Ca^{2+}-binding capacity with consequent loss of sarcolemmal structural integrity [5, 8]. This probably also attenuates the selective barrier function of the sarcolemma, giving rise to excessive Ca^{2+}-influx.

The data obtained with these experimental models of Ca^{2+}-overload show a divergence between the main groups of Ca^{2+}-antagonists, the drugs without slow-channel affinity being generally superior to the slow Ca^{2+}-channel blockers. This suggests that under these conditions Ca^{2+} enters the cell through domains other than the slow Ca^{2+}-channel. A similar mechanism has been suggested for the entry of Ca^{2+} in disease states such as ischaemia-reperfusion injury [5, 15, 16, 18, 19].

In view of these similarities, the pathological stimuli inducing shape change in isolated cardiomyocytes as described here have great potential for research in the field of Ca^{2+}-overload and pharmacological protection with Ca^{2+}-antagonists and related compounds. In addition, isolated cardiomyocytes can serve as a useful tool in many disciplines in order to extend our cardiological knowledge.

Acknowledgements

The authors acknowledge the technical assistance by Mrs. G. Verellen and Mr. P. Nieste and the secretarial work by Mrs. K. Donné and Mrs. L. Vanherle.

References

1. Powell, T. & Twist, V.W. (1976) *Biochem. Biophys. Res. Comm. 72*, 327-333.
2. Ver Donck, L., Pauwels, P.J., Vandeplassche, G. & Borgers, M. (1986) *Life Sci. 38*, 765-772.
3. Ver Donck, L., Liu, G.S., Vandeplassche, G. & Borgers, M. (1987) *Basic Res. Cardiol. 82*, 74-81.   [74.
4. Honerjäger, P. (1982) *Rev. Physiol. Biochem. Pharmacol. 92*, 2-
5. Borgers, M., Ver Donck, L. & Vandeplassche, G. (1988) *Ann. N.Y. Acad. Sci. 522*, 433-453.
6. McCord, J.M. (1985) *New Engl. J. Med. 312*, 159-163.
7. Ver Donck, L., Van Reempts, J., Vandeplassche, G. & Borgers, M. (1988) *J. Mol. Cell Cardiol. 20*, 811-823.
8. Langer, G.A. (1986) *J. Am. Coll. Cardiol. 8*, 65A-68A.
9. Vanhoutte, P.M. & Paoletti, R. (1987) *Trends Pharmacol. Sci. 8*, 4-5.
10. Jacobson, S.L. & Piper, H.M. (1986) *J. Mol. Cell Cardiol. 18*, 661-687.
11. Lamers, J.M. & Ruigrock, T.J.C. (1983) *Eur. Heart J. 4 (suppl. H)*, 73-79.
12. Fleckenstein, A., Janke, J., Döring, H.J. & Leder, O. (1975) in *Recent Advances in Studies on Cardiac Structure and Metabolism*, Vol. 1 (Fleckenstein, A. & Rona, G., eds.), University Park Press, Baltimore, pp. 21-32.
13. Eisner, D.A. (1986) *J. Cardiovasc. Pharmacol. 8 (suppl. 3)*, S2-S9.
14. Bigger, J.T. (1985) *J. Clin. Pharmacol. 25*, 514-521.
15. Bourdillon, P.D.V. & Poole-Wilson, P.A. (1981) *Cardiovasc. Res. 15*, 121-130.
16. Poole-Wilson, P.A., Harding, D.P., Bourdillon, P.D.V. & Tones, M.A. (1984) *J. Mol. Cell. Cardiol. 16*, 175-186; *see also art. #E-6, this vol.*
17. Miura, D.S. & Biedert, S. (1985) *J. Clin. Pharmacol. 25*, 490-500.
18. Borgers, M., De Clerck, F., Van Reempts, J., Xhonneux, R. & Van Neuten, J. (1984) *Int. Angio. 3 (suppl. 2)*, 25-31.
19. Borgers, M. (1985) in *Calcium Entry Blockers and Tissue Protection* (Godfraind, T., et al., eds.), Raven Press, New York, pp. 173-181.

#E-8

THE ROLE OF ION DEREGULATION IN CELL INJURY AND CARCINOGENESIS [†]

**B.F. Trump, I.K. Berezesky, M.W. Smith,
P.C. Phelps and K.A. Elliget**

Department of Pathology, University & The Maryland Institute
of Maryland School of Medicine,   for Emergency Medical
10 S. Pine Street   Services Systems

Baltimore, MD 21201, U.S.A.

*This article deals with the following aspects, reflected
in section headings and focused on calcium. -
#Calcium measurement: total; soluble/ionized.
#Changes in calcium during cell injury, investigated with
model cell systems (e.g. NHBE[*]); some influences of (e.g.)
Ca^{2+} antagonists, and some effects of injurious agents (e.g.
$HgCl_2$ or formaldehyde).
#The formation of cytoplasmic blebs following injury in relation
to calcium.
#Cell division, cell differentiation, tumour promotion and
carcinogenesis.*

*A concluding Summary refers to Fig. 8 as depicting consequences
of $[Ca^{2+}]_i$ augmentation from the extracellular space, mitochondria or the e.r., and showing injurious agents that modulate
these three sites, and compounds that may have a modifying
influence. Deregulation of, in particular, $[Ca^{2+}]_i$, $[Na^+]_i$
and $[H^+]_i$ is pertinent to cell injury (acute or chronic),
differentiation and proliferation and to carcinogenesis.*

[†]Contribution #2683 from the Cellular Pathobiology Laboratory.
[*]For free ions, e.g. Na^+ and $[Ca^{2+}]$ ([ ]*connotes* concentration),
subscript $_i$ = intracellular (cytosolic), $_e$ = extracellular.
(Ptd)InsP, a (phosphatidyl)inositol phosphate: e.g. $PtdInsP_2$
(the 4,5-diphosphate); AM, acetoxymethyl ester (of the probe
fura-2); BCECF, 2,7-bis-(2-carboxyethyl)-5-*(& -6)*carboxyfluores-
cein; CaM, calmodulin; CCCP, carbonyl cyanide *m*-chlorophenyl-
hydrazone; HBSS, Hank's buffered salt solution; IAA, iodoacetic
acid; NEM, *N*-ethylmaleimide; (NH)BE, (normal human) bronchial
epithelium ('normal' = culture outgrowths from immediate-autopsy
explants); PCMBS, *p*-chloromercuribenzenesulphonic acid; PTE,
proximal tubule epithelium (kidney); TGF, transforming growth
factor; TPA, 12-*O*-tetradecanoylphorbol-13-acetate; e.r., endoplasmic reticulum; p.m., plasma membrane *(authors' convention* ▬
ER, PM ▬ *altered by Ed.).*

This article explores the role of ion deregulation in cell injury - including lethal toxic injury as well as chronic injury - as it relates to control of differentiation, proliferation and carcinogenesis. Experiments in our laboratory have for many years associated increased total calcium, often in the form of calcium phosphate precipitates in mitochondria, with the lethal phase of cell injury (i.e. during the necrotic phase) [reviews: 1-7]. Furthermore, pathologists have recognized for some time that such precipitates, previously called dystrophic calcification, can be used as markers for areas of dead cells such as regions of toxic injury or for delineating the edges of infarcts in the heart, kidney or liver.

We have also noted early changes in $[Na^+]_i$, $[K^+]_i$, $[Cl^-]_i$, $[Mg^{2+}]_i$ and $[Ca^{2+}]_i$ [2-6] as well as changes in pH_i [8] in a variety of cell injuries. Many of these changes were seen to occur very early, long before cell death, and then to progress as the cells approached the point of irreversibility. It was, moreover, noted that cell death and calcium accumulation coincide in onset [9]. It has been our plan for some years to investigate $[Ca^{2+}]_i$ as this would be an entirely different phenomenon from that of the increases or changes observed in total calcium. However, only recently have techniques become available to measure $[Ca^{2+}]_i$ in living cells. At present, there are several fluorescent probes, e.g. fura-2, which can be used to determine changes in $[Ca^{2+}]_i$ in populations of cells exposed to injury. Cells in suspension can be studied using a spectrofluorimeter, and individual living cells attached to a substrate studied using digital imaging fluorescence microscopy coupled with video intensification microscopy. Many of the results presented in this article relate to the latter technique.

There is a well established relationship between cell injury, cell proliferation and carcinogenesis [reviews: 10-12]. In most model systems of carcinogenesis, the preneoplastic phases are accompanied by cell injury, inflammation and regeneration. With the development of new techniques, e.g. X-ray microanalysis and digital imaging fluorescence microscopy [13], it has become possible to evaluate ion content in dividing and neoplastic cells. Thus, neoplastic cells have increased total Na^+ and Cl^- and decreased K^+ [14, 15], and increased pH_i may be a signal for cell division [16]. Also, it has been shown in some cell types that increased $[Ca^{2+}]_e$ can serve as a stimulus for differentiation of normal but not initiated cells, indicating a relationship to the phenomenon of tumour promotion [17, 18]. We have, therefore, been investigating these relationships as part of our general study of ion regulation, cell injury and carcinogenesis.

CALCIUM MEASUREMENT

Recent measurement methods chosen by us for total Ca and $[Ca^{2+}]_i$ are now briefly discussed.

Soluble or ionized Ca^{2+}.- Using fluorescent probes as pioneered by R.Y. Tsien and colleagues [see first and later arts. in this vol. - *Ed.*], $[Ca^{2+}]_i$ can be measured in cell suspensions by spectrofluorimetry, or in monolayers by digital imaging fluorescence microscopy coupled with video image processing. Video imaging is especially attractive as it permits better analysis of dye leakage, individual cell-cell variation, and intracellular variation. We are currently using a Fluoroplex III fibre-optic dual-excitation advanced image analysis system (TN-8502; Tracor Northern, Middleton, NJ) for fluorescence microscopy analysis of single or multiple cell samples. This system was designed for *in vitro* cell study using various fluorescent probes, e.g. fura-2 for $[Ca^{2+}]_i$ and BCECF for pH_i measurement, and enables probes to be localized and quantitated during studies of time-dependent kinetic events. Specific-wavelength illumination is transferred to the epi-illuminator of an inverted microscope and cell images are transferred through an intensified video camera to the computer. Images at two different wavelengths can be collected and averaged simultaneously, and the ratios computed to provide a quantitated image of the $[Ca^{2+}]_i$ or pH_i distribution within the cell. Such a system may be coupled with various other procedures, including microinjection and gene transfection. As the field is being developed, probes are become available for $[Na^+]_i$, $[K^+]_i$, and other ions. The availability of such powerful image analysis has virtually revolutionized the field of calcium analysis and opened new doors for investigators, including experimental pathologists.

Total calcium is relatively simple to analyze using a number of different techniques, including atomic absorption spectroscopy for cell populations and X-ray microanalysis when imaging is desired in addition to measurement [15]. There are several examples in the literature where failure to recognize the distinction between total and ionized calcium has led to erroneous conclusions.

CHANGES IN CALCIUM DURING CELL INJURY

Experimental conditions utilized.- We have used various techniques to investigate the role of changes in $[Ca^{2+}]_i$ as a function of time in several cell-injury model systems that affect cell function through diverse mechanisms and lead to irreversible cell injury after periods of time ranging

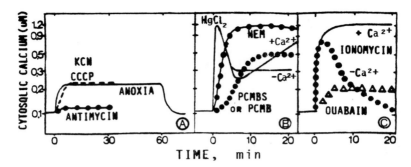

Fig. 1. Effect of different types of injury on $[Ca^{2+}]_i$ in rabbit PTE cultured (8-12 days) cells, loaded with fura-2/AM (5 µM, 60 min, 25°) and suspended by trypsinization. Spectrofluorimetric medium: HBSS + 10 mM HEPES/1.37 mM CaCl$_2$/1 mg per ml glucose (25°). **A**: effect of mitochondrial inhibitors (anoxia, 5 mM KCN, 4 µM CCCP, 5 µM antimycin). The $[Ca^{2+}]_i$ increase seen with each is independent of $[Ca^{2+}]_i$ and is presumably due to release from mitochondria. Reoxygenation of anoxic cells leads to recovery of $[Ca^{2+}]_i$ to control levels. **B**: effect of sulphydryl inhibition with 250 µM NEM, 1 mM PCMBS or 50 µM HgCl$_2$. The increased $[Ca^{2+}]_i$ seen with NEM and PCMBS is independent of $[Ca^{2+}]_e$ and is attributed mostly to release from the e.r.; HgCl$_2$ initially triggers Ca^{2+} release from e.r. followed by activation of a p.m. Ca^{2+}/H^+ pump and then by influx of $[Ca^{2+}]_e$ when 1.37 mM Ca^{2+} is present. **C**: effect of 5 µM ionomycin with 1.37 mM or <5 µM Ca^{2+} and of 500 µM ouabain. Ionomycin causes redistribution of Ca^{2+} from e.r. and mitochondria as well as $[Ca^{2+}]_e$ influx. In the absence of $[Ca^{2+}]_e$, $[Ca^{2+}]_i$ recovers in part by means of a p.m. Ca^{2+}/H^+-ATPase. The $[Ca^{2+}]_i$ rise following inhibition of Na^+/K^+-ATPase with ouabain is due to increased Na^+/Ca^{2+} activity in response to elevated $[Na^+]_i$. *From [19], by permission.*

up to 4 h. Our work on the evolution of events following cell injury has involved investigation of the role of $[Ca^{2+}]_e$, Ca^{2+}-entry blockers, and modifiers of calcium redistribution. The agents selected primarly target certain epithelial cells (PTE and BE), and include HgCl$_2$, ouabain, NEM, ionophores such as A 23187, CN$^-$, IAA, and several tumour promoters including formaldehyde, TPA, H$_2$O$_2$, TGF-β and serum (Figs. 1 & 2). These various agents were found to fit into the categories described below.

Agents causing the influx of calcium produce a $[Ca^{2+}]_i$ rise prior to cell death which is, in general, related to the $[Ca^{2+}]_e$ level. Influx studies have been performed with two agents: A 23187 and ionomycin. They effect rapid cell killing

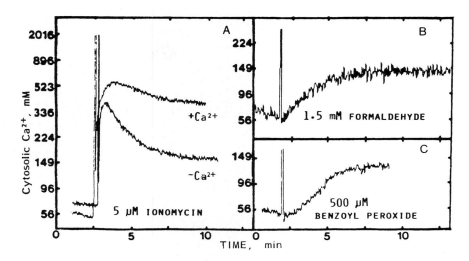

Fig. 2. Changes in $[Ca^{2+}]_i$ in NHBE cells loaded with fura-2 /AM (3 µM, 60 min, 25°) and then suspended by trypsinization in a modified HBSS containing 10 mM HEPES/110 µM $CaCl_2$. **A:** the $[Ca^{2+}]_i$ elevation resulting from exposure of cells to 5 µM ionomycin with Ca^{2+} present or (-Ca) reduced to <5 µM by EGTA (500 µM).   **B:** $[Ca^{2+}]_i$ elevation effected by 1.5 mM formaldehyde, identical if minus Ca^{2+} (+ 500 µM EGTA) *vs.* 110 µM Ca^{2+} — indicating an intracellular redistribution of Ca^{2+}. **C:** a similar $[Ca^{2+}]_i$ elevation by exposing cells to 500 µM benzoyl peroxide $\pm [Ca^{2+}]_e$. *From [19], by permission.*

and are highly dependent on $[Ca^{2+}]_e$. However, since ionophores are lipid-permeable, they can also cause redistribution of $[Ca^{2+}]_i$ from other intracellular sites (<u>e.g.</u> mitochondria and e.r.) prior to cell death.

Agents that effect calcium redistribution.- Many of the agents that we have examined do cause increases in $[Ca^{2+}]_i$ but also involve redistribution of $[Ca^{2+}]_i$ within the cell and, therefore, are not susceptible to down-regulation by decreasing $[Ca^{2+}]_e$. Such agents include H_2O_2, NEM, anoxia, CN^- alone, and CN^- + IAA. Ouabain may represent a special case in this category since it inhibits Na/K-ATPase and modifies Ca^{2+} extrusion at the cell membrane by means of Na^+/Ca^{2+} exchange.

Agents that influence both influx and redistribution.- In our laboratory, the principal agent used which is characteristic of this class is $HgCl_2$, a classic nephrotoxin. *In vitro* (at 50 µM) it causes redistribution of $[Ca^{2+}]_i$ within the cytoplasm, probably resulting from release from the e.r.

Fig. 3. Digitized fura-2 fluorescence-ratioed images illustrating $[Ca^{2+}]_i$ in rabbit PTE cells in primary cultures. Cells were grown on #1 coverslips which had been mounted over a $\frac{1}{2}$-inch hole drilled in the bottom of 35 mm plastic petri dishes, loaded with 1 µM fura-2/AM (30 min, 37°) and rinsed. The specially prepared petri dishes with cultures were then transferred to a Leitz Diavert inverted micro-scope equipped with epi-fluorescence and quickly scanned using a 40 × UV-Fluor objective. After selection of an area, cells were excited at 340 and 380 nm (emission 500 nM) and video images obtained _via_ an ISIT camera (Dage, Inc.). An Epyx board (Silicon Video) in an IBM/AT computer was used to digitize the video images. Calculations for $[Ca^{2+}]_i$ were performed as in [21]. **A**: zero-time control; **B**: 18 min following $HgCl_2$ treatment. Note the rises in $[Ca^{2+}]_i$ after $HgCl_2$ treat-ment as manifested by taller pixel heights and whiteness in colour of the cells. Scale = 20 µM. _From [22], by permission._

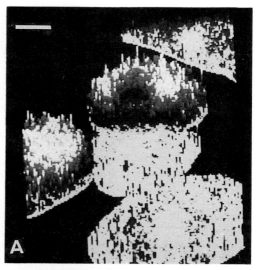

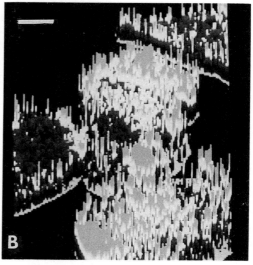

to the cytosol, followed by an influx of Ca^{2+}_e [20] (Fig. 3). Whereas the earlier redistribution component does not seem to correlate with cell death, the later influx of Ca^{2+}_e results in a killing time-course resembling that for Ca^{2+} ionophores. It is probable that many agents in animal and human disease exhibit both Ca^{2+} influx and redistribution.

CALCIUM-RELATED CYTOPLASMIC BLEB FORMATION FOLLOWING INJURY

The formation of cytoplasmic blebs at the cell periphery has long been thought to represent an early reaction to

injury [23]; but not till recently have these blebs received
attention.   In our experiments, it has become apparent that
this blebbing phenomenon occurs well before cell death, during
the reversible phase.   *In vitro*, blebs often pinch off,
detach and float away into the medium (Fig. 4), and *in
vivo* (as with kidney) they float into the tubule lumen.
Since blebs represent a major modification in cell shape,
we have been investigating what relationship they may have
to total calcium and $[Ca^{2+}]_i$, to changes in membrane function,
and to changes in the cytoskeleton. Using X-ray microanalysis,
we have noted in rat hepatocytes that, following bleb formation
after injury with A23187, total cell calcium increased over
the blebbed region [19] (Fig. 5). More recently, in studies
employing all of the injurious agents mentioned above and
in conjunction with phase, Nomarski and fluorescence microscopy
using probes such as fura-2, we have been able to characterize
the kinetics and significance of bleb formation in relation
to early cell injury, $[Ca^{2+}]_i$, and cell death [25] (Fig. 6).

The overall results of our studies to date suggest
that increased $[Ca^{2+}]_i$ precedes blebbing. It must be emphasized
that both increased $[Ca^{2+}]_i$ and bleb formation occur long
before cell death.   Conceivably the blebs, by pinching off,
may also be involved in the release of cytosolic enzymes
into the extracellular space.   For example, in the case
of liver cells, they may float into the serum, resulting
in increased liver-derived enzymes with no significant liver
necrosis.

It is evident from our studies that injuries which
cause large increases in $[Ca^{2+}]_i$ are associated with faster
bleb formation and, perhaps, with ultimately larger blebs
(Fig. 7).   It therefore appears that increased $[Ca^{2+}]_i$ is
intimately related to the formation of blebs, possibly through
alterations in proteins including CaM, CaM-associated proteins,
and cytoskeletal-associated proteins. In fact, in our experiments we have noted alterations in the immunocytochemical
staining pattern of actin filaments close to the time when
blebs form.

One · hypothesis for calcium-mediated, cytoskeletal-
associated changes is that perhaps Ca^{2+}-dependent proteinases
or calpains are being activated. With this in mind, we
have investigated inhibitors of calpains in connection with
bleb formation and have observed that both antipain and
leupeptin delay and diminish blebbing, perhaps by interacting
with actin-binding and/or spectrin-like proteins [26].

[continued on p. 448

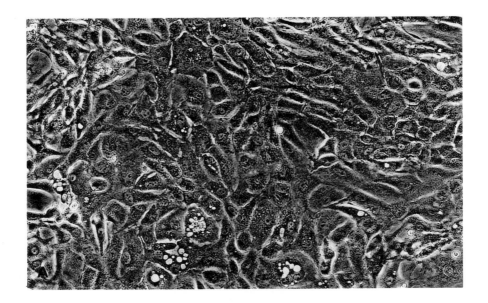

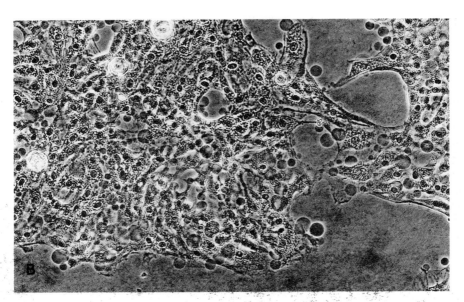

Fig. 4. Phase micrographs of rabbit PTE cells cultured for 9 days. **A**, untreated cells. **B**, cells exposed to 50 µM HgCl$_2$ for 7 min at 37° with 1.37 mM Ca^{2+}. Note the numerous blebs, many of which are free-floating. × 180. *From [24], by permission.*

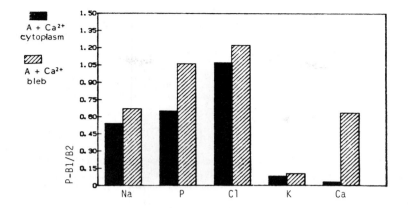

Fig. 5. Typical X-ray microanalysis results from a freeze–dried isolated hepatocyte following a 30–min treatment with A 23187 (**A**) in the
presence of extracellular Ca²⁺. Measurements made over the cytoplasm and a bleb. In the ratios expression P–B1/B2, P is the peak of
interest, B1 is the background count under the P, and B2 is the
continuum count between 5.50 and 6.50 keV after the continuum from
support film has been subtracted. Note the large increase in Ca in
bleb region *vs.* cytoplasm. (Other elemental changes: not elucidated.)

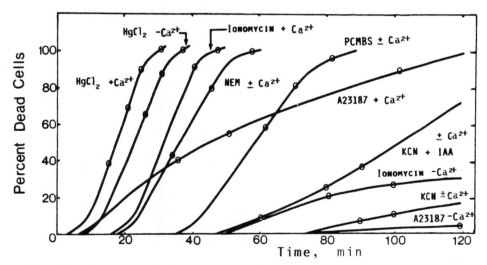

Fig. 6. Representative curves showing the loss of viability
in rabbit monolayer PTE cells during exposure at 37° to 50 µM
HgCl₂, 250 µM NEM, 1 mM PCMBS, 5 µM ionomycin, 10 µM A 23187,
4 µM CCCP, or 5 mM KCN alone or with 100 µM IAA: [Ca²⁺]ₑ normal
or low (as in Fig. 1), giving similar cell death for NEM, PCMBS
and KCN + IAA. Viability assessed by trypan blue. *From [25],
by permission (applicable to Fig. 7 also). (Fig. 5: from [19]).*

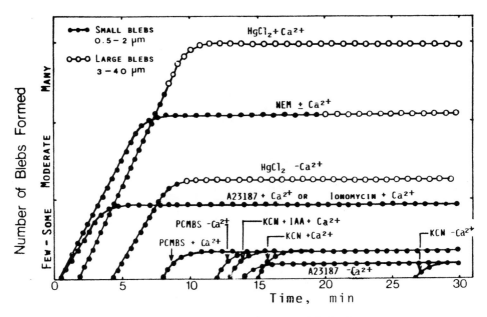

Fig. 7. Representative curves showing bleb formation, rate, number and size in rabbit monolayer PTE cells as affected by the agents in legend to Fig. 6, with normal or low $[Ca^{2+}]_e$ — which gave similar blebbing with NEM or PCMBS. *From [25].*

CELL DIVISION/DIFFERENTIATION; TUMOUR PROMOTION; CARCINOGENESIS

It is becoming increasingly evident that ion regulation may play an important role in these cell processes [reviews: 10-12]. X-ray microanalysis shows neoplastic cells to have higher Na^+ and Cl^-, and lower K^+, than normal cells [3, 14]. Although such ionic shifts are seen too in dividing cells, they are greater in neoplastic cells. However, as yet there are insufficient data to allow $[Ca^{2+}]_i$ comparison between normal and neoplastic cells. It is also known that in several systems increased pH_i results in stimulation of cell division, somewhat similar to the effect of growth factors: e.g. pH_i in marine eggs rises at the time of fertilization [27, 28]. Certainly in some systems this appears to be the result of stimulation of the Na^+/H^+ antiporter in the p.m., which in turn is activated by protein kinase C. As depicted in Fig. 8, protein kinase C can be activated directly in many cells by the phorbol ester TPA and also by diacylglycerol (DAG). DAG is part of the bifurcating pathway resulting from stimulation of a p.m. receptor whereby, through activation of a G protein, $InsP_3$ as well as DAG is released. $InsP_3$ can effect Ca^{2+} release from the e.r.-influencing protein kinase C as well as activating genes such as **fos** and **myc**. The G protein involved in this transduction

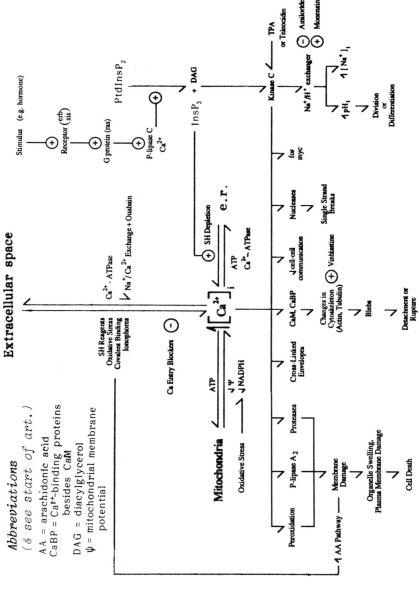

Fig. 8. Hypothesis (amplified in text) on the relationship between ion de-regulation and cell injury and carcinogenesis. *From Trump & Berezesky [11], by permission.*

Abbreviations (& see start of art.)

AA = arachidonic acid
CaBP = Ca^{2+}-binding proteins besides CaM
DAG = diacylglycerol
ψ = mitochondrial membrane potential

represents a **ras**-coated protein, which may be modified in neoplasia (Fig. 8).

Another line of investigation suggests the possibility that tumour promoters may exert selective action on normal *vs.* initiated cells, i.e. they may act by causing terminal differentiation in normal cells whilst being permissive to cell division in initiated cells. Thus, experiments in our laboratory have clearly shown that parameters such as TPA, H_2O_2 and formaldehyde induce terminal differentiation in NHBE cells, but do not alter cell division in a variety of human tumour cell lines [29, 30]. In the same context, it is perhaps even more striking that modification of the Ca^{2+} concentration in the medium can bring about the same results [31, 32]. It has also been observed that increased $[Ca^{2+}]_e$ results in terminal squamous differentiation in normal but not in tumour cell lines.

We have, therefore, been in the process of investigating the effects of such compounds on $[Ca^{2+}]_i$ regulation in normal BE (isolated from immediate autopsies) and in immortalized human BE. We have also compared Ca^{2+} regulation in a serum-susceptible strain to that in serum-resistant strains of the immortalized BE cell line (BEAS-2B) [32]. We have investigated a series of compounds or conditions known to exert differential effects on cell division, using spectrofluorimetry or digital imaging fluorescence microscopy [33]. The materials studied so far have included acrolein, acetaldehyde, formaldehyde, H_2O_2, TGF-β, TPA and calcium-free (dialyzed) serum.

The results of these studies were quite striking. After the addition of dialyzed serum, formaldehyde or H_2O_2 there was a rapid but reversible increase in $[Ca^{2+}]_i$. However, this reponse was not homogeneous within a cell population, but rather had definite regional differences. In contrast, addition of TGF-β or TPA did not result in such a Ca^{2+} signal, but TPA did modulate a serum-induced Ca^{2+} signal. For the present we infer that this may relate to direct activation of protein kinase C, thus bypassing the Ca^{2+} step. Further experimentation to clarify this point is in progress.

SUMMARY

Our current working hypothesis of the relationship between ion deregulation and cell injury and carcinogenesis [11] is illustrated in Fig. 8. Shown specifically are the many alterations which occur following influx of Ca^{2+} from the extracellular space, the mitochondria and the e.r., the injurious agents that modulate these three sites, and the compounds that can modify them.

From the pathways in Fig. 8 and the data presented above, it becomes quite obvious that there are many reasons to investigate the deregulation of ions, specifically $[Ca^{2+}]_i$, $[Na^+]_i$ and $[H^+]_i$, and the role they play in cell injury. Even more important is the understanding of how such ion deregulation relates to cell differentiation, proliferation and carcinogenesis. Since these same ions are often deregulated in acute cell injury, they could well serve as important links between acute cell injury and chronic effects on growth such as regeneration and neoplasia. Since recently developed technologies have provided the investigator with new topics for such studies, experimentation should result in data which provide a better understanding of these modulations.

Acknowledgements

The work was supported by NIH grants AM15440 and NO1CP51000.

References

1. Trump, B.F. & Ginn, F.L. (1969) in *Achievements in Experimental Pathology*, Vol. IV (Bajusz, E. & Jasin, E., eds.), Karger, Basel, pp. 1-29.
2. Trump, B.F., Croker, B.P. jr. & Mergner, W.J. (1971) in *Cell Membranes: Biological and Pathological Aspects* (Richter, G.W. & Scarpelli, D.G., eds.), Williams & Wilkins, Baltimore, pp. 84-128.
3. Trump, B.F., Berezesky, I.K., Chang, S.H., Pendergrass, R.E. & Mergner, W.J. (1979) *Scanning Electr. Micros. III*, 1-14 [SEM Inc., AMF O'Hare, Illinois 60666].
4. Trump, B.F., Berezesky, I.K., Laiho, K.U., Osornio, A.R., Mergner, W.J. & Smith, M.W. (1980) *Scanning Electr. Micros. III*, 437-462 [see ref. 3].
5. Trump, B.F., Berezesky, I.K. & Osornio-Vargas, A. (1981) in *Cell Death in Biology and Pathology* (Bowen, I.D. & Lockshin, R.A., eds.), Chapman & Hall, London, pp. 209-247.
6. Trump, B.F., Berezesky, I.K. & Phelps, P.C. (1981) *Scanning Electr. Micros. II*, 434-454 [see ref. 3].
7. Trump, B.F. & Berezesky, I.K. (1985) in *Current Topics in Membranes & Transport* (Shamoo, A.E., ed.), Academic Press, New York, pp. 279-319.
8. Pentilla, A. & Trump, B.F. (1974) *Science 185*, 277-278.
9. Trump, B.F., Berezesky, I.K., Sato, T., Laiho, K.U., Phelps, P.C. & DeClaris, N. (1984) *Envir. Health Perspectives 57*, 281-287.
10. Trump, B.F. & Berezesky, I.K. (1987) in *Non-genotoxic Mechanisms in Carcinogenesis*, Banbury Rep. 25, Cold Spring Harbor Laboratory, N.Y., pp. 69-79.
11. Trump, B.F. & Berezesky, I.K. (1987) *Carcinogenesis 8*, 1027-1031.

12. Trump, B.F. & Berezesky, I.K. (1987) in *Lung Carcinomas* (McDowell, E.M., ed.), Churchill Livingston, New York, pp. 162-174.

13. Arndt-Jovin, D.J., Robert-Nicoud, M., Kaufman, S.J. & Jovin, T.M. (1985) *Science 230*, 247-248.

14. Smith, N.R., Sparks, R.L., Pool, T.B. & Cameron, I.L. (1978) *Cancer Res. 38*, 1952-1959.

15. Cameron, I.L. & Smith, N.K.R. (1982) in *Ions, Cell Proliferation and Cancer* (Boynton, A.L., McKeehan, W.L. & Whitfield, J.F., eds.), Academic Press, New York, pp. 13-40.

16. Moolenaar, W.H. (1986) *Ann. Rev. Physiol. 48*, 363-376.

17. Hennings, H., Michaels, D., Cheng, C., Steinert, K., Holbrook, K. & Yuspa, S.H. (1980) *Cell 19*, 245-254.

18. Lechner, J.F. (1984) *Fed. Proc. 43*, 116-121.

19. Trump, B.F., Smith, M.W., Phelps, P.C., Regec, A.L. & Berezesky, I.K. (1988) in *Integration of Mitochondrial Function* (Lemasters, J.J. & Hackenbrook, C.R., eds.). Plenum, New York, pp. 437-444.

20. Smith, M.W., Ambudkar, I.S., Phelps, P.C., Regec, A.L. & Elliget, K.A. (1987) *Biochim. Biophys. Acta 931*, 130-142.

21. Grynkiewicz, G., Poenie, M. & Tsien, R.Y. (1985) *J. Biol. Chem. 260*, 3440-3450.

22. Trump, B.F. & Berezesky, I.K. (1989) in *Cell Calcium Metabolism* (Fiskum, C., ed.), Plenum, New York, pp. 441-449.

23. Trump, B.F., Goldblatt, P.J. & Stowell, R.E. (1962) *Lab. Invest. 11*, 986-1015.

24. Trump, B.F., Berezesky, I.K., Smith, M.W., Phelps, P.C. & Elliget, K.A. (1989) *Toxicol. Appl. Pharmacol. 97*, 6-22.

25. Phelps, P.C., Smith, M.W. & Trump, B.F. (1989) *Lab. Invest. 60*, in press.

26. Elliget, K.A., Phelps, P.C. & Trump, B.F. (1988) *J. Cell Biol. 107*, 394a.

27. Epel, D. (1980) *Ann. N.Y. Acad. Sci. 339*, 74-85.

28. Jaffee, L. (1980) *Ann. N.Y. Acad. Sci. 339*, 86-101.

29. Saladino, A.J., Willey, J.C., Lechner, J.F., Grafstrom, R.M., Laveck, M. & Harris, C.C. (1985) *Cancer Res. 45*, 2522-2527.

30. Willey, J.C., Saladino, A.J., Ozanne, C., Lechner, J.F. & Harris, C.C. (1984) *Carcinogenesis 5*, 209-215.

31. Trump, B.F., Smith, M.W., Miyashita, M., Phelps, P.C., Jones, R.T. & Harris, C.C. (1988) *J. Cell Biol. 107*, 72a.

32. Miyashita, M., Smith, M.W., Willey, J.C., Lechner, J.F., Trump, B.F. & Harris, C.C. (1989) *Cancer Res. 49*, 63-67.

33. Smith, M.W., Phelps, P.C., Jones, R.T., Trump, B.F. & Harris, C.C. (1989) *The Toxicologist, 9*, 160.

#ncE

NOTES and COMMENTS relating to

CELL PROTECTION; PATHOLOGY, INCLUDING CARDIOVASCULAR

'COMMENTS' are on pp. 479-482, starting with Forum discussions
on the preceding main articles, then on 'Notes'

#ncE.1

A Note on

ELEVATED $[Ca^{2+}]_i$ AS A LATE EVENT IN PATHOGENESIS IN CARDIOMYOCYTES AND HEPATOCYTES: MEASURMENTS WITH AEQUORIN IN SINGLE CELLS

A.P. Allshire, N.M. Woods and P.H. Cobbold

Department of Human Anatomy and Cell Biology,
University of Liverpool,
P.O. Box 147, Liverpool L69 3BX, U.K.

Blebbing of the sarcolemma (p.m.[*]) [1] and irreversible cell injury [2] have been attributed to a rise in $[Ca^{2+}]_i$. To examine this hypothesis we injected single cells with aequorin and related $[Ca^{2+}]_i$ to cell morphology. Studying histories of individual cells ensures that cell-cell hetero-geneity does not obscure the sequence of events. We chose aequorin as a reporter of $[Ca^{2+}]_i$ since its greater-than-linear response to Ca^{2+} resembles that of endogenous Ca^{2+} sensors such as calmodulin and troponin C, and may therefore be more relevant biologically than measurements of average $[Ca^{2+}]_i$ with probes which bind calcium in a 1:1 molar ratio.

We found that cardiomyocytes under anoxia, inhibited with cyanide and 2-deoxyglucose or uncoupled with CCCP, spon-taneously shorten before $[Ca^{2+}]_i$ begins to rise [3, 4]. This shape change probably reflects a rigor triggered by falling phosphorylation potential, and accelerates the fall by positive feedback through Ca^{2+}-activation of myosin ATPase [5]. Hepato-cytes uncoupled with CCCP developed hyaline blebs of the p.m.; only then did $[Ca^{2+}]_i$ rise, agreeing with observations using fura-2 [6].

Monitoring cell appearance and $[Ca^{2+}]_i$ in parallel shows that shortening of cardiomyocytes and blebbing of hepatocytes precede a rise in $[Ca^{2+}]_i$, although elevated $[Ca^{2+}]_i$ contributes to irreversible injury. The early events appear to depend on the phosphorylation potential in the cytosol.

---

[*]*Abbreviations.-* CCCP, carbonyl cyanide *m*-chlorophenylhydra-zone; $[Ca^{2+}]_i$, cytosolic free Ca^{2+} concentration; p.m., plasma membrane (*Ed.'s term for* sarcolemma & plasmalemma).

The idea that elevated $[Ca^{2+}]_i$ is fundamental to irreversible cell injury derives from observations of elevated cell calcium in tissue injury and necrosis, and the near-universal role of Ca^{2+} as an intracellular second messenger. Abnormal Ca^{2+} metabolism has been described in many pathogenesis contexts. However, the extent to which these abnormalities are the cause of the damage, rather than merely incidental to it, is unclear. We describe here an experimental approach whereby $[Ca^{2+}]_i$ was monitored in cardiomyocytes and hepatocytes - two cell types in which injury can readily be discerned by examination under a light microscope. The cells were subjected either to profound hypoxia or to treatment with a mitochondrial uncoupler or inhibitors of respiration and glycolysis.

Experimental rationale.- Even under apparently uniform conditions cells vary considerably in the time-course of their responses. Thus temporal and causal relationships between events in the pathological process become blurred. These relationships will be most apparent in the 'histories' of individual cells. We injected freshly isolated cells with aequorin (cf. Cobbold et al., #B-6, this vol.), and related $[Ca^{2+}]_i$-dependent aequorin luminescence to the cell's appearance during the pathogenic insult.

EXPERIMENTAL DESIGN

Fig. 1 outlines a system which enables us to measure aequorin light from a cell, with only 5-10 sec interruptions to monitor its appearance. Fig. 2 shows a part of the apparatus built for this purpose. The photomultiplier, a low-noise bi-alkali photocathode, is thermally insulated at 0-2° immediately over the top window of the incubation vessel containing the cell. Photoelectron counts are sampled at (usually) 50 msec intervals by a microcomputer, and stored for off-line analysis. The stainless steel incubation vessel has a top glass window and, at the bottom, another which comprises a long-pass filter (560 nm cut-off; >99% reflectance for vertically incident light) which reflects blue aequorin light but transmits the red light used to illuminate the cell. The cup is maintained at 37° and perfused with medium. An air-actuated shutter protects the photomultiplier while the appearance of the cell is being monitored. Illumination is from a red light-emitting diode (80 µW) via an angled fiberoptic over the upper window. The image passing through the lower window is focussed through a 60x objective lens and relayed via a CCD TV camera to an external monitor for recording by photgraphy or on video tape. The assembly comprising cup and optical system is constructed to slide out from under the phototube (as in Fig. 2). To load the

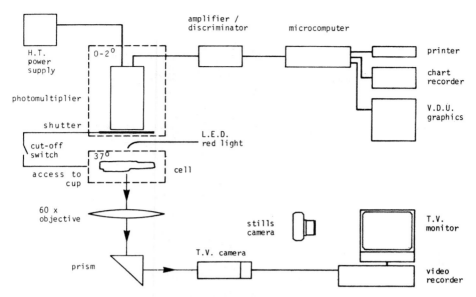

Fig. 1. Schematic outline of a system designed for monitoring aequorin luminescence and morphology in a single mammalian cell. X-Y translation of the cup and focus are motor-driven under remote control.

injected cell in its microslide into the cup, the top window (consisting of a circular glass cover-slip held in place by a film of liquid paraffin) is drawn aside. Once the assembly has been re-positioned beneath the photomultiplier, the shutter opened and the HT power supply (1.1 kV) switched on, cell-shape recording needs only 5-10 sec interruptions in recording aequorin light.

Software has been written by Dr K.S.R. Cuthbertson. The fractional rate of aequorin consumption is related to $[Ca^{2+}]_i$ on the basis of *in vitro* calibrations under appropriate conditions of $[Mg^{2+}]$, pH, temperature and ionic strength. Use of aequorin as a reporter of Ca^{2+} has been reviewed recently [7], and its advantages and limitations have been discussed along with a description of the injection procedure [5] (& Cobbold et al., this vol.: #B-6).

$[Ca^{2+}]_i$ AND CELL APPEARANCE

Fig. 3 shows the relationship between $[Ca^{2+}]_i$ and cell shape in a rat-ventricle myocyte under prolonged anoxia (for conditions see [4]; for maintenance of profound anoxia see [5]). Cardiomyocytes under anoxia or subjected to a 'chemical hypoxia' (2 mM CN^-, 5 mM 2-deoxyglucose) show considerable variation in their ability to withstand the insult, but

Fig. 2. Arrangement of the photomultiplier, cup and optical system. During experiments the photomultiplier chamber is closed off and equilibrated at 0-2°; the cup and optical assembly slide into place directly beneath the photomultiplier (1 cm clearance), and the whole apparatus is shielded against stray background light.

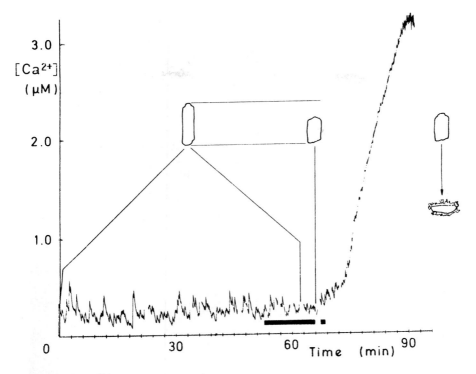

Fig. 3. $[Ca^{2+}]_i$ in a rat cardiomyocyte superfused with anoxic medium containing 1 mM Ca^{2+} but lacking glucose. Outlines depict cell shapes at the points indicated. Total aequorin light in the cell: 1.2×10^5 cts.; data exponentially smoothed on replotting. Time constants: 20 sec and 5 sec for aequorin consumption rates corresponding resp. to <560 and >560 nM $[Ca^{2+}]_i$.

finally shorten spontaneously by 30-40%. There is no measurable rise in $[Ca^{2+}]_i$ before shortening. Then within just a few minutes of shortening under control conditions (1 mM Ca^{2+} in the superfusate) $[Ca^{2+}]_i$ begins to rise as a net influx develops across the p.m. [4, 8]. We have suggested [5] that this shortening, apparently independent of $[Ca^{2+}]_i$, represents development of a rigor as the free-energy change of ATP hydrolysis in the cytosol falls to a critical level. Rigor would cause contraction and activate the myosin ATPase, thereby accelerating de-energization and limiting activity of the ATP-linked pumps responsible for ion homeostasis.

A $[Ca^{2+}]_i$-dependent contracture preceding loss of $[Ca^{2+}]_i$ homeostasis has also been seen in ferret papillary muscle injected with aequorin [9]. By contrast, $[Ca^{2+}]_i$ as measured with indo-1 was found to rise *before* onset of contracture in cultured chick embryo ventricle myocytes [10]; it was concluded that rising $[Ca^{2+}]_i$ contributed to onset of rigor through increased pumping of Ca^{2+}. It is unclear whether this discrepancy arises from the different measuring systems or from the conditions used. The main difficulties in interpreting aequorin data include (i) uncertainty about $[Mg^{2+}]_i$ and whether it changes, (ii) the sensitivity of aequorin to Ca^{2+} is altered inside cells, and (iii) the possibility that aequorin is not very sensitive to $[Ca^{2+}]_i$ in resting cells. The 5 mM Mg^{2+} used in our calibrations may prove too high when $[Mg^{2+}]_i$ is measured inside cardiomyocytes under appropriate conditions, in which case $[Ca^{2+}]_i$ levels will have been overestimated. That the other calibration conditions do not fully describe the cytosolic environment of injected aequorin cannot be discounted at this stage; however, our estimates of $[Ca^{2+}]_i$ generally agree quite well with those obtained by other methods. The lower sensitivity of aequorin to $[Ca^{2+}]_i$ at resting levels means that precise estimates and comparisons require integration of the small light signal over longer periods. However, this is rarely a serious limitation of the technique where $[Ca^{2+}]_i$ changes little over the period during which morphological changes occur.

In Fig. 3 $[Ca^{2+}]_i$ over the 1000 sec before shortening (mean 183 nM) does not differ significantly from that during an arbitrary 60 sec period immediately after shortening (258 nM; $P > 0.05$). On the basis of the in vitro calibration $[Ca^{2+}]_i$ was significantly below 350 nM for 78 sec and <400 nM for 109 sec after shortening was first detected ($P < 0.05$). Anoxic cardiomyocytes also shorten in nominally Ca^{2+}-free medium, although the $[Ca^{2+}]_i$ rise is delayed and attenuated [4]. The experiments described here therefore suggest that the anoxic contracture is at least partly independent of $[Ca^{2+}]_i$; yet the severity of the injury correlates with the extent of the $[Ca^{2+}]_i$ rise upon resumption of oxidative metabolism (the 'oxygen paradox'). To resolve the apparent discrepancy between aequorin and fluorescent indicator measurements, both techniques may need to be applied to the same system.

Hepatocytes incubated with the mitochondrial uncoupler CCCP (2 μM) develop large clear blebs (>3 μm) of the p.m. which progressively coalesce to form a 'halo' around the cell. Localized influx of Ca^{2+} at sites of blebbing should be accentuated in the aequorin signal. However, only in the latter stages of the process have we observed a rise in $[Ca^{2+}]_i$ (not shown). It appears that disruptions other than an elevated $[Ca^{2+}]_i$ - possibly involving the free energy of ATP hydrolysis in the cytosol - alter the cytoskeleton and cause volume changes. De-energization as the primary lesion would also lead to loss of $[Ca^{2+}]_i$ homeostasis through thermodynamic limitations on ATPase-linked ion pumps.

On the basis of parallel measurements of $[Ca^{2+}]_i$ and cell appearance, we conclude that an elevated $[Ca^{2+}]_i$ is inessential to, although linked with, the primary event in cell injury. De-energization probably triggers the injury; its severity and outcome are determined by the ensuing rise in $[Ca^{2+}]_i$. We are now extending this experimental approach by using the photoprotein pholasin to measure intracellular free radicals of oxygen [11]. The roles of $[Ca^{2+}]_i$ and oxygen radicals in pathogenic mechanisms can thereby be studied at the level of single cells.

Acknowledgements

This work was funded by the British Heart Foundation and the Wellcome Trust.

References

1. Jewell, S.A., Bellomo, G., Thor, H., Orrenius, S. & Smith, M. (1982) *Science 217*, 1257-1259.
2. Trump, B.F. & Berezesky, I.K.(1985) *Surv. Synth. Path. Res. 4*, 248-256. *[Cf. #E-8, this vol.]*
3. Cobbold, P.H. & Bourne, P.K. (1984) *Nature 312*, 444-446.
4. Allshire, A., Piper, H.M., Cuthbertson, K.S.R., & Cobbold, P.H. (1987) *Biochem. J. 244*, 381-385.
5. Allshire, A.P. & Cobbold, P.H. (1989) in *Adult Cardiac Myocytes* (Piper, H.M. & Isenberg, G., eds.), Vol. 1, CRC Press, Boca Raton, FL, Chapter 5, in press.
6. Lemasters, J.J., DiGuiseppi, J. & Nieminen, A-L. & Herman, B. (1987) *Nature 325*, 78-81.
7. Cobbold, P.H. & Rink, T.J. (1987) *Biochem. J. 248*, 313-328.
8. Allshire, A. & Cobbold, P.H. (1987) *Biochem. Soc. Trans. 15*, 960.
9. Smith, G.L. & Allen, D.G. (1988) *Circ. Res. 62*, 1223-1236.
10. Barry, W.H., Peeters, G.A., Rasmussen, C.A.F., jr. & Cunningham, M.J. (1987) *Circ. Res. 61*, 726-734.
11. Cotton, B., Allshire, A., Cobbold, P.H., Müller, T. & Campbell, A.K. (1988) *J. Mol. Cell Cardiol. 20 (Suppl. V)*, Abstract FR-2.

#ncE.2

A Note on

ALTERED Ca²⁺-ATPase ACTIVITY IN PLATELET MEMBRANES IN ESSENTIAL HYPERTENSION

J.M. Graham, R.B.J. Wilson and B.F. Robinson

Division of Biochemistry, Department of Cellular and
   Molecular Sciences and Department of Medicine,
St. George's Hospital Medical School,
Cranmer Terrace, London SW17 0RE, U.K.

Hypertension in man is sometimes observed as a consequence of some other disease – usually of a renal or endocrine nature; but in essential (primary) human hypertension (EH) no underlying disease can be detected. EH originates in a raised resistance to blood flow in the peripheral vessels, and $[Ca^{2+}]_i$ (the level of cytoplasmic Ca^{2+}) is a key factor in controlling both myocardial and arteriolar contractility. Not surprisingly, therefore, studies on EH and on the spontaneously hypertensive rat (SHR) have revealed abnormalities in calcium handling at the cellular level.

The control of $[Ca^{2+}]_i$ is very complex: recognized mechanisms include an ATP-dependent Ca^{2+} pump, Na^+/Ca^{2+} and H^+/Ca^{2+} exchangers, Ca^{2+} channels and Ca^{2+}-binding proteins such as calmodulin (CaM). The raised vascular smooth muscle contraction of EH could be caused by (i) increased local agonist levels leading to an opening of receptor-mediated Ca^{2+} channels; (ii) membrane depolarization; (iii) decreased transmembrane Ca^{2+}-pumping; or a release of Ca^{2+} from (iv) internal stores and/or (v) binding proteins. Other proposals involve a fundamental change in the concentration of some other ion to which the $[Ca^{2+}]_i$ alteration is secondary [1]. For example, Na^+/Ca^{2+} exchange is inhibited by an elevated $[Na^+]_i$ which could arise through depression of the Na-pump, maybe by some circulating ouabain-like inhibitor [2], or through a genetically determined defect [3]. Malfunctions in calcium handling at the molecular level observed in EH and in the rat (SHR) include the following.-
- SHR/EH: a $[Ca^{2+}]_i$ increase in platelets [4].
- SHR: impaired ATP-dependent Ca^{2+} efflux across mesenteric
         artery muscle membranes [5].
- SHR/EH: reduced ATP-dependent CaM-stimulated Ca^{2+} transport
         (erythrocytes) and Ca^{2+} binding (various membranes) [review: 6].
- SHR: reduced La^{3+} sensitivity of erythrocyte Ca^{2+}ATPase [7].

Because Ca^{2+} extrusion in vascular smooth muscle is thought to be dominated by the ATP-dependent Ca^{2+}-pump [8], we have investigated the Ca^{2+}-stimulated Mg^{2+}-dependent ATPase activity in membrane fractions isolated from the platelets of patients with EH and from a control group of subjects. We chose platelets because abnormalities in Ca^{2+} handling by these cells have already been demonstrated in EH, albeit not by all groups of workers, and because for study of the effect of treatment on the enzyme activity, a cell with a high turnover rate such as the platelet may be more relevant than the erythrocyte which has a relatively long half-life.

METHODS

There were 24 subjects in each group: *normotensives* (diastolic blood pressure <90 mm Hg) and hypertensives (>90 mm Hg on at least 3 separate occasions).
- Ethnic origin: Caucasian, *20* & 19; African, *2* & 4; Asian, *2* & 1.
- Sex: male, *22* & 21; female, *2* & 3.
- Age range: *25-63* (mean: *40*) & 25-64 (mean 49).

A platelet-rich fraction which contained ~95% of the total platelets, all the white cells and ~1% of the total erythrocytes was obtained from freshly drawn venous blood using a routine differential centrifugation method. CPD-adenine 1 (Travenol Labs., Thetford, Norfolk, U.K.) was used as anticoagulant (21 ml in 150 ml blood). Platelets were purified by free-flow electrophoresis [9], achieving better yield, purity and functional integrity than by any other method. (See Vol. 17, this series: #B-1, Graham et al.- *Ed.*).

Platelets were sedimented from the electrophoresis buffer by centrifugation (4,300 g, 10 min), resuspended in 10 ml 0.15 M NaCl/1 mM EDTA/5 mM Tris-HCl (pH 7.5), loaded onto two 36 ml linear 5-40% (v/v) glycerol gradients (in the same medium), and centrifuged at 1,500 g (30 min) and then at 10,000 g (10 min) [10]. Platelet pellets were resuspended quickly in 12 ml 0.25 M sucrose/5 mM Tris-HCl (pH 8.0) and homogenized with a Potter-Elvehjem homogenizer (6 strokes; 500 rpm). The homogenate was fractionated crudely by differential centrifugation: 4,300 g, 10 min; 20,000 g, 10 min; 100,000 g, 30 min.

The 4,300 g pellet which contained the bulk of the Na^+/K^+-ATPase activity was resuspended in 6 ml 40% (w/v) metrizamide and made part of a discontinuous gradient of 3 ml 40%/3 ml 30%/2 ml 25%/2 ml 20%/2 ml 15% metrizamide, and centrifuged at 80,000 g for 16 h. All gradient solutions contained Tris-HCl, pH 8.0. After harvesting the banded material, dilution with Tris-HCl, and centrifugation at 30,000 g for 20 min, it was resuspended in 0.25 M sucrose/5 mM

Table 1. Enzyme characterization of platelet membrane fractions: specific activities (μmol substrate/h per mg protein).

Fraction	NADPH–cyt. c red'ase	Na^+/K^+-ATPase	Mg^{2+}/Ca^{2+}-ATPase
A	0.27	0.12	0.24
B	<0.02	0.57	0.19
C	<0.02	0.41	0.39
Homogen.	0.05	0.07	0.16

Tris-HCl, pH 8.0, prior to enzyme analysis. NADPH-cytochrome c reductase was measured spectrophotometrically, and ATPase by the method of Avruch & Wallach [11].

RESULTS AND DISCUSSION

Characterization of platelet membrane fractions

Table 1 shows the enzyme composition of 3 of the major Ca^{2+}-ATPase-containing fractions isolated by differential and gradient centrifugation: **A** (the 100,000 **g** pellet); **B** and **C**, from the metrizamide gradient fractionation of the 4,300 **g** pellet. **A** contains very high activity of NADPH-cytochrome **c** reductase and also Ca^{2+}-stimulated Mg^{2+}-dependent ATPase: it is probably derived from the platelet e.r.* Fractions **B** and **C** were likewise rich in this activity, and also in Na^+/K^+-ATPase: whilst they might appear to be of p.m. origin, the reciprocal nature of the activities of these two ATPases in **B** and **C** make their origin less clear. Indeed the most active fraction in terms of Ca^{2+}-ATPase contained <15% of the total Na^+/K^+-ATPase. Hack et al. [12] also detected Ca^{2+}-stimulated ATPase in an internal membrane fraction but not in a p.m. fraction; but since the two membrane preparation methods differed widely, our fraction **C** may merely represent a membrane fraction contaminated by surface membrane, or may be a surface membrane domain which differs completely from those isolated by Hack et al. Although we have not attempted to further subfractionate **C** in order to test these two possibilities, it is noteworthy that Enyedi et al. [13] have reported that two Ca^{2+} pumps exist in platelets: one similar in properties to that of the sarcoplasmic reticulum, one like that of the erythrocyte membrane. Histochemically the Ca^{2+}/Mg^{2+}ATPase has been localized to both the e.r. and the open canalicular system near the surface [14].

Ca^{2+}-stimulated Mg^{2+}-ATPase in essential hypertension (EH)

Table 2 shows the basal Mg^{2+}-dependent ATPase and the Ca^{2+}-stimulated activity, at two Ca^{2+} concentrations (10 and 100 μM), of two platelet membrane fractions **A** and **C** from normal subjects and from EH patients. At both concentrations

* e.r., endoplasmic reticulum; p.m., plasma membrane.

Table 2 Ca^{2+}-stimulated Mg^{2+}-dependent ATPase in platelet membrane fractions from EH patients and normal subjects. Values are nmol ATP/h per mg protein, ±S.E.M.; Δ= increment.

Parameter		Normal	Hypertensive
Frac- Basal Mg^{2+}-ATPase		165 ±10	181 ±13
tion + Ca^{2+}: 10 µM		Δ 63 ±12	Δ 84 ±9
A 100 µM		Δ 150 ±18	Δ 226 ±18
100 µM & 1 mM EGTA		Δ 5 ±1	Δ 10 ±1
Frac- Basal Mg^{2+}-ATPase		166 ±11	170 ±10
tion + Ca^{2+}: 10 µM		Δ 50 ±7	Δ 91 ±7
C 100 µM		Δ 129 ±14	Δ 230 ±17
100 µM & 1 mM EGTA		Δ 4 ±1	Δ 12 ±1

the specific activity in **A** and **C** from EH patients was significantly raised *vs.* the normal subjects while the basal Mg^{2+}-dependent activity was unchanged. In both cases the maximum stimulation of the ATPase activity occurred between 20 and 30 µM Ca^{2+}, in accord with observations by other groups [15, 16]; but contrary to findings elsewhere, no pronounced inhibition of the activity at high concentrations of Ca^{2+} (>100 µM) was apparent, for reasons which are unclear. It is noteworthy, however, that Ca^{2+} concentrations as low as 3 µM and as high as 600 µM have been used by other workers to measure the Ca^{2+}-ATPase activity of platelets. As shown in Table 2 the enzyme in the presence of 100 µM Ca^{2+} was totally inhibited by EGTA (likewise with 10 µM, not shown).

No change in the 'K$_m$' for Ca^{2+} of the enzyme was observed in the EH patients. However, freezing the membranes to −170° almost completely abolished the hypertension-related increase in the Ca^{2+}-stimulated component of the ATPase activity. This process had no effect on the activity from the control subjects or on the basal Mg^{2+}-dependent ATPase from either the controls or the EH patients.

The change in the Ca^{2+}-stimulated Mg^{2+}-dependent ATPase of platelet membranes observed in EH is apparently in the V$_{max}$ of the enzyme rather than the K$_m$. This agrees with the data of Resink et al. [17]. The observations on the effect of freezing are, however, suggestive of a functional abnormality. This may be in the enzyme molecule itself, or it could reflect some change in the environment of the enzyme, rendering it more susceptible to freezing in the case of the EH platelet membranes.

Not all of the measured Ca^{2+}-stimulated Mg^{2+}-dependent ATPase may actually be involved in the transport of Ca^{2+} across the membrane. Some workers have used the La^{3+}-sensitive component of the enzyme as a measure of this transport activity. Interestingly, the La^{3+}-sensitivity of the enzyme in platelet membranes from EH patients is reduced by ~50%. This is in agreement with values on the spontaneously hypertensive rat (SHR).

In the light of the raised $[Ca^{2+}]_i$ observed in EH patients, it seems likely that the observed increase in the V_{max} of the Ca^{2+}-stimulated ATPase is a response to elevation of extracellular Ca^{2+}. The changed susceptibility to freezing and the reduction in La^{3+}-sensitivity suggest, however, that some fundamental difference does exist either in the enzyme or in some other components in the membranes of the platelets from EH patients. These changes in properties may reflect the inability of the enzyme to function effectively in reducing $[Ca^{2+}]_i$.

References

1. Blaustein, M.P. (1977) *Am. J. Physiol. 232*, C165–C173.
2. Poston, L., Sewell, P.B., Williamson, S.P., Richardson, P.J., Williams, R., Clarkson, E.M., MacGregor, G.A. & de Wardener, H.E. (1981) *Br. Med. J. 282*, 847–849.
3. Swales, J.D. (1982) *Biosci. Rep. 2*, 967–990.
4. Bruschi, G., Bruschi, M.E., Caroppo, M., Orlandini, G., Spaggiari, M. & Caratorta, A. (1985) *Clin. Sci. 68*, 179–184.
5. Kwan, C.Y., Bedbeck, L. & Daniel, E.E. (1979) *Blood Vessels 16*, 259–268.
6. Postnov, Y.V. & Orlov, S.N. (1984) *J. Hypertension 2*, 1–6.
7. Devynk, Y.V., Pernollet, M.G., Nunez, A.M. & Meyer, P. (1981) *Hypertension 3*, 397–403.
8. Carafoli, E. (1984) in *Calcium Antagonists and Cardiovascular Disease* (Opie, L.H., ed.), Raven Press, NY, pp. 29–41.
9. Wilson, R.B.J. & Graham, J.M. (1986) *Clin. Chim. Acta 159*, 211–217.
10. Barber, A.J. & Jamieson, G.A. (1970) *J. Biol. Chem. 245*, 6357–6365.
11. Avruch, J. & Wallach, D.F.H. (1971) *Biochim. Biophys. Acta 233*, 334–347.
12. Hack, N., Croset, M. & Crawford, N. (1980) *Biochem. J. 233*, 661–668.
13. Enyedi, A., Sarkadi, B., Folder-Papp, Z., Montostory, S. & Gardos, G. (1986) *J. Biol. Chem. 261*, 9558–9563.

14. Cutler, L., Rodan, G. & Feinstein, M.B. (1978) *Biochim. Biophys. Acta 542*, 357-371.
15. Steiner, B. & Luscher, E.F. (1985) *Biochim. Biophys. Acta 818*, 299-309.
16. Dean, W.L. (1984) *J. Biol. Chem. 259*, 7343-7348.
17. Resink, I.J., Tkachuk, V.A., Erne, P. & Buhler, R.R. (1986) *Hypertension 8*, 159-166.

#ncE.3

A Note on

ION-SELECTIVE MICROELECTRODES AS APPLIED TO THE ISOLATED PERFUSED RAT BRAIN DURING ISCHAEMIA AND REPERFUSION

D. Scheller*, **F. Tegtmeier, A. Bock, K. Dengler, E. Zacharias and M. Höller**

Janssen Research Foundation,
Raiffeisenstrasse 8, 4040 Neuss, F.R.G./W. Germany

In vitro perfusions of whole organs are useful in pharmacological and pathophysiological studies. Absence of systemic interferences, maintenance of the organ-specific circulatory system and strict control of experimental conditions are the advantages. Brain perfusions, however, are not so common. Subcellular preparations and tissue slices are mostly used in brain research. We describe here a brain perfusion technique as developed by Höller et al. [1] and its use for studying pathophysiological mechanisms of cerebral ischaemia. [See art. #A-3 in Vol. 18 for eicosanoid studies.- *Ed.*]. EEG recordings were performed and the ionic composition of the extracellular space was monitored by microelectrodes during and after global complete ischaemia.

METHODS

Male Wistar rats (200–220 g) were anaesthetized (1.5 g/kg urethan; 0.5 mg/kg atropine), the internal carotid arteries were cannulated, and the veins opened. For the pulsatory perfusion (280 cycles/min; mean pressure 135 mm Hg) a fluorocarbon emulsion was used, *viz.* 16% v/v FC 43 (3M, Neuss) in Pluronic F68 (Serva, Heidelberg), gassed with O_2/CO_2 (95:5). Bipolar EEG was recorded. Triple-barrelled microelectrodes were prepared from double-septum capillaries (R & D Scientific Glass Co., Spencerville, U.S.A.).

The capillaries were pulled and broken to achieve final tip diameters of 2-5 μm. Silanization of the ion-selective channels was performed with trimethylchlorosilane (Fluka). The inner channel was used as reference channel. H^+ or Ca^{2+} cocktail (Fluka) was sucked into one of the outer

---
* addressee for any correspondence

channels to a height of 100–150 μm after filling with the usual inner solutions. Ag/AgCl wires connected the electrode to an impedance change module. The DC-potential was measured between a 2 M KCl reference electrode (attached to the rat's nose) and the internal microelectrode reference channel. After drilling a hole into the skull, the dura was incised and the microelectrode was placed into the cortex (at 1000 μm depth). After 30–60 min of perfusion (equilibrium) at 32°, complete ischaemia was induced by clamping the cannula. After various ischaemic periods (max. 20 min) the clamps were opened and reperfusion was started. Flunarizine was introduced into the perfusion medium 30 min prior to ischaemia.

RESULTS AND DISCUSSION

During ischaemia, the ionic signals behaved as described for *in vivo* preparations [cf. 2, 3]: extracellular pH (pH$_e$) fell from 7.21 ±0.13 (±S.D.) in the control phase to 6.66 ±0.15 during ischaemia (Fig. 1). Extracellular Ca^{2+} (Ca^{2+}_e) increased from 1.11 ±0.03 to 1.29 ±0.09 mM during ischaemia. At 177 ±38.9 sec (n = 18) after induction of ischaemia, pH$_e$ manifested a sudden alkaline shift of 0.1–0.2 pH units. Simultaneously, Ca^{2+}_e fell to 0.098 ±0.083 mM. Extracellular K^+ (K^+_e) increased to 85.16 ±8.54 mM (not shown). These profound changes of extracellular ion composition indicate a sudden breakdown of ionic gradients across the cell membrane. They coincide with a sudden rapid DC negativation[†] [4].

The time lag from the onset of ischaemia until this terminal DC-negativation [5] may be used as a quantitative measure of cellular resistance to ischaemia (as first suggested by Bures & Buresova [6]).

Physiologically, this resistance can be improved by hyperglycaemia or hypothermia or reduced by hypoglycaemia [2]. Pre-ischaemic application of flunarizine, a cerebral protective drug [7], established a direct action on the brain: it increased this resistance *in vivo* (50 mg/kg i.p.) and *in vitro* (15 μg/ml medium), as shown by the following values (sec) for DC-latency.-

Anoxic rat:
  flunarizine, 330 ±26 (n = 8); control, 219 ±19 (n = 8).
Isolated perfused brain:
  flunarizine, 195 ±15 (n = 5); control, 165 ±33 (n = 30).

Cellular function, however, might further deteriorate during the recovery phase [8]. After ~1 h of reperfusion, power spectra showed that the EEG had completely recovered

[†] i.e. a negative DC deflection

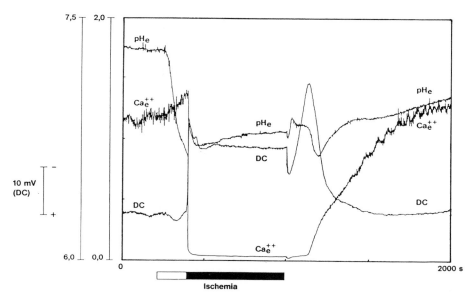

Fig. 1. Recorder traces for cortical pH_e (6.0-to-7.5 axis span), Ca^{2+}_e (mM) and DC-registration in isolated perfused rat brain (control). *Open bar*, period prior to DC-negativation; *closed bar*, ischaemic period after DC-negativation. For details see text.

in the isolated perfused brain.   The first indication of onset of recovery was a slight DC-positivation (Fig. 1) followed by a pronounced further negativation and a steep positivation. During 5–10 min pre-ischaemic levels were attained.   After a certain delay, but concomitantly with the steep DC-positivation, there commenced a recovery of K^+_e (paralleling the DC-positivation, and hence not illustrated separately) and, biphasically, of Ca^{2+}_e.   The DC-positivation in conjunction with the restoration of normal extra- and intra-cellular K^+ concentrations indicates the restoration of the membrane potential.   Together with this process, some of the Ca^{2+}_e which has entered the cell is rapidly released (fast recovery phase).   The rest of the Ca^{2+}_e could have been distributed into different organelles during ischaemia and/or very early recovery [9].   Its release to the extracellular space, therefore, is slower (slow recovery phase).

The pH_e restoration was even more complex.   A rapid acidification was followed by a rapid alkalinization; both processes were related to regeneration of membrane function. Since various factors contribute to pH-recovery (CO_2, Na^+/H^+-exchanger, anion transporters, buffer capacity), no further explanation can be offered.   It is noteworthy that (1) the recovery pattern is very constant and indicates that the above-mentioned factors contribute in a set sequence; (2) the pH_e changes seem not to be correlated with lactate changes [10, 11]. The effects of flunarizine on these ionic changes during the recovery phase were not statistically significant.

SUMMARY AND CONCLUSIONS

A system is described to perfuse rat brain with an artificial solution in which the extracellular ion changes both during and after complete ischaemia can be measured. The results obtained indicate the functional integrity of the model. Moreover, for the first time the direct effect of an antiischaemic/antianoxic drug on the brain could be shown. The model can now be used in the search for other cerebroprotective drugs. In addition, by applying such drugs, the mechanisms involved in the maintenance of the brain-specific ion environment and in pathophysiological compensations can be studied in more detail, independently of interferences by disturbed blood composition or circulatory disorders, after an anoxic insult.

References

1. Höller, M., Bruer, H. & Fleischhauer, K. (1983) *J. Pharmacol. Methods 9*, 19-32.
2. Hansen, A.J. (1985) *Physiol. Rev. 65*, 101-148.
3. Kraig, R.P., Ferreira-Filho, C.R. & Nicholson, Ch. (1983) *J. Neurophys. 49*, 831-850.
4. Lehmenkühler, A., Caspers, H. & Kersting, U. (1985) in *Ion Measurements in Physiology and Medicine* (Kessler, M., et al., eds.), Springer Verlag, Berlin, pp. 199-205.
5. Caspers, H., Speckmann, E-J. & Lehmenkühler, A. (1987) *Rev. Physiol. Biochem. Pharmacol. 106*, 127-178.
6. Bures, I. & Buresova, O. (1957) *Pflügers Archiv 264*, 325-334.
7. Van Reempts, J., Borgers, M., Van Dael, L., Van Eyndhoven, L. & Van de Ven, M. (1983) *Arch. Int. Pharmacodyn. 262*, 76-88.
8. Siesjo, B.K. (1981) *J. Cerebr. Flow Metab. 1*, 155-185.
9. Meldolesi, J., Volpe, P. & Pozzan, T. (1988) *Trends Neurosciences 11*, 449-452.
10. Tegtmeier, F., Scheller, D., Urenjak, J., Kolb, J., Peters, U., Bock, A. & Höller, M. (1989) in *Proc. 2nd Int. Symp. on Pharmacology of Cerebral Ischaemia*, Marburg 1988, (Krieglstein, I., ed.), Elsevier, Amsterdam, in press.
11. Paschen, W., Djuricic, B., Mies, G., Schmidt-Kastner, R. & Linn, F. (1987) *J. Neurochem. 48*, 154-222.

#ncE.4

A Note on

EFFECT OF HALOTHANE ON CALCIUM HOMEOSTASIS IN BLOOD PLATELETS FROM MALIGNANT HYPERTHERMIA PATIENTS

J.G. Hofmann and S. Fink

Institute of Pathological Biochemistry,
Medical Academy of Erfurt,
Nordhauser Str. 74, Erfurt 5010, G.D.R.

MH[*] is a disastrous metabolic syndrome which occurs in genetically predisposed individuals exposed to volatile anaesthetics and/or muscle relaxants such as halothane and succinylcholine. MH is a pharmacogenetic disorder inherited probably in an autosomal dominant fashion. The aetiology of MH has not yet been established, but it is generally believed that the basic abnormality of this disorder resides in skeletal muscle and the alterations in myoplasmic Ca^{2+} followed by an ATP depletion are involved in triggering an MH reaction.

We showed that adenine nucleotide metabolism in blood platelets is affected by halothane, more markedly in MH patients than in controls [1]. Accordingly, the effect of halothane on Ca^{2+} homeostasis was studied in blood platelets. The aim was to evaluate experimental conditions suitable for detecting the suggested defect in the regulation of $[Ca^{2+}]_i$.

The effect of halothane on Ca^{2+} homeostasis was studied by three different approaches: tracer experiments with $^{45}Ca^{2+}$ to follow the transport of Ca^{2+} across the platelet membranes, quin2-loaded cells to monitor changes in $[Ca^{2+}]_i$, and chlortetracycline-loaded cells to measure the Ca^{2+} release from the DTS. Furthermore, using a Ca^{2+} electrode the total Ca^{2+} content and the Ca^{2+} secretion were determined.

Halothane causes a transient increase in $[Ca^{2+}]_i$ to a peak, reached after 1 min, whose height depends on both the halothane concentration and $[Ca^{2+}]_e$. After reaching the peak value $[Ca^{2+}]_i$ declines rapidly; with low $[Ca^{2+}]_e$, $[Ca^{2+}]_i$ falls below the resting level.

---

[*]*Abbreviations.-* MH, malignant hyperthermia; DTS, dense tubular system (microsomal); MCC, see text. $[Ca^{2+}]_i$ (or $[Ca^{2+}]_e$) signifies the concentration of cytosolic (or external) free Ca^{2+}.

The total calcium content of the platelets from MH patients was the same as in controls. Half of the patients, however, had higher resting values for $[Ca^{2+}]_i$. In all patients the $[Ca^{2+}]_i$ transients after halothane addition differed from normals either in the peak height reached or in the velocity of decline or in both. To describe these effects the parameter MCC (mean $[Ca^{2+}]_i$) was defined [2].

Combining the results obtained by the three different methods with first results on the Ca^{2+} pumping system of the platelet plasma membrane, the suggestion can be made that susceptibility to MH is characterized by a generalized membrane defect. A fast halothane-induced Ca^{2+} release from the DTS seems to be of particular importance for adverse cellular reactions on exposure to the anaesthetic.

References

1. Hofmann, J.G., Büttner, C. & Till, U. (1989) *Biomed. Biochim. Acta 48*, 343-350.
2. Hofmann, J.G., & Schmidt. A.. eds. (1989) *Malignant Hyperthermia - An Update*, Volk und Gesundheit, Berlin.

#ncE.5

A Note on

ABSENCE OF A RELATIONSHIP BETWEEN $[Ca^{2+}]_i$
AND HEAT-INDUCED CELL KILLING

P.K. Wierenga and A.W.T. Konings

Department of Radiobiology, State University Groningen,
Bloemsingel 1, 9713 BZ Groningen, The Netherlands

It is not known how cell death results from hyperthermic treatment, which damages p.m.* and intracellular membranes [1]. Heat-induced impairment of membrane functions (<u>e.g.</u> lowered ATPase activity) may lead to raised $[Ca^{2+}]_i$, by net influx of Ca^{2+}_e and/or stored intracellular Ca^{2+}. To assess the possible role of $[Ca^{2+}]_i$ in hyperthermic cell killing, we examined the effect of heat on $[Ca^{2+}]_i$ and cell survival in two tumour lines differing in their Ca^{2+} requirement. We also investigated the effect of ionomycin, a Ca^{2+} ionophore, on $[Ca^{2+}]_i$ and cell kill.

For the asynchronously growing cells the medium, with foetal calf serum added to 10%, was RPMI 1640 (0.5 mM Ca^{2+}) for EAT cells and Joklik medium (0.05 mM Ca^{2+}) for HeLa S3 cells [2]. To load the cells with fura-2/AM for $[Ca^{2+}]_i$ measurement [3], they were incubated (10^{-6}/ml) at 37° for 30 min with the ester, then (using culture medium) washed and resuspended. The fluorimetric settings, with the ratio method [3] for calibration, were 340/380 nm and (emission) 505 nm. Cell survival after hyperthermia was determined by assaying the clonogenic ability of the plated-out cells [4].

$[Ca^{2+}]_i$ **dependence on [fura-2/AM].** - An apparent gradual rise in $[Ca^{2+}]_i$ was observed at 37° (not at room temp.) during the 60-min measurement period. Although no specific side-effects from fura-2/AM loading have been reported, we investigated the effect of increasing [fura-2/AM] on Ca^{2+} homeostasis, because a high cytosolic uptake of dye may perturb the resting $[Ca^{2+}]_i$. The apparent $[Ca^{2+}]_i$ in both cell types (as shown in Fig. 1, **a & b**) depends on [fura-2/AM], with no effect or (HeLa) only a slight effect on the slope of the $[Ca^{2+}]_i$ curve. For both types the lowest practicable [fura-2/AM], as adopted thereafter, was 1 μM - this being in the range of those reported in the literature.

Abbreviations. - p.m., plasma membrane(s); EAT, Ehrlich ascites tumour. Free calcium: intracellular (cytosolic), Ca^{2+}_i; extracellular, Ca^{2+}_e.

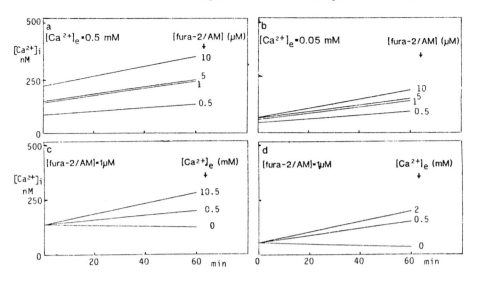

Fig. 1. $[Ca^{2+}]_i$ in EAT cells *(left)* and HeLa cells *(right)* as affected by [fura-2/AM] (**a**, **b**) and $[Ca^{2+}]_e$ (**c**, **d**). Using a K_d of 224, initial $[Ca^{2+}]_i$ was 133 ±22 (EAT) and 69 ±14 (HeLa) nM (±S.D.; n = 8).   Incubations were at 37°.

The rise in the $[Ca^{2+}]_i$ signal has three possible explanations: (1) net influx of $[Ca^{2+}]_e$, (2) Ca^{2+} release from intracellular stores, or (3) fura-2/AM leakage from the cells during the measurement. No increase in the $[Ca^{2+}]_i$ signal was seen in the absence of Ca^{2+}_e (Fig. 1, **c** & **d**), which excludes (2). Whilst (1) cannot at present be excluded, (3) was supported by the following evidence. We measured fluorescence in the cell supernatants at different times and temperatures, checking by a wavelength scan whether Ca^{2+}-fura-2/AM or Ca^{2+}-fura-2 was responsible. We found that no fura-2/AM is detectable but initially all intracellular fura-2/AM is hydrolyzed. At 37° and 44° (less at 20°) there is release of fura-2 into the supernatant (within a few min) with near-attainment of a steady-state (Fig. 2). Such leakage has been reported before [5]. Yet the $[Ca^{2+}]_i$ rise cannot be solely explained by the observed fura-2 leakage (so maybe there is Ca^{2+}_e influx too): at 37° the signal increase continues after 40 min and, moreover, at 20° fura-2 release is found that is not registered as an increase of $[Ca^{2+}]_i$ signal when the measurement is made on the total cell population.

By washing the samples at different time points the $[Ca^{2+}]_i$ signal was normalized - or even depressed below resting levels, which argues against Ca^{2+} influx. Despite

Fig. 2. Leakage of fura-2 from cells (identical for the two cell lines), at 20° (*), 37° (□) or 44° (+): assays on supernatants from cell centrifugation (800 **g**, 5 min) at different times. Monochromator nm settings: 340 and (emission) 505.

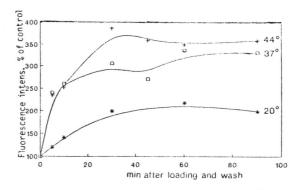

technical difficulties we prefer to measure $[Ca^{2+}]_i$ continuously, so avoiding the risks of cell damage through manipulations during hyperthermic treatments and of missing rapid $[Ca^{2+}]_i$ changes.

Hyperthermic $[Ca^{2+}]_i$ changes _vs._ $[Ca^{2+}]_e$.- In neither cell type did high $[Ca^{2+}]_e$ affect cell kill at 44° (60 min). During hyperthermia, with two different $[Ca^{2+}]_e$ levels, $[Ca^{2+}]_i$ in EAT cells hardly changed (Fig. 3, **a** & **c**), but showed a marked $[Ca^{2+}]_e$-dependent change in HeLa cells (**b** & **d**). Possibly the energy-dependent process, mainly involving a p.m. ATPase, that maintains intracellular Ca^{2+} homeostasis fails under hyperthermic conditons. The ATP content of EAT cells is unaffected by heat [6], but ATPase slightly decreases [7]. Our results agree with these findings. $[Ca^{2+}]_i$ increased if we inhibited energy metabolism by rotenone and iodoacetic acid (IAA). Whether in HeLa cells energy depletion or Ca-ATPase inactivation contributes to the $[Ca^{2+}]_i$ rise is unclear. We tried to mimic hyperthermia effects with ionomycin.⊗

Ionomycin (5 μM) at 37° caused, similarly to $[Ca^{2+}]_e$ elevation, a rapid (~3 min) but transient $[Ca^{2+}]_i$ rise in EAT cells and a sustained rise to a new steady-state level in HeLa cells. Hyperthermia with ionomycin present raised EAT-cell $[Ca^{2+}]_i$ only after 30 min, when hyperthermic cell death (unaffected by ionomycin in either cell type) was >90%; with HeLa cells an immediate $[Ca^{2+}]_i$ increase occurred.

Conclusion.- At 44°, or with ionomycin present, Ca^{2+} influx from the medium may occur. There are two reports on heat sensitivity _vs._ $[Ca^{2+}]_e$. The sensitivity of Reuber H35 rat hepatoma cells [8] or primary cultures of rat-hepatocytes [9] increased when $[Ca^{2+}]_e$ was raised. Also, in CHO cells $^{45}Ca^{2+}$ influx was unrelated to heat sensitivity [10].

⊗*Note by Ed.: henceforth some abridgement (two Figs. excluded).*

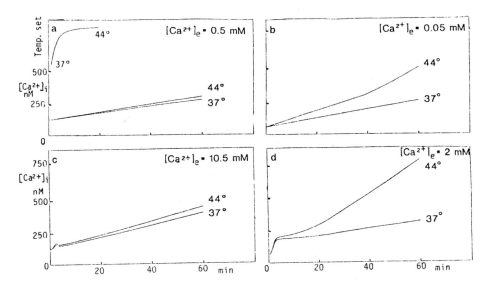

Fig. 3. Effect of hyperthermia (inset curve in **a**) on $[Ca^{2+}]_i$ in EAT *(left)* and HeLa *(right)* cells with normal (**a, b**) or raised (**c, d**) $[Ca^{2+}]_e$, this and the temp. being raised at 1 min. The data are representative of ~10 experiments.

In none of these reports were $[Ca^{2+}]_i$ measurements given. Our results show no relationship of $[Ca^{2+}]_i$ to heat sensitivity in EAT or HeLa S3 cells.

Acknowledgement.- We thank Drs P. Burgman and G. Stege for supplying the cells and performing the cell-survival studies.

References

1.   Konings, A.W.T. (1988) *Recent Results in Cancer Res. 109*, 9-21.
2.   Konings, A.W.T. & Ruifrok, A.C.C.(1985) *Radiat. Res. 102*, 86-98.
3.   Grynkiewicz, G., Poenie, M. & Tsien, R.Y. (1985) *J. Biol. Chem. 260*, 3440-3450.
4.   Jorritsma, J.B.M. & Konings, A.W.T. (1984) *Radiat. Res. 98*, 198-208.
5.   Malgaroli, A., Milani, D., Meldolesi, J. & Pozzan, T. (1987) *J. Cell Biol. 105*, 2145-2156.
6.   Pointec, G.E., Milner, J.A. & Cain, C.A. (1979) *Proc. Soc. Exp. Biol. Med. 140*, 597-599.
7.   Anghileri, L.J., Crone-Escanye, M.C., Marchal, C. & Robert, J. (1984) in *Hyperthermic Oncology*, Vol. 1 (Overgaard, J., ed.), Taylor & Francis, London, pp. 49-52.
8.   Wiegant, F., Kavelaars, A., Blok, F. & Linnemans, W. (1984) *as for 7.*, pp. 3-7.
9.   Malhotra, A., Kruuv, J. & Lepock, J.R. (1986) *J. Cell. Physiol. 128*, 279-284.
10. Vidair, C.A. & Dewey, W.C. (1986) *Radiat.Res. 105*, 187-201.

#ncE

COMMENTS related to

CELL PROTECTION; PATHOLOGY, INCLUDING
CARDIOVASCULAR

Comments on #**E-1**: C.A. Pasternak – PROTECTION BY $[Ca^{2+}]_e$
          #**E-2**: B.P. Morgan et al. – Ca^{2+} AND PORE FORMATION

D. Wermelskirchen asked C.A. Pasternak whether membrane leakage is causable merely by EGTA whereby extracellular Ca^{2+} is removed. **Reply.-** Brief exposure of cellular membranes to EGTA causes no detectable damage. Prolonged exposure to a high concentration results in a small leakage. **Reply to A.K. Campbell.-** Ca^{2+}-handling proteins have indeed been shown in bacteria, e.g. Ca^{2+}-binding proteins in *E. coli*. **B.P. Morgan's reply** to a question on MAC repair by Ca: the stimulus is likely to be from inside rather than outside the cell, since it is preceded by a $[Ca^{2+}]_i$ rise.

Comments on #**E-3**: P.W. Pauwels et al.- NEURON DEGENERATION
          #**E-4**: Yu.I. Gubskiy – MEMBRANE DAMAGE: Ca TRANSPORT

P.J. Pauwels, replying to S. Luciani.- We have not yet tried EGTA/Ca^{2+} buffer to find whether Ca^{2+} entry accounts for the effect of EGTA in high concentration (3 mM). **Reply to B.F. Trump.-** The effect of Ca^{2+}-entry blockers in protecting against anoxia in cardiac myocytes may not be generally applicable to any type of injury. **Answer to R. Pochet:** we have not yet studied pH_i variation in cell cultures without or with drug treatment. **T.J. Brown asked** how long the cells had to be in contact with EGTA for it to confer protection against the excitotoxic agent. **Reply:** only for as long as the cells are exposed to the Glu-like agonists.

A.P. Vieyra asked G.P. Gubskiy whether the increase, with antioxidant deficiency, in Ca efflux from pre-loaded microsomes occurs with ADP absent or present; the Ca^{2+}-ATPase data presented point to a non-specific pathway as distinct from pump involvement. **Answer.-** The Ca^{2+} leakage from peroxidized microsomes was tested in the absence of ADP, so we assume that the leakage took place via the non-specific pathways perhaps formed in conditions of free-radical pathology.

Comments on #**E-5**: K.E. Suckling – ATHEROSCLEROSIS STUDIES
          #**E-6**: P.A. Poole-Wilson – CARDIAC REPERFUSION
          #**E-7**: L. Ver Donck et al.- CARDIOMYOCYTE STUDIES

F.R. Maxfield asked K.E. Suckling whether the Ca^{2+}-channel blockers increase total smooth muscle receptors for LDL or

their expression on the surface. **Reply.**- The total number
is increased, by a process requiring protein synthesis. **Answer
to P.J. Pauwels:** the extent to which the effects of Ca^{2+}
blockers are specific has not been investigated. **V. Darley-Usmar
asked** whether, in the studies on cholesterol metabolism by
macrophages, its uptake was through the scavenger receptor.
Reply.- Most of the studies discussed concerned smooth muscle
cells.   It depends on which lipoproteins the cells are
presented with.   Macrophages are similar to other cells
in cholesterol-metabolizing machinery. Several receptors may
be involved.

    P.A. Poole-Wilson, answering L. Ver Donck.- The oxygen
free radicals that seem a major mediator of reperfusion
damage arise partly from various 'popular' sources such as
mitochondria, but mainly (depending on size) from the endothelium.   Scavengers provide protection, probably because
in the vessels a better re-flow is established, reasonably
attributable to reduced oxygen-radical production. **Pasternak
remarked** that K^+ leakage may be an unsuitable indicator
of membrane intactness in respect of leakage of ions ($^{45}Ca^{2+}$)
and low mol. wt. compounds. **Answer:** possibly so, but the
situation is actually more complex.

    Darley-Usmar asked Ver Donck how an effect of a compound
due to singlet-oxygen scavenging can be distinguished readily
from its Ca-overload effect. **Reply.**- The compounds   exerted
no effect on HDA production and so were not acting as   oxygen
scavengers. **F.R. Maxfield asked** whether the measurements with
fura-2 indicated a rise in $[Ca^{2+}]_i$ preceding the shape change
in isolated myocytes. **Answer.**- Our instrument lacked adequate
time resolution to determine clearly the timing of the $[Ca^{2+}]$
increase in relation to the shape change; but preliminary
data suggest an increase to ~500 nM before the shape change.

Comments on #**E-8:** B.F. Trump *et al.* - ION DEREGULATION ROLE
             #**ncE.1:** A.P. Allshire *et al.* - PATHOLOGICAL Ca^{2+} RISE
             #**ncE.2:** J.M. Graham *et al.* - HYPERTENSIVE CHANGES

    U.T. Rüegg asked B.F. Trump whether the protection by
low extracellular pH against necrosis applies to several
types of toxic influences. **Reply:** Yes! Low pH can protect
cells and tissues against damage due to various chemicals
as well as (*e.g.*) anoxia. **Comment by C.A. Pasternak.**- Low
pH may protect for reasons other than intracellular changes;
thus, H^+ has direct protective effects at the extracellular
p.m. face [see (1988) *J. Memb. Biol. 103*, 79].

    Trump, answering A.K. Campbell, could not explain the
discrepancy between the near-maximal Ca^{2+} fluorescent signal

in cell suspensions and the sub-maximum fluorescence in the imaged cells he studied.

Pasternak asked A.P. Allshire how intracellular Ca^{2+} can remain low during damage in the face of 1 mM extracellular Ca^{2+} when ATP levels have fallen dramatically.   **Answer.-** Contraction and the Ca^{2+} pump differ in ATP sensitivity; moreover, Ca^{2+} extrusion is maintained by Na^+/Ca^{2+} exchange.

J.M. Graham, answering Rüegg.- The need to take 150 ml blood samples precludes your good suggestion that the time course of blood-pressure change be compared with that of Ca^{2+}-ATPase activity after giving a Ca^{2+} antagonist to the patients. Likewise (**answer to T.J. Brown**) it would be impracticable to look at platelets from secondary hypertension patients in the hope of determining whether the observed changes are secondary to a blood-pressure increase rather than due to a primary defect in the p.m. **B.P. Morgan asked** whether, in view of the possibility that the observed effect is a result rather than a cause of the hypertension, platelet turnover after a blood-pressure rise had been investigated. **Reply:** not done; it would be interesting to establish whether newly formed platelets have higher enzyme activity than older platelets.   **M.B. Vallotton asked** whether dietary influences might affect platelet Ca^{2+}-ATPase.   **Reply.-** The subjects were always fasting, and we did not study dietary manipulation although this would be of interest (e.g. a vegetarian diet and a diet high in unsaturated fatty acids).

===========

SOME LITERATURE PERTINENT TO #E THEMES, *noted by Senior Editor* - with **bold type** for test material or other 'keyword'

Hepatocytes exposed to CCl_4 or $(Cl)_2C=CH_2$ manifested sustained rises in $[Ca^{2+}]_i$ as distinct from the transient rises that non-toxic hormonal agents produce. Besides quin-2, phosphorylase **a** activity served as a measure of $[Ca^{2+}]_i$.- Long, R.M. & Moore, L. (1987) *Biochem. Pharmacol. 36*, 1215-1221.

In mice given a low dose of CCl_4, Ca^{2+} sequestration by e.r. appeared to be an early step in **liver** injury.- Reitman, F.A., Berger, M.L. & Shertzer, H.G. (1988) *Biochem. Pharmacol. 37*, 2584-2586.

Cyanidol, a hepatoprotective flavonoid, "prevents the functional deterioration of rat **liver** mitochondria induced by Fe^{2+} ions", including their augmentation of lipid peroxidation and lowering of mitochondrial membrane potential.- Lukács, A. & Lukács, G.L. (1986) *Biochem. Pharmacol. 35*, 2119-2122.

Taurine pre-administration protected against adverse changes in chick **heart** due to re-perfusion with Ca^{2+} after omitting it ('calcium paradox'), including calcium accumulation.- Yamauchi-Takihara, K., Azuma, J., Kishimoto, S., Onishi, S. & Sperelakis, N. (1988) *Biochem. Pharmacol. 37*, 2651-2658.

Revascularization of ischaemic pig **heart**: a CaM antagonist (CGS 9343B) aided exocytosis of the excess calcium.- Das, D.K., Engelman, R.M., Prasad, M.R., Rousou, J.A., Breyer, R.H., Jones, R., Young, H. & Cordis, G.A. (1989) *Biochem. Pharmacol. 38*, 465-471.

In cultured **cardiomyocytes** calcium antagonists stimulate PG synthesis and prevent hypoxic effects.- Escoubet, B., Griffaton, G., Samuel, J-L. & Lechat, P. (1986) *Biochem. Pharmacol. 35*, 4401-4407.

Channel-binding studies with ^3H-nitrendipine on hamster **heart** homogenates and a membrane fraction ruled out a mere increase in binding sites as the cause of the Ca overload in dystrophy.- Howlett, S.E. & Gordon, T. (1987) *Biochem. Pharmacol. 36*, 2653-2659.

Heart membranes showed impaired sarcolemmal Ca^{2+} pump activity in diabetes mellitus.- Heyliger, C.E., Prakash, A. & McNeill, J.H. (1987) *Am. J. Physiol. 252*, H540-H544.

'Role of calcium in pathogenesis of acute **renal** failure' - an editorial review.- Humes, H.D. (1986) *Am. J. Physiol. 250*, F579-F589.

'Antioxidants as stabilizers of the Ca^{2+} transport system in **sarcoplasmic reticulum** membranes *in vivo*'.- Kagan, V.E., Ivanova, S.M., Murzakhmetova, M.K., Shvedova, A.A., Smirnov, L.D., Nadirov, N.K. & Voronina, T.A. (1986) *Bull. Exp. Med. (Engl. Transl.) 102*, 1526-1528.

'Inhibition of rat **heart** and **liver** microsomal lipid peroxidation by nifedipine': due evidently to its nitro group, which can yield free radicals when metabolically reduced. Nifedipine (a Ca^{2+}-channel blocker) possibly diverted electrons away from the NADPH-dependent lipid peroxidation path.-Engineer, F. & Sridhar, R. (1989) *Biochem. Pharmacol. 38*, 1279-1285.

POSTSCRIPT on experimental objectives and approaches

INTRACELLULAR CALCIUM — THE NEXT HORIZON

Anthony K. Campbell

Department of Medical Biochemistry,
University of Wales College of Medicine,
Heath Park, Cardiff CF4 4XN, U.K.

WHAT DO WE STILL NEED TO KNOW?

This author's survey of the past century of intracellular Ca^{2+} appears at the beginning of this book. Much has been learnt. A host of primary stimuli and pathogens have been identified which raise cytosolic free Ca^{2+}. Considerable advances have been made in our understanding of the role of InsP's in controlling it, and in the molecular biology of its regulation and action. We now also have the methodology and conceptual framework for elucidating its role in determining the threshold between cell recovery or death following injury. Yet many questions remain. No complete molecular scheme exists to take us from initiation to end-response, although this is close in muscle. The molecular basis of internal Ca^{2+} release, movement through receptor-operated channels and Ca^{2+} action is still poorly understood. Nor do we have a satisfactory unitary hypothesis rationalizing the relationship of Ca^{2+} oscillations and gradients with those of other intracellular signals. This requires a quantitative description of the chemisymbiosis between the signals and energy phosphate requisite to generate chemical thresholds and the eventual end-response in each cell [1].

THE NEED FOR NEW INDICATORS

We now have good fluorescent and bioluminescent indicators for cytosolic free Ca^{2+}, and sophisticated imaging devices for visualizing with time where changes in Ca^{2+} are occurring. However, we do not have a satisfactory way of monitoring free Ca^{2+} within particular organelles; nor do we have satisfactory ways of manipulating free Ca^{2+}, its binding proteins, and its phosphorylation sites in living cells. New techniques are needed for other signals also, applicable to single cells both *in vitro* and *in situ*. In examining living cells, indicators are particularly needed for: (1) cyclic nucleotides, InsP's and diacylglycerol; (2) energy phosphate, ATP and GTP; (3) the end-response in intact cells. It must be possible to measure these not only in populations but also in individual cells, where the site of 'action' can be localized using imaging microscopy.

We have developed a homogeneous ligand assay for cyclic nucleotides based on chemiluminescence energy transfer [2]. A chemi- or bio-luminescent cyclic nucleotide binds to a fluorescent antibody or binding protein or domain. The light emission reflects that of the fluor because this acts as an energy-transfer acceptor from the excited state generated by the chemiluminescent reaction. The transfer is radiation-less, _i.e._ it occurs not by direct transfer of a photon but rather by resonance. As the unlabelled cyclic nucleotide concentration rises, this displaces the labelled derivative and there is a shift towards the blue of the light emission, whose ratio measured simultaneously at two wavelengths gives the free nucleotide concentration.

A further exciting development is the ability to generate transgenic luminous cells expressing the gene for Ca^{2+}-activated photoproteins. We have also succeeded in incorporating the mRNA for obelin into human neutrophils (Fig. 1), where it is translated to form apoobelin [3]. Addition of the prosthetic group, coelenterazine, allows active photoprotein to form within the cells. We are also attempting to clone the gene for obelin using an expression vector, and to detect clones by light emission. One can detect obelin-mRNA readily (by adding rabbit reticulocyte lysate), and apoobelin or obelin down to 10^{-20} mol. Hence photoproteins should be usable to detect free Ca^{2+} gradients and oscillations in large numbers of individual cells, and circumvents the problem of consumption of the photoprotein when the free Ca^{2+} rises for prolonged periods. Photoproteins have particular advantages over fluors in the study of cell injury since they are less susceptible to leakage or intracellular redistribution.

THE NEED FOR NEW MANIPULATION TECHNIQUES

Nature and the pharmaceutical industry have provided us with a range of 'poisons' which can affect the regulatory pathway for Ca^{2+} at several sites: receptors, membrane channels, organelles, and binding proteins. What is needed now is the exploitation of genetic engineering to develop ways of inhibiting or activating specific gene expression or mRNA translation. Over-expression of calmodulin, for example, has already been achieved. Particular genes can be linked to specific promoters in order to control expression.

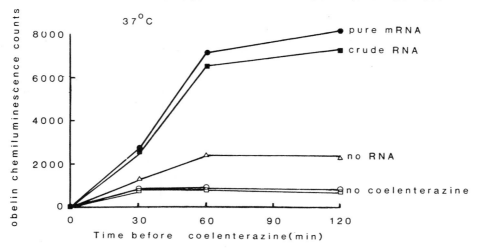

Fig. 1. Translation of obelin mRNA in human neutrophils. *From [3], courtesy of the Biochemical Journal.* SEE OVERLEAF.

A MODEL FOR EVOLUTION

So much for the last century and the present scene. But what about the era starting $3\frac{1}{2} \times 10^9$ years ago? – did life then begin in an environment very low in free Ca^{2+}? Ca^{2+} damages organelles, causes coagulation of protein and condenses nucleic acids. It is therefore proposed that the early prokaryocytes lived in a 'sea' where the free Ca^{2+} was sub-μM. Remember that all the rocks at this stage were igneous and hence the Ca^{2+}, while abundant in the earth's crust, was 'bio-unavailable'[4]. Climatic, chemical and biological forces resulted in a gradual rise in free Ca^{2+}. To prevent a 'Ca^{2+} holocaust' analogous to the 'oxygen holocaust' which occurred some $2\frac{1}{2} \times 10^9$ years ago, the primeval cells had to evolve Ca^{2+} pumps and internal stores. This enabled them to maintain their cytosolic free Ca^{2+} low. The large electrochemical gradient of Ca^{2+} generated as a result could then be exploited, and evolved a new function as a trigger for cell activation. A central feature of this evolutionary hypothesis is that throughout evolution there has been an intimate relationship between cell physiology and the need for cells to protect themselves from physical, chemical and biological attack. But how can we develop an experimental system to investigate this evolutionary hypothesis? One possible model is coelenterazine biolumines-cence [1].

Coelenterazine, an imidazolopyrazine, is responsible for the chemiluminescence of obelin and aequorin and occurs in six luminous phyla: Protozoa (Radiolaria), Cnidaria, Ctenophora, Mollusca, Arthropoda and Chordata; but only in the first three are Ca^{2+}-activated photoproteins found. Thus the same chemistry has evolved in six distinct phyla; yet there exist therein at least three versions of the biochemistry and moreover four distinct triggering mechanisms. Perhaps this intriguing phenomenon may shed some light on the apparent conflict between the principles of Darwinian–Mendelian evolution and the molecular pathway for the evolution of individual proteins. No doubt such controversies will continue to stretch our creative thinking, so that new techniques with new concepts can help us both understand and appreciate the natural world.

Acknowledgements

I thank the Director and staff of the Marine Biological Association Laboratory, Plymouth, and Dr P.J. Herring and Professor F. McCapra for their collaboration, and the MRC and the Royal Society's Brown and Maurice Hill bequest fund for financial support.

References

1. Campbell, A.K. (1988) *Chemiluminescence: Principles and Applications in Biology and Medicine*, Horwood/VCH, Chichester and Weinheim: pp. 548–554.
2. Campbell, A.K. & Patel, A. (1983) *Biochem. J. 216*, 185–194.
3. Campbell, A.K., Patel, A., Rasavi, L. & McCapra, F. (1988) *Biochem. J. 252*, 143–149.
4. Campbell, A.K. (1983) *Intracellular Calcium: its Universal Role as a Regulator*, Wiley, Chichester: pp. 4–6.

AMPLIFIED EXPLANATION of Fig. 1:

RNA was incorporated into human neutrophils using liposome fusion. Coelenterazine was added, and formation of Ca^{2+}-activated photoprotein detected by lysis of cells using Nonidet P40. Protein synthesis was inhibited either by removal of amino acids (Krebs *vs*. RPMI complete tissue culture medium) or by addition of puromycin.

Subject Index

This Index, patterned on those in previous vols. to aid back-consultation, focuses on cellular constituents (excluding gel-separated proteins), phenomena and processes. Listing is non-comprehensive for agents used, and lacking for use of techniques such as gel electrophoresis.

#Alerting to a superscript: ° denotes focus on Ca involvement.
#For major citations the page no. is represented (e.g.) '25-', the ensuing pp. being relevant too.

° denotes focus on Ca involvement

[continued

° denotes focus on Ca involvement

° denotes focus on Ca involvement

° denotes focus on Ca involvement

° denotes focus on Ca involvement